# AQUATIC ECOSYSTEM RESEARCH TRENDS

# AQUATIC ECOSYSTEM RESEARCH TRENDS

## GEORGE H. NAIRNE
### EDITOR

**Nova Science Publishers, Inc.**
*New York*

For permission to use material from this book please contact us:
Telephone 631-231-7269; Fax 631-231-8175
Web Site: http://www.novapublishers.com

### NOTICE TO THE READER

The Publisher has taken reasonable care in the preparation of this book, but makes no expressed or implied warranty of any kind and assumes no responsibility for any errors or omissions. No liability is assumed for incidental or consequential damages in connection with or arising out of information contained in this book. The Publisher shall not be liable for any special, consequential, or exemplary damages resulting, in whole or in part, from the readers' use of, or reliance upon, this material. Any parts of this book based on government reports are so indicated and copyright is claimed for those parts to the extent applicable to compilations of such works.

Independent verification should be sought for any data, advice or recommendations contained in this book. In addition, no responsibility is assumed by the publisher for any injury and/or damage to persons or property arising from any methods, products, instructions, ideas or otherwise contained in this publication.

This publication is designed to provide accurate and authoritative information with regard to the subject matter covered herein. It is sold with the clear understanding that the Publisher is not engaged in rendering legal or any other professional services. If legal or any other expert assistance is required, the services of a competent person should be sought. FROM A DECLARATION OF PARTICIPANTS JOINTLY ADOPTED BY A COMMITTEE OF THE AMERICAN BAR ASSOCIATION AND A COMMITTEE OF PUBLISHERS.

LIBRARY OF CONGRESS CATALOGING-IN-PUBLICATION DATA
Aquatic ecosystem research trends / editors, George H. Nairne.
   p. cm.
 Includes index.
 ISBN 978-1-60692-772-4 (hardcover)
 1. Aquatic ecology--Research. I. Nairne, George H.
 QH541.5.W3A6786 2009
 577.6--dc22

                                                    2008052987

*Published by Nova Science Publishers, Inc.* ✣ *New York*

# Errata for *Aquatic Ecosystem Research Trend,* George H. Nairne (Editor)

ISBN 978-1-60692-772-4

## The following equations were misprinted:

p. 120

**Eq. 1**

$$RBA = \frac{IC_{50}(\text{steroid hormone standard})}{IC_{50}(\text{test chemical})} \, x100$$

p129

**Eq. 2**

$$S = K_f C^{1/n}$$

where:  $S$ = amount of chemical sorbed to soil (mg kg$^{-1}$)

$K_f$ = Freundlich sorption coefficient

$C$ = equilibrium solution concentration (mg L$^{-1}$)

$1/n$ = index of nonlinearity

**Eq. 3**

$$S = K_d C$$

where:  $S$ = amount of chemical sorbed to soil (mg kg$^{-1}$)

$K_d$ = sorption coefficient

$C$ = equilibrium solution concentration (mg L$^{-1}$)

**Eq. 4**

$$K_{OC} = K_d / f_{oc}$$

where:  $K_{OC}$ = carbon normalized sorption coefficient

$K_d$ = sorption coefficient

$f_{OC}$ = fraction of total mass attributable to organic carbon

p. 133

**Eq. 5**

$$F_O / F = 1 + K_{OC}[colloids]$$

where:  $F_O$ = fluorescence without organic colloids

$F$ = fluorescence with organic colloids

[*colloids*] = the colloid concentration

p. 136

**Eq. 6**

$$C_r / C_f = 1 + K_p[colloids]$$

where:  $C_r$ = steroid hormone concentration in the retentate

$C_f$ = steroid hormone concentration in the permeate

[*colloids*] = the colloid concentration

# CONTENTS

# PREFACE

This book discusses various freshwater ecosystems and methods used for the purposes of water quality monitoring, including those utilizing aquatic macrophytes. Early-established management plans provide the first steps towards sustainable management of aquatic environments. Thus, the methods used for managing aquatic plants, including mechanical or manual harvesting, hand-pulling and biological control are discussed. The removal efficiency of nutrients and of metal by macrophytes is examined, as well as the potential use of these plants in wastewater treatment. This book surveys and critically reviews existing literature regarding the presence of steroid sex hormones in freshwater ecosystems, their sources and potential fate. Discharges from wastewater treatment plants and transport from agricultural operations are also examined. Finally, this book discusses the role of NF-kB inhibitors, an inducible transcription, in marine chemical ecology, as well as its role in biomedicine.

Chapter 1 - Submerged macrophytes respond to environmental fluctuations in nutrient concentrations. A general framework for the assessment of aquatic macrophyte communities has been developed in France, in part using phytosociology tools and water chemical composition. A bioindication scale based on assessment of the macrophyte communities for the water trophic level was established in both hard and soft streams. The relationships between water chemistry and aquatic bryoflora in Eastern France were studied in an attempt to survey the processes of acidification. A bioindication scale based on the degree of acidification was proposed using aquatic plant communities. Subsequently, research in Europe focused on the development of synthetic indices that allowed for the quantification of water quality. Three main macrophyte indices are available for assessing the ecological status of water bodies as tested in accordance with the aims of the Water Framework Directive "WFD". These are the Mean Trophic Rank (MTR), the Trophic Index of Macrophytes (TIM), and the French Biological Macrophyte Index for Rivers (IBMR). The MTR and IBMR indices could be useful for estimating the ecological status of rivers in accordance with the aims of the WFD. Ecological optima and ranges of submerged macrophytes are, amongst other factors, assumed to be influenced by ecoregion. Different macrophyte-based assessment systems have also been developed for application on lakes to fulfil the demands of the Water Framework Directive of the European Community. Aquatic macrophytes may be successfully used as ecological indicators for assessing and predicting environmental changes.

The metal and nutrient bioaccumulation ability of macrophytes was also investigated. The use of aquatic macrophytes for treatment of wastewater to mitigate a variety of pollution levels is a research area of great interest. The removal efficiency of nutrients and of metals by macrophytes has been discussed in this non-exhaustive review. The potential use of these plants in wastewater treatment was also explored.

Chapter 2 - Some invasive species are considered to cause "nuisance". In general, the priority of biological invasion control is to prevent new infestations from taking hold, especially in the cases of the fastest-growing and most disruptive species. Exotic macrophytes that do not rapidly increase in numbers are lower priorities. Controlling large infestations of plants with considerable environmental impact, such as *Ludwigia* spp. or Hydrocharitaceae (*Lagarosiphon major, Egeria densa, Elodea* species), is of the highest priority. There are several methods for managing aquatic plants: mechanical or manual harvesting, hand-pulling, biological control, changing the aquatic environment and chemical control. The management techniques chosen must be suited to the aquatic plant itself as well as the uses and function of the body water. Action is recommended only after careful analysis indicates that leaving the spreading species unchecked will result in greater damage than that caused by control efforts.

Early-established management plans provide the first steps towards sustainable management of aquatic environments. However, these efforts are compromised as long as invasive aquatic plant species continue to be sold to individuals. Stronger enforcement of existing laws coupled with an intensive public education campaign is needed to prevent further alien species introduction.

Chapter 3 – The authors present four mathematical models of algal blooms in aquatic ecosystems, and investigate the dynamics in the models. The models analyzed in this chapter are (i) a minimal nutrient-phytoplankton (NP) model, (ii) an algal-grazer model taking into account phenotypic plasticity in algae (cyanobacteria), (iii) the integrated model of the minimal NP model and the algal-grazer model, and (iv) a modified abstract version of the comprehensive aquatic simulation model (abs.-CASM). Bistability and oscillatory behaviors are exhibited in all the models as a function of nutrient inputs, or eutrophication levels. The authors also construct a reaction-diffusion-advection model by using the minimal NP model, and show that two-dimensional spatiotemporal patterns (patchiness) in algae can be exhibited by the combined effects of diffusion and advection by turbulent stirring and mixing, and biological interactions. The minimal NP model is further extended to a one-dimensional reaction-diffusion-advection model in the vertical direction of a lake, explaining the dynamics of both seasonal outbreak and diurnal vertical migration of cyanobacteria. Taking into account the dynamics of algal blooms, we propose effective lake restoration techniques by using the abs.-CASM. The authors address the fact that understanding the mechanisms and characteristics of the dynamical behaviors presented here are crucial to the management of lake ecosystems, in particular, to coping with regime shifts, nonlinear-discontinuous transitions, observed in lake ecosystems exposed to human stress. Finally the authors reexamine the validity of the present models and their analyses from mathematical viewpoints, and propose further mathematical insights into improving the models and their analyses in accordance with real aquatic ecosystems.

Chapter 4 - Natural and synthetic steroid sex hormones, including those administered to humans and livestock as pharmaceuticals, constitute an important class of environmental endocrine disruptors. Steroid sex hormones generally have high biological activities at low concentrations, and in river systems steroid sex hormones have been linked to various adverse effects on fish, including altered sex ratios, intersex fish, and diminished reproduction. In a national reconnaissance study conducted by the U.S. Geological Survey (USGS) from 1999 to 2000, steroid sex hormones were observed at various concentrations and frequencies in water samples from 139 stream sites across the United States. Steroid sex hormones also have been observed in rivers and river sediments in other countries throughout the world. Generally, the reported concentrations are in the low ng L$^{-1}$ (parts per trillion) range. However, in a 7-year, whole-lake experiment in northwestern Ontario, Canada, chronic exposure of fathead minnows to 17α-ethinylestradiol (the active ingredient in most oral contraceptive pills) at concentrations ranging from 5 to 6 ng L$^{-1}$ adversely affected gonadal development in males and egg production in females, and led to a near extinction of fathead minnows from the lake. Because steroid sex hormones and other endocrine disruptors have been detected in locations around the world at concentrations that could have adverse biological and ecological effects, it is important to understand their sources, the processes that transform them, and their ultimate fate in the environment. Therefore, the objective of this article is to survey and critically review existing literature regarding the presence of steroid sex hormones in freshwater ecosystems, their sources, and potential fate. In particular, the article will examine discharges from wastewater treatment plants, and transport from agricultural operations. The article also will consider how biodegradation, photodegradation, and sorption to sediments influence the fate of steroid sex hormones in freshwater ecosystems. Finally, the article will compile relevant information about the physical and chemical properties of selected steroid sex hormones, and conclude with suggestions for future research.

Chapter 5 - Patterns regarding elemental stoichiometry in stream ecosystems remain incompletely understood. The authors analyzed C, N, and P contents in water, benthic macroinvertebrates and their potential food resources in two reaches located upstream and downstream of a point source input in La Tordera stream (Catalonia, NE Spain). The nutrient contents in periphyton and mosses did not vary between the two reaches. The percentage of C and N in filamentous algae was also similar in the two reaches, but percentage of P was two times higher downstream than upstream of the point source. Stoichiometric ratios (C:nutrient) for CPOM, FPOM, and SPOM decreased considerably below the point source. Elemental contents and ratios were highly variable among macroinvertebrate taxa but did not differ significantly between the two reaches. Dipterans, caddisflies, and mayflies had similar elemental contents and stoichiometry, whereas C and N were lower in mollusks and P in beetles. Elemental contents and ratios of functional feeding groups were not significantly different between the two reaches. Predators had the higher C, N and P contents, and P was also high in filterers. Scrapers harbored the lowest elemental contents. Elemental imbalances between consumers and resources were reduced at the downstream reach relative to the upstream reach, reflecting lower nutrient retention efficiency by consumers at the altered reach. The understanding of the effects of stoichiometric constraints derived from human disturbances on in-stream processes provide valuable information that could pay for future research and the development of future management plans.

Chapter 6 - *Phragmites australis* (common reed) is a large perennial macrophyte found in aquatic ecosystems throughout the world. *P. australis* has many important roles in aquatic ecosystems. A functional allelopathic role, inhibiting cyanobacterial growth has been suggested for this species. To confirm this characteristic a series of assays using "rotting-reed solution (RRS)" against *Phormidium tenue* and *Microcystis aeruginosa* were performed. The results demonstrated the growth inhibition of both cyanobacteria. The degree of growth inhibition due to RRS was correlated to the dissolved organic carbon concentration of the RRS, which changed seasonally. *P. tenue* was more sensitive to the allelopathic effect of *P.*

*australis* than *M, aeruginosa*. To identify anti-cyanobacterial allelochemicals released from the reed, extracts of RRS were analyzed using GC/MS. Eight phenols and four fatty acids were identified. Of these, four phenols and one fatty acid were found to inhibit the growth of both cyanobacteria. The remaining compounds were species-specific in their inhibition or else exhibited no effect.

Other roles of *P. australis* include river water treatment, accumulation of sediment and plant transition. These were surveyed within a series of 10 year old constructed wetlands. The average total sediment accumulation in constructed wetlands with *P. australis* was 3.61 – 4.43 kg m$^{-2}$ yr$^{-1}$. This resulted in some land formation within the wetlands which may affect the plant community. The average community diversity index of four representative constructed wetland units was 0.88 in 2004, which was significantly higher than 10 years previously, indicating an increase in diversity. In addition, surveys during 2005 demonstrated that *P. australis* in these constructed wetlands had more gene types in comparison to *P. australis* populations in native wetlands.

Chapter 7 - Lead has been used since ancient times and its toxic effects are well documented. Although the introduction of unleaded gasoline has contributed enormously to the decrease of lead emissions to the environment, the metal is still one of the most commonly used in many industries.

For a long time, the activity of the enzyme δ-aminolevulinic acid dehydratase (ALA-D) has been recognized as a valuable and sensitive tool to assess the extent of lead exposure. Regarded as a specific biomarker, its usefulness has been investigated in different aquatic organisms, including fish. A great number of laboratory studies have proved a strong correlation between the inhibition of the enzyme activity and the lead content, encouraging its use for field monitoring programs. Previously, the authors had found variable levels of lead in liver and gills samples of the fish *Prochilodus lineatus* collected from a coastal area in the La Plata River, Argentina, where sewage effluents are still discharged without any kind of treatment. The aim of this work was to analyze ALA-D activity in blood and liver samples of *P. lineatus* to determine the extent of enzyme inhibition. Correlations between the enzyme activity and the levels of lead bioaccumulated by the fish were also investigated. The analyses were performed in adult organisms, collected each two or three months during June/2002 and May/2004. According to the data, any significant temporal trend could be observed. In addition, and perhaps the most surprising result, poor correlations between the enzyme activity and the lead content were found. In view of these results the authors decided to optimize the methodology to reactivate the enzyme, a technique already used for mammalian samples but not for fish. Only the enzyme from blood samples could be effectively reactivated. In all these samples, values of the reactivated enzyme were not significantly different among them, but significantly higher than those from the non-reactivated enzyme, demonstrating that ALA-D activity was actually inhibited. Good correlations were found between the percentages of enzyme inhibition and the blood lead levels. The results clearly show that for field studies, and even when no reference data are available, ALA-D activity may be regarded as a reliable biomarker of lead contamination.

Chapter 8 - Temperate port communities constitute a modified coastal environment, whose ecology has been poorly studied. Lately, the scientific interest on this subject has been increased for two main reasons: (1) the low ecological quality of their waters that severely influences nearby areas through the diffusion of organic matter, given that ports constitute a hot spot area of organic pollution and (2) the increased presence of allochthonous species, several of which proved to be invasive, that are probably transferred via coastal shipping. Accordingly, the biomonitoring of port communities has become a subject of priority. This study was aimed to assess the structure and function of the benthic communities that develop on artificial hard substratum at a temperate port, namely Thessaloniki Port, with high levels of commercial shipment. Sampling was carried out by diving at three sites and at three depth levels. Two separate assemblages were detected: the blocks of various Serpulids in the midlittoral and the upper sublittoral zone, and the beds of the common Mediterranean mussel *Mytilus galloprovincialis* in the sublittoral zone. The low ecological quality of port communities was evident in their structure, as most of the recorded species were opportunistic and tolerant to organic pollution. At a functional level, the fauna was classified at few trophic groups, among which suspension feeders dominated. An extensive comparison with similar, previously collected, data was also performed, and revealed an apparent degradation of the ecosystem. Algal-dominated communities that have been reported from the same area, are currently replaced by animal-dominated ones, possibly as a result of the extensive deployment of mussel-cultures at the nearby coastal zone that have caused a rapid expand of mussel populations, which now dominate on the harbor piers.

Chapter 9 - NF-κB is an inducible transcription factor found in virtually all types of vertebrate cells, as well as in some invertebrate cells. While normal activation of NF-κB is required for cell survival and immunity, its deregulated expression is characteristic of cancer, inflammation, and numerous other diseases. Hence, NF-κB has recently become one of the major targets in drug discovery.

Several marine organisms use NF-κB (or analogues thereof), NF-κB inducers, or NF-κB inhibitors as chemical defence mechanisms, for parasitic invasion, for symbiosis, or for larval development. In particular, a wide range of marine natural products have been reported to possess NF-κB inhibitory properties, and some of these marine metabolites are currently in clinical trials as anticancer or anti-inflammatory drugs.

In the present review, the authors discuss the role of NF-κB inhibitors in marine chemical ecology, as well as in biomedicine. The authors also describe synthetic modifications that have been made to a range of highly promising marine NF-κB inhibitors, including the macrolide bryostatin 1 isolated from the bryozoan *Bugula neritina*, the lactone-γ-lactam salinosporamide A isolated from the actinomycete *Salinispora tropica*, the alkaloid hymenialdisine isolated from various sponges, the sesquiterpenoid hydroquinone avarol isolated from the sponge *Dysidea avara*, and the sesterterpene lactone cacospongonolide B isolated from the sponge *Fasciospongia cavernosa*, to increase their bioactivity and bioavailability, to decrease their level of toxicity or to lower the risk of other detrimental side-effects, and to increase the sustainability of their pharmaceutical production by facilitating their chemical synthesis.

Chapter 10 - The global and anthropogenic climate change and variability, mainly global warming, is having measurable effects on ecosystems, communities, and populations. The combination of climate change and environmental degradation has created ideal conditions for the emergence, resurgence and spread of infectious diseases, and has led to growing concerns due to the effects of climate on health. Diverse environmental factors affect the distribution, diversity, incidence, severity, or persistence of diseases and other health effects - something that has been recognized for millennia. An important risk of climate change is its potential impact on the evolution and emergence of infectious disease agents. Evidence is indicating that the atmospheric and oceanic processes that occur in response to increased greenhouse gases in the broad-scale climate system may already be changing the ecology of infectious diseases. Ecosystem instabilities brought about by climate change and concurrent stresses such as land use changes, species dislocation, and increasing global travel could potentially influence the genetics of pathogenic microbes through mutation and horizontal gene transfer, giving rise to new interactions among hosts and disease agents. Recent studies have shown that climate also influences the abundance and ecology of pathogens, and the links between pathogens and changing ocean conditions, including human diseases such as cholera. *Vibrio cholerae* is well recognized as being responsible for significant mortality and economic loss in underdeveloped countries, most often centered in tropical areas of the world. Generally, *V. cholerae* is transmitted through contaminated food and water in communities that do not have access to proper sewage and water treatment systems, and is thus called, "the disease of poverty". During the last three decades, extensive research has been carried out to elucidate the virulence properties and the epidemiology of this pathogen. Within the marine environment, *V. cholerae* is found attached to surfaces provided by plants, filamentous green algae, copepods, crustaceans, and insects. The specific environmental changes that amplified plankton and associated bacterial proliferation and govern the location and timing of plankton blooms have been elucidated. Several studies have demonstrated that environmental non-O1 and non-O139 *V. cholerae* strains and *V. cholerae* O1 El Tor and O139 are able to form a three-dimensional biofilm on surfaces which provides a microenvironment, facilitating environmental persistence within natural aquatic habitats during interepidemic periods. Revealing the influence of climatic/environmental factors in seasonal patterns is critical to understanding temporal variability of cholera at longer time scales to improve disease forecasting. Recently, researchers have also been elucidating the environmental lifestyle of *V. cholerae*, and ecologically based models have been developed to define the role of environment, weather, and climate-related variables in outbreaks of this disease. This chapter provides current evidence for the influence of environmental factors on *Vibrio cholerae* dynamics and virulence traits of this organism, and the urgent need for action to prevent the consequences of climate change contributing to cholera.

Chapter 11 - Aquatic pollutant testing using biological assays is useful for ranking the toxicity of different chemicals and other stressors, for determining acceptable concentrations in receiving systems and for elucidating cause and effect relationships in the environment. This 'ecotoxicological' testing approach supplants previous approaches that indirectly estimated toxicity using chemical and physical surrogate measurements alone.

Nevertheless, many published aquatic pollution studies are restricted to examining the effects of a single toxicant on only a single species. Moreover, laboratory-based ecotoxicity tests often intrinsically suffer from a number of limitations due to their small-scale. For example, a major criticism of single-species bioassays is their failure to integrate and link toxicants (and other associated abiotic components) with higher scales of biological and ecological complexity (predation, competition, etc.). Many researchers have suggested that single-species toxicant testing has become so widely entrenched that it has hindered the development and greater use of testing at more ecologically-relevant scales. An improvement to single-species laboratory tests are microcosm and mesocosm studies using more complex and relevant measures to aquatic biotic communities.

Nevertheless, mesocosms still do not entirely simulate the ecosystem they come from, rather they mirror its general properties. As a result, there is increasing interest in correlating pollution measures from field surveys with measures of aquatic biotic community structure to determine a toxicant's scale of effect. However, field assessments, although extremely useful in determining site-specific impacts, may be limited by lack of experimental controls, too few or poorly-positioned regional reference sites and by confounding effects from impacts unrelated to the disturbance of concern.

A hegemony on the evolution of ecotoxicological science and practice is that the primary application of ecotoxicological data is regulatory. As ecological systems do not have a single characteristic of scale, "validation" of single-species toxicity assessments by higher ecological level assessments remains the highest standard for aquatic pollution studies. As such, multi-scale assessments at all of these scales are now being recognised as providing the highest reliability for environmental protection of lake ecosystems.

Chapter 12 - The investigation results for dynamics of $^{90}$Sr, $^{137}$Cs, $^{238}$Pu, $^{239+240}$Pu and $^{241}$Am content and distribution in components of the aquatic ecosystem located in the Chernobyl accident exclusion zone. The main data massif was obtained in the period of 1992 – 2004. Specific radionuclide activity data are present for bottom sediments, water, seston, macrozoobenthos (including bivalve molluscs), gastropods, higher aquatic plants, and fishes. The species specificity of radionuclide concentration by hydrobionts is studied, and the role of various groups of aquatic organisms in distribution of radioactive substances by the main components of the lake biocoenoses is estimated.

In: Aquatic Ecosystem Research Trends                    ISBN 978-1-60692-772-4
Editor: George H. Nairne                       © 2009 Nova Science Publishers, Inc.

*Chapter 1*

# MACROPHYTES AS INDICATORS OF THE QUALITY AND ECOLOGICAL STATUS OF WATERBODIES AND FOR USE IN REMOVING NUTRIENTS AND METALS

## *G. Thiébaut**

Laboratoire Interactions Ecotoxicité, Biodiversité Ecosystèmes (L.I.E.B.E.),
Université Paul Verlaine de Metz, UMR CNRS 7146,
Avenue Général Delestraint, 57070 Metz, France

## ABSTRACT

Submerged macrophytes respond to environmental fluctuations in nutrient concentrations. A general framework for the assessment of aquatic macrophyte communities has been developed in France, in part using phytosociology tools and water chemical composition. A bioindication scale based on assessment of the macrophyte communities for the water trophic level was established in both hard and soft streams. The relationships between water chemistry and aquatic bryoflora in Eastern France were studied in an attempt to survey the processes of acidification. A bioindication scale based on the degree of acidification was proposed using aquatic plant communities. Subsequently, research in Europe focused on the development of synthetic indices that allowed for the quantification of water quality. Three main macrophyte indices are available for assessing the ecological status of water bodies as tested in accordance with the aims of the Water Framework Directive "WFD". These are the Mean Trophic Rank (MTR), the Trophic Index of Macrophytes (TIM), and the French Biological Macrophyte Index for Rivers (IBMR). The MTR and IBMR indices could be useful for estimating the ecological status of rivers in accordance with the aims of the WFD. Ecological optima and ranges of submerged macrophytes are, amongst other factors, assumed to be influenced by ecoregion. Different macrophyte-based assessment systems have also been developed for application on lakes to fulfil the demands of the Water Framework Directive of the European Community. Aquatic macrophytes may be successfully used as ecological indicators for assessing and predicting environmental changes.

---

* Email: thiébaut@univ-metz.fr

The metal and nutrient bioaccumulation ability of macrophytes was also investigated. The use of aquatic macrophytes for treatment of wastewater to mitigate a variety of pollution levels is a research area of great interest. The removal efficiency of nutrients and of metals by macrophytes has been discussed in this non-exhaustive review. The potential use of these plants in wastewater treatment was also explored.

**Keywords:** aquatic plants - eutrophication - acidification - bioindication - indices -Water Framework Directive - monitoring - bioaccumulation - nutrients - metals – removal

# INTRODUCTION

The use of organisms that readily reflect the quality of the ecosystem is called bioindication. Macroinvertebrates, phytoplankton, and macrophytes are widely used for river quality assessment. The phytoplankton and the diatoms methods are well developed in the review of Prygiel and Haury (2006), which describes monitoring methods based on algae and macrophytes. In this work, we will focus only on the macrophytes. The term "aquatic macrophyte" is commonly used for all macroscopic forms of aquatic vegetation including algae, bryophytes, some pteridophytes, and many flowering plants (angiosperms). This assemblage contains extremely heterogeneous species, which survive in similar habitats, but result from fundamentally different evolutionary pathways.

Biological monitoring is perceived as a sensitive, economical, and easily comprehended means of determining the presence or effect of a pollutant in the environment. In many regards, biological approaches offer advantages over chemical techniques because they can deliver information on synergistic effects of disturbances or pollutions. Biological monitoring may offer alternatives to the chemical analysis of pollutants, but better results are obtained if a biological method is combined with chemical monitoring. Macrophytes are excellent indicators of water quality, where the main advantage of macrophytes is that they are relatively widespread and easy to identify. Most macrophytes are non-mobile. The perennial species are continuously exposed and are able to integrate environmental conditions over a longer period. They also may assimilate the cumulative effects of successive disturbances.

A range of methods based on macrophytes has been proposed in the literature. These methods can be applied to broad surveys or provide baseline information to assess possible future change. Macrophytes were first used in water quality assessment in relation to the saprobic system. Several indicators (Sladecek, 1973) included single macrophyte species to evaluate the degree of organic pollution. With increases in water eutrophication, methods have mainly focused on the evaluation of trophic levels. Approaches based on the whole community have been particularly developed in Germany and France, usually involving semi-quantitative estimates of abundance (the phytosociological approach). There has also been a rapid increase in the use of indices based on the sensitivity/tolerance of a species to a pollutant. The methodologies used by a number of research groups in Europe are broadly similar, which allows results to be compared between different regions. However, interpretation of the results is complicated, because the contribution of sediments and water to plant nutrition is dependent on the species of plants and the particular site. The intercalibration of indices based on macrophyte floristic composition in relation to river nutrient status is also under development, particularly in Europe, in accordance with the aims of the Water Framework Directive WFD (Directive 2000/60/EC).

In this non-exhaustive review, we will first focus on a discussion of mainly biological methods, including those utilizing aquatic macrophytes for water quality monitoring, but excluding those using plants in bioassays. Second, we will then describe the indices approach and, finally, the methods using the bioaccumulation capacities of aquatic plants. The second section will summarise the use of aquatic plants for removal of nutrient or metal pollution.

# 1. BIOLOGICAL METHODS

## 1.1. Methods Based on a Whole Community Approach

The assemblage of macrophyte species in streams might be an effective indicator of the integrated combination of stress and disturbance pressures influencing the plant habitat.

### 1.1.1. Aquatic Plant Communities as Bioindicators of Eutrophication

Submerged macrophytes respond to environmental changes in nutrient concentrations. A general framework on aquatic macrophyte communities has been developed since the beginning of the 20th century, partially with phytosociology tools and water chemical composition (Carbiener et al., 1990; Muller, 1990; Robach et al., 1996; Thiébaut and Muller, 1999; Kohler and Schneider, 2003).

These authors established a bioindication scale (based on the macrophyte communities) for assessing the trophic level in hard and soft waters.

Muller (1990) and Thiébaut and Muller (1999) established a bio-indicator scale for the degree of eutrophication and of the level of acidification, based on four aquatic macrophyte communities in soft waters (Table 1).

The A community was characterised by bryophytes (*Scapania undulata, Sphagnum* sp.) and *Potamogeton polygonifolius* in oligotrophic, acidic to slightly acidic, and poorly buffered streams.

The B community, defined by *P. polygonifolius* and the presence of *Ranunculus peltatus* and *Callitriche* species (*C. platycarpa, C. hamulata*) was found in oligotrophic waters with a higher buffer capacity than the A community. The C community was determined by the disappearance of *P. polygonifolius* and the presence of *Elodea* species *(E. canadensis, E. nuttallii)* and rare species in mesotrophic and neutral streams. The D community, with very high nutrient loading, was characterised by the abundance of *Callitriche obtusangula* and by the presence of *Amblystegium riparium, Fissidens crassipes*, and the development of filamentous algae. Upstream aquatic plant communities were composed of Group A, developed into Group B mid-stream, and become Group C or D downstream. Disturbed sites with very high nutrient loading were characterised by low vascular plant richness and the presence of filamentous algae and by the modification of the bioindication scale (Thiébaut and Muller, 1998).

Recently, Demars and Thiébaut (2008) demonstrated that floristic composition was more likely to indicate the role played by alkalinity than that of nutrients. They questioned the validity of current biomonitoring tools that used aquatic macrophytes to indicate the concentrations of inorganic N and P in soft waters.

**Table 1. The bioindication scale in the Northern Vosges Streams (Thiébaut et al., 1999 modified). Synoptic table of communities A to D based on frequency of species and physical-chemical characteristic. (V: 80–100%; IV: 60–80%; III: 40–60%; II: 20–40%; I: 0–20%). In brackets, the mean coefficient of abundance –dominance (Braun-Blanquet, 1964)**

| Plant community | A | B | C | D |
|---|---|---|---|---|
| *Potamogeton polygonifolius* | V(2) | V(1) | . | . |
| *Glyceria fluitans* | IV(2) | V(1) | V(+) | V(+) |
| *Sparganium emersum* | I(+) | III(1) | IV(1) | III(1) |
| *Callitriche stagnalis* | . | IV(+) | IV(+) | III(+) |
| *Callitriche hamulata* | . | III(1) | V(1) | V(2) |
| *Lemna minor* | . | II(+) | III (+) | IV(+) |
| *Ranunculus peltatus* | . | II(1) | II(2) | II(3) |
| *Callitriche platycarpa* | . | III(2) | III(2) | II(1) |
| *Berula erecta* | . | II(1) | I(1) | I(2) |
| *Potamogeton crispus* | . | . | I(1) | . |
| *Myriophyllum alterniflorum* | . | . | II (+) | . |
| *Potamogeton alpinus* | . | . | I(+) | . |
| *Potamogeton variifolius* | . | . | I(+) | . |
| *Elodea canadensis* | . | . | III(3) | I(1) |
| *Elodea nuttallii* | . | . | I(2) | III(3) |
| *Nastrurtium officinale* | . | . | III (1) | III (+) |
| *Oenanthe fluviatilis* | . | . | I(2) | I (+) |
| *Potamogeton berchtoldii* | . | . | I(2) | I(+) |
| *Callitriche obtusangula* | . | . | . | V (3) |
| pH | 6 +/- 0.2 | 6.5 +/- 0.2 | 6.9+/-0.3 | 6.7 +/- 0.4 |
| conductivity µS/cm | 56 +/- 13 | 48 +/- 5 | 74 +/- 18 | 80 +/- 21 |
| Alkalinity mg/l | 4.95+/-2.4 | 4.19 +/-1.1 | 5.78+/-1.1 | 6.3+/-1.7 |
| N-NH$_4^+$ µg/l | 53 +/- 19 | 47 +/- 6 | 111 +/- 99 | 142+/- 71 |
| P-PO$_4^{3-}$ µg/l | 25 +/-10 | 26 +/- 14 | 96 +/- 79 | 153 +/- 66 |
| N-NO$^{3-}$ mg/l | 0.6 +/- 0.2 | 0.3 +/- 0.2 | 0.5 +/- 0.2 | 0.7 +/- 0.3 |

The study of alkaline streams was based on a table of phytosociological releves for six plant communities, named A', B', C', CD', D', and E'. These six communities were classified according to the trophic scale (Carbiener et al., 1990). For the most part, the floristic composition found in the hard and soft waters differ and depend on water mineralization.

However, some species exhibited major differences in indicator values for the two scales; for example, *Callitriche obtusangula* was found in oligo-mesotrophic hard waters, while this same species was found in high trophic levels in soft waters (Robach et al., 1996). In the same way, *Elodea* species were not in the same trophic level in alkaline and soft waters (Thiébaut et al., 1997).

Therefore, caution must be taken when extrapolating local results to national or European scale. Ecological optima and ranges of submerged macrophytes are assumed to be influenced by nutrients, mineralization, and ecoregion.

Vanderpoorten et al., (2000) established that the distribution of bryophyte communities is first strongly related to mineralization and secondly to trophic level. Other factors besides trophic level might also influence the aquatic bryophyte assemblages and should be monitored in order to find the precise relationships between water quality and aquatic macrophytes (Vanderpoorten and Palm, 1998).

### 1.1.2. Aquatic Plant Communities as Bioindicators of Acidification

Acidification of surface waters is found in areas where acidic deposition is high and the catchments soil and bedrock are poor in minerals. During the last century, acidic deposition in Europe has increased markedly as a consequence of increased atmospheric emission of sulphur and nitrogen oxides, particularly from the burning of fossil fuels. However, acidification has been observed in a number of areas. During the last three decades, numerous studies have focused on the effects of acidification on aquatic biota (e.g., fish, macroinvertebrates, zooplankton, macrophytes, and algae). Most of these investigations have revealed deleterious effects on these populations due to low pH, leading to the loss of biodiversity. Despite the key role that aquatic macrophytes play in headwater stream functioning, aquatic bryophytes have received little attention as suitable tools for the evaluation of acidification (Tremp and Kohler, 1995; Thiébaut et al., 1998; Carbeillara et al., 2001). Most attention has focused on monitoring regional patterns of metal pollutions (for example: Satake et al., 1989). Chronic acidification of water bodies is readily detected, but short-term acidification events (for example, due to sporadic discharge or localised atmospheric contamination) are more difficult to identify.

The Vosges Mountains, located in northeastern France, are sensitive to acidification due to the prevalence of quartz-enriched sandstone and granite bedrock. Atmospheric deposition during the last few decades has resulted in surface water acidification (Probst et al., 1990a; Dambrine et al., 1998), particularly in catchments covered by coniferous forests (Probst et al., 1990b).

The relationships between water chemistry and aquatic bryoflora in the Vosges Mountains (Eastern France) were studied in an attempt to survey the processes of acidification. Thiébaut et al. (1998) found five groups of sites characterised by their floristic combinations and chemical value (Table 2). The first group concerns acidic streams (low pH, low cation concentrations) and includes *Marsupella emarginata*, *Jungermannia sphaerocarpa*, and *Dicranella heteromalla*. The last group was characterised by the absence of acidophilous species and the presence of the neutrophilous mosses including *Rhynchostegium riparoides*, *Chiloscyphus polyanthus*, *Dichodontium pellucidum*, and occasionally *Thamnobryum alopecurum*. *Scapania undula* appeared to be an acid-tolerant species, whereas the acid-sensitive *Rhynchostegium riparoides* did not tolerate high concentrations of aluminium (Thiébaut et al., 1998).

**Table 2. Classification of bryophyte community according to water quality from Thiébaut et al.(1998) modified**

| | Frequency % | | | | |
|---|---|---|---|---|---|
| | First group | 2nd group | Third group | Fourth group | Fitfh group |
| numbrer sites | 5 | 3 | 8 | 8 | 7 |
| *Dicranella heteromella* | 100 | | | | |
| *Marsupella emarginata* | 100 | 100 | 75 | | |
| *Jungermannia sphaerocarpa* | 20 | 100 | | | 14 |
| *Sphagnum auriculatum* | | 100 | 87.5 | | |
| *Racomitrium aciculare* | | 33 | 12.5 | | |
| *Rhynchostegium riparoides* | | | 37.5 | 100 | 85.7 |
| *Nardia compressa* | | | 12.5 | 25 | |
| *Fissidens pusillus* | | | 12.5 | 25 | |
| *Chiloscyphus polyanthos* | | | 25 | | 100 |
| *Brachytecium rivulare* | | | | 12.5 | |
| *Fissidens crassipes* | | | | | 28.6 |
| *Amblystegium fluviatile* | | | | | 28.6 |
| *Thamnobryum alopecurum* | | | | | 28.6 |
| *Dichodontium pellucidum* | | 33 | 12.5 | 25 | 42.9 |
| *Hyocomium armoricum* | 60 | 66 | 62.5 | 100 | 57 |
| *Scapania undulata* | 100 | 100 | 100 | 100 | 85.7 |
| Al µg/l | 428-798 | 458-881 | 114-177 | 90-165 | 49-122 |
| $Ca^{2+}$ mg/l | 1.30-1.80 | 1.40-2.30 | 2.30-2.50 | 3.00-3.60 | 4.2-8.5 |
| $Mg^{2+}$ (mg/l) | 0.53-0.71 | 0.56-1.50 | 1.00-1.09 | 1.22-1.56 | 1.64-3.38 |
| pH | 4.6-5.2 | 4.9-5.8 | 5.3-5.5 | 5.8-6.1 | 6.3-6.9 |

## 1.2. Water Quality Assessment by Macrophytes Indices

### 1.2.1. Biological Indices in Streams

In the 1980s, research focused on development of synthetic indices, allowing for the quantification of water quality. The three main macrophyte indices available for assessing eutrophication in flowing waters are the Mean Trophic Rank (MTR: Dawson et al., 1999; Holmes et al., 1999), the Trophic Index of Macrophytes (TIM: Schneider and Melzer, 2003), and the French Biological Macrophyte Index for Rivers (IBMR, Haury et al., 2006). Schneider (2007) indicated the limits of the macrophytes indices. Most indices require a certain number or biomass of indicator taxa present at a river site in order to ensure a reliable indication. By neglecting important indicator taxa a reliable indication of river trophic status might often be impossible due to the presence of too few indicator taxa. An adjustment of indicator species lists according to local conditions is required.

## 1.2.1.1. Mean Trophic Rank (MTR)

The Mean Trophic Rank (Dawson et al., 1999; Holmes et al., 1999) is a macrophyte-based method for assessing the trophic status of rivers that is particularly focused on the effects caused by phosphate enrichment and has been established in the United Kingdom. The MTR is not designed for classifying the ecological quality of rivers. For interpretation purposes only, Holmes et al. (1999) suggested the use of MTR boundary values to determine if the investigated site is (1)'unlikely to be eutrophic', (2) 'likely to be either eutrophic or at risk of becoming eutrophic' or (3) 'badly damaged by either eutrophication, organic pollution, toxicity, or physically damaged'.

Demars and Harper (1998) demonstrated the following: 1) that sites downstream of sewage producers did show a reduction in Mean Trophic Rank, and 2) that the effectiveness of the MTR method is limited at full catchment scale by low numbers of the indicator taxa at small upstream sites. Catchment-scale assessment of the plant community is probably best served by detailed phytosociological analysis and by the further development of the "habitat templet' approach" (Demars and Harper, 1998). However, numerous environmental and scientific institutions have used the MTR system across Europe. Szoszkiewicz et al. (2006) examined the relationships between several existing macrophyte metrics and nutrient enrichment in a Pan-European scale. Although most existing macrophyte metrics are useful, none can be applied at a pan-European scale in their current form. Attempts to redesign the Mean Trophic Rank (MTR) index by the addition of further species, and the re-scoring of existing species, have resulted in a considerable improvement in the relationship between MTR scores and nutrient variables. Thus, an enlarged core group of macrophyte species can form part of an improved pan-European macrophyte-based bioassessment system, although regional modifications may be required to adequately describe the nutrient status of certain stream types.

## 1.2.1.2. Trophic Index of Macrophytes (TIM)

The TIM is a macrophyte-based method for assessing the trophic status of rivers (Schneider and Melzer, 2003) that is applicable in Germany. Macrophyte indicator values are given on a scale from 1 to 4, with 1 indicating oligotrophic and 4 indicating polytrophic conditions. Thus, actual species indicator values range from 1.05 (*Chara hispida*, *Potamogeton coloratus*) to 3.20 (*Acorus calamus*) and lean towards the centre of the trophic gradient. An indication of the polytrophic category (index >3.50) is not possible. Within the polytrophic category, a depopulation of submerged macrophytes is mostly observed. To develop the Trophic Index of Macrophytes (TIM), the concentrations of soluble reactive phosphorus of both the water body and the sediment pore water were utilised (Schneider and Melzer, 2003).

## 1.2.1.3. Biological Macrophytes Index in Rivers (IBMR, AFNOR 2003)

The IBMR is a macrophyte-based method for determining the trophic status of rivers (Haury et al., 2006). This index has been normalised in France since 2003 and is now officially adopted for river macrophyte sampling through monitoring networks. This method has been overtaken by implementation of the European Water Framework Directive. To meet the new requirements proceeding from the application of this European regulation, adaptive measures have been intensively developed (Chauvin et al., 2008). The IBMR is applicable to shallow and deep running waters in continental France. This index has been detailed in the work of Chauvin et al. (2008). IBMR species values range from 0 to 20, with 0 indicating hypertrophic and 20 indicating oligotrophic conditions, respectively. Evaluation of the method in an acidic, lowland river and an alkaline mountain river validated the bio-indication scales based on macrophyte communities (Carbiener et al., 1990; Thiébaut and Muller, 1999). The IBMR effectively assessed not only trophic disruption, but also heavy organic pollutions induced by fish-farming (Haury et al., 2006). Although the French Trophic index appears to be a useful means to assess eutrophication and/or heavy organic pollution of running waters, IBMR was not effective in assessing disturbances, such as flood overflow or chemical pollution (Thiébaut et al. 2002; Thiébaut et al., 2006).

### 1.2.1.4. Intercalibration Methods

The purpose of Water Framework Directive WFD was to establish a European framework for the protection of surface waters, transitional waters, coastal waters, and groundwater (Directive 2000/60/EC). Macrophytes are one of the major groups of organisms (along with macroinvertebrates and diatoms) that the WFD prescribes for assessment of the ecological status of water bodies. Chauvin et al. (2005) compared the results of three French normalised indices; the first one is based on diatoms (Diatoms Biological Index "IBD"), the second one on macroinvertebrates (Biological Index "IBGN"), and the last one on macrophytes (River Macrophytes Biological Index "IBMR") calculated on different French rivers (Figure 1). The highest biological values were obtained with the macroinvertebrate index or "IBGN" index, which reflects the habitat quality and not the water chemical quality. The macrophyte index or "IBMR" indicates the water trophic level.

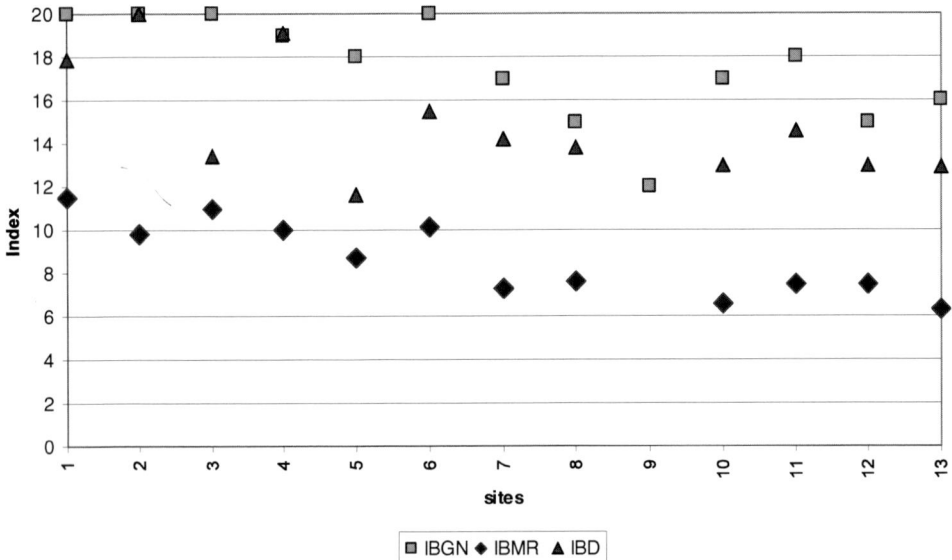

Figure 1. Comparison between three French indices in the STAR program. IBGN index:
Macroinverteberates index; IBD index: Diatoms index, IBMR index : Macrophyte index. Source:
Chauvin et al., (2005) modified. The macroinvertebrate index IBGN indicates a higher biological quality than the macrophyte index IBMR. IBGN reflects the quality of the habitat more than the water quality. The Diatoms Index IBD is similar to the IBGN.

The aim of a study by Staniszewski et al. (2006) was to compare different macrophytic indices and to quantify their variability in the STAR project. The MTR and IBMR indices could be useful for estimating the ecological status of rivers in accordance with the aims of the WFD.

Ecological optima and ranges of submerged macrophytes are, amongst other factors, assumed to be influenced by ecoregion. In order to test the influence of ecoregion factor within Europe, Schneider (2007) compared the species indicator values of three European macrophyte river trophic indices (MTR, IBMR, and TIM). Species indicator values for the English, French, and German bioindication methods were found to be significantly correlated. The most obvious difference between the three indicator systems is the number of included indicator taxa. Two species exhibited major differences in indicator values: *Callitriche hamulata* had a broader ecological amplitude in Germany and France than in the UK, where it is restricted to oligotrophic rivers, while *Ranunculus fluitans* has a broad ecological amplitude in the United Kingdom (UK), whereas the species is restricted to eutrophic rivers in Germany and France. Furthermore, if an index is to be applied in a different country from where it was developed, important local indicator taxa are probably neglected. Thus, minor differences in river trophic status might be overlooked. However, as individual species indicator values are not substantially different between countries, a rough classification of river trophic status is probably possible.

Moreover, Birk et al. (2006) underlined the lack of knowledge regarding pressure–impact relationships that may generally impede the intercalibration of macrophyte methods. The intercalibration of methods that specifically addresses eutrophication is possible, but there are deficiencies for the upcoming macrophyte intercalibration exercise (Birk et al. (2006).

### *1.2.2. Lake Biomonitoring by Macrophytes*

Aquatic macrophytes can serve as useful indicators of water pollution along the littoral zone of lakes. The first index is the Macrophyte Index "MI" (Melzer, 1999), which was developed in Bavaria and ranges from 1 (unpolluted) to 5 (heavily polluted). Many wastewater inflows or diffuse sources could be detected by utilizing "MI". However, investigations have shown that some species (*Ceratophyllum demersum, Myriophyllum spicatum, Potamogeton pectinatus*, and some *Chara*) exhibit different behaviours toward nutrient pollution in different areas (Melzer, 1999). Schneider (2007) compared the lake macrophytic index "MI" Index to the German trophic Index of Macrophytes "TIM" and found that the most important differences are due to the indicator taxa used and not differences in indicator values of individuals. However, three species exhibit a major difference in ecological optima between rivers and lakes: *Ranunculus circinatus* has a broader ecological amplitude in rivers, whereas this species is restricted to eutrophic lakes; *Myriophyllum spicatum* and *Nuphar lutea* show the opposite reaction. An adjustment of indicator species lists to local conditions is required.

Subsequently, different macrophyte-based assessment systems have been developed for application on lakes to fulfil the demands of the Water Framework Directive of the European Community (Schaumburg et al., 2004; Stelzer et al., 2005; Søndergaard et al., 2005; Wilby et al., 2006). Penning et al. (2008) tested a trophic index ("TI": Stelzer et al., 2005) and a lake trophic ranking ("LTR"; Willby et al., 2006) method on a Europe-wide scale. Changes in macrophyte communities due to eutrophication can be quantified using the "TI" and the "LTR" index.

However, caution must be exercised when translating the results to a judgement on the ecological status of a lake system. The result of the single index alone is not always sufficient to understand and predict how a lake macrophyte community changes along the eutrophication pressure gradient.

Combining the results of the eutrophication indices with additional environmental variables and other macrophyte community characteristics may help enhance the accuracy of status assessments as required for the WFD (Penning et al., 2008).

# 2. THE BIOACCUMULATION ABILITY OF PLANTS AND THEIR USE FOR NUTRIENT AND METAL REMOVAL

Aquatic macrophytes are biological filters, which perform purification water bodies by accumulating dissolved metals and toxins in their tissue. These macrophytes can reduce the concentrations of nutrient ions, such as P, N, and metals, significantly in a controlled environment, such as in a constructed wetland.

Constructed wetlands are the well-accepted and preferred ways to treat wastewater, particularly for the developing world. Treatment wetland systems can remove significant amounts of suspended solids, organic matter, nitrogen, phosphorous, trace elements, and microorganisms contained in wastewater (Kadlec and Knight, 1996). The constructed wetlands are also used as controlled environment for the removal of xenobiotics, such as pesticides (Cheng et al., 2002a).

## 2.1. The Bioaccumulation Ability of Macrophytes

In aquatic habitats, plant species of different groups serve as reliable indices for biological monitoring of pollution load. The level of a given contaminant in biological tissues is related to the concentration of this substance in the environment. Floating-leaved aquatic macrophytes potentially remove and recover the nutrient anions and metal cations from water and wastewaters (Kadlec, 2000). Floating Aquatic Macrophytes, because of their high productivity, high nutritive values, and ease of stocking and harvesting, are suitable in engineered wetlands to improve the quality of water (Sooknah and Wilkie, 2004). Submerged plants can maintain a clear water state by various buffering mechanisms, such as bicarbonates utilization and the luxuriant uptake of nutrient ions. Emergent plants influence metal storage indirectly by modifying the substratum through oxygenation, buffering pH, and adding organic matter (Dunabin and Bowmer, 1992).

### 2.1.1. Bioaccummulation of Nutrients

The accumulation of nutrients in macrophyte tissues depends on the physiological uptake capacity which is related to the species. *Chara* spp. can take up nitrogen as $NH_4^+$ and $NO_3^-$ from the water significantly (Vermeer et al., 2003) which causes a reduction in Characean biomass because of the increased levels of N in the cells (more than 2%). Submerged macrophyte communities exhibit phosphorus (P) removal mechanisms (Dierberg et al., 2002). Phosphorus is incorporated by plant cells in the form of phosphate (Pi) using transporter proteins, such as $H_2PO_4^-/H^+$ symporters in the plasma membrane (Schachtman et al., 1998; Smith et al., 2003). Phosphate can be translocated from shoots to roots or from roots to shoots, as has been demonstrated in laboratory studies (De Marte and Hartman, 1974; Eugelink, 1998). While shoot absorption is still a subject of debate, root uptake of Pi is commonly accepted as the mode that aquatic plants use to acquire Pi (Bole and Allan, 1978; Best and Mantai, 1979; Barko and Smart, 1981; Barko et al., 1988). Even plants with limited root systems have been shown to take up Pi via their roots (Barko and James, 1998). Therefore, nutrient uptake and storage by aquatic plants will depend upon species (biological type: free-floating, submerged, emergent). For example, Greenway and Wooley (1999) investigated nutrient bioaccumulation in 60 species. Although, emergent species had lower nutrient concentrations than aquatic plants, they had a greater biomass than submerged or free floating species and were therefore able to store more nutrients. Nitrogen concentrations were higher in the leaf/stem tissue compared to the root/rhizome, whereas phosphorus was higher in the root/rhizome tissue. The aims of numerous studies have been to measure the nutrient storage of aquatic and emergent plants (Table 3).

**Table 3. Removal of nutrients by aquatic macrophytes**

| | nutrient removal | References |
|---|---|---|
| Free floating macrophytes | | |
| *Eichhornia crassipes* | N, P | Boyd1976, Reddy et al. 1990; Xie and Yu 2003;Sooknah and Wilkie 2004 |
| *Ludwigia sp,* | N,P | Doyle et al. 2003, |
| *Salvinia herzogii* | N | Elankumaran et al.2003; Maine et al.2004; |
| *Pistia stratiotes* | NH4; NO3; P | Aoi andHayashi1996;Sooknah andWilkie 2004 |
| *Azolla spp,* | N | Lejeune et al.1999 |
| *Lemna minor* | N, P | Doyle et al. 2003, |
| *Lemna spp,* | N, P | DeBusk et al.1995; |
| Submerged macrophytes | | |
| *Hydrilla verticillata* | P | Bole and Allan1978; Barko and Smart1988 |
| *Elodea canadensis,* | N,P | Best 1977; Ozimek et al.1993, Eugelink et al.1998; Thiébaut and Muller 2003 |
| *Elodea nuttallii* | N,P | Ozimek et al.1993;Eugelink et al.1998; Thiébaut and Muller 2003 |
| *Ceratophyllum demersum* | N, P | Best 1977;Tracy et al.2003 |
| *Chara spp,* | N, P | Kufel and Kufel 2002; Vermeer et al, 2003 |
| *Myriophyllum sp,* | N,P | De Marte and Hartman1974;Best and Mantai 1979; |
| *Myriophyllum spicatum* | N,P | Bole and Allan 1978 |
| *Potamogeton spp,* | N | La-Montagne et al. 2003; Fritioff et al. 2005 |
| *Hygrophila polysperma* | N,P | Doyle et al,2003 |
| *Ipomea aquatica* | N, P | Sinha et al. 1996; Göthberg et al, 2002 |
| Emergent macrophytes | | |
| *Phragmites carca* | N; BOD | Juwarder et al,1995 |
| *Typha latifolia* | N; BOD,P | Juwarder et al,1995;Szymanowska et al. 1999 |
| *Phragmites australis* | P | Szymanowska et al. 1999 |

*Eichhornia crassipes* is capable of removing large amounts of nitrogen and phosphorus from water (Boyd, 1976). *Lemna* spp. have been widely studied for the uptake of P and N from the water by DeBusk et al. (1995). *Pistia stratiotes* is reported to reduce ammonium ions from the water as it utilises $NH_4$–N prior to $NO_3$–N as a nitrogen source and does not switch to utilization of $NO_3$–N until $NH_4$–N is entirely consumed (Aoi and Hayashi, 1996). Water ferns, including *Azolla filiculoides*, *Azolla caroliniana*, and *Salvinia molesta*, have great potential to affect water chemistry. Some of the most commonly submerged macrophytes used for nutrient removal include *Hydrilla verticillata* and *Ceratophyllum demersum*. Extensive studies of nutrient uptake from the water by *Chara* spp. have been performed by Kufel and Kufel (2002).

The nutrient uptake/storage ability varies according to the season (Thiébaut and Muller, 2003).

### 2.1.2. Bioaccumulation of Heavy Metals

Bioaccumulation of metal ions by aquatic macrophytes is also well documented (see for example Srivastava et al., 2008; Dhote and Dixit, 2008). Many free-floating, emergent, and submerged species have been identified as potential accumulators of heavy metals (Table 4). Submerged plants are useful in reducing heavy metal concentrations in water, as the biomass of their shoot can accumulate large amounts of heavy metals (Fritioff et al., 2005). Due to the ability to accumulate toxic substances, macrophytes indicate the presence of contaminants in the environment, even if they are present in very low concentrations. In many sensitive species, metal-induced morphological and structural changes may also be indicative of changes, which are specific to some metals. For example, in young differentiating leaves of *Elodea canadensis*, Cd interfered with the morphogenetic pattern, by inhibiting cell division and affecting cell enlargement. Cd hindered the division and the expansion of chloroplasts, also impairing organelle shape and thylakoid system arrangement. Cadmium, furthermore, greatly disturbed the cell wall organization, causing the demolition of the transfer cell-like frame. In differentiated leaves, Cd hastened the appearance of senescence symptoms in chloroplasts and worsened the cell wall alterations. Both young and mature leaves showed a decreased photosynthetic activity, not only ascribable to lowered chlorophyll contents (Dalla Vecchia et al., 2005). Elankumaran et al. (2003) established that morphology was dependent upon the duration of exposure and initial dose; morphological changes were observed at longer exposures and higher doses.

Heavy metals pose a serious threat to the aquatic environment as they do not degrade, but accumulate in aquatic macrophytes and enter into the food chain. Tolerance to metals in plants is correlated with an induced synthesis of a metal binding protein, binding of the metal to the cell wall, or compartmentation of metals in vacuoles. It has been reported that cysteine is one of the amino acids of cadmium binding protein. Any decrease in cellular cysteine is due to its incorporation in glutathione (GSH), which is a precursor for phytochelatin synthesis (Nussbaum et al., 1988).

The increased uptake of metals is caused by the increased concentration of metal ions in the surroundings of submerged aquatic macrophytes. Physical factors including pH, redox potential, temperature, and salinity in the surrounding aquatic environment largely affect metal uptake by submersed plants (Fritioff et al., 2005). Tripathi et al. (1995) demonstrated the light dependency of metabolic energy necessary for the transport of $Cd^{2+}$ in *Ceratophyllum demersum*.

**Table 4. Removal of of metals by aquatic macrophytes**

| | metal ions | References |
|---|---|---|
| Free floating macrophytes | | |
| *Eichhornia crassipes* | Pb,Cr, Cu, Cd, Zn, Mn, Ag, Ag | Muramoto and Oki 1983; Prakash et al.1987; Nor 1990, Pinto et al. 1987, Schneider and Rubio 1999; Maine et al. 2001, Sune et al. 2007; Tiwari et al. 2007;Sahu et al. 2007, |
| *Ludwigia sp,* | Cu, Zn,Pb, Cd, Ni, Hg | Lee et al. 1999; Pilon-Smith and Pilon 2002 |
| *Salvinia herzogii* | Cr, Cd | Schneider and Rubio 1999;Maine et al. 2001, 2004;Elankumaran et al.2003;Sune et al. 2007, |
| *Pistia stratiotes* | Cd,Hg,Cr | Maine et al. 2001,2004; Sune et al. 2007, |
| *Azolla spp,* | Hg, Cr | Kamal et al. 2004; Bennicelli et al. 2004 |
| *Lemna minor* | Ni, Cd, Zn, Al | Nasu and Kugomoto 1981, Noraho and Gaur1995; Maine et al. 2001, Goulet et al. 2005,Miretzky et al. 2006; |
| *Lemna spp,* | Pb | Gazi and Steven1999; SternbergandRahmani 1999;Maleva et al. 2004 |
| Submerged macrophytes | | |
| *Hydrilla verticillata* | Cu, Cd | Garg et al. 1997; Elankumaran et al. 2003; Bunluesin et al. 2007a |
| *Elodea canadensis,* | Cd,Pb, Cu, Ni,Cr, Co, Cu, Mn, Al, Zn | Mayes et al. 1977; Kähkönen and Kaireselo 1998; Kähkönen and Manninen 1998;Samecka-Cymerman and Kempers 2003; Maleva et al. 2004;Fritioff et al. 2005, Fritioff and Greger 2007, |
| *Elodea nuttallii* | Cd,Pb, Cu,Zn | Nakada et al. 1979 |
| *Ceratophyllum demersum* | Ni, Cr,Co, Zn,Mn,Pb, Cd, Hg, Fe | Ornes andSajwan1993; Szymanowska et al. 1999;Mal et al. 2002, Keshkinkan et al. 2004;Maleva et al. 2004; Bunluesin et al. 2007b |
| *Chara spp,* | | |
| *Myriophyllum sp,* | Cu, Zn,Pb, Cd, Ni | Lee et al. 1999;Sivaci et al. 2008, |
| *Groenlandia densa* | Cd, Cu | Kara and Zeytunluoglu, 2007, |
| *Potamogeton spp,* | Pb,Zn, Cu,Cd,Mn, Ni | Lee et al.1999; Kara and Zeytunluoghu 2007; Peng et al.,2008, Sivaci et al. 2008 |
| *Potamogeton natans* | Zn, Cu, Cd, Pb, Ni | Maleva et al. 2004; Fritioff et al. 2005; Fritioff and Greger 2006; |
| *Potamogeton lucens* | Pb,Zn, Cr, Cu,Ni, Cd,Mn | Schneider and Rubio, 1999; Maleva et al. 2004; Duman et al. 2006, |
| *Potamogeton pectinatus* | Cd,Pb,Cr, Ni, Zn,Cu, Mn | Demirezen and Aksoy, 2004;2006;Peng et al., 2008, |
| *Hygrophila polysperma* | Cu, Zn,Pb, Cd, Ni | Lee et al,, 1999 |
| *Hydromistia stolonifera* | Cd | Maine et al. 2001 |
| *Ipomea aquatica* | Hg | Götheberg et al. 2002 |
| Emergent macrophytes | | |
| *Typha latifolia* | Ni, Cr,Co, Zn,Mn,Pb, Cd, Hg, Fe | Szymanowska et al. 1999 |
| *Typha angustifolia* | Cd,Pb,Ni, Zn, Cu | Demirezen and Aksoy 2004 |
| *Phragmites australis* | Ni, Cr,Co, Zn,Mn,Pb, Cd, Hg, Fe | Szymanowska et al. 1999;Laing et al. 2003 |

Cadmium influx in these plants occurs more in the light than in dark. Moreover, Duman et al. (2006) have suggested seasonal changes in metal accumulation: Ni and Mn have a tendency accumulate in leaves in autumn, while Cr and Cd tend to accumulate in the shoot during summer.

The ability to accumulate metals varies according to the tissue (Vardanyan and Ingole, 2006). For example, Demirezen and Aksoy (2004) demonstrate that the organs of *Typha angustifolia* have a larger quantity of the measured elements than the *Potamogeton pectinatus*. Accumulation of most of the heavy metals was higher in the root system. Frittioff and Greger (2006) demonstrated that Zn, Cu, Cd, and Pb were absorbed by the leaves, stems, and roots, with the highest accumulation found in the roots. At elevated metal concentrations, the uptake of Cu, but not of Zn, Cd, or Pb, by the roots was somewhat limited, due to competition with other metals. Between 24% and 59% of the metal content was bound to the cell walls of the plant. Except in the case of Pb, the cell wall-bound fraction was generally smaller in stems than in leaves. No translocation of the metals to other parts of the plant was found, except for Cd, which was translocated from leaf to stem and vice versa. Recently, Fritioff and Greger (2007) showed that Cd is accumulated via direct uptake by both roots and shoots of *Elodea canadensis*.

Furthermore, the ability to accumulate metals varies according to the species. Demirezen and Aksoy (2006) found that the tissues of *Potamogeton pectinatus* accumulated heavy metals more than those of *Groenlandia densa* and Peng et al. (2008) demonstrated that the higher concentrations of Cd, Pb, Cu, Zn, and Mn were found in the leaves of *Potamogeton pectinatus* than in the leaves of *Potamogeton malaianus*. In one study, *Elodea canadensis* exhibited an unusual ability to remove Pb under high salinity conditions, while *Potamogeton natans* was found to have great promise for Zn and Cd removal (Fritioff et al., 2005).

Voluminous literature is available on the macrophytes used for water quality improvement and metal removal. In this work, we established a non-exhaustive review. Among the tested species, *Lemna minor* has been well investigated for the removal of heavy metals (Nasu and Kugimoto, 1981; Maine et al. 2001, Goulet et al., 2005). Researches have also shown the uptake of metals like Ni, Cd and Zn by various species of *Lemna* (Noraho and Gaur 1995; Miretzky et al., 2006). The Floating-leaved aquatic macrophytes, including *Pistia stratiotes*, have been extensively used to remove metals like Zn, Ni, and Cd from the water column (Sridahar, 1986). However, lower biomass of *Pistia stratiotes* was reported by Miretzky et al. (2006) when grown in water containing metal ions of Cd, Ni, Cu, Zn, and Pb as compared to *Spirodela intermedia* and *Lemna minor*. *Ipomoea aquatica* can accumulate higher contents of metals (Sinha et al., 1996; Göthberg et al., 2002) and *Ipomea carnea* has the potential to phyto-extract cadmium (Ghosh and Singh, 2005). *Salvinia molesta* has also been found to be well adapted for the removal of Cr (III) from wastewaters (Maine et al., 2004). *Azolla caroliniana* is helpful in the purification of waters polluted by Hg and Cr (Bennicelli et al., 2004). *Ceratophyllum demersum*, *Wolffia* spp., and *Hydrilla verticillata* have been used as markers to assess the level of heavy metal pollution in aquatic bodies (Keskinkan et al., 2004). Among the emergent species, *Phragmites australis* have been observed in removal of Zn from wastewater (Laing et al., 2003).

## 2.2. Nutrient and Metal Removal

### 2.2.1. Nutrient Mitigation

*Eichhornia crassipes* is reported to assimilate up to 777 mg $Nm^{-2}day^{-1}$ and 200 mg P $m^{-2}day^{-1}$ during the rooting and flowering stages as well as at high temperature conditions (DeBusk et al., 1995). However, phosphorous removal by the *Lemna* spp. is not significant and few papers have suggested employing this macrophyte for the removal of nutrients from polluted water (Al-Nozaily et al. 2000a, b; Cheng et al., 2002a,b).

Species such as *Najas guadalupensis, Ceratophyllum demersum, Chara* spp., and *Potamogeton illininoensis* have been reported to have the capability to remove different chemical species of P (e.g., total phosphorous, soluble reactive phosphorus, $PO_4^{3-}$) (Dierberg et al. 2002). *Chara* spp. has special affinity for N and *C. demersum* has it for P. Juwarkar et al. (1995) highlight the use of constructed wetlands for the removal of BOD, nitrogen, phosphorus, and pathogens from primary treated wastewater. BOD and nitrogen removal were 67%–90% and 58%–63%, respectively. Similarly, Greenway and Wooley (1999) demonstrated that the biochemical oxygen demand concentrations were reduced by 17%-89% and suspended solids concentrations by 14%-77%; reduction in total nitrogen concentrations ranged from 18% to 86%, ammonia nitrogen from 8% to 95%; and oxidised nitrogen from 55% to 98%. The constructed wetland treatment was found to be efficient in removal of BOD and N. This system, which is easy to operate and low cost, can provide an economically viable solution for wastewater management (Juwarkar et al., 1995).

### 2.2.2. Metal Removal

Studies on pollutant bioaccumulation in macrophytes are aimed at assessing removal efficiency of metals and bioaccumulation processes by aquatic macrophytes (Table 4). Chattopadhyay et al. (2002) examined the extent of toxic metal contamination of the east Calcutta wetland ecosystem. The concentrations of heavy metals in wastewater were reduced from 25% to 45% and total Cr was reduced by 95%, along the course of the 40 km journey of the composite wastewater from the sources to the river mouth via the wastewater carrying canal and the stabilization pond. Reductions of 65% Zn and 99% Cr in soil/sediment were recorded between sources and final discharge sites.

Sternberg and Rahmani (1999) reported that *Lemna minor* could remove 70%–80% Pb, using viable biomass. Peng et al. (2008) demonstrated that the average removal efficiencies of *Potamogeton pectinatus* and *Potamogeton malaianus* for Cd, Pb, Mn, Zn, and Cu were 92%, 79%, 86%, 67%, and 70%, respectively. These results indicated that *P. pectinatus* and *P. malaianus* had high capabilities for direct removal of heavy metals from the contaminated water and can be further utilised for pollution monitoring of these metals.

In conclusion, the potential use of aquatic macrophytes in wastewater treatment is worth further exploration.

## REFERENCES

Al-Nozaily F, Alaerts G, Veenstra S (2000a). Performance of duckweed-covered sewage lagoons—I. Oxygen balance and COD removal. *Water Res.* 34(10), 2727–2733.

Al-Nozaily F, Alaerts G, Veenstra S (2000b). Performance of duckweed-covered sewage lagoons—II. Nitrogen and phosphorus balance and plant productivity. Water Res 34(10), 2734–2741.

Aoi T, Hayashi T (1996). Nutrient removal by water lettuce (*Pistia stratiotes*). *Water Sci. Technol.* 34(7–8), 407–412.

Barko JW, James WF (1998). Effects of submerged aquatic macrophytes on nutrient dynamics, sedimentation, and resuspension. In: Sondergaard S, Sondergaard, C (eds) The structuring role of submerged macrophytes in lakes, Springer, Berlin Heidlberg New York, pp 197-214.

Barko JW, Smart RM (1981) Sediment-based nutrition of submersed macrophytes. *Aquatic Botany* 10, 339-352.

Barko JW, Smart RM, McFarland DG, Chen RL (1988). Interrelationships between the growth of *Hydrilla verticillata* (L.f.) Royle and sediment nutrient availability. A*quatic Botany* 32, 205-216.

Bennicelli R, Stezpniewska Z, Banach A, Szajnocha K, Ostrowski J (2004). The ability of *Azolla caroliniana* to remove heavy metals (Hg(II), Cr(III), Cr(VI)) from municipal waste water. *Chemosphere* 55,141–146.

Best EPH (1977). Seasonal changes in mineral and organic components of *Ceratophyllum demersum* and *Elodea canadensis. Aquatic Botany* 3, 337-348.

Best MD, Mantai KE (1979). Growth of *Myriophyllum*: sediment or lake water as the source of nitrogen and phosphorus. *Ecology* 59, 1075-1080.

Birk, S., T. Korte, D. Hering (2006). Intercalibration of assessment methods for macrophytes in lowland streams: direct comparison and analysis of common metrics. *Hydrobiologia* 566, 417-430.

Bole JB, Allan JR (1978). Uptake of phosphorus from sediment by aquatic plants, *Myriophyllum spicatum* and *Hydrilla verticillata*. Water Res 12, 353-358.

Boyd, C. E. (1976). Accumulation of dry matter N and P by cultivated water hyacinths. *Economic Botany, 30*(1), 51–56.

Braun-Blanquet, J. (1964). Pflanzensoziologie. Gründzüge der Vegetationskunde. Wien, Springer Verlag, 865p.

Bunluesin, S., Krutrachue, M., Pokethitiyook, P., Upatham, S., Lanza, G. R. (2007a). Batch and continuous packed column studies of cadmium biosorption by *Hydrilla verticillata* biomass. *Journal of Bioscience and Bioengineering, 103*(6), 509–513.

Bunluesin S., Pokethitiyook P., Lanza G.R., Tyson J.F., Kruatrachue M., Xing B., Upatham S., (2007b). Influences of Cadmium and Zinc interaction and humic acid on metal accumulation in *Ceratophyllum demersum. Water Air Soil Pollution*, 180, 225-235.

Carballeira, A., Vasquez, M.D., Lopez, J. (2001). Biomonitoring of sporadic acidification of rivers on the basis of release of preloaded cadmium from the aquatic bryophyte *Fontinalis antipyretica* Hedw. *Environmental Pollution* 111, 95-106.

Carbiener, R.,Trémolieres, M., Mercier, J.L., Ortscheit, A. (1990). Aquatic macrophyte communities as bioindicators of eutrophication in calcareous oligosaprobe stream waters (Upper Rhine plain, Alsace).*Vegetatio* 86, 71-88.

Chattopadhyay B., Chatterjee A. Mukhopadhyay S. K. (2002). Bioaccumulation of metals in the East Calcutta wetland ecosystem. *Aquatic Ecosystem Health and Management*, 5 (2), 191 – 203.

Chauvin C., Coudreuse J., Court E., Genin B., Germeur G., Haury J., Mignon A., Peltre M-C, Vandewalle C., (2005). Comparaison des réponses de l'indice biologique macrophytique en rivière et autres indices hydrobiologiques. Poster, 6 ème Conférence Internationale des Limnologues et Océanographes, Vaux-en-Vellin, 4-8 Juillet 2005.

Chauvin C., Peltre M-C, Haury J. (2008). La bio-indication et les indices mavcrophytiques, outils d'évaluation et de diagnostic de la qualité des cours d'eau. In : Plantes aquatiques d'eau douce : biologie, écologie et gestion, *Revue Ingénierie –EAT, numéro spécial*, 91-108.

Cheng S, Cifrek VZ, Grosse W, Karrenbrock F (2002a). Xenobiotics removal from polluted water by a multifunctional constructed wetland. *Chemosphere* 48, 415–418.

Cheng J, Bergmann BA, Classen JJ, Stomp AM, Howard JW (2002b). Nutrient recovery from swine lagoon water by *Spirodela punctata*. *Bioresour. Technol.* 81(1), 81–85.

Dalla Vecchia, F., La Rocca, N., Moro, I., De Faveri, S., Andreoli, C., Rascio, N. (2005). Morphogenetic, ultrastructural and physiological damages suffered by submerged leaves of Elodea canadensis exposed to cadmium. *Plant Science*, 168, 329-338.

Dambrine, E., Pollier, B., Poswa, A., Ranger, J., Probst, A., Viville, D., Biron, P. and Granier, A. (1998). Evidence of current acidification in spruce stands in the Vosges Mountains, north-eastern France. *Water, Air and Soil Pollution*, 105, 43– 52.

Dawson, F.H., J.R. Newman, M.J. Gravelle, K.J. Rouen and P. Henville, (1999). Assessment of the Trophic Status of rivers using macrophytes. Evaluation of the Mean Trophic Rank. RandD technical Report E39, *Environment Agency*, 178p.

De Marte JA, Hartman RT (1974). Studies on absorption of $^{32}$P, $^{59}$Fe and $^{45}$Ca by water-milfoil (*Myriophyllum excalbescens* Fernald). *Ecology* 55, 188-194.

DeBusk TA, Peterson JE, Reddy KR (1995). Use of aquatic and terrestrial plants for removing phosphorous from dairy waste waters. *Ecol. Eng.* 5, 371–390.

Demars B.O.L, Thiébaut, G. (2008). Distribution of aquatic plants in the Northern Vosges rivers:implications for biomonitoring and conservation. *Aquatic Conservation* 18, 619-632.

Demars, B.O.L.,Harper, D.M., (1998). The aquatic macrophytes of an English lowland river system: assessing response to nutrient enrichment. *Hydrobiologia*, 384, 75-88.

Demirezen D, Aksoy A. (2004). Accumulation of heavy metals in *Typha angustifolia* (L.) and *Potamogeton pectinatus* (L.) living in Sultan Marsh (Kayseri, Turkey). *Chemosphere* 56, 685–696.

Demirezen D, Aksoy A. (2006). Common hydrophytes as bioindicators of iron and manganese pollutions. *Ecological Indicators* 6 (2), 388-393.

Dhote, S., Dixit, S. (2008). Water quality improvement through macrophytes: a review. *Environmental Monitoring and Assessment*. In press.

Dierberg FE, DeBusk TA, Jackson SD, Chimney MJ, Pietro K (2002). Submerged aquatic vegetation-based treatment wetlands for removing phosphorus from agricultural runoff: response to hydraulic and nutrient loading. *Water Res.* 36, 1409–1422.

Directive, 2000/60/EC. Water Framework Directive of the European Parliament and of the Council of 23 October 2000.

Doyle RD, Francis MD, Smart RM (2003). Interference competition between *Ludwigia repens* and *Hygrophila polysperma*: two morphologically similar aquatic plant species. *Aquatic Botany* 77, 223–234.

Duman, F., Obali O., Demirezen, D. (2006). Seasonal changes of metal accumulation and distribution in shining pondweed (*Potamogeton lucens*), *Chemosphere* 65, 11, 2145–2151.

Dunbabin, J.S. Bowmer, K.H. (1992). Potential use of constructed wetlands for treatment of industrial wastewaters containing metals. *Sci. Total Environ.* 111, 151–168.

Elankumaran, R., Raj, M. B., and Madhyastha, M. N. (2003). Biosorption of copper from contaminated water by Hydrilla verticillata Casp. and Salvinia sp. *Green Pages, Environmental News Sources.*

Eugelink AH (1998) Phosphorus uptake and active growth of *Elodea canadensis* Michx. and *Elodea nuttallii* (Planch.) St. John. *Water Sci. Technol.* 37, 59-65.

Fritioff A, Kautsky L, Greger M (2005). Influence of temperature and salinity on heavy metal uptake by submersed plants. *Environ. Pollut.* 133 (2), 65–274.

Fritioff A., Greger M. (2007). Fate of Cadmium in *Elodea canadensis*. *Chemosphere* 67, 365-375.

Fritioff, A. Greger, M. (2006). Uptake and distribution of Zn, Cu, Cd, and Pb in an aquatic plant *Potamogeton natans*. *Chemosphere* 63, 220–227.

Garg P., Tripathi R.D., Rai U.N., Sinha S., Chandra P., (1997). Cadmium accumulation and toxicity in submerged plant *Hydrilla verticillata* (L.F.) Royle. *Environmental Monitoring and Assessment*, 47, 167-173.

Gazi NWR, Steven PKS (1999). Bioremoval of lead from water using *Lemna minor*. *Bioresour. Technol.* 70, 225–230.

Ghosh M, Singh SP (2005) A comparative study of cadmium phytoextraction by accumulator and weed species. *Environ. Pollut.* 133, 365–371.

Göthberg A, Greger M, Bengtsson B-E (2002). Accumulation of heavy metals in Water Spinach (*Ipomoea aquatica*) cultivated in the Bangkok region, Thailand. *Environ. Toxicol. Chem.* 21(9),1934–1939.

Goulet RR, Lalonde JD, Munger C, Dupuis S, Dumont G, Premont S, Campbell PGC (2005) Phytoremediation of effluents from aluminum smelters: a study of Al retention in mesocosms containing aquatic plants. *Water Res.* 39, 2291–2300.

Greenway M., Woolley A. (1999). Constructed wetlands in Queensland: *Performance efficiency and nutrient bioaccumulation Ecological Engineering*, 12(1), 39-55.

Haury J, Peltre M-C, Trémolieres M, Barbe J, Thiébaut G, Bernez I, Daniel H, Chatenet P, Muller S, Dutartre A, Laplace-Treyture C, Cazaubon A, Lambert-Servien E, (2006). A new method to assess water trophy and organic pollution : the Macrophyte Biological Index for Rivers (IBMR). Its application to different types of rivers and pollution. - *Hydrobiologia*. 570, 153-158.

Holmes, N. T. H., Newman, J. R., Chadd, S., Rouen, K. J., Saint, L., Dawson, F. H. (1999) Mean Trophic Rank: A users manual. RandD Technical Report E38, *Environment Agency of England and Wales*, Bristol, UK.

Juwarker, A. S., Oke, B., Juwarkar, A., Patnaik, S. M. (1995). Domestic wastewater treatment through constructed wetland in India. *Water Science Technology*, London 32(3), 291–294.

Kadlec R, Knight R (1996) Treatment wetlands. *Lewis Publishers*, Boca Raton.

Kadlec RH (2000) The inadequacy of first-order removal models. *Ecol. Eng.* 15, 105–119.

Kähkönen M.A, Kaireselo, T. (1998). The effects of nickel on the nutrient fluxes and on the growth of *Elodea canadensis*. *Chemosphere* 37, 8, 1521-1530.

Kähkönen M.A, Manninen, P.K.G. (1998). The uptake of nickel and chromium from water by *Elodea canadensis* at different nickel and chromium exposure levels. *Chemosphere,* 36, 6, 1381-1390.

Kamal M, Ghalya AE,Mahmouda N, Cote R (2004) Phytoaccumulation of heavy metals by aquatic plants. *Environ. Int.* 29, 1029–1039.

Kara Y., Zeytunluoglu A. (2007). Bioaccumulation of toxic metals (Cd and Cu) by *Groenlandia densa* (L.) *Fourr. Bull Environ Contam Toxicol.*, 79, 609-612.

Keskinkan O, Goksu MZL, Basibuyuk M, Forster CF (2004) Heavy metal adsorption properties of a submerged aquatic plant (*Ceratophyllum demersum*). *Bioresour. Technol.* 92,197–200.

Kohler A., Schneider S. (2003) Macrophytes as bioindicators. Archiv für Hydrobiologie, *Supplement*, 147, 17–31.

Kufel L, Kufel I (2002) *Chara* beds acting as nutrient sinks in shallow lakes—a review. *Aquat Bot.* 72, 249–260.

Laing, G. Du., Tack, F. M. G., Verloo, M. G. (2003). Performance of selected destruction methods for the determination of heavy metals in reed plants (*Phragmities australis*). *Analytica Chimica Acta* 497 (1-2), 191-198.

LaMontagne JM, Jackson LJ, Barclay RMR (2003) Compensatory growth responses of *Potamogeton pectinatus* to foraging by migrating trumpeter swans in spring stop over areas. *Aquatic Botany* 76, 235–244.

Lee C.L., Wang T.C., Hsu C.H, Chiou A.A., (1999). Heavy Metal Sorption by aquatic plants in Taiwan. *Bull. Environ. Contam. Toxicol.*, 61, 497-504.

Lejeune A, Cagauan A, van Hove C (1999) Azolla research and development: recent trends and priorities. *Symbiosis* 27,333–351.

Maine M A, Duarte M, Suñé N (2001) Cadmium uptake by floating macrophytes. *Water Resour.* 35:2629–2634.

Maine MA, Suné NL, Lagger SC (2004) Chromium bioaccumulation: comparison of the capacity of two floating aquatic macrophytes. *Water Res.* 38,1494–1501.

Mal T.K., Adorjan P., Corbett A.L., (2002). Effect of copper on growth of an aquatic macrophyte, *Elodea canadensis. Environmental Pollution* 120, 307-311.

Maleva M.G., Nekrasova G.H., Bezel V.S. (2004). The responses of hydrophytes to Environmental Pollution with Heavy Metals. *Russian Journal of Ecology*, 35, 4, 230-235.

Mayes, R.A., MacIntosh, A.W., Anderson V.L. (1977). Uptake of cadmium and lead by a rooted aquatic macrophyte (*Elodea canadensis*). *Ecology*, 58, 1176-1180.

Melzer, A. (1999). Aquatic macrophytes as tools for lake management. *Hydrobiologia* 395-396(0): 181-190.

Miretzky P, Saralegui A, Cirelli AF (2006) Simultaneous heavy metal removal mechanism by dead macrophytes. *Chemosphere* 62(2), 247–254.

Muller S. (1990).Une séquence de groupements végétaux bio-indicateurs d'eutrophisation croissante des cours d'eau faiblement mineralisés des Basses Vosges gréseuses du Nord C. R. *acad. Sci. Paris* 310,(serie III): 509-514.

Muramoto, S., Oki, Y. (1983). Removal of some heavy metals from polluted water by water hyacinth (*Eichhornia crassipes*). *Bulletin of Environmental Contamination and Toxicology* 30, 170–177.

Nakada M., Fukaya, K., Takesshita, S., Wada, Y., (1979). The accumulation of heavy metals in the submerged Plant (*Elodea nuttallii*). *Bull. Environm. Contam. Toxicol.*, 22, 21-27.

Nasu Y, Kugimoto M (1981) Duckweed as an indicator of water pollution. I. The sensitivity of *Lemna pancicostata* to heavy metals. *Arch. Environ. Contam. Toxicol.* 10,159–169.

Nor, Y. M. (1990). The absorption of metal ions by *Eichhornia crassipes*. *Chemical Speciation and Bioavailability* 2, 85–91.

Noraho N, Gaur JP (1995) Effect of cations, including heavy metals, on cadmium uptake by *Lemna polyrhiza* L. *Biometals* 8, 95–98.

Nussbaum S., Schmutz D., Brunold, C. (1988). Regulation of assimilatory sulfate reduction by cadmium in *Zea mays* L, *Plant Physiol.* 88, 1407–1410.

Ornes, W. H.Sajwan, K. S.(1993). Cadmium accumulation and bioavailability in Coontail (*Ceratophyllum demersum* L.) plants. *Water, Air, and Soil Pollution*, 69 (3-4), 291-300.

Ozimek T, Van Donk E, Gulati RD (1993) Growth and nutrient uptake by two species of *Elodea* in experimental conditions and their role in nutrient accumulation in a macrophyte-dominated lake. *Hydrobiologia* 251:13–18.

Peng K, Luo C, Lou L, Li X, Shen Z. (2008). Bioaccumulation of heavy metals by the aquatic plants *Potamogeton pectinatus* L. and *Potamogeton malaianus* Miq. and their potential use for contamination indicators and in wastewater treatment. *Sci. Total Environ.*, 392 (1), 22-29.

Penning, W.,Dudley, B., Mjelde, M., Hellsten, S., Hanganu, J., Kolada, A., van den Berg, M., Poikane, S., Phillips, G., Willby, N., Ecke, Frauke (2008). Using aquatic macrophyte community indices to define the ecological status of European lakes. *Aquatic Ecology*, 42 (2), 253-264.

Pilon-Smits E, Pilon M (2002) Phytoremediation of metals using transgenic plants. *Crit. Rev. Plant. Sci.* 21(5), 439–456.

Pinto, C. L. R., Caconia, A., Souza, M. M. (1987). Utilization of water hyacinth for removal and recovery of silver from industrial waste water. Water, *Science and Technology* 19(10), 89–101.

Prakash O, Mehroira I, Kumar P (1987). Removal of cadmium from water by water hyacinth. *J. Environ. Eng.* 113, 352–365.

Probst, A., Dambine, E., Viville, D., Fritz, B. (1990a). Influence of acid atmospheric inputs on surface water chemistry and mineral fluxes in a declining spruce stand within a small granitic catchment (Vosges, France). *Journal of Hydrology*, 116, 101-124.

Probst, A., Massabuau, J.C., Probst, J.L., Fritz, B. (1990b). Acidification des eaux de surface sous l'influence des précipitations acides: rôle de la végétation et du substratum, conséquence pour les populations de truites. Le cas des ruisseaux des Vosges. *Compte Rendu de l'Académie des Sciences de Paris*, 311, 405– 411.

Prygiel J., Haury J. (2006). Monitoring methods based on algae and macrophytes. In: Biological Monitoring of Rivers Edited by G.Ziglio, M.Siligardi and G.Flaim, *John Wiley and Sons*, 155-170.

Reddy KR, Agami M, Tucker JC (1990). Influence of phosphorus on growth and nutrient storage by water hyacinth (*Eichhornia crassipes* (Mart.) Solms) plants. *Aquatic Botany* 37, 355–365.

Robach, F., Thiébaut G, Trémolières M., Muller S (1996). A reference system for continental running waters: plant communities as bioindicators of incrising eutrophication in alkaline and acidic water in north-east France. *Hydrobiologia* 340, 67-76.

Sahu RK, Naraian R, Chandra V (2007). Accumulation of metals in naturally grown weeds (aquatic macrophytes) grown on an industrial effluent channel. *Clean* 35(3), 261–265.

Samecka-Cymerman, A.,Kempers, A. J. (2003). Biomonitoring of Water Pollution with *Elodea canadensis*. A Case Study of Three Small Polish Rivers with Different Levels of Pollution. *Water, Air, and Soil Pollution*, 145, 139-153.

Satake, K., Nishikawa, M., Shibata, K., (1989). Distribution of aquatic bryophytes in relation to water chemistry of the acid riverAkagawa, Japan. *Archiv für. Hydrobiology.* 116, 299-311.

Schachtman DP, Reid RJ, Ayling SM (1998). Phosphorus uptake by plants: From soil to cell. *Plant Physiol.* 116, 447-453.

Schaumburg, J., Schranz Ch., Hofmann G., Stelzer D., Schneider S., Schmedtje U., (2004). Macrophytes and phytobenthos as indicators of ecological status in German lakes - a contribution to the implementation of the Water Framework Directive. *Limnologica* 34, 302-314.

Schneider, I.A.H., Rubio J., (1999). Sorption of Heavy metal ions by the non living Biomass of Freshwater Macrophytes. *Environ. Sci. Technol.*, 33, 2213-2217.

Schneider, S, Melzer, A. (2003). The Trophic Index of Macrophytes (TIM)- a new Tool for indicating the trophic state of running waters. *Internat. Rev. Hydrobiol.* 88, 49-67.

Schneider, S. (2007). Macrophyte trophic indicator values from a European perspective. Limnologica - *Ecology and Management of Inland Waters.* 37 (4), 281-289.

Sinha S, Rai UN, Chandra P (1996) Metal contamination in aquatic etables *Trapa natans* L. and *Ipomea aquatica* Forsk. In: Proceedings of conference on progress in crop sciences from plant breeding to growth regulation, 17–19 June, Mosonmagyarovar, Hungary.

Sivaci A., Elmas E., Gümüs, Sivaci E.R., (2008). Removal of cadmium by *Myriophyllum heterophyllum* Michx. and *Potamogeton crispus* L. and its effects on pigments and total phenolic compounds. *Arch. Environ. Conttam. Toxicol.*, 54, 612-618.

Sladecek, V., (1973). System of water quality from the biological point of view. Arch. Hydrobiol., Beiheft 7, *Ergebnisse der Limnologie* (I–IV), 1–218.

Smith FW, Mudge SR, Rae AL, Glassop D (2003). Phosphate transport in plants. *Plant Soil* 248, 71-83.

Søndergaard M, Jeppesen E, Peder JJ, Lildal SA (2005) Water Framework Directive: ecological classification of Danish lakes. *J. Appl. Ecol.* 42(4):616–629.

Sooknah RD, Wilkie AC (2004) Nutrient removal by floating aquatic macrophytes cultured in anaerobically digested flushed dairy manure wastewater. *Ecol. Eng.* 22(1),27–42.

Sridahar M (1986). Trace element composition of *Pistia stratiotes* in a polluted lake in Nigeria. *Hydrobiologia* 131, 273–276.

Srivastava J., Gupta A, Chandra H. (2008). Managing water quality with aquatic macrophytes Reviews in Environmental Science and Bio/Technology. In press.

Staniszewski, R., Szoszkiewicz K., Zbierska J., Lesny J., Szymon Jusik S., Clarke R.T (2006). Assessment of sources of uncertainty in macrophyte surveys and the consequences for river classification. *Hydrobiologia* 566(1), 235-246.

Stelzer D, Schneider S, Melzer A (2005) Macrophyte based Assessment of lakes—a contribution to the implementation of the European Water Framework Directive in Germany. *Int. Rev. Hydrobiol.* 90(2), 223–237.

Sternberg SPK, Rahmani GNH (1999) Bioremoval of lead from water using *Lemna minor*. Bioresour Technol 70:225–230. Re: from Lemna Corporation (1992) Harvesting equipment makes the difference. Lemna Corporation's Retention Times.

Sune, N., Sanchez, G., Caffaratti S., Maine, M.A., (2007). Cadmium and chromium removal kinetics from solution by two aquatic macrophytes. *Environmental pollution*, 145, 467-473.

Szoszkiewicz K, Ferreira T, Korte T., Baattrup-Pedersen A, Davy-Bowker J., O'Hare M, (2006). European river plant communities: the importance of organic pollution and the usefulness of existing macrophyte metrics. *Hydrobiologia* 566(1): 211-234.

Szymanowska A., Samecka- Cymerman A., Kempers A.J., (1999). Heavy metals in three lakes in West Poland. *Ecotoxicology and Environmental Safety*, 43, 21-29.

Thiébaut, G., Muller, S. (1998). The impact of eutrophication on aquatic macrophyte diversity in weakly mineralised streams in the Northern Vosges Mountains. (N-E, France). *Biodiversity and Conservation* 7, 1051-1068.

Thiébaut, G., Muller, S. (1999). A macrophyte communities sequence as an bioindicator of eutrophication and acidification levels in weakly mineralised streams in North-Eastern France. *Hydrobiologia* 410, 17-24.

Thiébaut G, Muller S (2003). Linking phosphorus pools of water, sediment and macrophytes in running waters. *Ann. Limnol.* 39, 307-316.

Thiébaut, G., Guérold F, Muller, S. (2002). Are trophic indices based on macrophyte communities pertinent tools to monitor water quality ? *Water Research* 36, 3602-3610.

Thiébaut G., Tixier G., Guérold F., Muller S., (2006). Comparison of different biological indices for the assessment of rivers quality: application to the river Moselle (France). *Hydrobiologia*, 570, 159-164.

Thiébaut G., Rolland T., Robach F., Trémolières M., Muller S., (1997). Quelques conséquences de l'introduction de deux espèces de macrophytes, *Elodea canadensis* Michaux et *Elodea nuttallii* St. John, dans les écosystèmes aquatiques continentaux: exemple de la Plaine d'Alsace et des Vosges du Nord (Nord-Est de la France). *Bulletin Fran9ais de la Pêche et de la Pisciculture*, 344/345, 441-452.

Thiébaut, G., Vanderpoorten A, Guérold F, Boudot J-P, S. Muller S. (1998). Bryological patterns and streamwater acidification in the Vosges mountains (N-E France): An analysis tool for the survey of acidification processes. *Chemosphere 36* (6), 1275-1289.

Tiwari, S., Dixit, S., Verma, N. (2007). An effective means of bio-filtration of heavy metal contaminated water bodies using aquatic weed *Eichhornia crassipes*. *Environmental Monitoring and Assessment*, 129, 253–256.

Tracy M, Montante JM, Allenson TE, Hough RA (2003). Long-term responses of aquatic macrophyte diversity and community structure to variation in nitrogen loading. *Aquatic Botany* 77, 43–52.

Tremp H., Kohler, A., (1995). The usefulness of macrophytes monitoring-systems, exemplified on eutrophication and acidification of running waters. *Acta botanica Gallica* 142, 541-550.

Tripathi RD, Rai UN, Gupta M, Yunus M, Chandra P (1995). Cadmium transport in submerged macrophyte *Ceratophyllum demersum* L. in presence of various metabolic inhibitors and calcium channel blockers. *Chemosphere 31*(7), 3783–3791.

Vanderpoorten, A., G. Thiébaut, M. Trémolières, S. Muller, (2000). A model for assessing water chemistry by using aquatic bryophyte assemblages in north-eastern France. *Verhandlungen des Internationales Verein für Limnologie* 27, 807-810.

Vanderpoorten, A., Palm R. (1998). Canonical variables of aquatic bryophyte combinations for predicting water trophic level. *Hydrobiologia* 386(1), 85-93.

Vardanyan L.G. Ingole B.S., (2006). Studies on heavy metal accumulation in aquatic macrophytes from Sevan (Armenia) and Carambolim (India) lake systems. *Environ. Int.* 32, 208–218.

Vermeer CP, Escher M, Portielje R, de Klein JJM (2003) Nitrogen uptake and translocation by *Chara*. *Aquatic Botany* 76, 245–258.

Willby N, Pitt J, Phillips G (2006). Summary of approach used in LEAFPACS for defining ecological quality of rivers and lakes using macrophyte composition. *Draft Report* January 2006.

Xie Y, Yu D (2003). The significance of lateral roots in phosphorus (P) acquisition of water hyacinth (*Eichhornia crassipes*). *Aquatic Botany* 75, 311–321.

In: Aquatic Ecosystem Research Trends
Editor: George H. Nairne

ISBN 978-1-60692-772-4
© 2009 Nova Science Publishers, Inc.

*Chapter 2*

# MANAGEMENT OF INVASIVE AQUATIC PLANTS IN FRANCE

## *G. Thiébaut*[*1] *and A. Dutartre*[2]

[1]Laboratoire Interactions Ecotoxicité, Biodiversité Ecosystèmes (L.I.E.B.E.),
Université Paul Verlaine de Metz, UMRCNRS 7146,
Avenue Général Delestraint, 57070 Metz, France
[2]Cemagref, Unité de Recherches Réseaux, Epuration et Qualité des Eaux, 50,
Avenue de Verdun, 33612 CESTAS Cedex. alain.dutartre@bordeaux.cemagref.fr

## ABSTRACT

Some invasive species are considered to cause "nuisance". In general, the priority of biological invasion control is to prevent new infestations from taking hold, especially in the cases of the fastest-growing and most disruptive species. Exotic macrophytes that do not rapidly increase in numbers are lower priorities. Controlling large infestations of plants with considerable environmental impact, such as *Ludwigia* spp. or Hydrocharitaceae (*Lagarosiphon major, Egeria densa, Elodea* species), is of the highest priority. There are several methods for managing aquatic plants: mechanical or manual harvesting, hand-pulling, biological control, changing the aquatic environment and chemical control. The management techniques chosen must be suited to the aquatic plant itself as well as the uses and function of the body water. Action is recommended only after careful analysis indicates that leaving the spreading species unchecked will result in greater damage than that caused by control efforts.

Early-established management plans provide the first steps towards sustainable management of aquatic environments. However, these efforts are compromised as long as invasive aquatic plant species continue to be sold to individuals. Stronger enforcement of existing laws coupled with an intensive public education campaign is needed to prevent further alien species introduction.

**Keywords**: invasive aquatic plants, freshwaters, France, management

---

[*] Email: thiebaut@univ-metz.fr

# INTRODUCTION

It is widely acknowledged that biological invasions are a major cause of biodiversity loss. In freshwater, Invasive Aquatic Plants (IAP) attract considerable attention [67]. The characteristic conditions of inland waters such as internal connectivity, high seasonal and spatial variability, and extensive interfaces (shorelines) of contact with terrestrial ecosystems make them especially vulnerable to the invasion of IAP. IAP present considerable challenges for conservationists and managers; these plants cause significant damage in freshwater, such as increased risk of flooding due to occupation of the riverbeds by high densities of plants, physicochemical impacts (i.e., oxygen depletion in dense beds), and impacts on aquatic fauna and native plants (competition for resources). In addition, they present problems for water usage (exploitation, fishing, swimming, sailing, drainage, irrigation). For example, *Ludwigia grandiflora*, *Myriophyllum aquaticum* and *Egeria densa* form dense monospecific stands that restrict water movement, trap sediment and cause fluctuations and water quality deterioration.

Since the review presented by Nichols [48], the field of IAP management has expanded significantly. Selection of management techniques is based on site use intensity as well as economic, environmental, and technical constraints [18]. In France, many management actions are partially financed by public administration at different levels (ranging from local to national), which facilitates their application given the high cost of treatments.

# 1. INVASIVE AQUATIC PLANTS IN FRANCE

IAP can be aggressive in their native countries, countries in which they have been introduced or both. *Myriophyllum spicatum* (a common species in France) and *Trapa natans* (a rare and locally protected species in France) are two clear examples. The number of European aquatic plants that have invaded North America is greater than the number of North American aquatic plants that have invaded Europe [59]. Nonindigenous plant species may be widespread in one part of the introduced area and restricted in another part. For example, *Crassula helmsii* is a rapidly spreading aggressive species that is damaging aquatic ecosystems in Great Britain, Belgium, Germany, and Ireland [66], whereas previously it was found only occasionally in aquatic environments in the northwestern part of France.

France is a good place to study the distribution of IAP into Europe because it is characterized by three bio-geographical areas (Atlantic [A], Continental [C], and Mediterranean [M]). Distribution of plants varies with bio-geographical zone. For example, the diploid *L. peploides* colonizes mainly the Mediterranean region of France, while the polyploid *L. grandiflora* predominates in the other regions [13].

This does not aim to be an exhaustive review of IAP in metropolitan France, but rather an overview of the highest priority species and the potential need for control of IAP (Table 1). Twenty species were selected, and several sources with information on IAP were consulted [2, 23, 44].

It is less of a priority to control IAP that are not rapidly increasing in numbers, proliferating in undisturbed habitats, or interfering in areas recovering from disturbances. However, controlling large infestations of plants with considerable environmental impact, such as *Ludwigia* spp. or Hydrocharitaceae (*Lagarosiphon major*, *Egeria densa*), is of the highest priority. The IAP most in need of control include four widespread invasive species belonging to the Hydrocharitacae family (*Egeria densa, Elodea canadensis, Elodea nuttallii, Lagarosiphon major*).

**Table 1. Origin, date and vector of the introduction of the most invasive or potentially invasive aquatic plants in France. ?: unknown date of introduction. Vectors: (a) escapes from aquaria, (b) by seagoing ships (ballast water, timber trade), (c) intentionally introduced,(d) natural expansion, ?: unknown vector**

|  | Family | Origin Area | First found | Vectors |
|---|---|---|---|---|
| *Azolla filiculoides* Lam. | Azollaceae | N.Am, S. Am, Australia | 1880 | a |
| *Crassula helmsii* (Kirk) Cockayne, (Hook.f.) Ostenf. non N.E.Br., (Hook.f.) Hook.f. | Crassulaceae | Australia, New-Zealand | before 2006 | c |
| *Cabomba caroliana A.Gray var. caroliana* | Cabombacae | S.Am | 2005 | a |
| *Egeria densa* Planchon | Hydrocharitaceae | S.Am | 1961 | a |
| *Eichhornia crassispes* (Mart.) Solms | Pontederiaceae | S.Am | before 1993 | c |
| *Elodea canadensis* Michaux | Hydrocharitaceae | N.Am | 1845 | a/b |
| *Elodea ernstiae* H. St. John | Hydrocharitaceae | S.Am | 1959 | a |
| *Elodea nuttallii* (Planchon) H. St. John | Hydrocharitaceae | N.Am | 1959 | a |
| *Hydrilla verticillata* (L.f.) Royle | Hydrocharitaceae | Australia, Asia, Africa | before 1997 | a/c |
| *Hydrocotyle ranunculoides L.f* | Apiacaea | N.Am | ? | c |
| *Lagarosiphon major* (Ridley) Moss | Hydrocharitaceae | S. Af | 1940 | a |
| *Lemna minuta (Kunth)* | Lemnacaeae | N. and S. Am | 1965 | d |
| *Lemna turionifera* Landolt | Lemnacaeae | N. Am | 1992 | d? |
| *Ludwigia grandiflora subsp. hexapetala* (Hook. and Arn.) Nesom and Kartesz | Onagraceae | S.Am | 1820-1830 | c |
| *Ludwigia peploides subsp. montevidensis* (Spreng.) Raven | Onagraceae | S.Am | 1820-1830 | c |
| *Myriophyllum aquaticum* (Velloso) Verdcourt | Haloragaceae | S.Am | 1880 | a |
| *Myriophyllum heterophyllum* Michaux | Haloragaceae | N.Am | before 1997 | a/c |
| *Pistia stratiotes* L. | Araceae | S. Am | 2002-2003 | a/c |
| *Pontederia cordata* L. | Pontederiaceae | N. Am | 2006 | c |

Three more IAP (*Ludwigia grandiflora, L. peploides, Myriophyllum aquaticum*) are amphibious species [12, 15, 20, 29, 35, 50, 51]. The *Ludwigia* species are now considered to be the most invasive aquatic plants in France. The high morphological plasticity of these plants has allowed them to tolerate fluctuations in habitat characteristics (e.g., water level, water current, and soil moisture) and to efficiently colonize bodies of water (Figure 1). Of the 567 sites investigated in France by Dutartre et al. [25], *Ludwigia* sp. were found in rivers with low water velocity in summer (29%), in shallow wetlands (20%), in ditches and channels (20%), in ponds and on lake shores (13%), in oxbows (9%), and in wet meadows (4%).

Many IAP were introduced at the end of the 19th century and the beginning of the 20th century (Table 1). Of all the invasive or potentially invasive species listed here, 55% came from South America, 40% came from North America and 15% came from Australia or New-Zealand (Table 1). Most invasive plant species arrived in France as a result of human intervention (e.g., aquarium plants, ornamental use). Among the well-known examples of aquarium plants are certain Hydrocharitaceae species (*Egeria densa, Elodea sp., Lagarosiphon major, Hydrilla verticillata*) and some other taxa (*Myriophyllum aquaticum*). Plants that escaped from outdoor ponds (*Cabomba caroliana, Eichhornia crassipes, Ludwigia sp.* and *Pistia stratiotes)* can easily colonize freshwater environments (Table 1).

The IAP *Althernanthera philoxeroides, Cabomba caroliana, Crassula helmssii, Eichhornia crassipes, Hydrilla verticillata, Hydrocotyle ranunculoides, Myriophyllum heterophyllum, Pistia stratiotes*, and *Pontederia cordata* have spread worldwide, but have only recently been introduced in France [16, 21, 33, 65; Table 2].

Figure 1. Some morphological adaptations of *Ludwigia sp* in France. (picture Dutatre A).

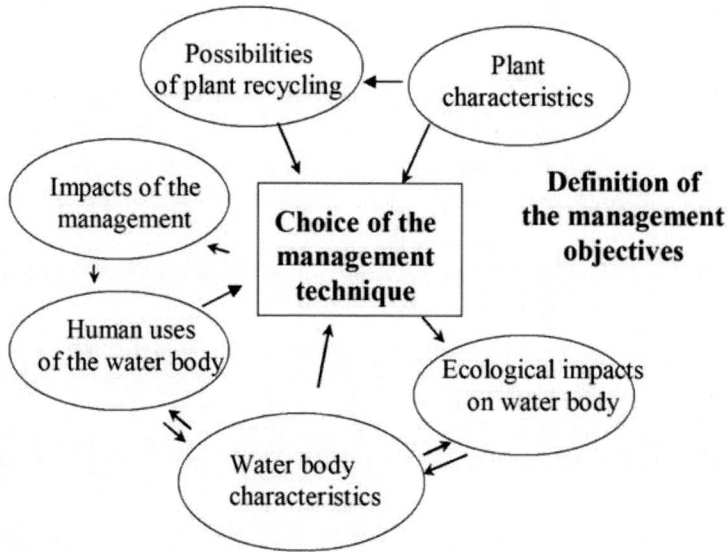

Figure 2. Choice elements for management techniques of aquatic plant (adapted from Dutartre, 2002).

## Table 2. Potential high priority IAP for control
## and their Distribution in Europe and in France

| | Distribution | | |
|---|---|---|---|
| | Europe | First found | France |
| | widespread | in France | |
| *Althernanthera philoxeroides* (Martius) Griseb | ? | 1971 | South West of France |
| *Crassula helmsii* (Kirk) Cockayne, (Hook.f.) Ostenf. non N.E.Br., (Hook.f.) Hook.f. | Belgium, Denmark,Germany, The Netherlands, the UK | before 2006 | North-Pas de Calais, Brittany |
| | Europe | First found | France |
| | widespread | in France | |
| Cabomba caroliana A.Gray var. caroliana | ? | 2005 | Burgundy ( close to Dijon) |
| Eichhornia crassispes (Mart.) Solms | South Europe, Portugal | before 1993 | Moselle river, |
| Hydrilla verticillata (L.f.) Royle | the UK, Belgium | before 1997 | South West of France |
| Hydrocotyle ranunculoides L.f | Belgium,Italy, Netherlands, Portugal, Spain, the UK | ? | south of Paris, Normandie, Corsica, North West, Dombes area ( near Lyon) |
| Myriophyllum heterophyllum Michaux | Ireland, Germany, Belgium | before 1997 | |
| Pistia stratiotes L. | ? | 2002 | Moselle river, Bordeaux, Landes-Pyrénées-Atlantiques |
| Pontederia cordata L. | ? | 2006 | Ardennes |

These species, sold in the aquarium trade, are potentially invasive. *Althernanthera philoxeroides, Myriophyllum heterophyllum* and *Pontederia cordata* are now present in very few sites with no apparent colonization dynamics. *Eichhornia crassipes, Hydrilla verticillata* and *Pistia stratiotes* are observed in some sites, but in a sporadic manner (they disappear quickly at the end of summer). *Cabomba caroliana, Crassula helmssii,* and *Hydrocotyle ranunculoides* were recently found in some sites and exhibited visible dynamics of colonization.

Although the time-lag phenomenon (during which a given population remains small and geographically restricted) was long in France for *Ludwigia grandiflora* subsp. *Hexapetala* and *Ludwigia peploides* subsp. *Montevidensis* (about 1.5 centuries) [20] and for Hydrocharitaceae such as *Elodea* species (*Elodea □anadensis, Elodea nuttallii*), the invasion of aquatic environments can be extremely rapid. Once a highly invasive species arrives, it is difficult to prevent its rapid spread. Many IAP are freely dispersed by water or animals to new ecosystems. Thus, prevention is the most effective and economically least-expensive environmental response. The difficulties of fighting IAP mean investment in prevention is likely to be the most successful response to biological invasions. However, an early warning system to trigger rapid interventions is difficult to put into practice because it requires the involvement of field observers and alert managers, which is not always possible. Much greater effort should be expended to avoid the introduction of IAP into novel environments.

## 2. MANAGEMENT OF IAP

The management techniques chosen must be appropriate to both the IAP and the uses and function of the body water. Some invasive species are considered to cause "nuisance", wherein the degree of nuisance is judged in relation to the aim of water body management (such as transportation, recreation, fishery management, or conservation). The degree of invasiveness of a species is a misleading indicator for management. For example, some species (such as the non-indigenous Lemna sp. Or the water-fern Azolla filiculoides, which are found at fewer sites than the Elodea species) are highly invasive and continuing to spread. However, typically no management is required except in cases of colonization of dense networks of drainage ditches or wetlands in western (e.g., Marais Poitevin) and northwestern France.

High-priority, high-intensity use sites such as beaches might justify high-cost management techniques such as harvesting, while low-intensity use areas might remain untreated if resources are low or be marked for less expensive techniques. Likewise, areas with higher concentrations of plants, such as boat launches, should receive more attention than areas with no plants or with acceptable levels of infestation (Figure 3a).

The action plan should take into account the plants' actual and potential impacts on ecosystem functioning as well as the indigenous species and communities present, particularly if rare and/or ecologically important species are targeted for conservation. Action is recommended only after careful analysis indicates that leaving the spreading species unchecked will result in greater damage than that caused by the control effort.

Figure 3a. Harvesting *Elodea* species in the La Plaine reservoir (picture Thiébaut G).

Before acting, three preliminary steps must take place [18]:

*First,* one must define the management objectives, waterbody and IAP characteristics, human uses and the nuisances caused by the invasive species.

*Second,* one must consider the choice of techniques (single or combined), taking into account the expected results and potential impacts.

*Finally,* one must define the timetable of the management program (multi-year organization and funding).

After acting, two further assessments are required: analysis of efficiency and analysis of the ecological impacts of management.

The use of techniques varied both spatially and temporally. For example, Marais Poitevin, the second-largest wetland in the west of France, has a dense network of waterways (more than 4000 km) and was invaded by the *Lemna* species 30 years ago and by the *Ludwigia* sp. Almost 20 years ago. Since 1992, the management of *Lemna sp., Spirodela sp.* And *Wolffia sp.* has consist of plant removal and its subsequent usage as green manure (Pipet, personal communication). Three combined techniques for managing *Ludwigia sp.* were tested from 1994 to 2006: herbicide, mechanical extraction and hand-pulling [26]. Hand-pulling, the first removal operation in the initial stages of the invasion, was also used in other sites invaded by *Ludwigia grandiflora* and *Myriophyllum aquaticum* in southwest France (Figure 3b). When these species became well-established, mechanization was necessary [20, 22].

No management technique will work for all situations in a management program. Some techniques are more expensive but will better control dense populations in larger areas. For small nuisance plant populations or new colonies, or in biological conservation areas, hand-pulling may actually be the best approach.

**Table 3. Management of the highest Invasive Aquatic plants (IAP)**

| 1: not allowed now, no literature data<br>2: tested in laboratory in France | | Afil | Ccar | Eden | Ecan | Enut | Hran | Lmaj | Lmin | Lgra | Lpep | Maqu |
|---|---|---|---|---|---|---|---|---|---|---|---|---|
| Biological techniques | Grass Carp[1] | | | x? | x | x | | x ( in 1965) | x? | x | | |
| | Macroinvertébrates[2] | | | | x | x | | | | | | |
| | Pasture | | | | | | | | | x | x | |
| Chemical techniques | Herbicide | | | x | | | | x | | x | x | x |
| Manual & Mechanical techniques | Cutting | | | x | | | | x | | x | x | x |
| | Removal | x? | | | | | | | x | | | |
| | Harvest | x? | x | x | x | x | x | x | | x | x | x |
| | Hand-pulling | | | x? | x | x | | x | | x | x | x |
| Physical techniques | Dredging | x | | x | x | x | | | | x | x | x |
| | Drawdown | | | | x | | | | | x | x | |
| | Salt management | | | | | | | | | x | x | |
| | Light attenuation[2] | | | | x | | | | x | | | |
| No action | | x | | | | | | | | | | |

Figure 3b. Mechanical extraction of dense beds of *Ludwigia grandiflora* in the pond « Etang du Turc » in southwest France. (picture Dutatre A).

Generally, the dominant management techniques used in Europe are mechanical harvesting and the introduction of grass carp; the latter is also used to control native plants [66]. Herbicides are used in some European countries to fight IAP, but restrictions often apply. The management techniques used in France are illustrated in Table 3.

## 2.1. Biological Management Techniques

Many exotic and native organisms have been used for biological control programs [30] in North America and Australia; however, current operational and research and development efforts center on a few in particular: grass carp (or white amour, *Ctenopharyngodon idella*) and introduced insects.

These techniques were not used in the field in France, with the exception of grass carp. Introduction of grass carp is no longer allowed, but in the past it was a popular control agent for aquatic plants (natives and non-indigenous species), especially in small ponds or isolated bodies of water. There are many concerns about using grass carp, including the difficulty of controlling where and what they eat, the escape of carp from the managed system, the impact of their feeding on non-target plant and animal species, and the difficulty of removing them when control is no longer needed [8]. The introduction of biological organisms such as macroinvertebrates or grass carp could lead to consumption of native plants. For example, grass carp may not be suitable for natural water bodies with multiple human uses [10] due to their feeding choices: invasive species were often not the first plants to be eaten. For example, the consumption rate of the native species *Callitriche platycarpa* by the two gammarid species was significantly higher than that of the introduced *Elodea nuttallii*. Similarly, *Gammarus roeseli* (the naturalized species) showed higher daily food consumption than *G. fossarum* (the native species) for each plant species, and, in particular, for the native species [61].

Biocontrol agents (*Lymnea stagnalis, Gammarus species*) were tested in the laboratory for *Elodea* species regulation [6, 7, 61]. *Elodea nuttallii* was found to have slightly higher palatability than *Elodea canadensis* for *Lymnea stagnalis* [7]. Snail herbivory can influence the outcome of competition between the *Elodea* species. However, snails are probably not an effective biological agent control for *Elodea nuttallii* [6]. Although *Elodea* species are often the preferred food for waterfowl or crayfish [40], they are avoided by many insect herbivores [46]. *E. nuttallii* is avoided by herbivorous larvae of *Acentria ephemerella* [28]. No herbivore damage of apical meristems was observed for *Elodea canadensis* because *Acentria ephemerella* larvae also avoid feeding on this species' leaves below the apical tips [36]. In Lake La Plaine (in northeast France), the spread of *E. canadensis* may also be explained by the abundance of this aquatic lepidopteran in dense mats of the *Elodea* species. *Altica lythri* Dawn, a Chrysomelidae with phytophagous larva and adults, has been observed in some sites in France on *Ludwigia* sp.

In 2005, two sites of dense beds of *Ludwigia grandiflora* of the Marais d'Orx, a well-colonized wetland in southwest France, we observed many traces of consumption by this coleopteran. Laboratory surveys have shown important consumption by larvae and adults [49]. On a large scale, such insects are not efficient biological control agents, but it is possible that their consumption contributed to the long-term naturalisation of *Ludwigia* in France. Some other phytophagous insects have also been observed on invasive plants (for example the lepidoptera *Parapoynx statiotata* on *Lagarosiphon major* [17]] but had only small and localized impacts.

Pastures with cows or horses were also used to eradicate *Ludwigia* species in wetlands, but in many cases, the results of these experimentations were weak or inconsequential.

## 2.2. Chemical Management Techniques

Although chemical treatment can replace or enhance manual removal operations, it has been used in France only as a last resort when water use and environmental considerations make it possible. The response of different plant species to different herbicides is a function of the properties of both the plant and the herbicide.

The use of herbicides to control aquatic plants has been common for several decades. It has also been, for many years, the object of criticism regarding its short and medium-term toxicological risks, as well as indirect risks such as, for example, oxygen depletion of the water.

At present, France and Europe tend to focus on reducing the use of herbicides. In France in 2007, only commercial products based on two herbicidal molecules (dichlobénil and glyphosate) are homologated for the "destruction" of aquatic or semi-aquatic plants (http://e-phy.agriculture.gouv.fr), whereas in 2001, five active molecules, utilized in nine commercial products, were allowed [18].

Herbicides are sometimes used in conjunction with other management techniques to improve their effectiveness. For example, glyphosates have been used with mechanical and manual control techniques to manage *Ludwigia sp.* in the Marais Poitevin [26].

## 2.3. Mechanical and Manual Management Techniques

Mechanical management methods have been widespread in attempts to control aquatic plants. Larger-scale control efforts require more mechanization For example, 40% of *Lagarosiphon major* mats (40 ha) have been harvested yearly since 1990 to permit summer recreational activities in a pond in southwest France (Etang Blanc). The volume of plants extracted per year is about 10,000 to 11,000 m$^3$. These harvests don't seem to affect the biomass production of the plant; the biomass of *L. major* at this site is on the order of 1 kg of dry matter per m² [26].

Although cutting is relatively rapid, it leaves large mats of plants that can not only spread the plant, but also create a floating obstacle, wash up on shorelines, and cause water-quality problems through decomposition. Because of these problems, cutting operations are typically combined with plant removal. One neglected aspect of such harvesting operations is the disposal of plant material. The hydrophyte is generally more than 90% water and can be used as green manure or compost, often in conjunction with other plants. Composting is also an option for amphibious plants such as *Myriophyllum aquaticum* or *Ludwigia* sp. However, before using *Ludwigia sp.,* viable seeds must be destroyed [26].

Although harvesting is environmentally superior to using herbicides, we must take into account the fact that harvesting removes large numbers of macroinvertebrates, semi-aquatic vertebrates, forage fishes, and even adult gamefish from the environment. For example, a study of the harvesting of dense beds of *Ceratophyllum demersum* in a recreational water body showed that the majority of fish captured were the young-of-the-year; the amount of captured fish remained small and corresponded to weak impacts on the fish population [24]. However, it is likely that in a lake or pond, the quantities of fish extracted are significantly more important. The harvester acts as a large, nonselective predator "grazing" in the littoral zone.

In addition, harvesting can resuspend bottom sediments into the water column, releasing nutrients and other accumulated compounds. Mechanical control could cause a reduction in the biomass of indigenous plants. *Elodea canadensis* appears to be less susceptible to cutting-based weed control measures than *Myriophyllum spicatum*, a native species [1]. Sabbatini and Murphy [54] showed that *Elodea. canadensis* has a strong tolerance for management based on disturbance, such as cutting. Mechanical management may allow *Elodea* to spread to new areas because this management breaks up the plant. Management such as harvests favours downstream plant propagation. However, harvests can be quite useful in areas where IAP are already established or when the species disperses into areas unfavourable to its survival [9]. The use of nets to recover floating stem cuttings can reduce the dispersal of these plant parts.

However, not all secondary effects of harvesting are negative. Removal of large amounts of plants can improve the oxygen balance of littoral zones and rivers, particularly in shallower water [42].

Harvesting is very expensive. Dutartre et al. [25] give one example: the partially publicly financed (costs of 38,000 €) treatment of four hectares of *Ludwigia grandiflora* and *Myriophyllum aquaticum* in France using mechanical extraction. The yearly cost of the harvesting of *Lagarosiphon major* in Etang Blanc is more that 60,000€, from both public and private financing [26].

## *Hand-Pulling*

Hand-pulling techniques are most appropriate for localized nuisance problems with both non-indigenous and native plants. This method is efficient in first stages of colonisation by exotic plants or as a complementary action after the mechanical extraction of plants; in the second case, hand-pulling permits the extraction of stem fragments, which are vectors of the spread of some species. After several hand-pullings, the spread of *Elodea nuttallii* and of *E. canadensis* is area-restricted. Figure 4 illustrates the decrease in biomass after three hand-pulling campaigns. The maximum biomass for *Elodea nuttallii* was obtained in June 2005. The three treatments induced a drastic reduction in *E. nuttallii* in the stream, resulting in less than 2% remaining by the end of the study.

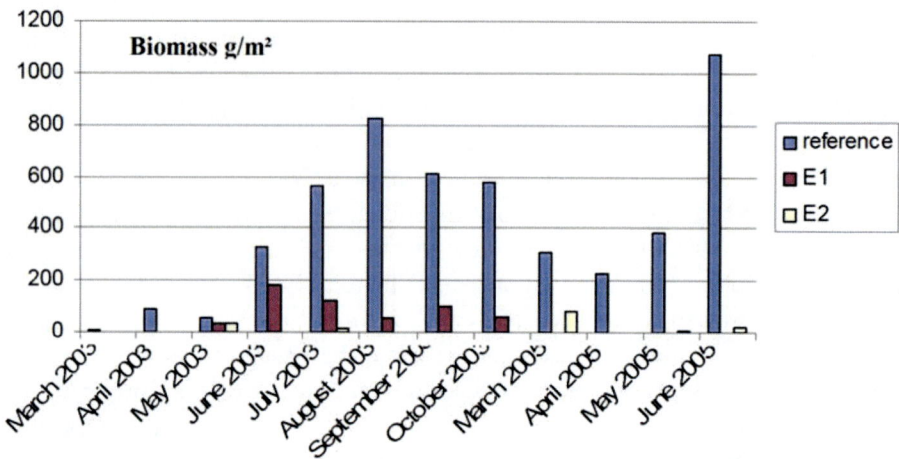

Figure 4. Evolution of the biomass of *Elodea nuttallii* at reference site and two additional sites (E1; E2) managed by hand-pulling in the Fakensteinbach river. Site E1: 1 manual extraction in February 2003; Site E2; 3 manual extractions in February 2003, May 2003 and March 2005. reference site: no manual extraction.

This technique is also used at some sites for management of *Ludwigia* species, such as some ponds in southwest France [19] and in the largest channels in Marais Poitevin [27, 50]. In this second case, hand-pulling was used in conjunction with herbicides (until 1995) and mechanical extraction. A chronological analysis (Figure 5) of the results of this management plan demonstrates the regularity of interventions permitted to control the plant developments in this part of the wetlands. The tonnage of extracted plants increased until 1998, and then decreased to under one hundred tons up to 2003. Two quantitative parameters of analysis used in these works illustrate this success.

First, the linear of river and channel banks maintained by the methods of management passed of some kilometres in 1994 to reach more than 1000 by 2007. Second, the number of workdays increased substantially until 2000, reaching about 1200 days; from that point, it evolved little, reaching 1400 days in 2006.

A team of twelve seasonal workers, working under the direction of a permanent agent of the IISBN, is hired during six months of every year to accomplish this . The cost of this management is important: about 210,000 € (out investments) in 2007. It is, however, in line with the scale of the geographical dimensions and human uses of this site.

tons

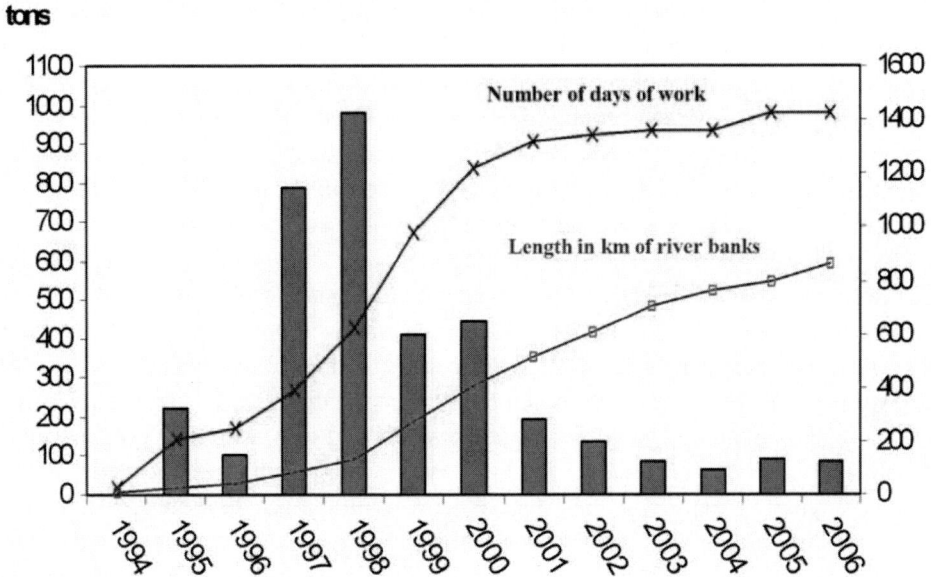

Figure 5. Management of *Ludwigia* sp in the "Marais Poitevin", 1994 – 2006: Evolution of extracted plant quantities, length of river banks and number of days of work (adapted from IIBSN report, 2007).

The main disadvantage of this technique is that it is expensive and arduous (Figure 6). This method may be applicable in bodies of water where no herbicide can be tolerated, such as in a lake used as a municipal drinking water supply or running waters.

Figure 6. Hand-pulling of *Ludwigia grandiflora* (southwest France). (picture Dutatre A).

## 2.4. Physical Management Techniques

Several physical techniques are commonly used, including dredging, drawdown and light attenuation.

By opening more diverse habitats and creating depth gradients, dredging may also create more diversity in the plant community [47]. The results of dredging can be long-term ; the biomass of *Potamogeton crispus* in Collins Lake, New York, remained significantly lower than pre-dredging levels ten years after dredging [64]. Due to the cost, environmental impact, and problems presented by disposal, dredging should not be used alone for aquatic plant management. Rather, it is best used as a multi-purpose lake remediation technique. Dredging is efficient for removing organic sediment (a nutritional source for rooted plants) and seed banks from the water bodies.

Drawdown is another effective aquatic plant management technique that alters the plant's environment. Essentially, the water body has water removed until a pre-determined depth. It is best if this depth includes the entire depth range of the target species. To be effective, drawdown needs to be at least one month long to ensure thorough drying [11]. This method was tested over one month last winter in the northeast part of France (the Vosges Mountains) for controlling the *Elodea* species in a reservoir. This method is inexpensive and has significant environmental effects that may interfere with the use and intended function (e.g., power generation or drinking water supply) of the water body during the drawdown period. Lastly, species respond very differently to drawdown and often not in a consistent fashion [11]. To manage amphibious plants such as *Myriophyllum aquaticum* or *Ludwigia* species, the duration of this technique must be longer than one month, preferably in winter. In shallow bodies of water with wet organic sediments, it is possible that these plants may survive the drawdown and colonise a larger surface; particular attention must be taken for this type of intervention. Drawdown may provide an opportunity for the spread of highly invasive aquatic plants in the Mediterranean climate of southeast France.

The most basic environmental manipulation for plant control is light reduction or attenuation. This, in fact, may have been the first physical control technique. Although light manipulation techniques may be useful for most IAP in narrow streams or small ponds in France, in general, these techniques are of limited applicability. The growth of *E. canadensis*, contrary to that of *E. nuttallii*, is affected by low light intensity [4]. This method was tested only in the laboratory.

## 2.5. Ecological Restoration

Changes in management practices, especially those with the potential to directly impact the river, are important for helping manage exotic macrophytes. This includes practices such as maintaining and enhancing the condition of riparian vegetation and storm water and nutrient management. The use of nutrient inactivation for the control of aquatic vascular plants is still in the research and development phase. Reduction of nutrients entering the river or in the ponds is important in the fight against exotic macrophytes species as these species grow best in high nutrient waters. For example, *Elodea nuttallii* and *E. canadensis* have a wide amplitude with regard to nutrient levels [52, 53, 58, 62].

The last type of biological management technique, native plant restoration, is an ecological approach to managing a desired plant community. The basic idea is that restoring a native plant community should be the end goal of most aquatic plant management programs [48, 57]. Disturbances favour the introduction of exotic species. In communities that have only recently been invaded by exotic species, a propagule bank that can restore the native community probably exists. However, in communities that have harbored highly invasive aquatic plants for a long period of time (e.g., more than ten years), native plants may have to be reintroduced.

In the wetlands along the Mediterranean Sea in southern France, management with salt water has been used successfully to eradicate salt-sensitive *Ludwigia peploides* (Grillas 2004).

## 2.6. No Action

Getting rid of one invasive species by management may not solve the problem, since it has also been shown that this can make way for other introduced species [66]. There are examples in Europe in which a second or third introduced species has been competing successfully with a previous introduction.

### *Example of Elodea Species*

Native to North America, *Elodea canadensis*, first recorded in the early 19th century in the British Isles [55], is now naturalised. *Elodea canadensis* became a persistent invasive species following its naturalisation, choking waterways before declining to its present, less-abundant (but still common) level [3, 63]. *E. nuttallii*, native to North America, was first found in Belgium in 1939 and spread into the Alsatian Plain (northeast France) by the end of the 1950s [31]. It has been colonizing numerous ponds and streams in many metropolitan areas of France for the past 30 years [60]. *Elodea nuttallii* is replacing *E. canadensis* at many sites [3, 43, 63]. Although *Elodea canadensis* and *E. nuttallii* have been spreading for several years in eastern France, these species are more problematic in other European countries. For example, *E. nuttallii* was classified as one of the top ten invasive species in Germany (F. Klingenstein, pers. com.); colonies have been expanding in Lake Leman in Switzerland since 1993 [14], as well as in numerous ponds, reservoirs and streams in Brittany [56], Belgium (G. Verniers, pers. com.), and Sweden [39]. In contrast, there has been a noticeable decline in *Elodea canadensis* in European freshwaters [56] and in *E. nuttallii* in Japan [45]. The *Elodea nuttallii* populations in Japan exhibited a genetic uniformity that made them vulnerable to attack by fungi or pathogens and could explain the decline in *E. nuttallii*. Recent investigations in the Alsatian Plain revealed that the distribution of *E. nuttallii* has been stable over the last few decades, despite local fluctuations in abundance [34].

Furthermore, the palatability of *Elodea* species is low [6, 7, 61]. Herbivores are probably not an effective biological agent control for *Elodea nuttallii* [6, 28]. The growth of *E. canadensis* is affected by low light intensity, unlike *Elodea nuttallii* [4], so light reduction is not an alternative management method.

*Elodea nuttallii* possesses a high tolerance for dehydration, so summer drawdown would not be efficient in the control of this IAP [5]. An experimental study on the impact of winter drawdown on *Elodea* species suggested no effect of low temperature combined with desiccation on the plant's phenology (unpublished data). Mechanical management may allow *Elodea* to spread to new areas because it breaks up the plant.

Both *Elodea* species showed similar resistance to currents, while fragment regeneration and colonisation were only slightly higher in *Elodea nuttallii* than in *E. canadensis* [7]. Hand-pulling is an effective method, but is limited to small mats. So what can we do? One solution is to wait and see; maybe the *Elodea* populations will decline in the future.

Further work is needed to improve our knowledge of the ecological risk of management on biodiversity and ecosystem function. The risk of adverse side-effects for users of the water and for the health of the ecosystem must always be taken into account.

## 3. POLITICAL TOOLS

National strategies must be implemented as a good management tool (preferably as quickly as possible) once the presence of a new IAP is detected. In inland and littoral waters, the European Water Framework Directive is an extremely useful tool requiring the restoration of degraded water masses to good ecological status by 2015. The first European workshop showed that different countries have their own vision about how much the presence of IAP influences the determination of the final state of a mass of water. This process would have to be normalized within the European Union Invasive Alien Species are a priority at the continental level, and their control must be legislated at different scales, ranging from the European community as a whole to the local scale [32, 37, 38]. In a recent European Conference on Invasive Alien Species[1] in Spain in January 2008, the general principals were:

–   Prevention is the first and most desirable response.
–   To mitigate the effects of IAP, there must be investments in the creation and implementation of early warning systems and immediate action mechanisms.
–   Political commitment and the development and implementation of a specific legal framework for the prevention and fight against IAP at the community and national levels is a priority.
–   Good practices must be created that are focused on prevention, and national committees devoted exclusively to IAP must be established.
–   Coordination between research centres and governmental entities should be encouraged.
–   Citizen participation should also be encouraged.
–   Environmental education must be considered a basic tool in the fight against IAS.

In France, the current national legislation seems to be insufficient; legislators must consider the development of regulations and economic means that make the IAP real and effective development possible. As of April 2007, introduction and trade of only both *Ludwigia*s species has been forbidden; other ornamental invasive species continue to be sold in garden centers.

---

[1]   http://www.fundacionbiodiversidad.info/eei/pdf/PUBLICACION%20FINAL/Version%20Ingles/Versioningles.pdf

# 4. RECOMMENDATIONS

In general, the priority of biological invasion control is to prevent new infestations from taking hold, especially for the fastest-growing and most disruptive species. Although only a fraction of non-indigenous species become established, and only a small proportion of these pose a direct threat to human health or are otherwise invasive [68], the number of invasive species is large and continuing to grow because of increasing global movement of humans and goods. The prevention approach would include greater investment in prevention: preventing organisms from entering a particular pathway, and preventing organisms that are transported from being released or escaping alive.

The major pathways through which species are intentionally and unintentionally introduced into Europe is the first step in a risk analysis of invasion [41]. For example, non-indigenous species potentially available for sale must be subject to risk analysis. Lodge et al. defined the invasion risk associated with a pathway as *"a function of the number of the nonindigenous species transported, the number of individuals of each species transported, the characteristics of the species (including their environmental tolerances), the number and characteristics of their hitchhiking species(including, parasites, and other associated organisms) and the likehood and frequency that a species and associated hitchhikers would be released or escape into an environment suitable for the species to thrive (either initially or through secondary transport)"* [41]. The feasibility of this and the cost of eradication should be taken account.

Early detection, as well as acting quickly, is the second step for IAP control. The lag time between establishment and spread associated with many invading populations provides an opportunity for detection and eradication. For species distributed worldwide, eradication efforts must proceed quickly if there is to be a substantial probability of success. Effective eradication must be based on evidence from other countries rather than "proof" of invasiveness [67]. The recent campaign to eradicate water primrose from southern England promises to be the first example of successful eradication of an introduced aquatic plant in the UK (Defra, 2007 in 67). In France, eradicating *Hydrocotyle ranunculoides,* localized in northwest France and in the Ile de France (Conservatoire Botanique de Bailleul, in 26) should become a high priority. Initial action was realized through dredging a channel of about 3 kilometers entirely colonized by this species, which gives good results without regrowth of *L. ranunculoïdes* (B. Toussaint, personal communication). This species is widespread in the UK and in Belgium. A similar recommendation could be applied to limit the spreading of *Crassula helmssii.*

A basic recommendation is to foster good ecological conditions in water masses as a self-defence measure for aquatic ecosystems against the colonising pressure of IAP. The water bodies most vulnerable to biological invasions are often subject to disturbances such as high or fluctuating nutrient-loading, water level or flow regime alteration, toxic contaminants (pesticide, heavy metals) and morphological alteration of the riverbed. Specific emphasis on hydromorphological as well as nutrient-related pressures, combined with the delivery of programmes on a catchment scale, should support the environmental improvements needed to sustain reduction of IAP [67].

At some point, it becomes necessary to urge the participation of all local water users. To this effect, training those responsible for plant management is necessary. One positive action would be to have periodic working forums bringing together all users and administrators.

Environmental education is also an essential tool in the fight against IAP.

## CONCLUSION

There is no single cure-all solution to aquatic plant problems, no single best choice. None of these techniques is without flaws or potential environmental impacts. It is up to each management group to select the most appropriate techniques for their situation given a set of social, political, economic and environmental conditions.

To complete the essential steps of preventive management and reduce the control activities for aquatic plants to what is necessary, efforts must be carried on to improve the management practices. We should insist on the assessment of a given method's impacts, coordination of undertaken actions, circulation of information between all the partners, and intensification of a training program of all these actors of aquatic plant regulation practices [26].

Management plans established early on were the first steps towards sustainable management of aquatic environments. However, these efforts are compromised as long as invasive aquatic plant species continue to be sold to individuals or unintentionally imported without control. Stronger enforcement of existing laws and control coupled with an intensive public education campaign is needed to prevent further spreading of IAP.

## REFERENCES

[1]    Abernethy, V. J., Sabbatini, M.R., Murphy, K.J. 1996. Response of *Elodea canadensis* Michx. and *Myriophyllum spicatum* L. to shade, cutting and competition in experimental culture. *Hydrobiologia* 340, 219-224.

[2]    Aboucaya, A. 1999. Premier bilan d'une enquête nationale destinée à identifier les xénophytes invasifs sur le territoire métropolitain français (Corse comprise). *Bulletin de la Société Botanique du Centre-Ouest* 19, 463-482.

[3]    Barrat-Segretain, M.-H. 2001. Invasive species in the Rhône River floodplain (France) : replacement of *Elodea canadensis* Michaux by *E. nuttallii* St. John in two former river channels. *Archiv für. Hydrobiologie.* 152, 237-251.

[4]    Barrat-Segretain, M.-H. 2005. Growth of *Elodea canadensis* and *Elodea nuttallii* in monocultures and mixture under different light and nutrient conditions. *Archiv für. Hydrobiogie.* 161, 133-144.

[5]    Barrat-Segretain, M.-H. and B. Cellot 2007. Response of invasive macrophyte species to drawdown: The case of *Elodea* sp. *Aquatic Botany* 87(4), 255-261.

[6]    Barrat-Segretain, M.-H. and Lemoine, D. 2007. Can snail herbivory influence the outcome of competition between *Elodea* species. *Aquatic Botany* 86, 157-162.

[7]     Barrat-Segretain, M.-H., Elger, A., Sagnes, P. and Puijalon, S. 2002. Comparison of three life-history of invasive *Elodea canadensis* Michx and *Elodea nuttallii* (Planch). H. St. John. *Aquatic Botany* 74, 299-313.

[8]     Bonar S.A., Vecht, S.A., Bennett, C.R., Pauley, G.B., Thomas, G.L. 1993. Capture of grass carp from vegetated lakes. *Journal of Aquatic Plant Management* 31, 168-174.

[9]     Bowmer, K.H., Jacobs, S.W.L. and Sainty, G.R. 1995. Identification, Biology and Management of *Elodea canadensis*, Hydrocharitaceae. *Journal of Aquatic Plant Management* 33, 13-19.

[10]    Codhant H., and Dutartre A., 1992. Utilisation de la carpe chinoise comme moyen de contrôle biologique des macrophytes aquatiques. Revue bibliographique. 1099-1107. In ANPP, 15° Conférence du COLUMA, Journées Internationales sur la lutte contre les mauvaises herbes, Versailles, 2-4 décembre 1992, 1274 p.

[11]    Cooke, G.D. 1980. Lake level drawdown as a macrophyte control technique. *Water Resources Bulletin* 16, 317-322.

[12]    Damien, J. P. 2002. Control of *Ludwigia grandiflora* communities in the Brière Regional Nature Park. Pages 341-344 in Proceedings of the 11th EWRS *International Symposium on Aquatic Weeds,* Moliets Maâ, France.

[13]    Dandelot, S. 2004. Les *Ludwigia* spp. invasives du Sud de la France: Historique, Biosystématique, Biologie, Ecologie. Thesis, Université d'Aix-Marseille III, Marseille, France.

[14]    Demierre, A. and Perfetta, J. 2002. Gestion du faucardage des macrophytes sur les rives genevoises du Léman (Suisse). Pages 345-347 in Proceedings of the 11th EWRS International Symposium on Aquatic Weeds, Moliets Maâ, France.

[15]    Di Nino, F., Thiébaut, G. and Muller, S. 2005. Response of *Elodea nuttallii* (Planch.) H. St. John to manual harvesting in the north-eastern France. *Hydrobiologia* 551, 147-157.

[16]    Dupont, P. 1989. Une découverte inédite d'Emile Contre: Althernanthera philoxeroides dans la vallée de la Garonne. *Bulletin de la Société Botanique du Centre-Ouest* 20, 27-28.

[17]    Dutartre A., 1979. Recherches préliminaires sur *Lagarosiphon major* (Ridley) Moss., Hydrocharidacée, dans le lac de Cazaux-Sanguinet-Biscarrosse. 1979, D.E.A. Université de Bordeaux III, Laboratoire de Botanique, 73 p.

[18]    Dutartre, A., 2002. Panorama des modes de gestion des plantes aquatiques : nuisances, usages, techniques et risques induits. *Ingénieries* - E A T, n° 30, p. 29 – 42.

[19]    Dutartre, A., 2004. De la régulation des plantes aquatiques envahissantes à la gestion des hydrosystèmes. Ingénieries - E A T, n° *spécial Ingénierie écologique*, p. 87–10.

[20]    Dutartre, A. and Oyarzabal, J. 1993. Gestion des plantes aquatiques dans les lacs et les étangs landais. *Hydroécologie Appliquée* 5(2), 43-60.

[21]    Dutartre, A, Chauvin, C. and Grange, J. 2006. Colonisation végétale du canal de Bourgogne à Dijon. Bilan 2006 – Propositions de gestion. *Voies Navigables de France*, Cemagref. 87 p.

[22]    Dutartre, A., Haury, J. and Jicorel, A. 1999. Succession of *Egeria densa* in a drinking water reservoir in Morbihan (France). *Hydrobiologia* 415, 243-247.

[23]    Dutartre, A., Haury, J. and Planty-Tabacchi, A-M. 1997. Introductions of aquatic and riparian macrophytes into continental French hydrosystems: an attempt at evaluation. *Bulletin Français de la Pêche et de la Pisciculture* 344-345, 407-426.

[24] Dutartre, A., Pipet, N. and Bachelier, E., 2005. Suivi de l'impact de la moisson mécanique des plantes aquatiques sur les populations piscicoles. Synthèses des expérimentations 2002-2003 sur le plan d'eau de Noron (Deux Sèvres). *Rapport*, Cemagref, 33 p.

[25] Dutartre, A., Dandelot, S., Haury, J., Lambert, E., Le Goffe, P. and Menozzi, M.J., 2007, Programme de recherche Invasions Biologiques. Les jussies : caractérisation des relations entre sites, populations et activités humaines. Implications pour la gestion. Rapport final. Cemagref, 128 p.

[26] Dutartre, A., Peltre, M-C, Pipet, N., Fournier, L., and Menozzi, M-J. 2008. Régulation des développements de plantes aquatiques. In Ingénieries - *E A T, n° spécial*, 135-154.

[27] Dutartre, A., Charbonnier, C., Dosda, V, Fare, A. Lebougre, C., Saint Macary, I. and Touzot, O. 2002. Primary production of Ludwigia spp. In the South West of France. Pages 23-26 in Proceedings of the 11th EWRS International Symposium on Aquatic Weeds, Moliets Maâ, France.

[28] Erhard, D., Pohnert G., and Gross, E. 2007. Chemical Defense in *Elodea nuttallii* Reduces Feeding and Growth of Aquatic Herbivorous Lepidoptera. *Journal of Chemical Ecology* 33, 1646-1661.

[29] Fournier, L. J. Oyarzabal, J. 2002. Management of aquatic invasive plants in lakes and ponds (Landes, France). Pages 287-290 in Proceedings of the 11th EWRS International Symposium on Aquatic Weeds, Moliets Maâ, France.

[30] Gallagher, J.E. and Haller, W.T. 1990. History and development of aquatic weed control in the United States. *Reviews of Weed Science* 5, 115-192.

[31] Geissert F., Simon M. and Wolff P., 1985. Investigations floristiques et faunistiques dans le Nord de l'Alsace et quelques secteurs limitrophes. *Bulletin de l'Association Philomatique d'Alsace- Lorraine*, 21, 111-127.

[32] Genovesi P. and Shine C., 2004. Stratégie européenne relative aux espèces exotiques envahissantes. Editions du Conseil de l'Europe. *Sauvegarde de la nature*, 137, 74 p.

[33] Georges, N. 2004. L'herbe à l'alligator (*Althernanthera philoxeroides* (Martius) Grisebach) atteint le département du Tarn-et-Garonne. Le Monde des Plantes 484, 1-3.

[34] Greulich, S. and Trémolières, M. 2006. Present distribution of the genus *Elodea* in the Alsatian Upper Rhine floodplain (France) with a special focus on the expansion of *Elodea nuttallii* St. John during recent decades. *Hydrobiologia* 570(1), 249-255.

[35] Grillas, P. 2004. Bilan des actions de gestion de Ludwigia grandiflora et L. peploides (jussies) dans les espaces protégés du Languedoc-Roussillon. Pages 148-152 in Patrimoines naturels, editors. Plantes invasives en France - Etat des connaissances et propositions d'actions. Museum National d'Histoire Naturelle, Paris, France.

[36] Gross, E. M., Johnson R. L., and Hairston, J. N. G. 2001. Experimental evidence for changes in submersed macrophyte species composition caused by the herbivore Acentria ephemerella (Lepidoptera). *Oecologia* 127, 105-114.

[37] Hulme P. E., 2007. Biological Invasions in Europe: Drivers, Pressures, States, Impacts and Responses. Issues Biodiversity Under Threat; in *Environmental Science and Technology*, No. 25. 56 – 80

[38] Hulme P. E., Bacher S., Kenis M., Klotz S., Kühn I., Minchin D., Nentwig W., Olenin S., Panov V., Pergl J., Pysek P., Roques A., Sol D., Solarz W.and Vilà M., 2008. Grasping at the routes of biological invasions: a framework for integrating pathways into policy. *Journal of Applied Ecology*, 45 (2), 403–414

[39] Larson D. 2007. Non-indigenous Freshwaters Plants. Patterns, Processes and Risk Evaluation. Doctoral Thesis, *Swedish University of Agricultural Sciences*, Uppsala.

[40] Lodge DM. 1991. Herbivory on freshwater macrophytes. Aquatic Botany 41, 195-224.

[41] Lodge, D.M., S. Williams, H. MacIsaac, K. Hayes, B. Leung, L. Loope, S. Reichard, R.N. Mack, P.B. Moyle, M. Smith, D.A. Andow, J.T. Carlton, and A. McMichael. 2006. Biological invasions: recommendations for policy and management [Position Paper for the Ecological Society of America]. *Ecological Applications* 16:2035-2054.

[42] Madsen, J.D., Adams MS. and Ruffier, P. 1988. Harvest as a control for sago pondweed (*Potamogeton pectinatus* L.) in Badfish Creek, Wisconsin: Frequency, efficiency and its impact on stream community oxygen metabolism. *Journal of Aquatic Plant Management* 26, 20-25.

[43] Mériaux, J.-L. 1979. *Elodea nuttallii* St. John, espèce nouvelle pour le Nord de la France. *Bulletin de la Société Botanique Nord France* 32, 30-32.

[44] Muller, S., Aboucaya, A., Affre, L., Cassan, S., Decocq, G., Dinger, F., Dutartre, A., Fournier, L., Gavory, L., Grillas, P., Largier, G., Maillet, J., Médail, F., Moiroud, L., Mony, C., Oyarzabal, J., Pénélon, L., Pénin, D., Quertier, P., Saïd, S., Sinnassamy, J-M., Suehs, C., Tabacchi, E., Thiébaut, G., Toussaint, B. and Yavercovski, N. 2004. Plantes invasives en France. Etat des connaissances et propositions d'actions. In Patrimoines naturels, editors. Plantes invasives en France - Etat des connaissances et propositions d'actions. *Museum National d'Histoire Naturelle*, Paris, France.

[45] Nagasaka, M., Yoshizawa, K., Ariizumi, K., and Hirabayashi, K. 2002. Temporal changes and vertical distribution of macrophytes in Lake Kawaguchi. *Oecologia* 3, 107-114.

[46] Newman, R. M. 1991. Herbivory and detritivory on freshwater macrophytes by invertebrates: a review. *Journal of The North American Benthological Society* 110, 89-114.

[47] Nichols, S.A. 1984. Macrophyte community dynamics in a dredged Wisconsin lake. *Water Resources Bulletin* 20, 573-576.

[48] Nichols, S.A. 1991. The interaction between biology and the management of aquatic macrophytes. Aquatic Botany 41, 225-252.

[49] Petelszyc, M., Dutartre, A., and Dauphin, P., 2006. La jussie (*Ludwigia grandiflora*) plante-hôte d'*Altica lythri* Aubé (Coleoptera Chrysomelidae) : Observations *in situ* dans la Réserve Naturelle du Marais d'Orx (Landes) et en laboratoire. Bulletin de la société linnéenne de Bordeaux 141, 34, 221 – 228.

[50] Pipet, N. 2002. Management of *Ludwigia* spp in Marais Poitevin (West of France). Pages 389-392 in Proceedings of the 11th EWRS International Symposium on Aquatic Weeds, Moliets Maâ, France.

[51] Rebillard, J.-P., Dutartre, A., Far A., and Ferroni, J.-M. 2002. Management of the development of aquatic plants in the Adour-Garonne river basin (South West of France). Pages 307-310 in Proceedings of the 11th EWRS International Symposium on Aquatic Weeds, Moliets Maâ, France.

[52] Robach, F., Hajnsek, I., Eglin, I. and Tremolières, M. 1995. Phosphorus sources for aquatic macrophytes in running waters: water or sediment ? *Acta botanica Gallica* 142, 719-731.

[53] Rolland, T. and Tremolières, M. 1995. The role of ammonium in the distribution of two species of *Elodea*. *Acta Botanica Gallica* 142, 733-739.

[54] Sabbatini, M.R. and Murphy, K.J. 1996. Submerged plant survival strategies in relation to management and environmental pressures in drainage channel habitats. *Hydrobiologia* 340 , 191-195.

[55] Simpson, D. A. 1984. A short history of the introduction and spread of *Elodea* in the British Isles. *Watsonia* 15, 1-9.

[56] Simpson, D. A. 1990. Displacement of *Elodea canadensis* Michx. by *Elodea nuttallii* (Planch.) St. John in the British Isles. *Watsonia* 18, 173-177.

[57] Smart, R.M. and Doyle, R. 1995. Ecological Theory and the Management of Submersed Plant Communities. Information Exchange Bulletin A-95-3, US Army Engineer Waterways Experiment Station, Vicksburg, Mississippi.

[58] Thiébaut, G. 2005. Does competition for phosphate supply explain the invasion pattern of *Elodea* species? *Water Research* 39, 3385-3393.

[59] Thiébaut, G. 2007a. Invasion success of exotic aquatic plants in their native and introduced ranges. A comparison between their invasiveness in North America and in France. *Biological Invasions* 9 (1) : 1-12.

[60] Thiébaut 2007b Non-indigenous aquatic and semi-aquatic plant species in France, pp. 209-229. In: Gherardi F. (Ed.). *Biological invaders in inland waters: Profiles, distribution and threats*, vol. 2, chap. 11. Berlin: Springer. (Springer Series in Invasion Ecology). ISBN: 978-1-4020-6028-1

[61] Thiébaut, G. and Gierlinski, P. 2007. Gammarid (Crustacea: Amphipoda) herbivory on native and alien freshwater macrophytes.In: Tokarska-Guzik, B., Brock, J.H., Brundu, G.,Child, L., Daehler, C.C. and Pyšek, P. (Eds). *Plant Invasions: Human perception, ecological impacts and management*, pp. 333-34 Backhuys Publishers, Leiden, The Netherlands.

[62] Thiébaut, G. and Muller, S. 2003. Linking phosphorus pools of water, sediment and macrophytes in running waters. *International Journal of Limnology* 39: 307-316.

[63] Thiébaut, G., Rolland, T., Robach, F., Trémolières, M. and Muller, S. 1997. Quelques conséquences de l'introduction de deux espèces de macrophytes, *Elodea canadensis* Michaux et *Elodea nuttallii* St. John, dans les écosystèmes aquatiques continentaux: exemple de la Plaine d'Alsace et des Vosges du Nord (Nord-Est de la France). *Bulletin Français de la Pêche et de la Pisciculture* 344/345, 441-452. OK pour ref.

[64] Tobiessen, P., Swart, J. and Benjamin, S. 1992. Dredging to control curly-leaf pondweed: A decade later. *Journal of Aquatic Plant Management* 30,71-72.

[65] Vivant, J. 2005. Plantes signalées dans les Landes et Pyrénées Atlantiques en 2004. *Le monde des plantes* 486, 6-10.

[66] Wallentinus, I. 2002. Introduced marine algae and vascular plants in European aquatic environments. In E. Leppäkoski, S. Gollasch and S. Olenin (Eds;). *Invasive Aquatic Species of Europe: Distribution, Impacts and Management*, pp 27-52, Kluwer Academic Publishers, Dordrecht, The Netherlands.

[67] Wilby, N.J. 2007. Managing invasive aquatic plants: problems and prospects. Aquatic Conservation : *Marine and Freshwater ecosystems* 17, 659-665.

[68] Williamson, M.,1996. *Biological invasions*. Chapman and Hall, London, UK.

In: Aquatic Ecosystem Research Trends
Editor: George H. Nairne

ISBN 978-1-60692-772-4
© 2009 Nova Science Publishers, Inc.

*Chapter 3*

# MATHEMATICAL MODELS FOR DYNAMICS AND MANAGEMENT OF ALGAL BLOOMS IN AQUATIC ECOSYSTEMS

## *Takashi Amemiya[1], Hiroshi Serizawa[1], Takashi Sakajo[2,3] and Kiminori Itoh[4]*

[1]Graduate School of Environment and Information Sciences, Yokohama National University, 79-7 Tokiwadai, Hodogaya-ku, Yokohama 240-8501, Japan
[2]Department of Mathematics, Hokkaido University, Kita 10 Nishi 8, Kita-ku, Sapporo 060-0810, Japan
[3]PRESTO, Japan Science and Technology Agency
[4]Graduate School of Engineering, Yokohama National University, 79-7 Tokiwadai, Hodogaya-ku, Yokohama 240-8501, Japan

## ABSTRACT

We present four mathematical models of algal blooms in aquatic ecosystems, and investigate the dynamics in the models. The models analyzed in this chapter are (i) a minimal nutrient-phytoplankton (NP) model, (ii) an algal-grazer model taking into account phenotypic plasticity in algae (cyanobacteria), (iii) the integrated model of the minimal NP model and the algal-grazer model, and (iv) a modified abstract version of the comprehensive aquatic simulation model (abs.-CASM). Bistability and oscillatory behaviors are exhibited in all the models as a function of nutrient inputs, or eutrophication levels. We also construct a reaction-diffusion-advection model by using the minimal NP model, and show that two-dimensional spatiotemporal patterns (patchiness) in algae can be exhibited by the combined effects of diffusion and advection by turbulent stirring and mixing, and biological interactions. The minimal NP model is further extended to a one-dimensional reaction-diffusion-advection model in the vertical direction of a lake, explaining the dynamics of both seasonal outbreak and diurnal vertical migration of cyanobacteria. Taking into account the dynamics of algal blooms, we propose effective lake restoration techniques by using the abs.-CASM. We address that understanding the mechanisms and characteristics of the dynamical behaviors presented here are crucial to the management of lake ecosystems, in particular, to coping with regime shifts, nonlinear-discontinuous transitions, observed in lake ecosystems exposed to human stress. Finally we reexamine the validity of the present models and their analyses from mathematical viewpoints, and propose further mathematical insights into improving the models and their analyses in accordance with real aquatic ecosystems.

## INTRODUCTION

Eutrophication causes serious environmental problems such as algal blooms in aquatic ecosystems (Scheffer 1998; Sigee 2005). In the progress of eutrophication, the aquatic ecosystem is thought to be changed from the clear-water to the turbid-water states, in which algal blooms in midsummer become an annual event. The change to a degraded turbid state is usually referred to as a regime shift, abrupt and discontinuous transitions from one state to a completely different alternative state in terms of ecosystem structures and functions (Scheffer et al. 2001; Scheffer et al. 2003). Besides the regime shifts, aquatic ecosystems exhibit various kinds of irregular behaviors originating from the nonlinear interactions among organisms. These include oscillations in the abundances of organisms, and inhomogeneous distribution of phytoplankton on the water surface.

The spatial distribution of phytoplankton on the water surface often shows inhomogeneous patterns called patchiness. The so-called patchiness with stretched and curled structures is characteristically observed in seas or oceans, although the more obscure patterns appear in lakes. It is known that reaction-diffusion-advection equations can simulate various kinds of pattern formation including patchiness, when the diffusion and advection terms are added to the ordinary differential equations exhibiting a limit cycle oscillation (Medvinsky et al. 2002). Consequently, the fundamental model that describes the basic behaviors of aquatic ecosystems requires both bistability and limit cycle oscillations within a realistic range of parameters without diffusion and advection terms. These two are the minimum conditions for fundamental models in aquatic ecosystems.

In this chapter, we firstly present a minimal two-component model with nutrients and phytoplankton that describes the basic behaviors of phytoplankton in aquatic ecosystems. This fundamental model as presented in section 3.1, referred to as the minimal NP model, can show both bistability and limit cycle oscillations in its ordinary differential equation form. Then, the model is extended to reaction-diffusion-advection equations to simulate two characteristic phenomena concerned with algal blooms: one is plankton patchiness as shown in section 3.2, and the other is annual and diurnal vertical migration of cyanobacteria as shown in section 3.3.

In section 3.4, we focus on biology of algae, and model the morphorogical changes in algae, especially colony formation in *Microcystis*. Cyanobacteria or blue-green algae such as *Microcystis*, *Anabaena* and *Oscillatoria* are known to form colonies or filaments under natural circumstances (Reynolds et al. 1987; Dokulil and Teubner 2000; Oberholster et al. 2004; Ozawa et al. 2005; Yoshinaga et al. 2006). *Microcystis* exists as colonial forms, while *Anabaena* and *Oscillatoria* exists as filamentous forms. Such morphological changes from unicells to the filaments or colonies are thought to be one of the most important defensive strategies, i.e., phenotypic plasticity, to protect these algae from grazing by herbivorous zooplankton. This mechanism could be an adaptive reaction for survival that has developed through evolutionary processes (Jakobsen and Tang 2002; Tang 2003).

A number of studies suggest that defensive morphologies can be induced by zooplankton grazing. As for *Microcystis*, it is reported that grazing by the protozoan flagellate *Ochromonas* can induce colony formation, whereas grazing by the metazoan copepod *Eudiaptomus graciloides*, the cladoceran *Daphnia magna* and the rotifer *Brachionus calyciflorus* can not (Burkert et al. 2001; Yang et al. 2006). The mechanism for inducing defensive morphologies is thought to be triggered by the emission of chemical cue in grazing activities of zooplankton (Jakobsen and Tang 2002; Tang 2003; Lürling 2003a; Lürling 2003b; van Holthoon et al. 2003). We show by using a phenotypic plasticity model that abrupt increase in biomass of colonial form of algae occurs when nutrient loading crosses a threshold value.

As a result of the formation of colonial and filamentous forms, the scum of algae covering the surface of lakes during summertime degrades water qualities and the odor gives a nuisance to humans. For instance, Lake Sagami and Lake Tsukui (Kanagawa Prefecture, Japan) have suffered from these cyanobacteria every summer season since the 1970s. Microscopic photos reveal that the dominant species of algal blooms is *Microcystis*, while *Anabaena* and *Oscillatoria* are also observed. Two photos in Figure 1 are a view of algal blooms in Lake Sagami (a) and a microscopic image of *Microcystis* (b), which is responsible for algal blooms in these lakes.

The present phenotypic plasticity model as shown in section 3.4 is composed of two trophic levels with three components: unicellular algae, colonial algae, and their grazer. On the other hand, the minimal NP model as presented in section 3.1 is composed of nutrient and algae. Because the two models cover different parts of trophic levels, it is possible to connect these two models and unify them into a four-component model of two trophic levels. In section 3.5, the unified four-component model is presented as an integrated aquatic model.

Figure 1. Algal blooms in Lake Sagami (September, 2006). Lake Sagami and Lake Tsukui are located in Kanagawa Prefecture, Japan. Two lakes have suffered from algal blooms every summer since the 1970s. The main component of algal blooms is *Microcystis*, while *Anabaena* and *Oscillatoria* are also observed. Two photos are a view of algal blooms in Lake Sagami (a) and a microscopic image of *Microcystis* (b). The colony in (b) is about 500 μm in diameter. Permission is granted to reproduce the figure by Ecological Research.

Taking into account the dynamics of algal blooms obtained by the analyses of the present mathematical models, we consider aquatic-ecosystem management in section #-6. In terms of ecosystem management, ecological resilience (Holling 1973), referred to as resilience hereafter, is a key concept for understanding not only the degraded process but also the restoration process in lake ecosystems. Resilience is introduced to describe the properties of bistable or alternative stable states in ecosystems, and is defined by "the amount of change the system can undergo and still retain the same controls on function and structure" (Holling 1973, 1996). Resilience is commonly illustrated by using a sigmoidal bifurcation curve of an ecosystem model (Scheffer et al. 2001). The horizontal axis is a control parameter such as nutrient loading in lakes, and the vertical axis is a state variable such as biomass of algae in the bifurcation curve. Using a bifurcation diagram (Scheffer et al. 2001), resilience is visually represented by the basins of attractions of a stable state or an attractor. It should be noted that there are two kinds of resilience in ecosystems: resilience of natural state and that of a degraded state. In the case of eutrophication of lakes, resilience of both clear-water and turbid-water states exist.

The magnitude of resilience of a clear-water state in lakes decreases with increasing the nutrient loading. Ecosystems with decreased resilience are vulnerable to external perturbations, and tend to shift to an alternative state. We will argue the importance of resilience in ecosystems management in section 3.6.

Biomanipulation is a lake-restoration technique that was firstly introduced in 1975 (Shapiro et al. 1975), focusing on trophic-cascade top-down effects. A number of results of biomanipulations including both successful and unsuccessful cases are summarized (Hansson et al. 1998; Drenner and Hambright 1999).

At present, the concept of resilience and biomanipulation is thought to be closely linked; successful baiomanipulations can be realized by manipulating the abundance of organisms or materials in order to overcome the basins of attractions of turbid-water state. However, resilience is generally embedded in a multivariable-parameter space in ecosystem models or real ecosystems, and thus may be a highly complex object (Amemiya et al. 2005). We will come back this point when discussing the effect of biomanipulation using a graphical representation of resilience in section 3.6.

Finally in section 3.7, we critically reexamine the present models and their analyses from mathematical viewpoints. While we have tried to make the mathematical models as simple as possible for the sake of easy analysis, we need to study more on the mathematical models to confirm how they are better compared to other models. In what follows, we will discuss what direction do mathematical analysis proceed from now.

# ALGAL BLOOMS:
## MATHEMATICAL MODELS, DYNAMICS, AND MANAGEMENT

## 3.1. Minimal NP Model

### 3.1.1. Scheffer Model

The first step of this chapter is construction of the fundamental aquatic model. The simpler model is preferable as a fundamental model. Algebraically simple models are suitable for examining the generic behavior of the system (Scheffer 1991). It is known that two variables are sufficient to describe both bistability and limit cycle oscillations. For example, Scheffer (1991) presented a following minimal two-component model with phytoplankton and zooplankton:

$$\frac{dP}{dt} = \mu P \left( 1 - \frac{P}{K} \right) - f_P \frac{P}{H_P + P} Z, \tag{1}$$

$$\frac{dZ}{dt} = \eta f_P \frac{P}{H_P + P} Z - f_Z \frac{Z^2}{H_Z^2 + Z^2} - m_Z Z. \tag{2}$$

Two variables $P$ and $Z$ represent the biomasses of phytoplankton and herbivorous zooplankton, whose unit is g/m$^3$, and $t$ is time measured in days. The meanings and values of the parameters are summarized in Table 1.

The Scheffer model (1), (2) assumes the logistic growth of phytoplankton and the Holling type II functional response in zooplankton grazing on phytoplankton. Without the fish predation term (the second term of the right side of the equation (2)), the local kinetics of the model shows a typical phytoplankton-zooplankton limit cycle oscillation. Moreover, the addition of a non-dynamical fish predation term with the Holling type III functional response gives the possibility of bistable behaviors.

Thus, it is certain that the Scheffer model (1), (2) meets the requirement of being a fundamental aquatic model, showing both bistability and limit cycle oscillations. However, the most serious problem is that the Scheffer model cannot avoid the paradox of enrichment. That is, the model shows continuous increase in the amplitude of limit cycle oscillations with the increase in the carrying capacity $K$, as shown in Figure 2. In this model, the minimum values of the phytoplankton and zooplankton biomasses approach to zero with the increase in the carrying capacity, which means that the possibility of species extinction increases as well. This kind of phenomena is hardly seen in the natural environment. The paradox of enrichment cannot be avoided in the Scheffer model.

**Table 1. Parameters in Scheffer model (1)-(2)**

| Parameters | Meanings | Values | Units |
|---|---|---|---|
| $K$ | Carrying capacity | 0-20.0 | g/m$^3$ |
| $\mu$ | Maximum growth rate of phytoplankton | 0.5 | day$^{-1}$ |
| $f_P$ | Maximum feeding rate of zooplankton on phytoplankton | 0.5 | day$^{-1}$ |
| $H_P$ | Half-saturation constant of phytoplankton | 4.0 | g/m$^3$ |
| $\eta$ | Assimilation coefficient in zooplankton feeding on phytoplankton | 0.5 | |
| $f_Z$ | Maximum feeding rate of fish on zooplankton | not used | g m$^{-3}$ day$^{-1}$ |
| $H_Z$ | Half-saturation constant of zooplankton | not used | g/m$^3$ |
| $m_Z$ | Mortality rate of zooplankton | 0.1 | day$^{-1}$ |

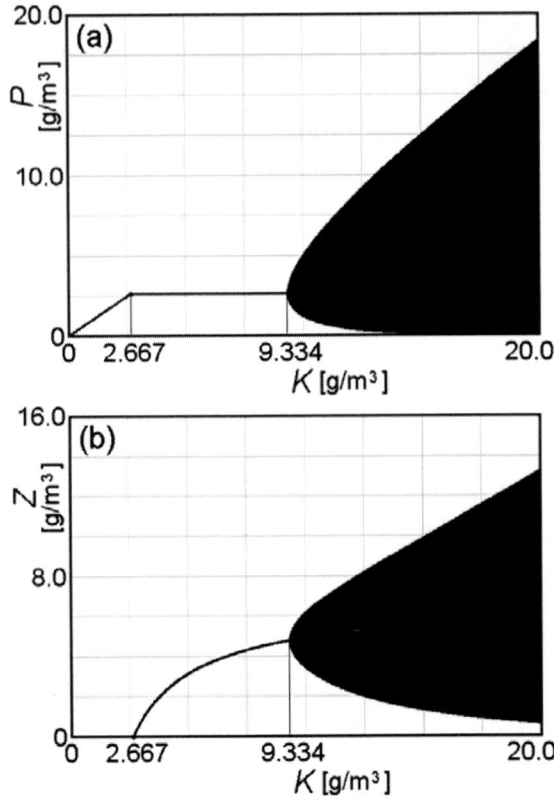

Figure 2. Paradox of enrichment in Scheffer model (1)-(2). These are the bifurcation diagrams of the phytoplankton biomass $P$ (a) and the zooplankton biomass $Z$ (b) for the parameters in Table 1 as a function of the carrying capacity $K$. The solid lines represent the stable state. The black areas in $K>9.334$ g/m³ represent limit cycle oscillations. The amplitude of oscillations is continuously increasing, which means that the possibility of species extinction is increasing as well.

### 3.1.2. Minimal NP Model

In this section, we propose a new minimal two-component model with nutrients and phytoplankton, which not only exhibit both bistability and limit cycle oscillations but can avoid the paradox of enrichment (Serizawa et al. 2008a). This model is named the minimal NP model. The mean-field model represented by ordinary differential equations is described as follows:

$$\frac{dN}{dt} = I_N - k\mu \frac{N}{H_N + N} P - m_N N, \tag{3}$$

$$\frac{dP}{dt} = I_P + \mu \frac{N}{H_N + N} P - f_P \frac{P}{H_P + P} - m_P P. \tag{4}$$

Two variables $N$ and $P$ denote the nutrient concentration (the unit is mmol/m³) and the phytoplankton biomass (the unit is g/m³), which are functions of actual time $t$ (the unit is day). The minimal NP model (3), (4) includes the effect of nutrient uptake by phytoplankton as a Holling type II functional response. Moreover, the implicitly assumed zooplankton grazing on phytoplankton is incorporated as a non-dynamical term, which is also formulated by the Holling type II functional response. The mean-field model without the diffusion and advection terms shows both bisatability and limit cycle oscillations as a couple of parameters are changed. The typical parameter values are listed in Table 2. The model shows bistability for the parameter set I and a limit cycle oscillation for the parameter set II, respectively.

The use of the Holling type II functional response in zooplankton grazing on phytoplankton is crucial for the minimal NP model (3), (4), although it is formulated as a non-dynamical term. This type of functional response has been broadly adopted to describe zooplankton predation on phytoplankton in many theoretical studies (Holling 1973; Scheffer 1991; Petrovskii and Malchow 1999; Medvinsky et al. 2002). It is also reported that the Holling type II functional response shows good accordance with experimental data (Mullin et al. 1975; Gentleman et al. 2003).

The removal term of nutrients $m_N N$ is added to evaluate the losses except for the uptake by phytoplankton, such as sedimentation to a deep bottom or an outflow from the system. However, this term has another meaning in terms of mathematical modeling. Suppose that phytoplankton is completely extinct. Without the remocal term, the nutrient concentration $N$ continues to increase unlimitedly. We can avoid this unfavorable situation by adding the removal term.

### 3.1.3. Dependence of Minimal NP Model on Parameters

Figure 3 shows the bifurcation diagrams of the nutrient concentration $N$ (a) and the phytoplankton biomass $P$ (b) as a function of the input rate of nutrients $I_N$ for the parameter set I in Table 2. The solid lines represent stable states, while the broken lines represent unstable states. Obvious bistability is observed in a parameter range of $0.187 \leq I_N \leq 0.426$ mmol m$^{-3}$ day$^{-1}$, showing sigmoid curves characteristic of bistability. Each bifurcation diagram consists of two stable branches. The first branch in $0 \leq I_N \leq 0.426$ mmol m$^{-3}$ day$^{-1}$ is referred to as a nutrient-dominated (ND) branch, in which phytoplankton do hardly exist, while nutrients increase almost linearly with $I_N$. Meanwhile, the other branch in $I_N \geq 0.187$ mmol m$^{-3}$ day$^{-1}$ is referred to as a phytoplankton-dominated (PD) branch, where nutrients decrease gradually, while phytoplankton increases dramatically.

### Table 2. Parameters in minimal NP model (3)-(4)

| Parameters | Meanings | Set I | Set II | Units |
|---|---|---|---|---|
| $I_N$ | Input rate of nutrients | 0-0.6 | 0-0.3 | mmol m$^{-3}$ day$^{-1}$ |
| $k$ | Nutrient content in phytoplankton | 0.2 | 0.2 | mmol/g |
| $H_N$ | Half-saturation constant of nutrients | 0.2 | 0.2 | mmol/m$^3$ |
| $m_N$ | Removal rate of nutrients | 0.075 | 0.005 | day$^{-1}$ |
| $\mu$ | Maximum growth rate of phytoplankton | 0.5 | 0.5 | day$^{-1}$ |
| $I_P$ | Input rate of phytoplankton | 0.01 | 0.01 | g m$^{-3}$ day$^{-1}$ |
| $f_P$ | Maximum feeding rate of zooplankton on phytoplankton | 2.0 | 2.0 | g m$^{-3}$ day$^{-1}$ |
| $H_P$ | Half-saturation constant of phytoplankton | 4.0 | 4.0 | g/m$^3$ |
| $m_P$ | Mortality rate of phytoplankton | 0.05 | 0.05 | day$^{-1}$ |

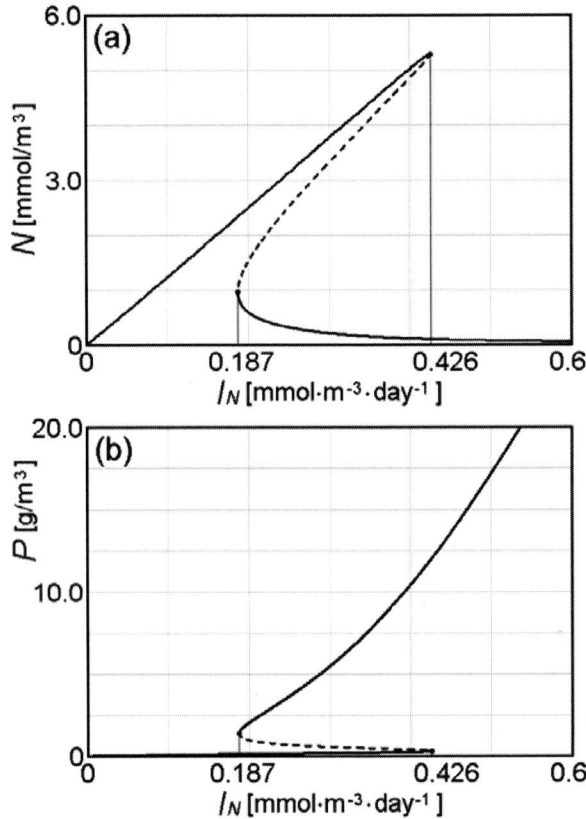

Figure 3. Bistability in minimal NP model (3)-(4). These are the bifurcation diagrams of the nutrient concentration $N$ (a) and the phytoplankton biomass $P$ (b) for the parameter set I in Table 2 as a function of the nutrient input rate $I_N$. The solid lines represent the stable state, while the broken lines represent the unstable state. The sigmoid curves characteristic of bistability are observed in a region $0.187 \leq I_N \leq 0.426$ mmol m$^{-3}$ day$^{-1}$. The stable state in $0 \leq I_N \leq 0.426$ mmol m$^{-3}$ day$^{-1}$ is referred to as a nutrient-dominated (ND) state, and the stable state in $I_N \geq 0.187$ mmol m$^{-3}$ day$^{-1}$ is referred to as a phytoplankton-dominated (PD) state, respectively.

The bifurcation diagrams of $N$ (a) and $P$ (b) exhibiting limit cycle oscillations are shown in Figure 4 as a function of $I_N$, using the parameter set II in Table 2. Two Hopf bifurcation points $I_{N0}=0.058$ mmol m$^{-3}$ day$^{-1}$ and $I_{N1}=0.18$ mmol m$^{-3}$ day$^{-1}$ appear, between which limit cycle oscillations occur. The system converges to equilibrium stable states in the region $0 < I_N \leq 0.058$ mmol m$^{-3}$ day$^{-1}$ or $I_N \geq 0.18$ mmol m$^{-3}$ day$^{-1}$, while it shows limit cycle oscillations in the region $0.058 < I_N < 0.18$ mmol m$^{-3}$ day$^{-1}$.

Figure 5 shows the parameter regions related to different behaviors in terms of three parameters, the input rate of nutrients $I_N$, the removal rate of nutrients $m_N$ and the maximum feeding rate of zooplankton $f_P$. The system behaviors are classified into six categories: (i) the convergence to a stable ND state (blue region), (ii) the convergence to a stable PD state (green region), (iii) the limit cycle oscillation (red region), (iv) bistability between (i) and (ii) (magenta region), (v) bistability between (i) and (iii) (cyan region), and (vi) divergence (yellow region).

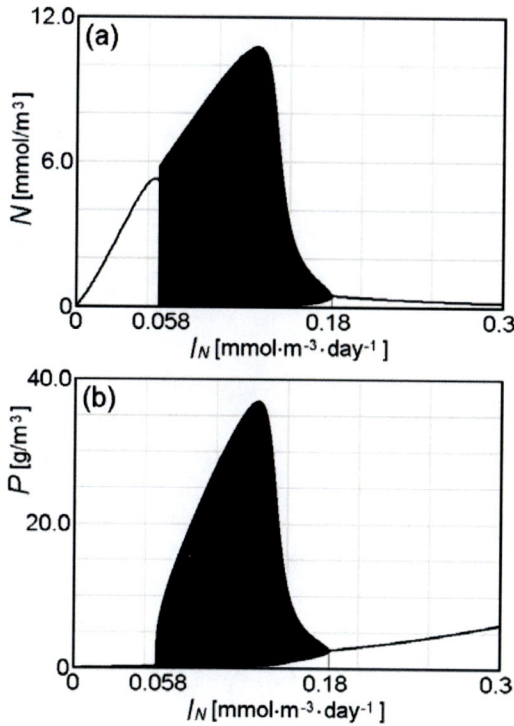

Figure 4. Limit cycle oscillations in minimal NP model (3)-(4). These are the bifurcation diagrams of the nutrient concentration $N$ (a) and the phytoplankton biomass $P$ (b) for the parameter set II in Table 2 as a function of the nutrient input rate $I_N$. The solid lines represent the stable state. Limit cycle oscillations occur in a region $0.058 < I_N < 0.18$ mmol m$^{-3}$ day$^{-1}$, where $I_{N0}=0.058$ mmol m$^{-3}$ day$^{-1}$ and $I_{N1}=0.18$ mmol m$^{-3}$ day$^{-1}$ specify two Hopf bifurcation points. Different from the Scheffer model in Figure 2, limit cycle oscillations are not continuing with the increase in $I_N$, which means that the paradox of enrichment could be avoided in the present model.

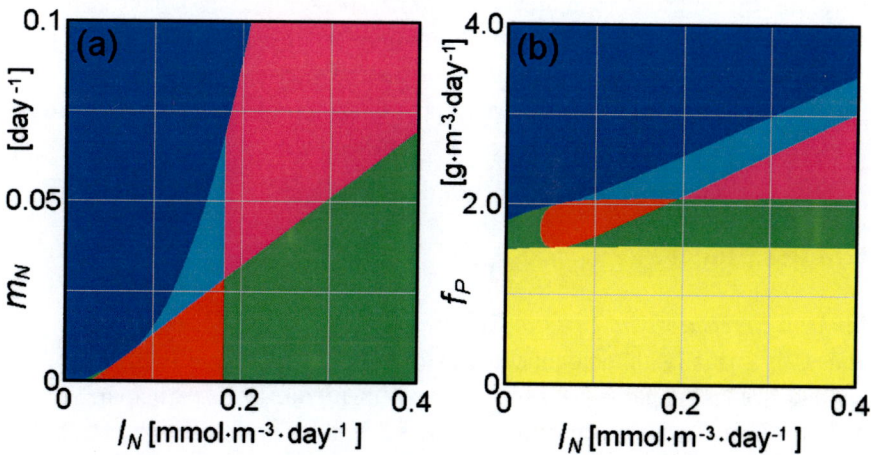

Figure 5. Dependence of minimal NP model (3)-(4) on parameters. The minimal NP model shows six different behaviors depending on the parameter values, as written in the text. The value of $m_N=0.005$ day$^{-1}$ in (b).

### 3.1.4. Avoidance of Paradox of Enrichment

The input of nutrients is one of the most critical external factors that can promote eutrophication in aquatic ecosystems. Figure 4 clearly shows the transition of the phytoplankton biomass $P$ from convergence to the stable equilibrium state with a small amount of phytoplankton to limit cycle oscillations and return to convergence without oscillation, as the nutrient input $I_N$ is increased. Within the limit cycle oscillation region in $0.058 < I_N < 0.18$ mmol m$^{-3}$ day$^{-1}$, the amplitude varies as eutrophication progresses, showing a peak near $I_N = 0.13$ mmol m$^{-3}$ day$^{-1}$. Meanwhile, within the convergence region in $I_N \geq 0.18$ mmol m$^{-3}$ day$^{-1}$, the value of $P$ increases gradually with an increase in the nutrient input $I_N$.

The paradox of enrichment has been a controversy among theoretical ecologists. It predicts that an increase in the nutrient input can destabilize a prey-predator system and even lead to the extinction of species (Rosenzweig 1971; Petrovskii et al. 2004). It has been shown theoretically in some models that a progress of enrichment or eutrophication leads to population oscillations of increasing amplitude, as shown in Figure 2 (Scheffer 1998). As a result, the minimum value of the population density decreases and species extinction becomes more probable due to stochastic environmental perturbations (Petrovskii et al. 2004). Although the extinction of a prey-predator community following system eutrophication has been observed in some laboratory experiments (Luckinbill 1974; Bohannan and Lenski 1997), this kind of phenomenon is not commonly seen in nature (McCauley and Murdoch 1990; Petrovskii et al. 2004).

In order to avoid the paradox of enrichment, various kinds of mechanisms have been proposed. For example, it is claimed that the paradox can be resolved by the transition to spatiotemporal chaos (Petrovskii et al. 2004) or by inducible defenses (Vos et al. 2004a). Scheffer (1998) also claimed that the presence of inedible algae such as large cyanobacterial colonies and spatial heterogeneity are potential stabilizing mechanisms.

The minimal NP model (3), (4) presented here does not exhibit the paradox of enrichment. As shown in Figure 4, the limit cycle oscillations in this model are restricted in a certain range of the nutrient input level ($0.058 < I_N < 0.18$ mmol m$^{-3}$ day$^{-1}$), and a further increase in the nutrient input makes the system return to the convergence to a stable equilibrium state. Considering that an increase in the nutrient input is equivalent to an increase in the carrying capacity, our nutrient-phytoplankton model is supposed to be one of the counter-examples of the paradox for ecological modeling.

## 3.2. Patchiness Model

### 3.2.1. Plankton Patchiness in Aquatic Ecosystems

Patchiness is spatially inhomogeneous distributions of plankton, which is observed in lakes, seas and oceans of various sizes, as shown in Figure 6. In particular, stretched and curled structures characteristic of patchiness patterns seem to be clearly observed in seas and oceans (Figure 6 (b) and (c)). In fact, this kind of pattern can emerge in a large scale region from one to hundreds of kilometers, where the lateral stirring and mixing play a more important role than the vertical circulation (Martin 2003).

Turing (1952) demonstrated that reaction-diffusion equations could form inhomogeneous patterns in his historical work. However, Turing pattern formation requires different diffusivities among the compnents. The diffusivity of the activator should be less than that of

Figure 6. Algal blooms in Lake Tsukui, Japan (a), the Black Sea (b) and the North Atlantic Ocean (c). The credits of (b) and (c) belong to NASA Goddard Space Flight Center.

the inhibitor. Moreover, the distribution of Turing patterns is stationary and spatially fixed. These restrictions are inconsistent with the field observations that the diffusivities of passive tracers such as nutrients and plankton are equal and that the movement of patterns is a general phenomenon (Martin 2003). The application of the Turing's approach to pattern formation in aquatic ecosystems confronts these difficulties.

In this section, we adopt the different approach from the Turing's one to simulate patchiness pattern formation, using reaction-diffusion-advection equations extended from the minimal NP model (3), (4). We assume equal diffusivities for both components, which is the usual case for nutrients and phytoplankton. In this case, the necessary condition of pattern formation is limit cycle oscillations in the mean-field model without diffusion and advection terms. If the parameter values are chosen from the limit cycle oscillation region, the corresponding reaction-diffusion equations or reaction-diffusion-advection equations can show spatial pattern formations by the combined effects of diffusion, advection by turbulent stirring and mixing, and biological interactions.

The studies to simulate the processes of patchiness formation seem to be classified into two groups depending on what is regarded as a major cause of these phenomena. One group focuses on biological factors such as non-linear growth and grazing interaction between phytoplankton and zooplankton (Medvinsky et al. 2002). In contrast, another group emphasizes the effects of physical factors such as currents, eddies and turbulent stirring induced by them (Abraham 1998).

One of the representative studies of the first group was conducted by Medvinsky et al. (2002). They have revealed a different route from the Turing's one to attain inhomogeneous pattern formation, using the Scheffer model with equal diffusivities. In spite of neglecting turbulent stirring, they have succeeded in reproducing patchiness-like patterns developed from regular spirals. In their simulations, the initial spiral pattern formation is followed by its collapse, resulting in spatiotemporal chaos in the entire region. Further, it is shown that these processes are almost independent of the initial distributions of the components. They claim that biological factors play an essential role in the emergence of plankton patchiness (Medvinsky et al. 2002).

In the second group, the main cause of patchiness pattern formation is considered to be physical factors (Abraham 1998; Neufeld et al. 2002; Tzella and Haynes 2007). For example, Abraham (1998) has revealed that the maturation time of zooplankton induces fine structures in the spatial distributions of the zooplankton population, using a three-component model whose state variables are the carrying capacity and population densities of phytoplankton and zooplankton.

We intend to construct the algebraically simpler model than preceding ones that can reproduce patchiness patterns. Our model is two-component reaction-diffusion-advection equations with diffusion terms of equal diffusivities and advection terms induced by simple turbulent stirring. Within a certain range of the nutrient input level, the model shows various types of pattern formation. However, a further increase in nutrient loading makes the inhomogeneous patterns disappear, showing homogeneous distribution of phytoplankton.

### 3.2.2. Diffusion and Advection

There are two types of diffusion. The first type of diffusion is usual diffusion in the context of molecular hydrodynamics, which is thought to be a tendency toward homogeneity originating from the second law of thermodynamics. Therefore, it is effective without any physical water flow. This type of diffusion is represented by the second order differential for position coordinates. On the other hand, the second type of diffusion is diffusion by advection. Advection is the substantial water flow such as currents and eddies, and diffusion by advection is passive movement by advection. Therefore, it is ineffective within perfectly calm water. This type of diffusion is represented by the first order differential for position coordinates. The actual mixing of nutrients or phytoplankton is caused by the combination of these two effects. In this chapter, the first type of diffusion is simply called diffusion, and the second type of diffusion is called advection.

Turbulent stirring is considered to be a crucial factor in creating patchiness patterns in aquatic ecosystems. In this section, we use a two-dimensional turbulent flow developed by referring to the seeded-eddy model (Dyke and Robertson 1985; Abraham 1998). The stream function $\psi$ and fluid velocity $V$ are described as follows:

$$\psi(x,y) = s\sum_i \sigma_i \exp\left\{-\frac{(x-x_i)^2 + (y-y_i)^2}{r_0^2}\right\}, \quad \sigma_i = 1 \; or \; -1, \quad V = (V_x, V_y) = \left(-\frac{\partial \psi}{\partial y}, \frac{\partial \psi}{\partial x}\right). \quad (5)$$

The velocity field is composed of 100 eddies, the half of which rotate clockwise, while the other half rotate counter-clockwise. In our model, the center of each eddy $(x_i, y_i)$ is randomly distributed within the domain. For simplicity, we use a constant value of radius $r_0$ for all eddies without considering a distribution of varying eddy sizes. The scaling constant $s$ is introduced in order to adjust the maximum velocity $V_{max}$.

Figure 7 shows the velocity field $V$ used in this section. Except for Figure 11 (a), the domain in our simulations is a 200×200 km square, that is, a half length of each side is $L=100$ km. The radius of each eddy is $r_0=10$ km, and the maximum velocity is $V_{max}=0.15$ km/day. The velocity field is stationary and remains temporally unchanged, and also meets the periodic boundary conditions.

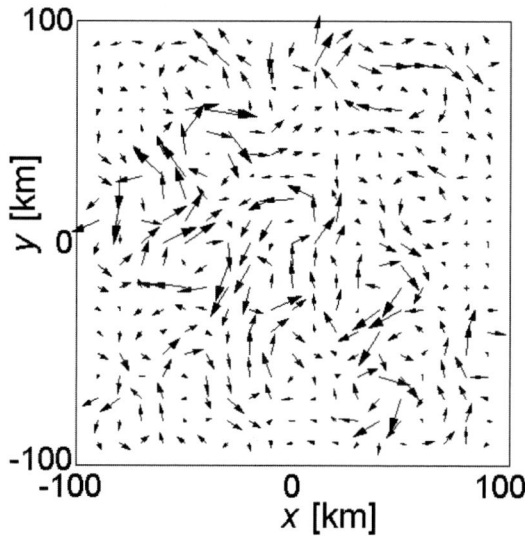

Figure 7. Velocity field by turbulent stirring. The velocity field is constructed by the superimposition of 100 eddies with the radius $r_0=10$ km. The maximum velocity is $V_{max}=0.15$ km/day. The velocity field meets the periodic boundary conditions.

### 3.2.3. Reaction-Diffusion-Advection Equations

The previous minimal NP model (3), (4) is extended to reaction-diffusion-advection equations to simulate patchiness pattern formation (Serizawa et al. 2008a). Therefore, the state variables are the nutrient concentration $N$ and the phytoplankton density $P$, and mathematical representation that describes the biological interaction is the same as that of the minimal NP model. For simplicity, two parameters $I_P$ and $m_P$ are omitted, which represent the input and mortality rates of phytoplankton, respectively. The present model, referred to as the patchiness model in this chapter, is described as follows:

$$\frac{\partial N}{\partial t} = D\nabla^2 N - \nabla \cdot (V N) + I_N - k\mu \frac{N}{H_N + N}P - m_N N, \qquad (6)$$

$$\frac{\partial P}{\partial t} = D\nabla^2 P - \nabla \cdot (V P) + \mu \frac{N}{H_N + N}P - f_P \frac{P}{H_P + P}. \qquad (7)$$

$$\nabla = \left(\frac{\partial}{\partial x}, \frac{\partial}{\partial y}\right), \quad \nabla^2 = \frac{\partial^2}{\partial x^2} + \frac{\partial^2}{\partial y^2}.$$

Two variables $x$ and $y$ denote the position coordinates, the unit of which is km. The parameter $D$ is the diffusion coefficient, and we assume that the lateral diffusivities are equal for both components. The first terms of the right side in equations (6) and (7) are referred to as diffusion terms, whereas the second terms are referred to as advection terms.

The parameter values used in the simulations are listed in Table 3. The units of variables and parameters are changed as mmol/m$^3$ → mmol/m$^2$, g/m$^3$ → g/m$^2$, and so on, reflecting the use of two-dimensional reaction-diffusion-advection equations.

In our simulations of the partial differential equation system (5)-(7), the initial distributions of $N$ and $P$ are given by the following equations, respectively:

$$N(x, y, 0) = N_1\left(1 + \sin\frac{\pi}{L}x\right), \qquad (8)$$

$$P(x, y, 0) = P_1\left(1 + \sin\frac{\pi}{L}y\right). \qquad (9)$$

### Table 3. Parameters in patchiness model (5)-(7)

| Parameters | Meanings | Values | Units |
|---|---|---|---|
| $I_N$ | Input rate of nutrients | 0.02-0.16 | Mmol m$^{-2}$ day$^{-1}$ |
| $k$ | Nutrient content in phytoplankton | 0.2 | mmol/g |
| $H_N$ | Half-saturation constant of nutrients | 0.2 | mmol/m$^2$ |
| $m_N$ | Removal rate of nutrients | 0.015 | day$^{-1}$ |
| $\mu$ | Maximum growth rate of phytoplankton | 0.5 | day$^{-1}$ |
| $f_P$ | Maximum feeding rate of zooplankton on phytoplankton | 3.6 | g m$^{-2}$ day$^{-1}$ |
| $H_P$ | Half-saturation constant of phytoplankton | 8.0 | g/m$^2$ |
| $D$ | Diffusion coefficient | 0.02, 0.5 | km$^2$/day, m$^2$/day |
| $V_{max}$ | Maximum velocity | 0.15 | km/day, m/day |
| $L$ | Half length of square domain side | 100 | km, m |

For simplicity, two parameters $I_P$ and $m_P$ in the minimal NP model (3)-(4) are omitted in the patchiness model (5)-(7). The units of $D$, $V_{max}$ and $L$ are changed to m$^2$/day, m/day or m in the simulation of Figure 11 (a).

Here, $N_1$ and $P_1$ denote the values of two components at the unstable fixed point $F_1$, which is shown in Table 4. The distributions of both $N$ and $P$, as well as their first-order derivatives, are continuous throughout the boundaries. In addition, the gradients of two components are perpendicular to each other. These initial conditions are used together with the periodic boundary conditions.

**Table 4. Stability analyses in corresponding mean-field model of (6)-(7)**

|  | $(N, P)$ | Eigenvalues | Stability | Fixed points |
|---|---|---|---|---|
| $F_0$ | $(6.0, 0.0)$ | $0.034, -0.015$ | Unstable | Saddle point |
| $F_1$ | $(0.814, 0.969)$ | $0.005 \pm 0.078i$ | Unstable | Repeller |

The corresponding mean-field model of (6)-(7) generates two unstable fixed points, the saddle point $F_0$ and the repeller $F_1$, within the region $N \geq 0$ and $P \geq 0$ for the parameters in Table 3. $I_N = 0.09$ mmol m$^{-2}$ day$^{-1}$.

The initial conditions are chosen so that the spatial distributions do not show any spatial symmetry. The initial distributions represented by (8), (9) seem to be one of the simplest forms that satisfies this condition and also consistent with the periodic boundary conditions. It should be noted that the appearance of the pattern is not induced by the initial conditions. In fact, the process of patchiness pattern formation is almost independent of the initial conditions. However, it is at least necessary to include the values at the unstable fixed point $F_1$, which functions as a seed of irregular pattern formation (Petrovskii and Malchow 1999).

### 3.2.4. Bifurcation Diagrams

Prior to the exploration of spatial pattern formation, it is useful to investigate the corresponding ordinary differential equation system. Stability analysis shows that this system has two unstable fixed points $F_0$ and $F_1$ in a realistic range of two components ($N \geq 0$ and $P \geq 0$) under the conditions given in Table 3. As summarized in Table 4, these are the saddle point $F_0$ and the repeller $F_1$, and a limit cycle is formed around $F_1$.

Figure 8 shows the bifurcation diagrams of the nutrient concentration $N$ (a) and the phytoplankton density $P$ (b) as a function of the input rate of nutrients $I_N$. In the case that the maximum zooplankton feeding rate $f_P < \mu H_P$, there exists a transcritical point $I_{Ntc}$, which is described by

$$I_{Ntc} = \frac{m_N H_N f_P}{\mu H_P - f_P}. \tag{10}$$

In the region $I_N \leq I_{Ntc}$, phytoplankton cannot survive, while nutrients linearly increase with $I_N$. When the parameter values are given in Table 3, $I_{Ntc} = 0.027$ mmol m$^{-2}$ day$^{-1}$. Meanwhile, in the region $I_N > I_{Ntc}$, nutrients and phytoplankton coexist. Two Hopf bifurcation points $I_{N0} = 0.049$ mmol m$^{-2}$ day$^{-1}$ and $I_{N1} = 0.125$ mmol m$^{-2}$ day$^{-1}$ are specified, and the system shows limit cycle oscillations in the intermediate region $I_{N0} < I_N < I_{N1}$, while it converges to an equilibrium stable state in the region $I_{Ntc} < I_N \leq I_{N0}$ or $I_N \geq I_{N1}$.

### 3.2.5. Temporal Variation and Dependence on Input Rate of Nutrient

Figure 9 shows the temporal change of the spatial distribution of phytoplankton for the parameters in Table 3. The input rate of nutrients $I_N$=0.09 mmol m$^{-2}$ day$^{-1}$. As shown in Figure 9 (b), the typical patchiness pattern formation with stretched and curled structures is completed within 270 days.

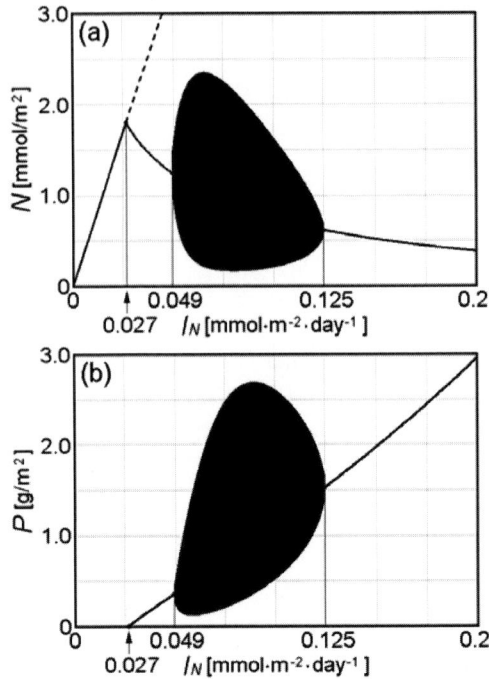

Figure 8. Limit cycle oscillations in corresponding mean-field model of (6)-(7). These are the bifurcation diagrams of the nutrient concentration $N$ (a) and the phytoplankton biomass $P$ (b) for the parameters in Table 3 as a function of the nutrient input rate $I_N$. The solid lines represent the stable state, while the broken line represents the unstable state. Limit cycle oscillations occur in a region $0.049 < I_N < 0.125$ mmol m$^{-2}$ day$^{-1}$, where $I_{N0}$=0.049 mmol m$^{-2}$ day$^{-1}$ and $I_{N1}$=0.125 mmol m$^{-2}$ day$^{-1}$ specify two Hopf bifurcation points. Phytoplankton cannot exist in the region $I_N \leq 0.027$ mmol m$^{-2}$ day$^{-1}$, where $I_{Ntc}$=0.027 mmol m$^{-2}$ day$^{-1}$ is a transcritical point.

One of the characteristic features seen in Figure 9 is its periodic behavior. Obvious similarity is recognized between two images of Figure 9 (b) and (f), indicating that similar patterns appear periodically without decaying to homogeneous distributions. The period is estimated at about 90 days, which is also thought to be equal to the period of the limit cycle oscillation.

Eutrophication is the main factor that promotes algal blooms in aquatic ecosystems. Therefore, understanding the correlation between the nutrient loading level and the spatial patterns of phytoplankton blooms is important for predicting the degradation of aquatic ecosystems. The correlation between these two factors in the patchiness model (5)-(7) is examined by changing the input rate of nutrients $I_N$.

Figure 10 shows the simulation results of the spatial patterns for different values of $I_N$ at $t$=360 days. When $I_N$=0.2 mmol m$^{-2}$ day$^{-1}$, no pattern appears, corresponding to the total extinction of phytoplankton (Figure 10 (a)). On the other hand, when $I_N$=1.6 mmol m$^{-2}$ day$^{-1}$, the homogeneous distribution of a large quantity of phytoplankton occurs, reflecting a mass occurrence of phytoplankton blooms (Figure 10 (f)).

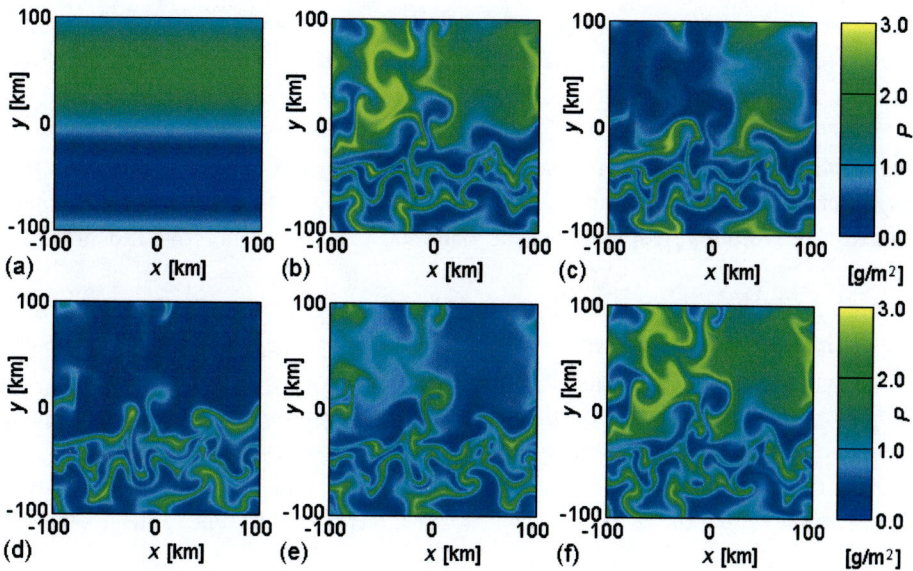

Figure 9. Temporal change of spatial distribution of phytoplankton in patchiness model (5)-(7). Obvious similarity between two images (b) and (f) indicates that the temporal variation of the system shows periodicity in an about 90-day cycle, which is also the period of the limit cycle oscillation. $I_N$=0.09 mmol m$^{-2}$ day$^{-1}$. (a) $t$=0 day, (b) $t$=270 days, (c) $t$=288 days, (d) $t$=312 days, (e) $t$=336 days, (f) $t$=360 days.

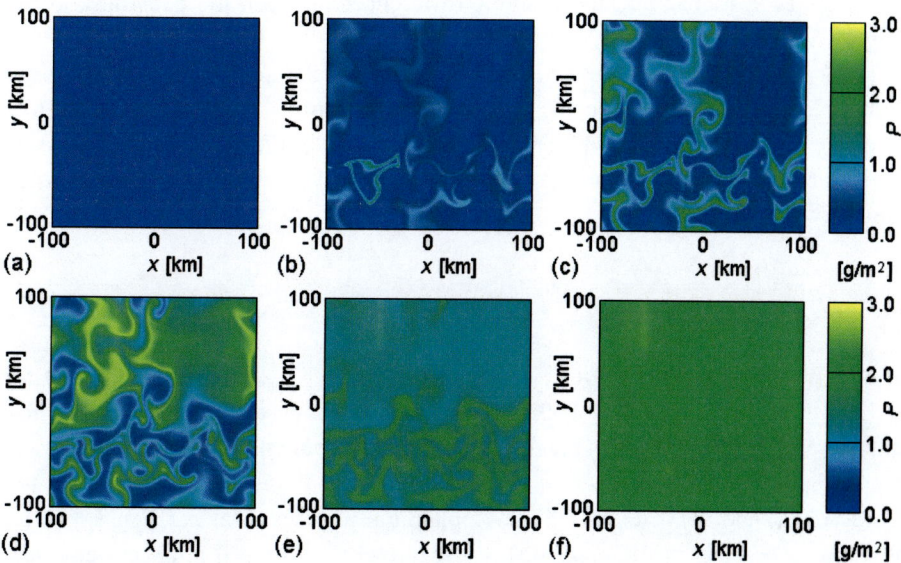

Figure 10. Dependence of spatial distribution of phytoplankton on nutrient input rate in patchiness model (5)-(7). $t$=360 days. (a) $I_N$=0.02 mmol m$^{-2}$ day$^{-1}$, (b) $I_N$=0.05 mmol m$^{-2}$ day$^{-1}$, (c) $I_N$=0.07 mmol m$^{-2}$ day$^{-1}$, (d) $I_N$=0.09 mmol m$^{-2}$ day$^{-1}$ (same as Figure 9 (f)), (e) $I_N$=0.125 mmol m$^{-2}$ day$^{-1}$, (f) $I_N$=0.16 mmol m$^{-2}$ day$^{-1}$.

The cases within a limit cycle oscillation region are more interesting. For example, the cases in $I_N$=0.5 and 0.7 mmol m$^{-2}$ day$^{-1}$ show filamentous patterns (Figures 10 (b) and (c)), which means the inhomogeneous pattern with locally concentrated and thinly stretched structures. Meanwhile, the case in $I_N$=0.9 mmol m$^{-2}$ day$^{-1}$ seems to be a typical example of a patchiness pattern, in which stretched and curled patterns are widely dispersed (Figure 10 (d)). This figure seems to replicate real plankton patchiness as seen in Figure 6 (b) and (c). The patchiness pattern becomes obscure gradually with an increase in $I_N$ (Figure 10 (e)), reaching a uniformly distributed homogeneous pattern (Figure 10 (f)). Consequently, as the nutrient input $I_N$ is increased, the system behavior changes from the entire extinction of phytoplankton to formation of filamentous patterns, patchiness patterns and homogeneous distributions.

According to Rietkerk et al. (2004), there is a sequence of vegetation patterns in arid ecosystems that can predict the level of resource availability. Vegetation patterns are changed from homogeneous cover to gaps, labyrinths or stripes and spots, as the resource (such as water or nutrients) input is decreased. Our simulation results show a similar sequence of plankton distribution, such as homogeneous cover, patchiness patterns, and filamentous patterns, as the nutrient input is decreased, indicating that the spatial distributions of phytoplankton can function as an indicator to evaluate the eutrophication level in aquatic ecosystems.

### 3.2.6. Dependence on Diffusion and Advection

We can recognize stretched and curled structures in the images of phytoplankton blooms, as seen in Figure 6. These features of patchiness patterns are supposed to be due to the coupled effects of diffusion and advection such as stretching by currents and turbulent stirring by eddies. However, there seem to be subtle differences between the patchiness patterns in lakes (Figure 6 (a)) and those in seas or oceans (Figure 6 (b) and (c)).

Figure 11 seems to explain these differences, in which the dependence of spatial patterns of the phytoplankton density $P$ on the diffusion coefficient $D$ and the maximum turbulent velocity $V_{max}$ is examined. It should be noted that the velocity due to diffusion $V_D$ is estimated by the square root of the product of the maximum growth rate $\mu$ and the diffusion coefficient $D$, as follows:

$$V_D = \sqrt{\mu D}. \tag{11}$$

Comparing with the maximum velocity of turbulent stirring $V_{max}$, we can consider that Figure 11 (a), where $V_D$=0.5 m/day and $V_{max}$=0.15 m/day, shows the case in which lateral diffusion is dominant ($V_D > V_{max}$). On the other hand, advection by turbulent stirring is considered to be dominant in Figure 11 (b), where $V_D$=0.1 km/day and $V_{max}$=0.15 km/day ($V_D < V_{max}$). It seems that Figures 11 (a) and (b) resemble the phytoplankton blooms in lakes (Figure 6 (a)) and in seas or oceans (Figure 6 (b) and (c)), respectively. Stretched and curled structures characteristic of plankton patchiness in seas or oceans are clearly reproduced in Figure 11 (b), which suggests that turbulent stirring and mixing are essential for patchiness formation in marine ecosystems.

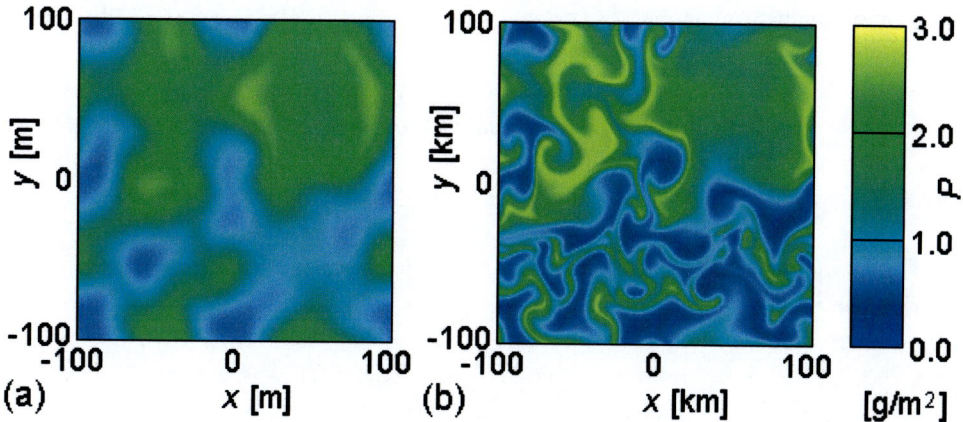

Figure 11. Dependence of spatial distribution of phytoplankton on diffusion coefficient in patchiness model (5)-(7). $I_N$=0.09 mmol m$^{-2}$ day$^{-1}$. $t$=360 days. (a) $D$=0.5 m$^2$/day, $V_{max}$=0.15 m/day, (b) $D$=0.02 km$^2$/day, $V_{max}$=0.15 km/day (same as Figure 9 (f) and Figure 10 (d)).

We suppose that either diffusion or advection mainly determines the phytoplankton distributions on the water surface. The contribution of advection becomes more significant as the water area is expanded. As a result, the non-diffusive stirring and mixing become major factors in creating plankton patchiness in seas or oceans. In contrast, the phytoplankton distribution is principally dominated by lateral diffusion in freshwater lakes of small or intermediate sizes. The dominance of lateral diffusion seems to obscure the spatial patterns of phytoplankton.

## 3.3. Vertical Migration Model

### 3.3.1. Annual Life Cycle of Cyanobacteria

Cyanobacteria have dominated various kinds of aquatic ecosystems on Earth. This group has developed some adaptive mechanisms throughout its evolutionary history, which include phenotypic plasticity between unicellular and colonial or filamentous morphologies (Yang et al. 2006) as will be shown in section 3.4, production of toxic materials such as microcystins or anatoxins (Watanabe et al. 1996) and an ability to migrate vertically by means of buoyancy regulation (Reynolds et al. 1987). By these adaptive mechanisms, cyanobacteria have succeeded to occupy areas with various environmental conditions ranging from freshwater lakes to brackish water estuaries. In this section, we focus on vertical migration of cyanobacteria, and present the vertical migration model, which is another application of the minimal NP model (3), (4) to reaction-diffusion-advection equations.

The annual life cycle of cyanobacteria is as follows. These cyanobacteria overwinter as vegetative colonies on the bottom sediment until early summer. Then, some of the colonies rise into the upper water column in response to increased light intensity. Successful recruitment of benthic cells to the water column requires a preceding phase of high water clarity and light penetration to the lake bottom, which possibly cause the photo-activation of gas-vesicle synthesis. Other seasonal change that triggers the activation of dormant cells includes increasing water temperature (Sigee 2005).

The recruitment of cyanobacteria to the water column results in an algal bloom during midsummer. With the coming of autumn, thermal stratification collapses, and colonial cyanobacteria fall to the bottom, where they remain viable until next early summer. This is the seasonal succession of life stages in colonial cyanobacteria. They are present in the water column over a limited period, spending most of the year in a dormant state on the lake sediments (Sigee 2005).

### 3.3.2. Buoyancy Regulation Mechanism of Cyanobacteria

During the algal bloom in midsummer, cyanobacteria repeat diurnal vertical migration within a water column. For photosynthetic micro-organisms such as cyanobacteria, the primary requirement is to remain within near-surface layers where sunlight is abundant (Reynolds et al. 1987). However, photosynthetic productivity is governed not only by light but also by nutrients, which tend to be depleted in the light-rich epilimnion. Therefore, it is also advantageous for cyanobacteria to move from the nutrient-deficient epilimnion down to the metalimnion or the hypolimnion in order to take up nutrients. Buoyancy regulation of cyanobacteria is thought to have evolved to allow access to both the epilimnion and the hypolimnion (Brookes and Ganf 2001; Sigee 2005).

Buoyancy regulation is a mechanism to float upwards and sink downwards alternately within a water column. These movements can be achieved by using vesicles, intracellular gas-filled spaces, and there are many reports that suggest cyclic diurnal migration of cyanobacteria, rising up during the night and sinking down during the day (Ibelings et al. 1991; Sigee 2005). The process of buoyancy regulation is sensitive to changes in light intensity so as to allow cyanobacteria to migrate within a day.

Two explanations are conceivable for the mechanism of buoyancy regulation. One is related to control of vesicle numbers and the other is due to extra-vesicle processes (Reynolds et al. 1987). According to the former explanation, floating and sinking of cyanobacteria are caused by synthesis and turgor-collapse of gas-vesicles. However, the mechanisms based on these processes are less than feasible, because it takes more than a day to synthesize vesicles. Gas-vesicle synthesis takes place on time scales of days to weeks (Brookes et al. 2000; Brookes and Ganf 2001), and thus the process is too slow for diurnal migration (Reynolds et al. 1987). These vesicles are also too strong to be collapsed by usual turgor-pressure (Walsby 1994). In fact, extra-vesicle processes, which operate outside the vesicles, are more likely to influence the vertical migration of cyanobacteria.

Staying within the light-rich epilimnion promotes carbon fixation, resulting in accumulation of photosynthetic products such as carbohydrates. Cyanobacteria accumulate these carbohydrates in the form of glycogen, which functions as ballast. As a result, cyanobacteria become heavier and begin to sink downwards when in the light. In contrast, a reduction in glycogen increases buoyancy of cyanobacteria, enabling them to return to the light-rich epilimnion for photosynthesis. It is known that control of this ballast can be accomplished within several hours (Reynolds et al. 1987), and therefore a short-term adjustment and a rapid reversal between floating and sinking become possible (Ibelings et al. 1991).

### 3.3.3. Mathematical Model

Algal blooms caused by cyanobacteria are characterized by two features with different time scales: one is seasonal outbreak and collapse of a bloom and the other is diurnal vertical migration. In order to simulate both phenomena, we present a two-component mathematical model with nutrients and cyanobacteria, developed from the minimal NP model (3), (4) (Serizawa et al. 2008b).

The model is a set of one-dimensional reaction-diffusion-advection equations, and temporal changes of these two variables are regulated by the following five factors: (i) annual variation of light intensity, (ii) diurnal variation of light intensity, (iii) annual variation of water temperature, (iv) thermal stratification within a water column and (v) the buoyancy regulation mechanism.

The seasonal change of cyanobacteria biomass is mainly controlled by factors, (i), (iii) and (iv), among which annual variations of light intensity and water temperature directly affect the maximum growth rate of cyanobacteria. The latter also contributes to formation of the thermocline during the summer season. Thermal stratification causes a reduction in vertical diffusion and largely prevents mixing of both nutrients and cyanobacteria between the epilimnion and the hypolimnion. Meanwhile, the other two factors, (ii) and (v), play a significant role in diurnal vertical migration of cyanobacteria. A key mechanism of diurnal migration is buoyancy regulation due to gas-vesicle synthesis and ballast formation.

To simplify the model, lateral homogeneity is assumed, meaning that two state variables are described as functions of time and depth. We assume that the depth of the water column $z_B$ is 20 m and that the vertical position of the thermocline $z_T$ is 8 m below the water surface. It is also assumed that one year consists of 365 days, and all the simulations start at midnight on December 31.

## Light Intensity

Annual and diurnal variations of incident light intensity at the surface at noon on each day $I_S$ follows the formula:

$$I_S(t) = -\frac{1}{2}\left\{(I_{S\,max} + I_{S\,min})\cos 2\pi t + (I_{S\,max} - I_{S\,min})\cos\frac{2\pi(t-t_0)}{365}\right\}. \quad (12)$$

$I_S$ is a function of time $t$ measured in days since January 1. The vertical distribution of light intensity $I$ at noon on each day is described as follows:

$$I(t,z) = \begin{cases} I_S(t)\exp\left(-\alpha_W z - \alpha_P \int_0^z P dz'\right) & (I_S \geq 0), \\ 0 & (I_S < 0). \end{cases} \quad (13)$$

Here, the variable $z$ represents the depth within the water column. The annual variation of light intensity at the surface is shown in Figure 12 (a). The meanings and values of parameters used in the vertical migration model are listed in Table 5.

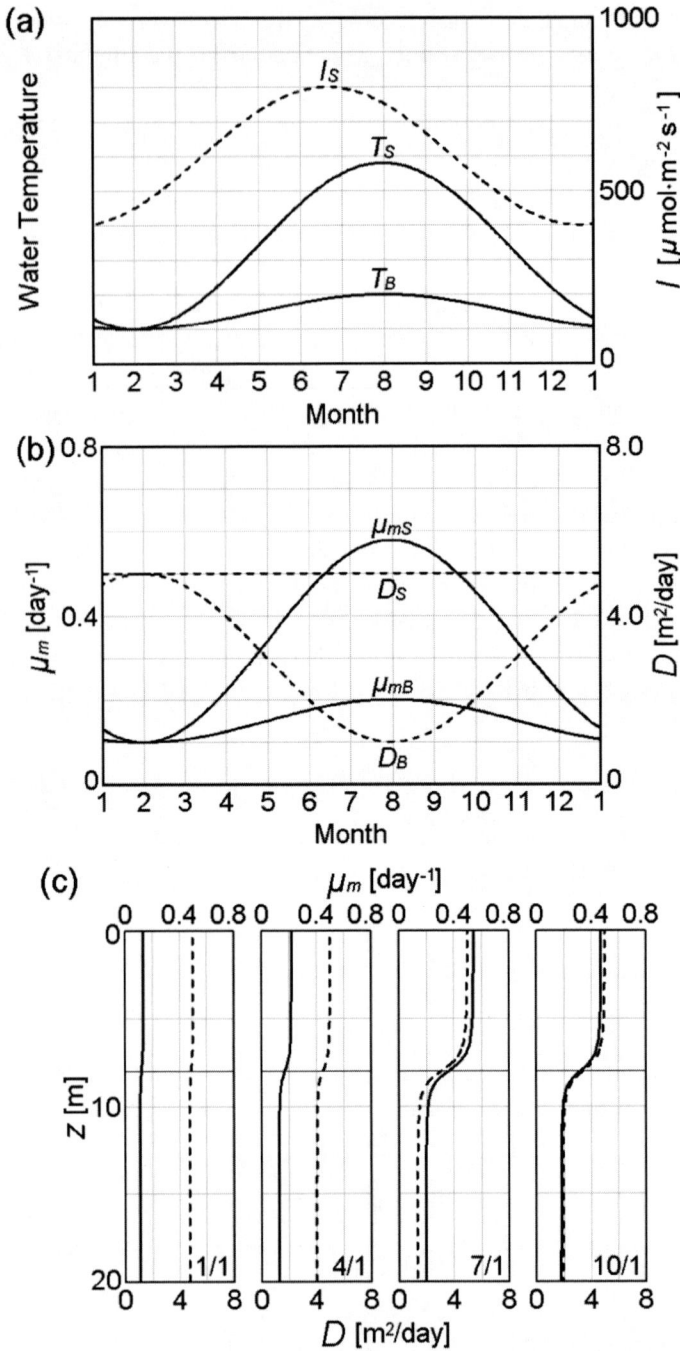

Figure 12. (a) Annual variations of water temperature at the surface (solid line: $T_S$) and at the bottom (solid line: $T_B$), and light intensity at the surface (broken line: $I_S$) in vertical migration model (12)-(28). The unit of water temperature is arbitrary. (b) Annual variations of maximum growth rate of cyanobacteria at the surface (solid line: $\mu_{mS}$) and at the bottom (solid line: $\mu_{mB}$), and turbulent diffusivity at the surface (broken line: $D_S$) and at the bottom (broken line: $D_B$). (c) Annual variations of vertical distributions of maximum growth rate (solid lines) and turbulent diffusivity (broken lines). The horizontal lines at $z_T=8$ m indicate the depth of the thermocline. Permission is granted to reproduce the figure by Limnology.

## Table 5. Parameters in vertical migration model (12)-(28)

| Parameters | Meanings | Values | Units |
|---|---|---|---|
| $I_{Smax}$ | Maximum light intensity at the surface at noon on the summer solstice | 800 | $\mu mol\,m^{-2}\,s^{-1}$ |
| $I_{Smin}$ | Maximum light intensity at the surface at noon on the winter solstice | 400 | $\mu mol\,m^{-2}\,s^{-1}$ |
| $d_0$ | Difference between Jan. 1 and the winter solstice | -10 | day |
| $\alpha_W$ | Absorption coefficient by water | 0.23 | $m^{-1}$ |
| $\alpha_P$ | Absorption coefficient by cyanobacteria | 0.06 | $m^2/g$ |
| $\mu_{mSmax}$ | Maximum growth rate of cyanobacteria at the surface on the warmest day | 0.58 | $day^{-1}$ |
| $\mu_{mSmin}$ | Maximum growth rate of cyanobacteria at the surface on the coldest day, Jan. 31 | 0.1 | $day^{-1}$ |
| $\mu_{mBmax}$ | Maximum growth rate of cyanobacteria at the bottom on the warmest day | 0.2 | $day^{-1}$ |
| $\mu_{mBmin}$ | Maximum growth rate of cyanobacteria at the bottom on the coldest day, Jan. 31 | 0.1 | $day^{-1}$ |
| $d_1$ | Difference between Jan. 1 and the coldest day | 30 | day |
| $H_I$ | Half-saturation constant for light intensity | 20.0 | $mmol\ m^{-2}\ s^{-1}$ |
| $H_N$ | Half-saturation constant for nutrient concentration | 0.2 | $mmol/m^3$ |
| $D_S$ | Vertical turbulent diffusivity at the surface | 5.0 | $m^2/day$ |
| $D_{Bmax}$ | Vertical turbulent diffusivity at the bottom on the coldest day, Jan. 31 | 5.0 | $m^2/day$ |
| $D_{Bmin}$ | Vertical turbulent diffusivity at the bottom on the warmest day | 1.0 | $m^2/day$ |
| $r$ | Reciprocal of decay time for actual growth rate | 3.0 | $day^{-1}$ |
| $V_m$ | Scale factor for velocity of cyanobacteria | 250.0 | m/day |
| $F_0$ | Ballast factor for neutral buoyancy | 0.1 | |
| $I_N$ | Input rate of nutrients | 0.5 | $mmol\ m^{-3}\ day^{-1}$ |
| $k$ | Nutrient content in cyanobacteria | 0.75 | mmol/g |
| $m_N$ | Removal rate of nutrients | 0.05 | $day^{-1}$ |
| $f_P$ | Maximum feeding rate of zooplankton on cyanobacteria | 0.6 | $g\ m^{-3}\ day^{-1}$ |
| $H_P$ | Half-saturation constant for cyanobacteria | 4.0 | $g/m^3$ |
| $N_B$ | Nutrient concentration at the bottom | 40 | $mmol/m^3$ |
| $w_T$ | Vertical extent of the thermocline | 1 | m |
| $z_T$ | Depth of the thermocline | 8 | m |
| $z_B$ | Depth of the water column | 20 | m |

## Growth Rate of Cyanobacteria

In the present model, we schematically assume that water temperatures both at the surface and at the bottom vary as shown in Figure 12 (a). We also assume that the maximum growth rates of cyanobacteria vary in the same manner as water temperature. Then, the maximum growth rates of cyanobacteria at the water surface $\mu_{mS}$ and at the bottom $\mu_{mB}$ are formulated as follows:

$$\mu_{mS}(t) = \frac{1}{2}\left\{(\mu_{mS\,max} + \mu_{mS\,min}) - (\mu_{mS\,max} - \mu_{mS\,min})\cos\frac{2\pi(t - t_1)}{365}\right\}, \qquad (14)$$

$$\mu_{mB}(t) = \frac{1}{2}\left\{(\mu_{mB\,max} + \mu_{mS\,min}) - (\mu_{mB\,max} - \mu_{mB\,min})\cos\frac{2\pi(t - t_1)}{365}\right\}. \qquad (15)$$

Figure 12 (b) shows annual variations of the maximum growth rate at the surface $\mu_{mS}$ and at the bottom $\mu_{mB}$.

Next, we consider the dependence of the maximum growth rate on depth $z$. Denoting the vertical extent and the depth of the thermocline as $w_T$ and $z_T$, we can formulate the maximum growth rate $\mu_m$, as follows:

$$\mu_m(t,z) = \mu_{mB}(t) + \{\mu_{mS}(t) - \mu_{mB}(t)\}\frac{1}{2}\left(1 - \frac{z - z_T}{\sqrt{w_T^2 + (z - z_T)^2}}\right). \qquad (16)$$

Vertical profiles of the maximum growth rate are shown in Figure 12 (c) for January 1, April 1, July 1 and October 1.

The actual growth rate $\mu$ is described by a product of the maximum growth rate $\mu_m$ and the minimum function $f$, as follows (Yoshiyama and Nakajima 2002; Huisman et al. 2006):

$$f(I(t,z), N(t,z)) = \min\left(\frac{I(t,z)}{I(t,z) + H_I}, \frac{N(t,z)}{N(t,z) + H_N}\right). \qquad (17)$$

$$\mu(t,z) = \mu_m(t,z)f(I(t,z), N(t,z)). \qquad (18)$$

The minimum function $f$ gives the smaller value of two Monod functions for light intensity $I$ and nutrient concentration $N$.

**Turbulent Diffusivity**

The thermocline not only affects the vertical distribution of the maximum growth rate of cyanobacteria, but also largely prevents vertical diffusive mixing within the water column. The turbulent diffusivity at the bottom $D_B$ is a function of time $t$, which is represented as follows:

$$D_B(t) = \frac{1}{2}\left\{(D_{B\,max} + D_{B\,min}) + (D_{B\,max} - D_{B\,min})\cos\frac{2\pi(t - t_1)}{365}\right\}. \qquad (19)$$

Annual variations of turbulent diffusivities at the surface $D_S$ and at the bottom $D_B$ are also shown in Figure 12 (b). The value of $D_S$ is constant throughout the year.

The depth dependence of the turbulent diffusivity $D$ follows the representation:

$$D(t,z) = D_B(t) + (D_S - D_B(t))\frac{1}{2}\left(1 - \frac{z - z_T}{\sqrt{w_T^2 + (z - z_T)^2}}\right).$$ (20)

Vertical profiles of the turbulent diffusivity are also shown in Figure 12 (c).

## Buoyancy Regulation Mechanism

Denoting the vertical velocity and the cell density of cyanobacteria by $V$ and $\rho_P$, we can derive the following relationship between $V$ and $\rho_P$ from Stokes' law by comparing the densities of cyanobacteria $\rho_P$ and water $\rho_W$ (Reynolds et al. 1987; Visser et al. 1997; Wallace and Hamilton 2000).

$$V(t,z) \propto \rho_P(t,z) - \rho_W.$$ (21)

Differentiating this formula (21) with respect to time leads to the following relationship:

$$\frac{\partial}{\partial t}V(t,z) \propto \frac{\partial}{\partial t}\rho_P(t,z) \propto \mu(t,z).$$ (22)

We assume that the differential of the cell density is proportional to the actual growth rate $\mu$.

## Ballast Factor

Buoyancy of cyanobacteria is controlled by carbohydrate accumulation and loss, which affects the cell density $\rho_P$. Here, we define the ballast factor $F$, which is a function of time $t$ and depth $z$, as follows:

$$F(t,z) = \int_0^\infty \mu(t-\tau,z)e^{-r\tau}d\tau = \int_0^\infty \mu_m(t-\tau,z)[f(I(t-\tau,z), N(t-\tau,z))]e^{-r\tau}d\tau.$$ (23)

We assume that the ballast factor $F$ is determined by previous light and nutrient histories and that the contribution of the past growth rate decays exponentially in the progress of time. Therefore, the overall contribution of previous histories is represented by integration of the product of the actual growth rate $\mu$ and an exponential decay factor $\exp(-r\tau)$, where $r$ is the reciprocal of decay time and $\tau$ is the variable for integration.

Finally, the following representation describes vertical movement of cyanobacteria:

$$V(t,z) = V_m \times \{F(t,z) - F_0\}.$$ (24)

The constant $V_m$ is the scale factor for the velocity of vertical migration and the parameter $F_0$ represents the ballast factor for neutral buoyancy, i.e., the equilibrium between floating and sinking.

**Reaction-Diffusion-Advection Equations**

The spatiotemporal dynamics of the nutrient concentration and the cyanobacteria biomass is described by the following partial differential equations (Yoshiyama and Nakajima 2002; Fennel and Boss 2003; Huisman et al. 2006):

$$\frac{\partial N}{\partial t} = \frac{\partial}{\partial z}\left(D\frac{\partial N}{\partial z}\right) + I_N - a\mu P - m_N N, \tag{25}$$

$$\frac{\partial P}{\partial t} = \frac{\partial}{\partial z}\left(D\frac{\partial P}{\partial z}\right) - \frac{\partial}{\partial z}(VP) + \mu P - f_p\frac{P}{H_p + P}. \tag{26}$$

Two dependent variables $N$ and $P$ represent the nutrient concentration and the cyanobacteria biomass. Time and depth-dependent $D$ and $V$ specify the vertical turbulent diffusivity and the vertical velocity of cyanobacteria, respectively.

Since the value of $\mu$ changes periodically with time, this is a forced oscillatory system. The system oscillates at two different frequencies, i.e., yearly and daily, due to annual and diurnal variations of light intensity and water temperature.

We assume the following boundary conditions at the surface ($z$=0) and at the bottom ($z$=$z_B$) of the water column:

$$\frac{\partial N}{\partial z} = 0, \quad \frac{\partial P}{\partial z} = \frac{V_m}{D(t,0)}\{(F(t,0) - F_0\}P \quad (z = 0). \tag{27}$$

$$N = N_B, \quad \frac{\partial P}{\partial z} = 0 \qquad\qquad (z = z_B). \tag{28}$$

Here, $N_B$ denotes the nutrient concentration at the bottom, which does not vary throughout the year. Otherwise, the usual zero-flux boundary conditions are imposed.

The model gradually settles into yearly periodic oscillations with superimposed daily oscillations, independent of initial conditions. The results shown here are those during the second year after the one-year start up. It is confirmed that the system shows repetitive oscillations after the first year.

### 3.3.4. Annual Variation of Cyanobacteria

Figure 13 shows annual variations of the nutrient concentration (broken line) and the cyanobacteria biomass (solid line), using diurnally averaged values. The values of both components are also spatially averaged within the layers between 0 and 2 m depth. A sudden increase in the cyanobacteria biomass $P$ occurs during June, showing a peak at the beginning of July. The maximum value of $P$ at the peak is more than 6.5 g/m$^3$. The value of $P$ begins to reduce in July, and a gradual decrease continues until the end of the year.

Meanwhile, the nutrient concentration $N$ begins to decrease during June because of uptake by cyanobacteria. The value of $N$ shows a minimum in the end of July, with a gradual recovery of concentration continuing until the end of the year.

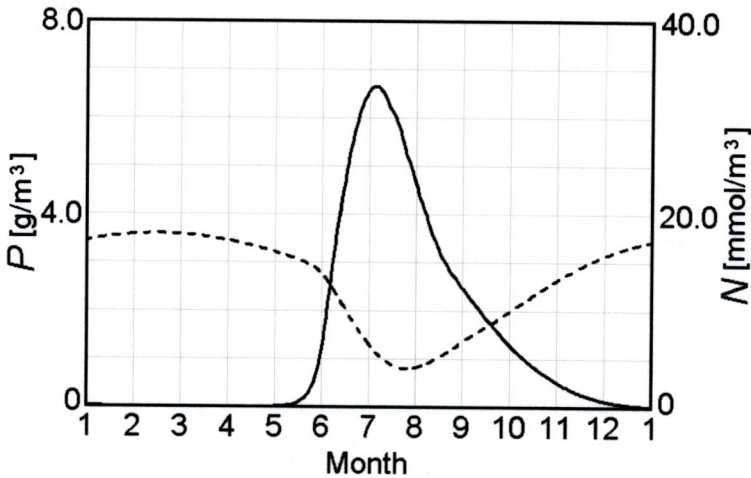

Figure 13. Annual variations of nutrient concentration (broken line) and cyanobacteria biomass (solid line) in vertical migration model (12)-(28). Both values are diurnally averaged and also vertically averaged in layers between 0 and 2 m depth.

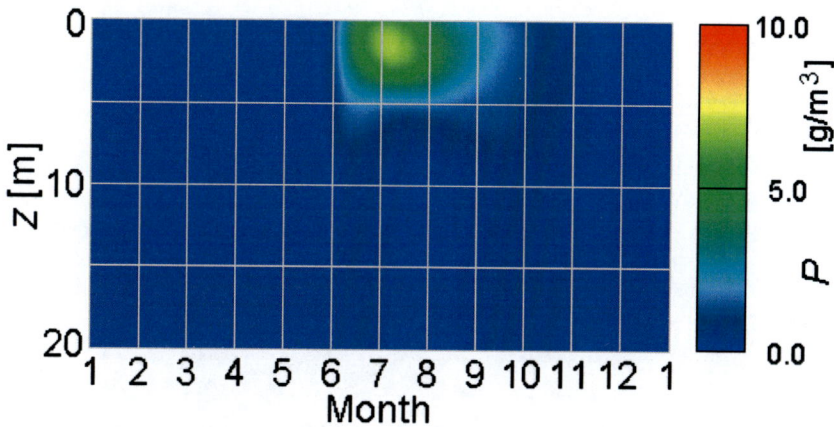

Figure 14. Annual variation of the depth profile of cyanobacteria biomass in vertical migration model (12)-(28). A year consists of 365 days.

Referring to the observations by Takamura and Yasuno (1984) in hypertrophic Lake Kasumigaura, Japan, we define an algal bloom as the state in which the amount of cyanobacteria biomass accumulated at the surface exceeds $1.0 \times 10^5$ cells/ml, i.e., $1.0 \times 10^{11}$ cells/$m^3$. This value corresponds to 4.0 g/$m^3$, considering that the dry weight of a cyanobacterium cell $m_{cell}$ is $40.0 \times 10^{-12}$ g/cell (Long et al. 2001). Judging from the value of $P$ in Figure 13, we can claim that the algal bloom occurs during about two months from the middle of June to the beginning of August.

The annual variation of the depth profile of cyanobacteria biomass $P$ is shown in Figure 14. The values of $P$ are diurnally averaged. Mass occurrence of cyanobacteria during a few months between June and August is obviously observed above the thermocline, particularly in the layers between 0 and 4 m. In contrast, cyanobacteria are at low levels in the layers below the thermocline throughout the year.

### 3.3.5. Diurnal Variation of Cyanobacteria

Figure 15 shows diurnal variations of light intensity at the surface $I_S$ (broken lines) and the cyanobacteria biomass $P$ in different layers (solid lines). The period is from noon July 10 to noon July 11, in the midst of the algal bloom. Four solid lines labeled 0, 1, 2 and 3 show the cyanobacteria biomass averaged within the layers between 0 and 2 m, 2 and 4 m, 4 and 6

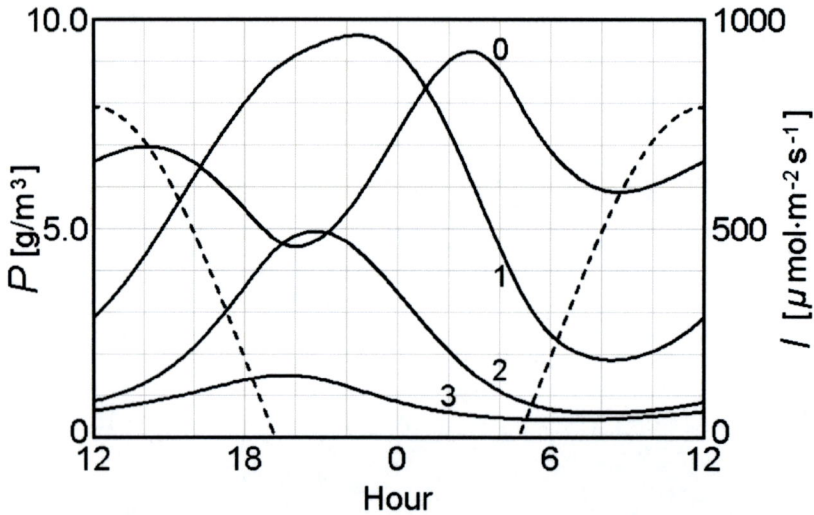

Figure 15. Diurnal variations of light intensity (broken lines) and cyanobacteria biomass (solid lines) from noon July 10 to noon July 11 in vertical migration model (12)-(28). The value of $P$ is vertically averaged in layers between 0 and 2 m (0), 2 and 4 m (1), 4 and 6 m (2), and 6 and 8 m (3) depth.

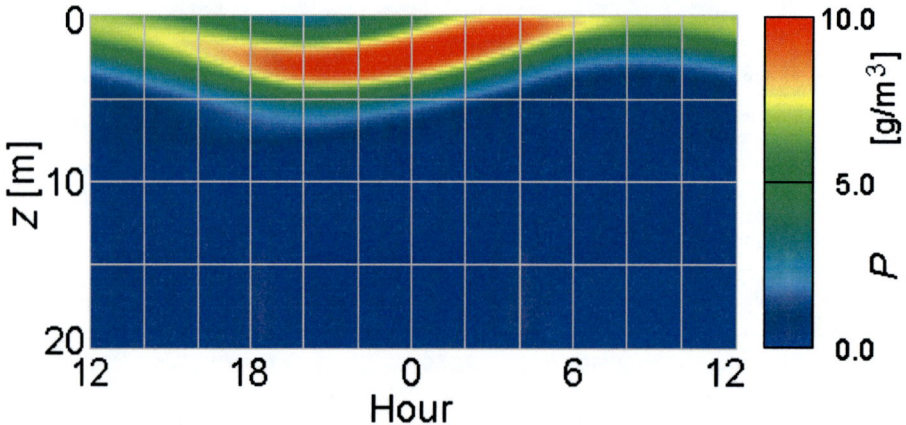

Figure 16. Diurnal variation of the depth profile of cyanobacteria biomass from noon July 10 to noon July 11 in vertical migration model (12)-(28).

m, and 6 and 8 m below the surface, respectively. The movement of the peak in the order of lines labeled 3, 2, 1 and 0 corresponds to the ascent of cyanobacteria.

The diurnal variation of the depth profile of the cyanobacteria biomass $P$ is shown in Figure 16. The period is the same as that in Figure 15. The cyanobacteria peak concentrations provide evidence of descent and ascent, corresponding to vertical migration.

### 3.3.6. Alteration of Floating and Sinking above the Thermocline

Figure 17 shows the diurnal variation of the ballast factor $F$ (solid lines) from noon July 10 to noon July 11 in the layers between 0 and 4 m (line 0), 4 and 8 m (line 1), and 8 and 20 m (line 2) depth, respectively. The two lines labeled 0 and 1 show the variation of $F$ above the thermocline, while the line labeled 2 shows that below the thermocline. The ballast factor for neutral buoyancy $F_0=0.1$ is also indicated (broken line).

The ballast factor $F$ is defined as an integration of the actual growth rate. Comparing the vertical distribution of $F$ with $F_0$ in Figure 17, we find that sinking is dominant from the late morning to the evening in near-surface layers above 4 m depth, whereas floating is dominant from midnight to the early morning (line 0). In other regions below 4 m depth, floating is dominant throughout the day (lines 1 and 2).

The ballast factor for neutral buoyancy $F_0$ is a key parameter that regulates cyanobacteria distributions. In the case of $F=F_0$, cyanobacteria are suspended within the water column. A ballast factor smaller than $F_0$ ($F<F_0$) induces floating, while that larger than $F_0$ ($F>F_0$) results in sinking. In the present model, we adopt the value of $F_0=0.1$, though the value of $F_0$ should be derived from Stokes' law referring to the density of water $\rho_W$. The choice of the parameter value can be justified for the following reasons. As shown in Figure 17, the highest value of the ballast factor $F$ is about 0.14 in near-surface layers (line 0), while the lowest value is almost 0 in deep layers below the thermocline (line 2). The value of $F_0=0.1$ lies within this range (0-0.14), which means that both floating and sinking can take place within the water column.

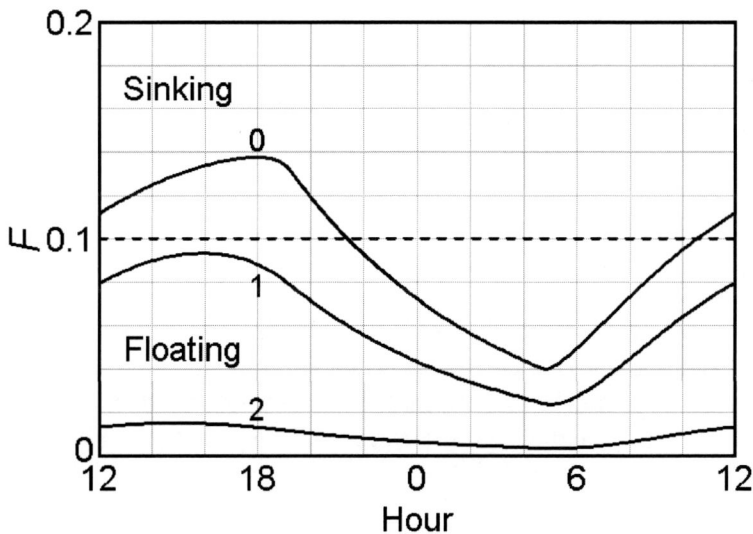

Figure 17. Diurnal variation of the ballast factor (solid lines) from noon July 10 to noon July 11 in vertical migration model (12)-(28). The value of $F$ is vertically averaged in layers between 0 and 4 m (0), 4 and 8 m (1), and 8 and 20 m (2) depth. The ballast factor for neutral buoyancy $F_0=0.1$ is also indicated (broken line).

When $F_0=0.1$ as in the present study, the value of the difference between $F$ and $F_0$ ranges from about -0.1 to 0.04. Considering that the scale factor is $V_m=250$ m/day, the values of the vertical velocity of cyanobacteria range from -25 m/day (floating) to 10 m/day (sinking). These values show a good agreement with the vertical velocity of *Microcystis* estimated at -30 m/day to 10 m/day by Reynolds et al. (1987).

Further attention should be paid to the parameter $r$, which denotes the reciprocal of the decay time for the contribution of the actual growth rate to ballast formation. Our model adopts $r=3$, meaning that the contribution of the actual growth rate reduces to $1/e$ in 8 hours (=1/3 day). It should be noted that this value is comparable with the lifetime of ballast, which was estimated by Reynolds et al. (1987) to be 4 to 12 hours.

## 3.4. Phenotypic Plasticity Model

### 3.4.1. Mathematical Model

In this section, we aim at modeling the fundamental structures of phenotypic plasticity or defensive morphologies of cyanobacteria induced by zooplankton predation. As explained in introduction, this mechanism must be one of the most effective strategies for survival developed by cyanobacteria. We restrict our model to the bitrophic system with unicellular and colonial algae and herbivorous zooplankton. Among these organisms, colonial algae are the main components of algal blooms.

We use the similar model to the one developed by Vos et al. (2004a; 2004b) with almost the same parameter values. However, some original viewpoints are adopted in our model. Firstly, the effect of the increase in the mean colony size is incorporated in our model, because it could be more difficult for zooplankton to ingest large colonies than to do small colonies. We assume that the colony size is distributed by the binomial form as a function of the total number of cells that constitute the colonies. Considering that the number is proportionally related to the biomass, the mean colony size is also in proportion to the colonial algae biomass. In addition, we assume that the increase in the mean colony size causes the increase in the handling time of zooplankton on colonial algae. This size effect is one of the key mechanisms in our mathematical model (Serizawa et al. 2008c).

Secondly, we introduce the source terms to the mathematical model. These source terms not only represent the inputs of the components from the outside but also bring the stabilization effect on the system. For example, the food of zooplankton is not restricted to one species. If one species is depleted, zooplankton must forage other species as alternate foods. Such effect as alteration or compensation should stabilize the system. We intend to introduce these stabilization effects to the system by adding source terms to the differential equations.

The present mathematical model is a system of the following ordinary differential equations:

$$\frac{dP}{dt} = I_P + \mu P\left(1 - \frac{P+Q}{K}\right) - c\frac{Z^2}{Z^2+g^2}P + c\frac{g^2}{Z^2+g^2}Q - \frac{v_P P}{1+v_P h_P P + v_Q h_Q(1+k_Q Q)Q}Z - m_P P, \quad (29)$$

$$\frac{dQ}{dt} = I_Q + \mu Q\left(1 - \frac{P+Q}{K}\right) + c\frac{Z^2}{Z^2+g^2}P - c\frac{g^2}{Z^2+g^2}Q - \frac{v_Q Q}{1+v_P h_P P + v_Q h_Q(1+k_Q Q)Q}Z - m_Q Q, \quad (30)$$

$$\frac{dZ}{dt} = I_Z + \eta \frac{v_P P + v_Q Q}{1 + v_P h_P P + v_Q h_Q (1 + k_Q Q) Q} Z - f_Z \frac{Z^2}{H_Z^2 + Z^2} - m_Z Z. \tag{31}$$

Three variables $P$, $Q$ and $Z$ denote the biomasses of unicellular algae, colonial algae and herbivorous zooplankton, respectively. The unit of these variables is $g/m^3$. The variable $t$ represents time measured in days. We assume that the local growths of both unicellular and colonial algae follow the logistic equation with the same maximum growth rate $\mu$. A parameter $K$ is the carrying capacity for the total algal biomass, which includes both unicellular and colonial algae. The input rates of unicellular algae, colonial algae and zooplankton are represented by three parameters $I_P$, $I_Q$ and $I_Z$, respectively. The source terms are introduced only for unicellular algae and zooplankton, and the input of colonial algae are not considered in the present simulations, therefore, $I_Q = 0$.

### Table 6. Parameters in phenotypic plasticity model (29)-(31)

| Parameters | Meanings | Values | Units |
|---|---|---|---|
| $K$ | Carrying capacity of environment | 0-24.0 | $g/m^3$ |
| $\mu$ | Maximum growth rate of algae | 0.5 | $day^{-1}$ |
| $c$ | Scaling parameter for inducible effect | 1.0 | $day^{-1}$ |
| $g$ | Half-saturation constant for inducible effect | 0.1 | $g/m^3$ |
| $I_P$ | Input rate of unicellular algae | 0.1 | $g\,m^{-3}\,day^{-1}$ |
| $v_P$ | Search rate on unicellular algae | 0.5 | $m^3\,g^{-1}\,day^{-1}$ |
| $h_P$ | Handling time on unicellular algae | 0.5 | $day$ |
| $m_P$ | Mortality rate of unicellular algae | 0.2 | $day^{-1}$ |
| $I_Q$ | Input rate of colonial algae | 0 | $g\,m^{-3}\,day^{-1}$ |
| $v_Q$ | Search rate on colonial algae | 0.5 | $m^3\,g^{-1}\,day^{-1}$ |
| $h_Q$ | Handling time on colonial algae at $k_Q=0$ | 0.5 | $day$ |
| $k_Q$ | Increasing rate of handling time on colonial algae | 1.0 | $m^3/g$ |
| $m_Q$ | Mortality rate of colonial algae | 0.2 | $day^{-1}$ |
| $I_Z$ | Input rate of zooplankton | 0.1 | $g\,m^{-3}\,day^{-1}$ |
| $\eta$ | Assimilation coefficient on algae | 0.4 | |
| $f_Z$ | Maximum feeding rate of fish on zooplankton | 0.84 | $g\,m^{-3}\,day^{-1}$ |
| $H_Z$ | Half-saturation constant of zooplankton | 3.0 | $g/m^3$ |
| $m_Z$ | Mortality rate of zooplankton | 0.15 | $day^{-1}$ |

The actual handling time of zooplankton on colonial algae is represented as $h_Q(1+k_Q Q)$, where $Q$ is the biomass of colonial algae.

The efficiency of inducible morphological changes is estimated by the parameters $c$ and $g$ (Vos et al. 2004a). The parameter $c$ denotes the scaling constant for the inducible effect, and $g$ denotes the half-saturation constant for the same effect. The selective feeding of zooplankton on unicellular and colonial algae constitutes another important part of the model. These functional responses are described by the following parameters. That is, $v_P$ and $v_Q$ are the search rates of zooplankton on unicellular and colonial algae, $h_P$ is the handling time of zooplankton on unicellular algae, $h_Q$ is the constant related to the handling time of zooplankton on colonial algae, and $\eta$ is the assimilation coefficient of zooplankton on algae.

As mentioned above, our assumption is that the handling time of zooplankton on colonial algae is proportionally related to the biomass of colonial algae $Q$. In order to estimate the increasing rate of the handling time on colonial algae, a new parameter $k_Q$ is introduced to the model. Therefore, the actual handling time of zooplankton on colonial algae is represented as $h_Q(1+k_Q Q)$. As for the other parameters, $m_P$, $m_Q$ and $m_Z$ denote the mortality rates of unicellular algae, colonial algae and herbivorous zooplankton, respectively. The meanings and values in the present model are given in Table 6.

Figure 18 is a schematic diagram that illustrates the structure of morphological changes in the phenotypic plasticity model (29)-(31). Arrows indicate the routes of morphological changes.

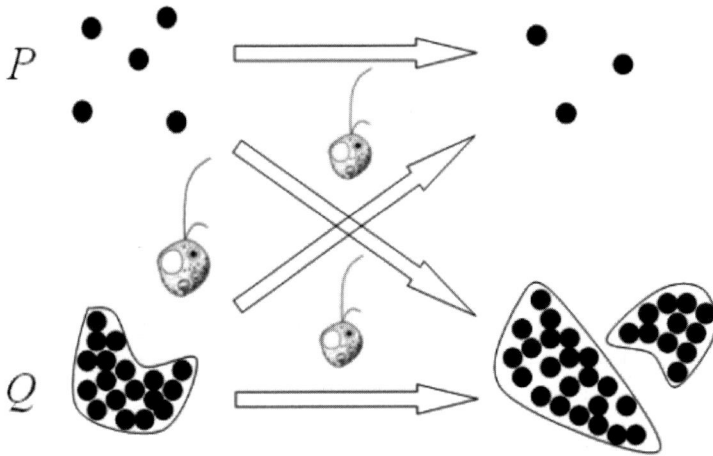

Figure 18. Structure of morphological changes in phenotypic plasticity model (29)-(31). Arrows indicate the routes of morphological changes. The change from the unicellular to the colonial type dominates when the density of zooplankton is high, while otherwise the change from the colonial to the unicellular type dominates. Besides, both types of algae show the logistic growth independently. Permission is granted to reproduce the figure by Ecological Research.

### 4.4.2. Bifurcation Diagrams with Constant Handling Time ($K_q=0$)

We begin with the analyses of the simplified case with a constant handling time, i.e., $k_Q=0$. Figure 19 shows the bifurcation diagrams of the biomasses of unicellular algae $P$ (a), colonial algae $Q$ (b) and herbivorous zooplankton $Z$ (c) as a function of the carrying capacity $K$. In each figure, three bifurcation curves are drawn corresponding to $h_Q=0.5$, 1.5 and 2.5.

For all cases of $h_Q=0.5$, 1.5 and 2.5, the biomasses of three components $P$, $Q$ and $Z$ show monotonous increase with the increase in the carrying capacity $K$, although the increasing rates represented by the gradient of the bifurcation curve are quite different depending on both the value of $h_Q$ and the components. As for unicellular and colonial algae, the increasing rates seem to increase with the increase in $h_Q$. Particularly in the case of $h_Q=2.5$, the biomasses of both algae show considerable increase with the increase in $K$. Meanwhile, the

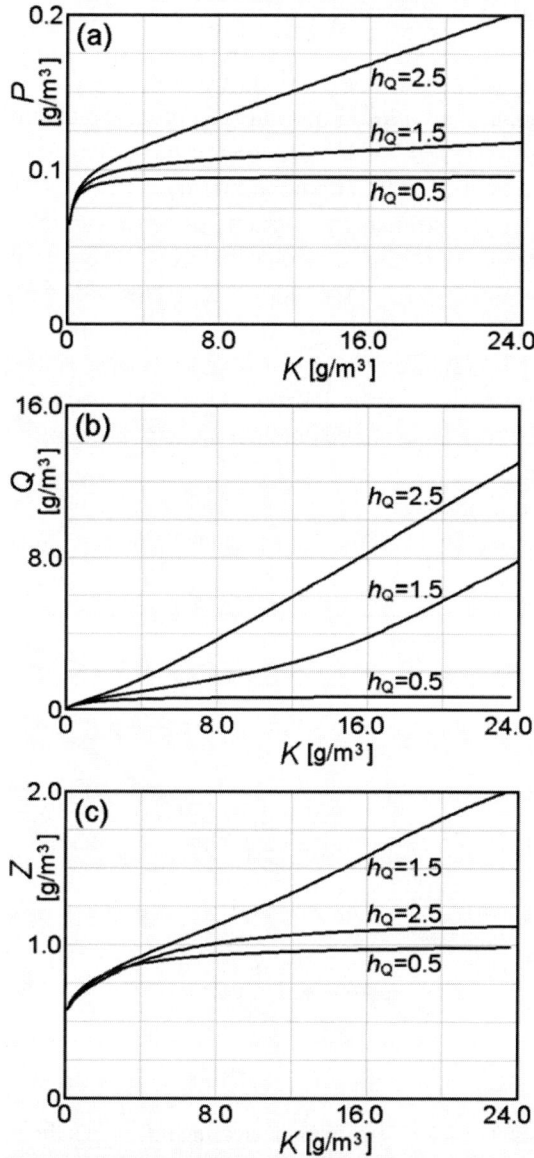

Figure 19. Dependence of bifurcation diagrams on handling time on colonial algae in phenotypic plasticity model (29)-(31). These are the bifurcation diagrams of unicellular algae $P$ (a), colonial algae $Q$ (b) and herbivorous zooplankton $Z$ (c) as a function of the carrying capacity $K$. The solid lines represent the stable equilibrium states. The handling time of zooplankton on colonial algae $h_Q$ is constant ($k_Q$=0).

case of zooplankton shows different aspects. The increasing rate in $h_Q$=1.5 is larger than those in $h_Q$=0.5 and 2.5. However, it seems that no bistable behavior is observed when $k_Q$=0.

### 4.4.3. Bifurcation Diagrams with Variable Handling Time ($K_q \neq 0$)

Next, we investigate the cases in which $k_Q \neq 0$. In the current phenotypic plasticity model, the handling time on colonial algae is represented as $h_Q(1+k_Q Q)$, which means that it increases with the biomass of colonial algae $Q$ at the increasing rate $k_Q$. Figure 20 shows the dependence of the bifurcation diagram of $Q$ on $k_Q$ at $h_Q$=0.5. Two types of monostability occur corresponding to $k_Q$=0.0 and 2.0. When $k_Q$=0.0, the biomass of colonial algae $Q$ is saturated soon with the increase in the carrying capacity $K$, while $Q$ continues to increase when $k_Q$=2.0. On the other hand, typical bistability is observed when $k_Q$=1.0. In summary, as the parameter $k_Q$ moves from 0 to 2, the behaviors of the system show a gradual change from monostability to bistability and a return to monostability.

Figure 21 is a set of the bifurcation diagrams of three variables $P$, $Q$ and $Z$ that show typical bistability as a function of the carrying capacity $K$. The parameter values are given in Table 6 ($h_Q$=0.5 and $k_Q$=1.0). The sigmoid curves characteristic of bistability are observed in a range of 8.16≤$K$≤16.25 g/m$^3$. Two stable branches are observed in each bifurcation diagram, which are the first branch in the region $K$≤16.25 g/m$^3$ and the second branch in the region $K$≥8.16 g/m$^3$. In the regime shift from the first branch to the second branch, the biomasses of both unicellular algae $P$ and colonial algae $Q$ increase, while that of herbivorous zooplankton $Z$ decreases.

On the basis of these observations, we can recognize that the first branch specifies the clear state, while the second branch does the turbid state. Within the clear-state branch, all $P$,

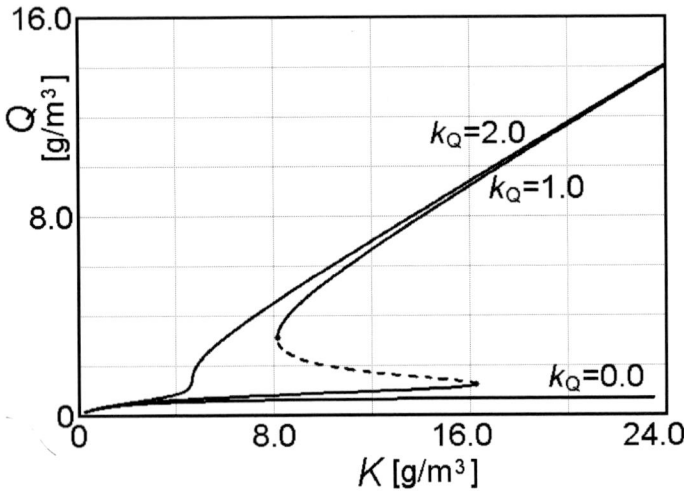

Figure 20. Dependence of bifurcation diagram on increasing rate of handling time on colonial algae in phenotypic plasticity model (29)-(31). The solid lines represent the stable equilibrium states, while the broken lines represent the unstable states. The actual handling time of zooplankton on colonial algae is represented as $h_Q(1+k_Q Q)$, where $k_Q$ is the increasing rate of the handling time on colonial algae and $h_Q$ is the handling time of zooplankton on colonial algae at $k_Q$=0.0. When $k_Q$=0.0 or 2.0, the system shows monostable behaviors, while it shows bistability when $k_Q$=1.0.

$Q$ and $Z$ increase with the increase in $K$. Within the turbid-state branch, on the other hand, the biomasses of $P$ and $Q$ increase drastically with the increase in $K$, while $Z$ decreases gradually.

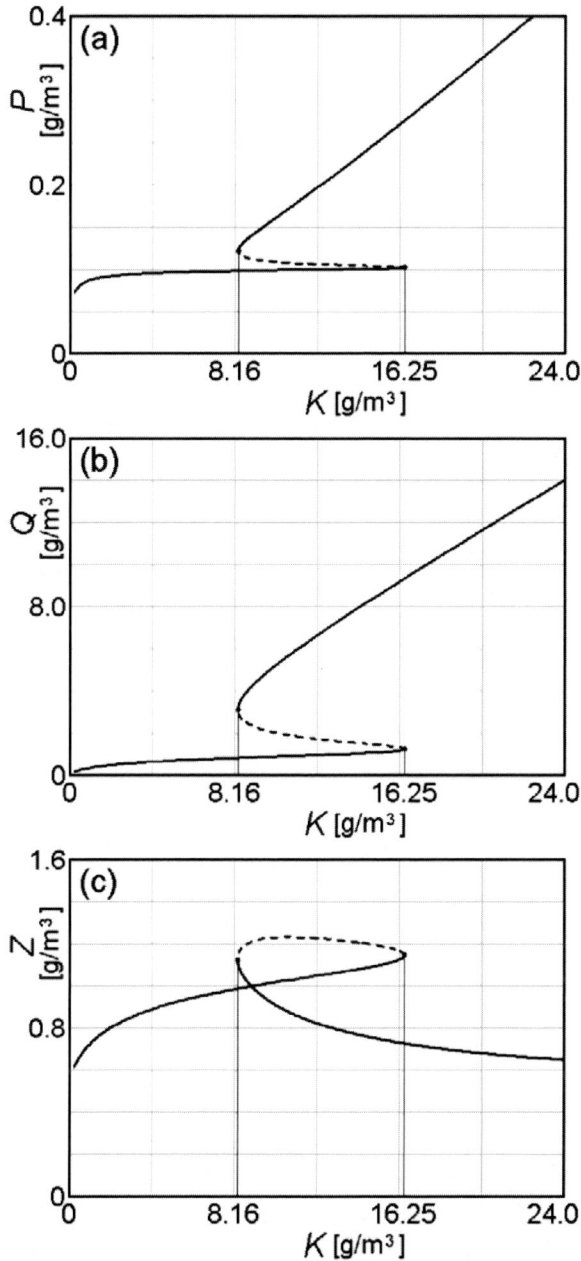

Figure 21. Typical bistability in phenotypic plasticity model (29)-(31). These are the bifurcation diagrams of unicellular algae $P$ (a), colonial algae $Q$ (b) and herbivorous zooplankton $Z$ (c) when $h_Q=0.5$ and $k_Q=1.0$. The solid lines represent the stable equilibrium states, while the broken lines represent the unstable states. Under the conditions given in Table 6, the sigmoid curve characteristic of bistability is observed in a range of $8.16 \leq K \leq 16.25$ g/m$^3$. The stable branch in $K \leq 16.25$ g/m$^3$ represents the clear state, while that in $K \geq 8.16$ g/m$^3$ does the turbid state.

Different from the case in the constant handling time ($k_Q=0$), the bifurcation diagrams with the variable handling time ($k_Q \neq 0$) shows bistability. Considering that the only difference is the way to deal with the handling time on colonial algae, it is likely that the occurrence of bistability is due to the introduction of the variable handling time.

In the bifurcation diagram of Figure 21 (b), the biomass of colonial algae $Q$ on the turbid-state branch varies from about 3.1 g/m³ to 9.3 g/m³ within the bistable range of $8.16 \leq K \leq 16.25$ g/m³. In the previous section, we defined that the critical value that marks the algal blooming state is 4.0 g/m³. It should be noted that the critical value 4.0 g/m³ is included within the range of $3.1 \leq Q \leq 9.3$ g/m³, which could afford credibility to the model.

## 3.5. Integrated Aquatic Model

### 3.5.1. Unification of Minimal NP Model and Phenotypic Plasticity Model

Synthesizing previous studies, we present an integrated aquatic model that unifies the minimal NP model and the phenotypic plasticity model. The food web structure of the integrated aquatic model is illustrated in Figure 22. The main part of the model is composed

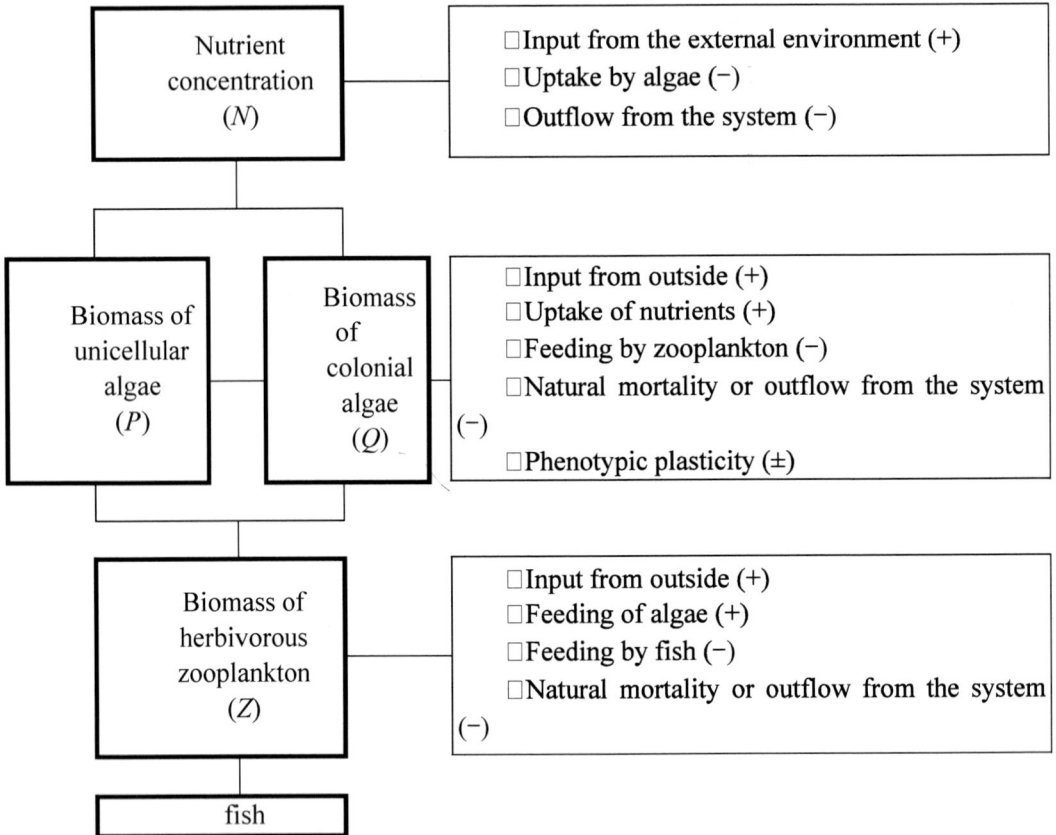

Figure 22. Food web structure of integrated aquatic model (32)-(35). (+) means the increase in components, while (-) does the decrease.

## Table 7. Parameters in integrated aquatic model (32)-(35)

| Parameters | Meanings | Values | Units |
|---|---|---|---|
| $I_N$ | Input rate of nutrients | 0-0.6 | mmol m$^{-3}$ day$^{-1}$ |
| $k$ | Nutrient content in phytoplankton | 0.2 | mmol/g |
| $H_N$ | Half-saturation constant of nutrients | 0.2 | mmol/m$^3$ |
| $m_N$ | Removal rate of nutrients | 0.1 | day$^{-1}$ |
| $\mu$ | Maximum growth rate of algae | 0.5 | day$^{-1}$ |
| $c$ | Scaling parameter for inducible effect | 1.0 | day$^{-1}$ |
| $g$ | Half-saturation constant for inducible effect | 0.1 | g/m$^3$ |
| $I_P$ | Input rate of unicellular algae | 0.1 | g m$^{-3}$ day$^{-1}$ |
| $v_P$ | Search rate on unicellular algae | 0.5 | m$^3$ g$^{-1}$ day$^{-1}$ |
| $h_P$ | Handling time on unicellular algae | 0.5 | day |
| $m_P$ | Mortality rate of unicellular algae | 0.2 | day$^{-1}$ |
| $I_Q$ | Input rate of colonial algae | 0 | g m$^{-3}$ day$^{-1}$ |
| $v_Q$ | Search rate on colonial algae | 0.5 | m$^3$ g$^{-1}$ day$^{-1}$ |
| $h_Q$ | Handling time on colonial algae at $k_Q=0$ | 0.5 | day |
| $k_Q$ | Increasing rate of handling time on colonial algae | 1.0 | m$^3$/g |
| $m_Q$ | Mortality rate of colonial algae | 0.2 | day$^{-1}$ |
| $I_Z$ | Input rate of zooplankton | 0.1 | g m$^{-3}$ day$^{-1}$ |
| $\eta$ | Assimilation coefficient on algae | 0.4 | |
| $f_Z$ | Maximum feeding rate of fish on zooplankton | 0.8 | g m$^{-3}$ day$^{-1}$ |
| $H_Z$ | Half-saturation constant of zooplankton | 3.0 | g/m$^3$ |
| $m_Z$ | Mortality rate of zooplankton | 0.15 | day$^{-1}$ |

of nutrient and two trophic levels with four components, among which unicellular algae and colonial algae are correlated with each other by phenotypic plasticity within a same trophic level. Besides, an additional third trophic level of fish is included in the model, which is described as a non-dynamical fish predation term on zooplankton in the mathematical model. In Figure 22, the symbol (+) marks the factor by which the number of component increases, while the symbol (-) does those by which the number decreases.

Unification of the minimal NP model and the phenotypic plasticity model leads to the four-component integrated aquatic model described as follows:

$$\frac{dN}{dt} = I_N - k\mu \frac{N}{H_N + N}(P+Q) - m_N N, \tag{32}$$

$$\frac{dP}{dt} = I_P + \mu \frac{N}{H_N + N}P - c\frac{Z^2}{Z^2+g^2}P + c\frac{g^2}{Z^2+g^2}Q - \frac{v_P P}{1+v_P h_P P + v_Q h_Q (1+k_Q Q)Q}Z - m_P P, \tag{33}$$

$$\frac{dQ}{dt} = I_Q + \mu \frac{N}{H_N + N}Q + c\frac{Z^2}{Z^2+g^2}P - c\frac{g^2}{Z^2+g^2}Q - \frac{v_Q Q}{1+v_P h_P P + v_Q h_Q (1+k_Q Q)Q}Z - m_Q Q, \tag{34}$$

$$\frac{dZ}{dt} = I_Z + \eta \frac{v_P P + v_Q Q}{1+v_P h_P P + v_Q h_Q (1+k_Q Q)Q}Z - f_Z \frac{Z^2}{H_Z^2+Z^2} - m_Z Z. \tag{35}$$

Four state variables $N$, $P$, $Q$ and $Z$ represent the nutrient concentration and the biomasses of unicellular algae, colonial algae and herbivorous zooplankton, respectively. The unit of $N$ is mmol/m$^3$, while those of $P$, $Q$ and $Z$ are g/m$^3$. Another variable $t$ is time measured in days.

In the integrated aquatic model (32)-(35), the eutrophication level is estimated by the nutrient loading, i.e., the input rate of nutrients $I_N$, as is the case in the minimal NP model. The input of colonial algae is neglected as in the phenotypic plasticity model (29)-(31), that is, $I_Q=0$. The meanings and values in the integrated aquatic model are listed in Table 7.

### 3.5.2. Bistability in Bifurcation Diagrams

Figure 23 shows the bifurcation diagrams of the integrated aquatic model (32)-(35) for the nutrient concentration $N$ (a), the biomasses of unicellular algae $P$ (b), colonial algae $Q$ (c) and herbivorous zooplankton $Z$ (d), as a function of the nutrient input rate $I_N$. Obvious bistability appears with the progress of eutrophication in the integrated aquatic model, as well as in the minimal NP model and the phenotypic plasticity model. The bistability region exists in $0.229 \leq I_N \leq 0.433$ mmol m$^{-3}$ day$^{-1}$, and the stable branch in $I_N \leq 0.433$ mmol m$^{-3}$ day$^{-1}$ is identified as the clear-state branch, while that in $I_N \geq 0.229$ mmol m$^{-3}$ day$^{-1}$ is identified as the turbid-state branch.

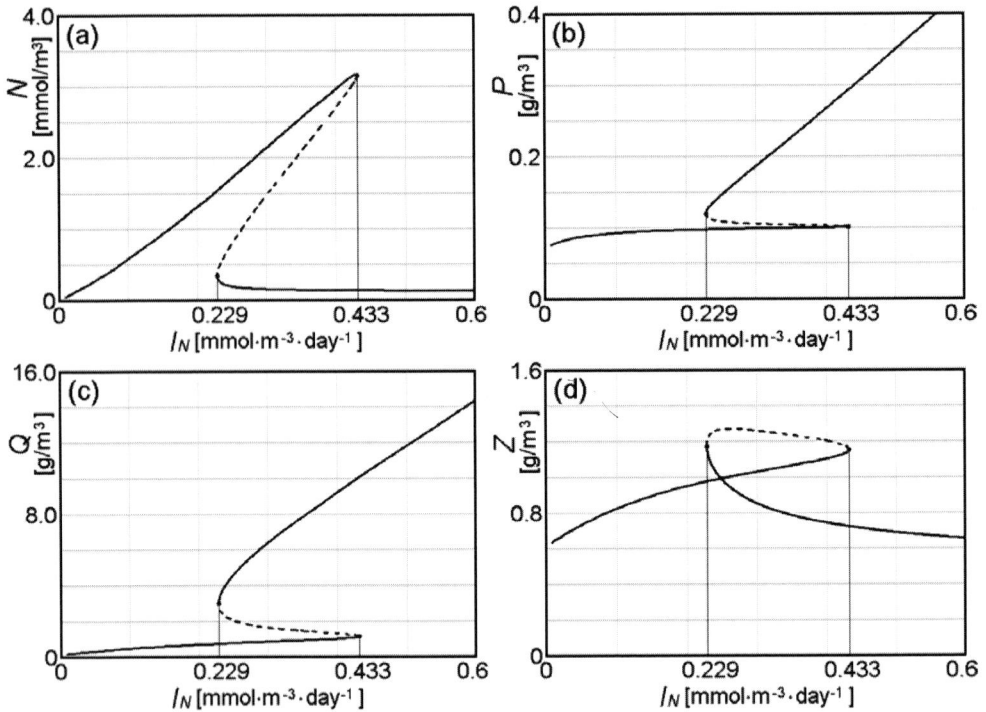

Figure 23. Typical bistability in integrated aquatic model (32)-(35). These are the bifurcation diagrams of nutrients $N$ (a), unicellular algae $P$ (b), colonial algae $Q$ (c) and herbivorous zooplankton $Z$ (d) as a function of the nutrient input rate $I_N$. The solid lines represent the stable equilibrium states, while the broken lines represent the unstable states. Under the conditions given in Table 7, the sigmoid curve characteristic of bistability is observed in a range of $0.229 \leq I_N \leq 0.433$ mmol m$^{-3}$ day$^{-1}$. The stable branch in $I_N \leq 0.433$ mmol m$^{-3}$ day$^{-1}$ represents the clear state, while that in $I_N \geq 0.229$ mmol m$^{-3}$ day$^{-1}$ does the turbid state.

As is stated previously, the ecosystem state in which the biomass of phytoplankton exceeds 4.0 g/m$^3$ is defined as the algal blooming state in this chapter. In the case of the integrated aquatic model (32)-(35), the bifurcation diagram of colonial algae $Q$ (Figure 23 (c)) shows that the value of $Q$ varies from 3.1 g/m$^3$ to 10.0 g/m$^3$ within the bistability region $0.229 \leq I_N \leq 0.433$ mmol m$^{-3}$ day$^{-1}$. It should be noted that the critical value 4.0 g/m$^3$ is included within the variable range of $Q$ also in this case. These observations suggest that the integrated aquatic model is one of the candidates that can describe synthetically the population dynamics of aquatic ecosystems in which algal blooms are flourishing.

It is interesting to investigate the influence of fishes on colonial algae in the integrated aquatic model (32)-(35). Figure 24 shows the dependence of the bifurcation diagram of $Q$ on $f_Z$. Originally, the parameter $f_Z$ denotes the maximum feeding rate of fish on zooplankton. However, considering that it is used in the non-dynamical term, $f_Z$ is thought to represent the product of the maximum feeding rate and the fish biomass. This means that the value of $f_Z$ increases with the increase in the fish biomass.

We can recognize two characteristic features in Figure 24. One of the features is the movement of the bifurcation range of the nutrient input rate $I_N$ to the lower side, which finally results in the disappearance of bistability and the shift to monostability, as is seen in the case of $f_Z$=1.3. Another feature is the considerable increase in the colonial algae biomass $Q$ with the increase in $f_Z$. For instance, when $I_N$=0.2, the value of $Q$ in $f_Z$=1.3 is much larger than that in $f_Z$=0.5 or $f_Z$ =0.8. These features of the bifurcation diagram show that the increase in the fish number could induce serious algal blooms with large quantity of colonial algae. Mass occurrence of colonial algae could be due to the indirect effects of two steps: the first step is the reduction of zooplankton by fish predation, and the second step is the subsequent decline of zooplankton predation on phytoplankton that causes the increase in colonial algae.

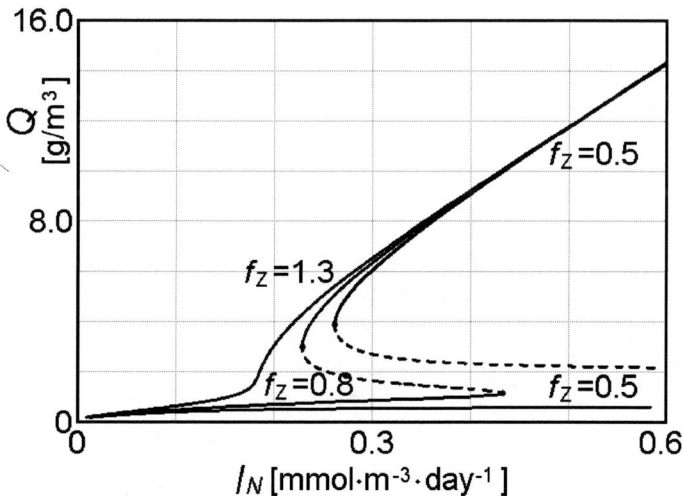

Figure 24. Dependence of bifurcation diagram on maximum feeding rate of fish on zooplankton in integrated aquatic model (32)-(35). The solid lines represent the stable equilibrium states, while the broken lines represent the unstable states. $f_Z$ is a parameter which denotes the maximum feeding rate of fish on zooplankton. In this model, the parameter $f_Z$ increases with the increase in the fish biomass, because the fish predation term is formulated as a non-dynamical term. It is shown that the bistable region of $I_N$ moves to the lower side as $f_Z$ is increased, resulting in disappearance of bistability, that is, transition to monostability.

### 5.3. Extension to Reaction-Diffusion-Advection Equations

Finally, we present the integrated aquatic model represented in the form of reaction-diffusion-advection equations. Denoting the right side of the equations (32)-(35) as $F(N,P,Q,Z)$, $G(N,P,Q,Z)$, $H(N,P,Q,Z)$ and $I(N,P,Q,Z)$, respectively, we can formulate the following three-dimensional reaction-diffusion-advection equations:

$$\frac{\partial N}{\partial t} = D_N \left( \frac{\partial^2 N}{\partial x^2} + \frac{\partial^2 N}{\partial y^2} + \frac{\partial^2 N}{\partial z^2} \right) - \frac{\partial}{\partial x}(V_x N) - \frac{\partial}{\partial y}(V_y N) + F(N,P,Q,Z), \tag{36}$$

$$\frac{\partial P}{\partial t} = D_P \left( \frac{\partial^2 P}{\partial x^2} + \frac{\partial^2 P}{\partial y^2} + \frac{\partial^2 P}{\partial z^2} \right) - \frac{\partial}{\partial x}(V_x P) - \frac{\partial}{\partial y}(V_y P) - \frac{\partial}{\partial z}(V_B P) + G(N,P,Q,Z), \tag{37}$$

$$\frac{\partial Q}{\partial t} = D_Q \left( \frac{\partial^2 Q}{\partial x^2} + \frac{\partial^2 Q}{\partial y^2} + \frac{\partial^2 Q}{\partial z^2} \right) - \frac{\partial}{\partial x}(V_x Q) - \frac{\partial}{\partial y}(V_y Q) - \frac{\partial}{\partial z}(V_B Q) + H(N,P,Q,Z), \tag{38}$$

$$\frac{\partial Z}{\partial t} = D_Z \left( \frac{\partial^2 Z}{\partial x^2} + \frac{\partial^2 Z}{\partial y^2} + \frac{\partial^2 Z}{\partial z^2} \right) - \frac{\partial}{\partial x}(V_x Z) - \frac{\partial}{\partial y}(V_y Z) + I(N,P,Q,Z). \tag{39}$$

Here, $x$ and $y$ denote the horizontal position coordinates, and $z$ denotes the vertical position coordinate. Four constants $D_N$, $D_P$, $D_Q$ and $D_Z$ are the diffusion coefficients of nutrients, unicellular algae, colonial algae and herbivorous zooplankton, respectively. Furthermore, the two-dimensional horizontal velocity field by eddies $V_{Eddy}$ derived from the flow function $\Psi$ and the one-dimensional vertical velocity field originating from buoyancy regulation $V_{Buoy}$ are described as follows:

$$V_{Eddy} = (V_x, V_y, 0) = \left( -\frac{\partial \psi}{\partial y}, \frac{\partial \psi}{\partial x}, 0 \right), \quad V_{Buoy} = (0, 0, V_B). \tag{40}$$

We can perform more detailed analyses of the population dynamics and three-dimensional movement of colonial algae in algal blooms by using the three-dimensional version of the integrated aquatic model (36)-(40).

## 3.6. Abs.-CASM: Abstract Version of the Comprehensive Aquatic Simulation Model

### 3.6.1. Characteristics of abs.-CASM

Here we employed a different model as mentioned above, the abstract version of the Comprehensive Aquatic Simulation Model (CASM) (DeAngelis et al. 1989), abbreviated as abs.-CASM hereafter, to investigate the bistable behavior in lakes due to nutrient loading, and also to propose a method of lake- ecosystem management. The abs.-CASM is a set of five-variable autonomous differential equations and sufficiently simple for the mathematical analysis, however retains important characteristics of lake ecosystems such as material (phosphorus) circulation from detritus to water, species interactions between three trophic levels, and the effect of human activities (phosphorous loading) on lake ecosystems. Thus this model is preferable for studying lake-restoration techniques utilizing interactions of materials and organisms in lakes. Such a lake-restoration technique that focuses on trophic-cascade top-down effects is called biomanipulation (Shapiro et al. 1975). We will show that effective biomanipulation is possible by simultaneously manipulating appropriate variables to restore lakes from turbid-water to clear-water state.

### 3.6.2. Mathematical Model

The model is a set of five-variable differential equations for a three-trophic food chain proposed by DeAngelis et al. (1989):

$$\frac{dN}{dt} = I_N - r_N N - \frac{\gamma\, r_1 NX}{k_1 + N} + \gamma\, d_4 D,$$

(41)

$$\frac{dX}{dt} = \frac{r_1 NX}{k_1 + N} - \frac{f_1 X^2 Y}{k_2 + X^2} - (d_1 + e_1)X,$$

(42)

$$\frac{dY}{dt} = \frac{\eta\, f_1 X^2 Y}{k_2 + X^2} - \frac{f_2 Y^2 Z}{k_3 + Y^2} - (d_2 + e_2)Y,$$

(43)

$$\frac{dZ}{dt} = \frac{\eta\, f_2 Y^2 Z}{k_3 + Y^2} - (d_3 + e_3)Z,$$

(44)

$$\frac{dD}{dt} = \frac{(1-\eta) f_1 X^2 Y}{k_2 + X^2} + \frac{(1-\eta) f_2 Y^2 Z}{k_3 + Y^2} + d_1 X + d_2 Y + d_3 Z - (d_4 + e_4)D,$$

(45)

where $N$ is the dissolved nutrient (phosphorus), $X$ the biomass of phytoplankton, $Y$ the biomass of zooplankton, $Z$ the biomass of zooplanktivorous fish, and $D$ is the detrial biomass. The parameter $I_N$ stands for the input rate of the limiting nutrient from the environment, representing the main cause of anthropogenic eutrophication. The other parameters are the followings: $r_N$, loss rate of the nutrient; $\gamma$, ratio of nutrient mass to biomass; $r_1$, maximum growth rate of phytoplankton; $k_i$, half-saturation constant of nutrient (i=1) or biomass (i=2,3); $d_i$, death rate of organisms (i=1-3) or decomposition rate of detritus (i=4); $e_i$, removal rate of organisms (i=1-3) or detritus (i=4); $f_i$, feeding rates; $\eta$, assimilation efficiency. The details of the original model can be found in the literature (DeAngelis et al. 1989). It is noted that the uptake kinetics of the nutrient is described by a Michaelis-Menten-Monod (Holling type-II) function, and that interactions of organisms are modeled by a Holling type-III function. The type-III functional response assumes that predation rate is not proportional to prey density but very low when prey are scarce, which may be the case where predator activity finds out large concentrations of prey, or prey hides itself behind in an obstacle.

**Table 8. Parameters used in the abs.-CASM model (41)-(46)**

| parameters | values | units | meaning of the parameters | ref. |
|---|---|---|---|---|
| Control parameters | | | | |
| $IN$ | 0 - 1.5 | [mg m-2 day-1] | input rate of limiting nutrient | |
| Constants | | | | |
| $rN$ | 0.005 | [day-1] | loss rate of limiting nutrient | 1) |
| $r1$ | 0.3 | [day-1] | maximum growth rate of phytoplankton | 1) |
| $k1$ | 0.005 | [g m-2] | half-saturation constant of nutrient | 1) |
| $k2$ | 6 | [(g m-2)2] | half-saturation constant of phytoplankton biomass | 2) |
| $k3$ | 2 | [(g m-2)2] | half-saturation constant of zooplankton biomass | 1) |
| $f1$ | 2 | [day-1] | feeding rate of zooplankton | 1) |
| $f2$ | 10 | [day-1] | feeding rate of zooplanktivorous fish | 1) |
| $\gamma$ | 0.02 | [-] | ratio of nutrient mass to biomass | 1) |
| $\eta$ | 0.5 | [-] | assimilation efficiency | 1) |
| $d1$ | 0.1 | [day-1] | death rate of phytoplankton | 1) |
| $d2$ | 0.1 | [day-1] | death rate of zooplankton | 1) |
| $d3$ | 0.5 | [day-1] | death rate of zooplanktivorous fish | 1) |
| $d4$ | 0.1 | [day-1] | decomposition rate of detritus | 1) |
| $e1$ | 0.001 | [day-1] | removal rate of phytoplankton from the system | 1) |
| $e2$ | 0.001 | [day-1] | removal rate of zooplankton from the system | 1) |
| $e3$ | 0.001 | [day-1] | removal rate of zooplanktivorous fish from the system | 1) |
| $e4$ | 0.001 | [day-1] | removal rate of detritus from the system | 1) |

1) DeAngelis et al., 1989. 2) Amemiya et al., 2005.

### 3.6.3. Bifurcation Diagrams

A numerical linear stability analysis for the five-variable model was carried out with the input rate ($I_N$) of nutrient being used as a bifurcation parameter. The parameters used in the present analysis are listed in Table 8.

The linear stability analysis yielded monostable or bistable behavior depending on the values of the parameters. When using the original parameter values reported by DeAngelis et al. (1989), we obtained only monostable behavior (not shown). However, bistable behavior was exhibited as shown in Figure 25, when a more realistic value of six (6) than the original value of sixty (60) for the half-saturation constant of phytoplankton biomass ($k_2$) was used (Bowie et al. 1985; Amemiya et al. 2005). Nearly the same bistable behavior could be exhibited for different values of parameters as listed in Table 1: for instance, by changing one of the parameter values in the following range $k_2$=5.7-7.2, $r_1$=0.14-0.16, and $f_1$=6.2-6.5, while leaving the other parameters at the original values as listed in Table 8.

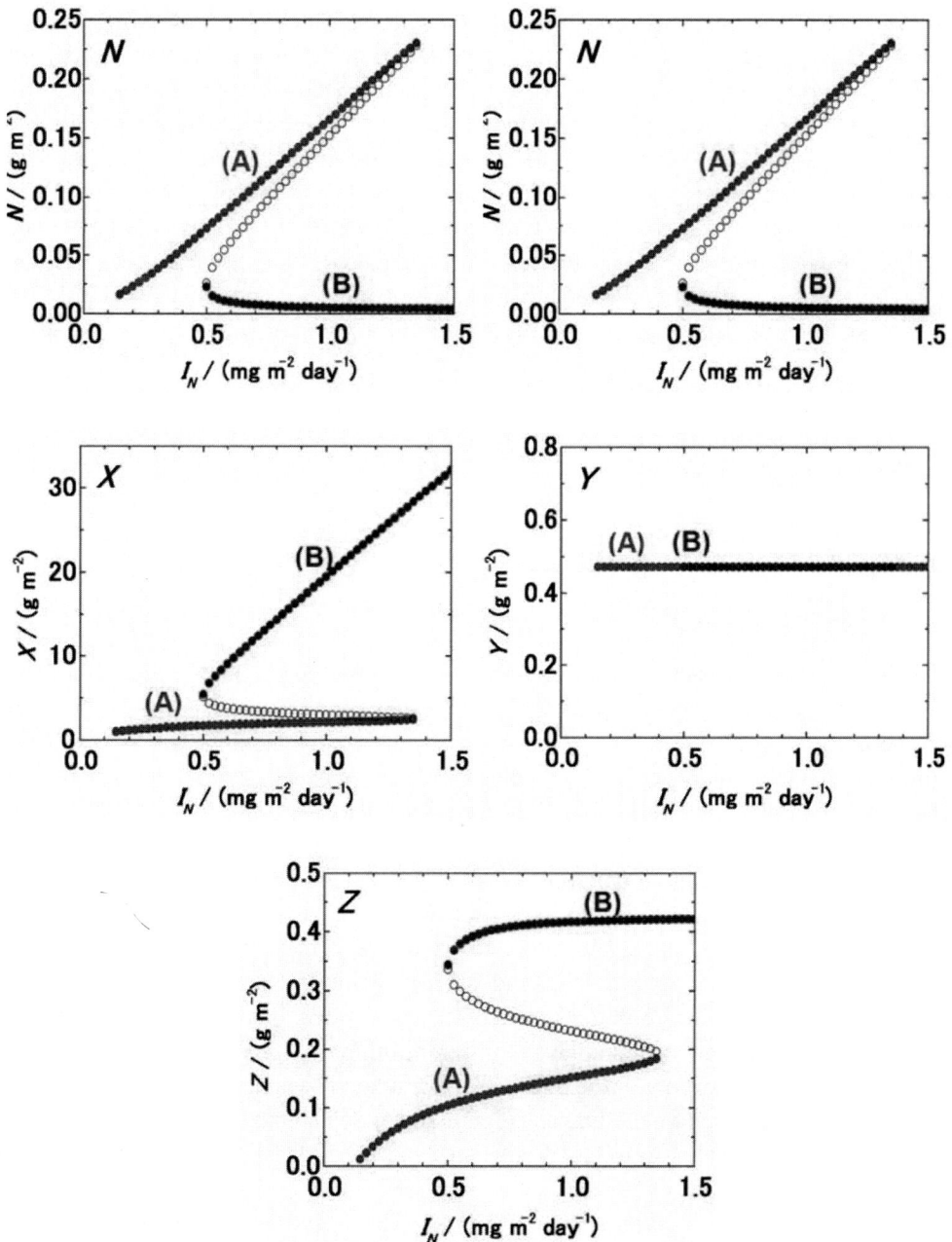

Figure 25. Local stability of the steady-state solutions of abs.-CASM as a function of $I_N$, the rate of nutrient loading. $N$ is the amount of dissolved nutrient, $X$ the biomass of phytoplankton, $Y$ the biomass of zooplankton, $Z$ the biomass of zooplanktivorous fish, and $D$ the detrial biomass. Filled circles are stable steady states, and represent clear (A) state and (B) turbid state, respectively. Open circles are unstable steady states. The parameter values are listed in Table 3-1. (This figure is re-used from Ecology and Society 10(2):3 2005. Permission is granted from the authors).

Though we did not examine all the parameters, we could also find bistable behavior by changing the values of other parameters such as $k_3$, $r_1$, and $f_1$ instead of $k_2$. Thus we can say that the abs.-CASM reproduce bistable behavior robustly. However, we should note that the above bistable behavior was not found if we changed the Holling type-III functional responses to a type-II functional response, as far as we investigated. The nonlinearity of the type-III functional responses appears to be indispensable for exhibiting bistable behavior in the abs.-CASM. This characteristic is different from that of the phenotypic plasticity model presented in section 3.4. The type-III functional response assumes that the feeding interactions between species decreases quickly at low prey abundances. At present we can not explain the discrepancy of the different functional responses between the two models for exhibiting bistable behavior. Under the limitation permitted, the present abs.-CASM can be used as a lake-model exhibiting alternative-stable-states (Scheffer et al. 1993, 2001, 2003), and also can be employed for proposing a method of restoration of lakes.

### 3.6.4. Modification of abs.-CASM

The abs.-CASM was found to exhibit different bifurcation diagrams by slightly changing the differential equation of $Z$ as

$$\frac{dZ}{dt} = \frac{\eta\, f_2 Y^2 Z}{k_3 + Y^2} - (d_3 + e_3)(Z - Z^*),\qquad(46)$$

while the other equations remained the same (Amemiya et al. 2006). This modification generalizes the abs.-CASM by allowing the predators, zooplanktivorous fish, to maintain a low equilibrium level $Z=Z^*$ ($>0$) even when its prey, zooplankton, are scarce. Such a modification can be understood if we consider that alternative food sources that are not included in the model might be available for the top predator.

The present modified abs.-CASM exhibits different bifurcation diagrams depending on the value of $Z^*$ as shown in Figure 26. It is noted that the same constant parameter values as listed in Table 8 except for $d_2=0.05$ was used in Figure 26. In the diagrams, stable and unstable steady states and stable oscillatory solutions are plotted. A folded bifurcation curve with a saddle node, an unstable fold, and a Hopf point is shown in Figure 26(a) with $Z^*=0.0002$. Stable oscillatory solutions were obtained between the unstable fold and the Hopf point on the unstable branch. A similar but different bifurcation curve was obtained with the value of $Z^*=0.0006$; a portion of one of the unstable branch became stable, and a saddle node and a Hopf point appeared on the branch (Figure 26(b)). When the value of $Z^*$ was increased to 0.0009, oscillatory solutions were disappeared, and a folded bifurcation curve with two saddle nodes was obtained (Figure 26(c)).

A folded bifurcation curve as shown in Figure 26(c) has been used to explain regime shifts between clear-water state and turbid-water state in lakes (Scheffer et al. 1993, 2001, 2003). However, stable states may either be steady or oscillatory (non-steady) state, and thus any abrupt and discontinuous shifts can be called regime shifts. The present model exhibits three kinds of bistable behaviors, indicating that small changes in the parameter value of $Z^*$ can also induce a shift of bistable behavior. We have also confirmed that such divergent bistable behavior can be obtained by changing the value of other parameters.

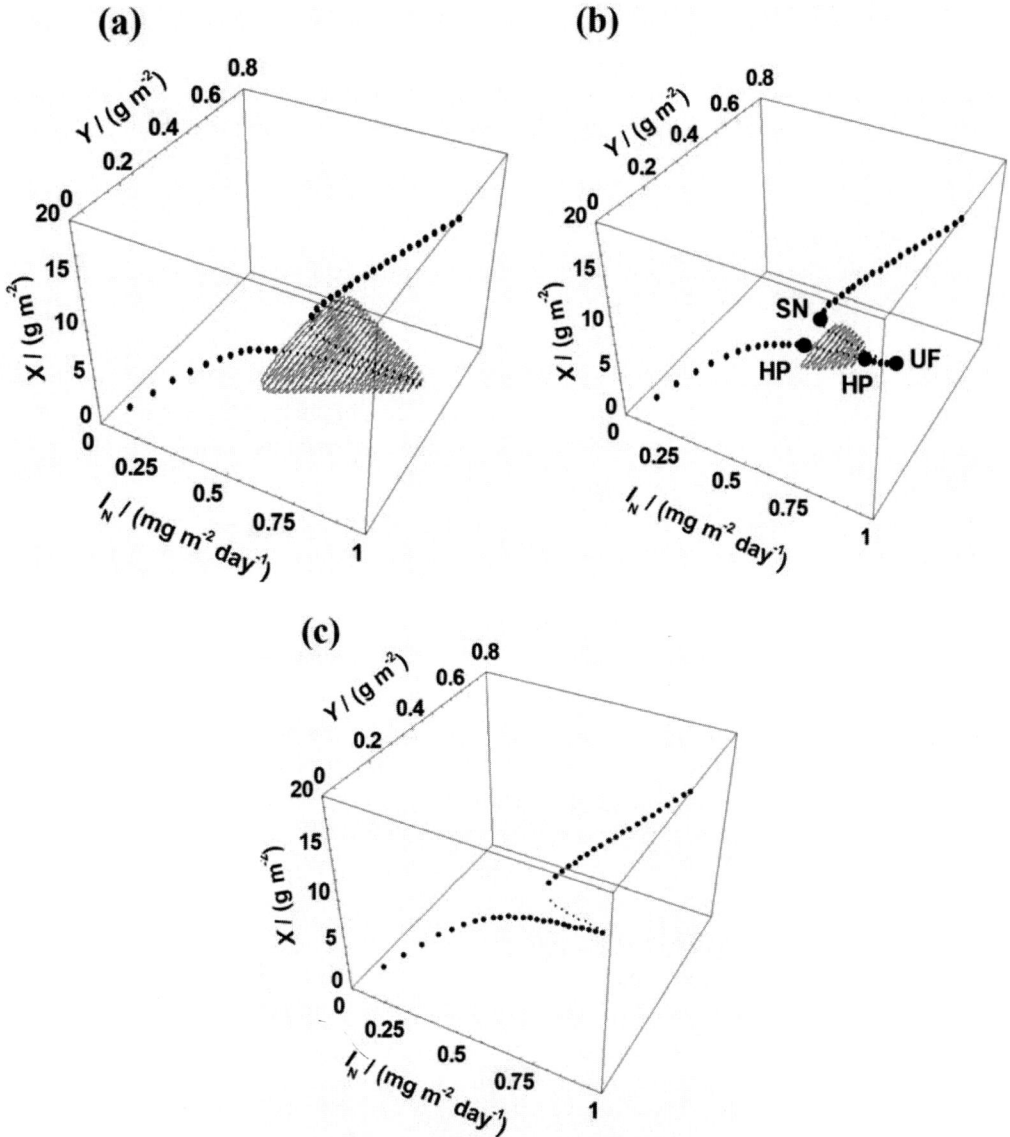

Figure 26. Bifurcation diagrams for a modified abs.-CASM with different values of $Z^*$: (a) 0.0002, (b) 0.0006, and (c) 0.0009. The other parameters values are listed in Table 3-1 except for $d_2$=0.55. Filled circles and small dots show the stable and unstable steady states, respectively, and lines show the limit-cycle oscillations. SN represents saddle node, HP is Hopf bifurcation point, and UF is unstable fold. (This figure is a part of Fig.1 in Ecological Modelling, 206, 54-62, 2007. Permission is granted form the authors.)

These results imply that different regimes can be induced by small changes in the natural (not anthropogenic) parameters such as $Z^*$ and $d_2$, which are difficult to control in real lakes. Nevertheless, such naturally induced regime shifts do not occur in the present model if the nutrient loading is small, indicating that controlling anthropogenic impacts is still crucial to prevent this kind of regime shifts.

Persistent switching of clear and turbid states has been observed in lakes with time scale of weeks to years (Walker and Meyers, 2004). The analyses of the present model suggest that lakes may exhibit such a switching by a threshold phenomenon; the system-state is changed from one state to the other when the nutrient input moves back and forth crossing the bifurcation points. On the other hand, the modified abs.-CASM exhibits autonomous oscillations with a period of weeks (Amemiya et al. 2006), indicating that lakes may change their states by the internal dynamics.

### 3.6.5. Biomanipulation

Biomanipulation is a lake-restoration technique that uses a pulse-like perturbation of the abundance of organisms. For a long-term success of biomanipulation, the external nutrient loading is said to be reduced below 0.5-2.0 (g m$^{-2}$ year$^{-1}$) in the case of phosphorous (Jeppesen et al., 1990), suggesting the existence of alternative stable states around that level of nutrient loading in real lakes. Our simulation of biomanipulation by controlling one (or two) state variable(s) agree with the findings, and give new insights into how the concept of resilience can be used in lake restoration.

Simulations of biomanipulation were carried out using the original abs.-CASM in a range of $I_N$ exhibiting bistable states as shown in Figure 25. Firstly we manipulated one of the five variables over the whole range of the bistable region, and found that only manipulations of $X$ and $D$ succeeded, whereas the others failed to switch the states from turbid to clear (Amemiya et al. 2005). Namely, decreasing the abundance of $X$ or $D$ (phytoplankton or detritus) from the algal-bloomed state changed all the variables from the original state to the alternative stable state, indicating a transition from a turbid to a clear state. On the other hand, manipulations of $N$, $Y$, or $Z$ did not shift the state, i.e., all the variables returned to the original stable state after a spike-like response at the beginning of the manipulation.

We also carried out similar simulations by manipulating two of the variables; we selected each of the ten possible combinations ($(X, Y)$, $(Y, Z)$, etc.) of the variables $N$, $X$, $Y$, $Z$, $D$, and manipulated the two variables synchronously. The system was found to switch from turbid to clear or *vice versa* only when the variable $X$ or $D$, or both, were involved in the manipulations. Typical results are shown in Figure 27. Quasi-basins of attractions are indicated in the figure when each set of the two variables were manipulated while the other three variables were fixed at the turbid or clear steady states. One can see from Figure 27(a) that manipulation of only variable $Z$ cannot switch the system from turbid to clear states. The basins of attractions obtained by linear stability analysis differ from those obtained by the present manipulations as indicated by the arrows in the figure. This discrepancy originates from the fact that bifurcation curves are obtained by the local stability of steady states whereas the biomanipulations are conducted away from the steady states.

In the case of $X$ and $D$ manipulation, the diagrams are shown in Figure 27(b): manipulation from turbid to clear state, and in Figure 27(c): manipulation from clear to turbid state. In terms of the lake restoration, where manipulation is started from turbid states, the quasi-basins of attraction of the clear state was the largest in the case of the combination of $X$ and $D$ in the present model. These results reflect that the full vector of the manipulated variables in the five-dimensional space determines the quasi-basins of attraction.

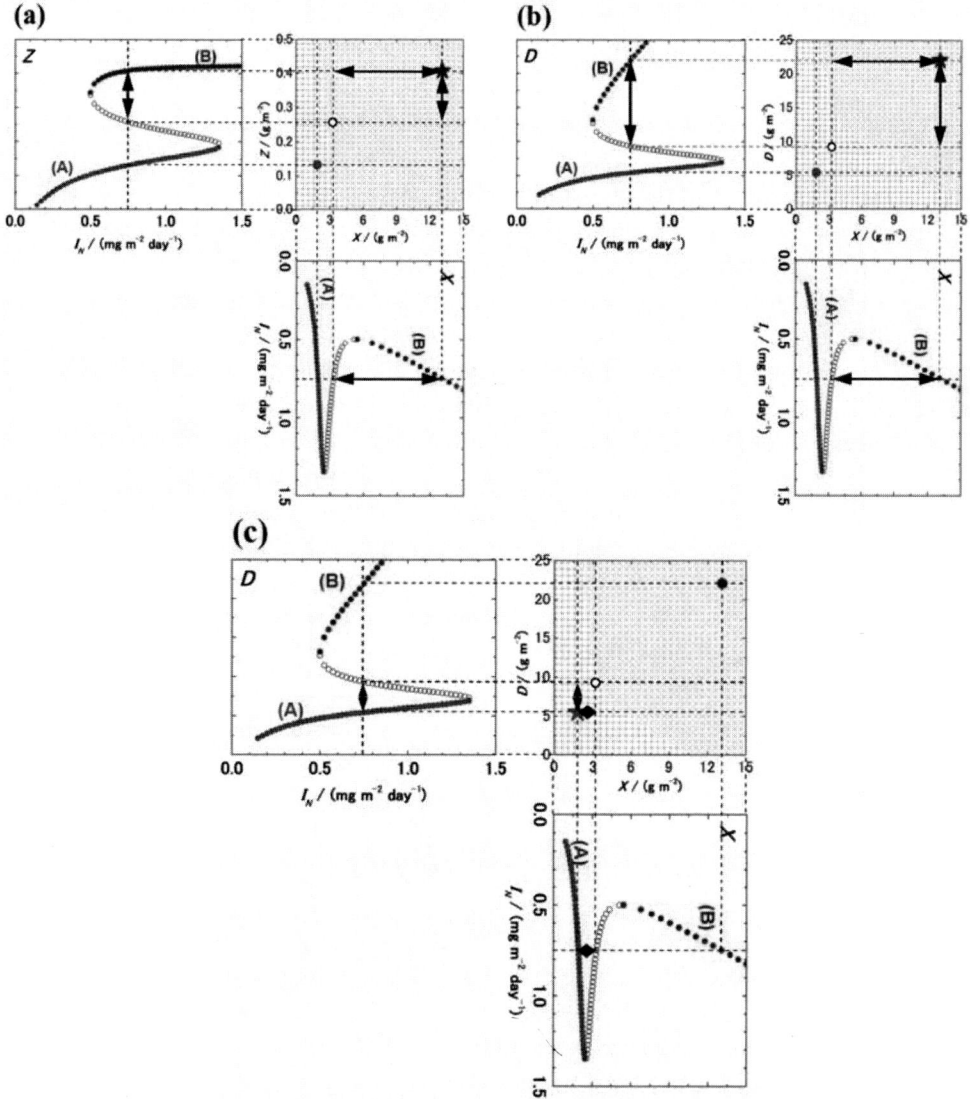

Figure 27. Phase diagrams (quasi-basins of attractions) spanned by two of the five variables in abs.-CAMS. (a) $X$ and $Z$ manipulation from a turbid state (B), $X$ and $D$ manipulation from a turbid state (B), and (c) $X$ and $D$ manipulation from a clear state (A). All the variables were initially set to either the turbid or the clear states (red or blue filled star). Two of the variables were manipulated from the initial state to a certain state within the phase diagram. The regions represented by red and blue crosses represent basins of attractions of the turbid and the clear states, respectively. (This figure is re-used from Ecology and Society 10(2):3, 2005. Permission is granted from the authors.)

The present simulations of biomanipulations show that the success of switching the turbid states to clear states depends on the structure of the basins of attraction in the five-variable space. In multivariable system or natural systems, resilience is embedded in the full multi-dimensional space. We have proposed representations of resilience by quasi-basins of attractions as shown in Figure 27. An implication of the present results of biomanipulations is that multivariable manipulations may be effective for the practical restoration of lakes showing alternative states. The present results suggest that a direct reduction in the abundance of phytoplankton and detritus is the most effective for switching lakes from turbid to clear states. The present result supports, for instance, the regional plan for conservation of lake water quality in Japan (Chiba prefecture 2001); direct removal of algae and dredging of benthic mud have been and will be carried out for the restoration of eutrophic lakes.

### 3.6.6. Implications for Aquatic Ecosystem Management

The analyses of the original and a modified version of abs.-CASM show that not only the changes in the anthropogenic parameters such as nutrient loading but also those in natural parameters in lakes may induce regime shifts. One of the important implications obtained from the present study is that uncontrollable regime shifts induced by the changes in the natural parameters may not occur if the nutrient loading is small, indicating that controlling anthropogenic impacts is still crucial to prevent regime shifts to a degraded state.

Real lakes can originally be regarded as oscillatory systems due to either their internal dynamics or seasonal forcing. Under such oscillatory conditions, the timing of biomanipulation may affect the results of temporal behaviors even if the strength of the manipulation is nearly the same. Nonlinear dynamics shows that the sensitivity of oscillatory systems to pulse-perturbation depend on the phase of oscillations. Much more effective biomanipulations may possible when pulse-perturbations of multi-variables are applied at an appropriate phase of the lake-systems, a situation utilizing the resonant phenomena.

## 3.7. Perspectives for the Mathematical Models and Their Mathematical Analysis

### 3.7.1. Effect of Advection – Their Quantitative/Qualitative Studies with Controllable Fluid Models

In our mathematical models (6), (7), (25), (26) and (36)-(39), we have introduced the advection terms to represent the effect of flow fields on the basic ecological processes. However, we didn't couple it with the equations of fluids such as the Navier-Stokes equations and we have made some heuristic assumptions on the flow. That is to say, in the reaction-diffusion-advection equations for the minimal NP model (6) and (7), we use the seeded-eddy model as a background flow. We approximate the two-dimensional turbulent flow with some rotational eddies whose centers and sizes are randomly chosen, and fixed at the same position throughout the evolution. However, in the real flows, these coherent eddies can interact with each other and they produce more complicated unsteady flow patterns, which possibly affect the ecological processes. The vertical velocity field appearing in the equations (26) is derived from the Stokes law, which is proportional to the cell density of cyanobacteria. In the model (36)-(39), the three-dimensional flow field (40) does not satisfy the incompressibility condition, which is somewhat unrealistic and artificial in aquatic ecosystems.

These specific flows are certainly helpful to confirm that the basic dynamics found in the minimal mean-field ecological models persists under the influence of background flow field. However, in the real aquatic ecosystems, the flow is usually unsteady and the complex flow pattern sometimes has a strong influence on the basic ecological dynamics. So we need to investigate the effect of advection in more quantitative and qualitative ways by coupling the fluid equations with the reaction-diffusion-advection ecological equations. One candidate for the flow is two-dimensional turbulence, but the theory of two-dimensional turbulence is far from being complete for the time being. Another possible flow is the chaotic advection induced by some simple periodic flows. See the review paper by Aref (Aref 2002) for many references. It is interesting to see how the chaotic mixing enhances or inhibits the ecological processes.

Let us finally mention the importance of the topography of the flow domains in the study of aquatic ecosystems. The aquatic regions such as lakes, swamps and inland seas usually have many islands and artificial obstacles with complicated boundaries, which are mathematically described as multiply connected regions. Thus theoretical investigation of the two-dimensional incompressible and inviscid flows in multiply connected regions coupled with the ecological models is essential to see the relation between the topography of the flow domain and the ecological pattern arising in the aquatic ecosystems. It would be possible to construct the inviscid and incompressible lateral flows in such multiply connected regions through an explicit representation of the hydrodynamic Green function given by Crowdy et al. (Crowdy et al. 2005).

### 3.7.2. Fundamental Mathematical Reduction in the Ecological Model

It is needless to say that mathematical models should be as simple as possible for the sake of analysis, but they should also represent some of basic dynamics occurred in the aquatic ecosystems. In this sense, bifurcation analysis for the integrated minimal NP model (32)-(35), although it is a simple four-component model, has successfully revealed that the existence of bistability of clear state and turbid state and limit cycle oscillations play a key role in understanding of the abrupt transition between the two states in aquatic ecosystems and gives some useful information on the ecosystem managements. On the other hand, the NP model is too simple to represent essential complexity of the ecosystems, which thus arises another fundamental question to be answered theoretically. How the reduction to the simple component model is justified in the real ecosystems where there exist complicated interactions between elements. Generally speaking, the more elements we take into considerations in the mathematical model, the more steady states we can find, which gives rise to multi-stable states dynamics. Nevertheless the bistable states become dominant in the real aquatic ecosystems. So we need to answer how these two states are chosen among the multi-stable states, which is realizable with some powerful mathematical tools developed in the theory of nonlinear dynamical systems.

One possible mathematical approach to this problem is applying a reduction theory to the N components model,

$$\frac{dP_i}{dt} = I_{P_i} + \sum_{j \neq i}^{N} A_{ij} F(P_i, P_j) - m_{N_i} P_i, \qquad for \ \ i = 1,...,N, \tag{47}$$

in which $I_{P_i}$, $m_{N_i}$ and $A_{ij}$ denote the input/birth rate, the mortality rate for the component $P_i$ and the component of an interaction matrix between components $P_i$ and $P_j$ respectively. Let us note that the component $A_{ij}$ is determined from the food-chain structure for the aquatic ecosystem we are dealing with. The nonlinear interaction term $F(P_i, P_j)$ is given by either the Holling type II or type III responses,

$$F(P_i, P_j) = \frac{P_i}{H_{N_i} + P_i} P_j \quad or \quad \frac{P_i^2}{H_{N_i} + P_i^2} P_j \tag{48}$$

At first, we try to obtain all equilibrium states of the model, and then we investigate their stability for an appropriate parameter region. For each of equilibrium state, we compute the distribution of eigenvalues for the linearized equation that determines the local linear stability of the equilibrium. Depending on the distribution of the eigenvalues, we may use a mathematical reduction theory such as the normal forms and the center manifold theory. Bifurcation analysis for this N component model is the next step. Although the bifurcation diagram for the multi-equilibiria-states is expected to be very complicated, it may be sometimes successful to describe the evolution of the system if we can obtain global structures such as heteroclinic or homoclinic connections between the equilibria. As a result of this nonlinear analysis, we may be able to explain why the minimal two or three components model is sufficient to describe the bistable states in the aquatic ecosystems. Furthermore, we can expect that this analysis could reveal the difference in the nonlinear responses (48) from the mathematical point of view.

Finally let us point out another completely different approach toward understanding the bistability in the ecosystems. That is to say, developing a statistical theory for the N component model (47) is important. It is equivalent to state that we have to describe global attractors for the N-component system, and we then investigate the attractors behave in a statistical sense. We define a certain "stable" state in the ecosystems as an ensemble average for the orbits in terms of the global attractors for the system. And the transition between the clear state and the turbid state may be described as a transition between the global attractors of the dynamical system (47).

### 3.7.3. An Approach from the Theory of Nonlinear Partial Differential Equations

The mathematical analysis in the present chapter is rely on the theory of nonlinear dynamical systems for the ordinary differential equations defined in the finite dimensional phase space. Namely, we first investigate the stability of the equilibria and bifurcation diagram for the equilibria in the ODE models such as the minimal NP models and then we describe some of the properties in the extended reaction-diffusion-advection (PDE) models by numerical means. In the process, we implicitly assume that the reaction term is dominant and the diffusion and advection terms work as weak "perturbation" terms in the PDE model. However, strictly speaking, the assumption does not hold for the PDE system. For example, it is not trivial to conclude that there exists an equilibrium state for the PDE model in the neighborhood of the equilibrium state found in the ODE model. To make the present theory more complete from the mathematical point of view, we need to deal with the PDE model by using the theory of nonlinear partial differential equations.

# CONCLUSION

In the present chapter, we have proposed some simple mathematical models to understand basic dynamical aspects observed in the aquatic ecosystems. Owing to their simplicity, mathematical analysis of the models through bifurcation analysis and numerical analysis indicates that the existence bistable states of clear-state and turbid-state, limit-cycle oscillations and basin of attraction play a significant role in the description of key notions in ecological sciences such as "resilience", "paradox of enrichment", "patchiness", "phenotypic plasticity" and so on. In particular, folded-bifurcation curves obtained as a function of nutrient inputs in all the models confirm regime shifts from clear-water to turbid-water states due to eutrophication. The critical points of nutrient levels for the regime shifts are realistic ranges, indicating that these models can be used for the prediction of regime shifts and management of aquatic ecosystems.

Not only the mathematical models capture some of dynamical properties occurred in the aquatic ecosystems, but they also provide us with implications to biomanipulations and ecosystem management. Bistability or alternative-stable-state theory is the guiding principle for the successful biomanipulations. Although direct field observations of alternative stable states in aquatic ecosystems due to eutrophication may be difficult, many field experiments of biomanipulations suggest the existence of the alternative stable states. One of the practical difficulties for the observation of the alternative stable states may be that it takes long periods of time to discern the histeretic behavior due to eutrophication and reverse-eutrophication processes. Further, spatial inhomogeneities in real aquatic ecosystems and many kinds of environmental noise and fluctuations in nature tend to hide the bistable characteristics in the systems. Thus, bistable ranges as a function of nutrient levels seem to be smaller in real systems than in the model systems presented in this chapter. Nonetheless, we believe that the concept of alternative stable stats and resilience are very useful concept for the management of aquatic ecosystems.

Advanced mathematical insights reveal further improvements of the mathematical models and their analyses presented in this chapter. First, consideration of more realistic fluid dynamics such as unsteady flow patterns arising from the interaction of many eddies can help us to understand a strong influence of flows on the basic ecological dynamics. Such a complex flow dynamics may also be exhibited by the topography of aquatic ecosystems that typically have many islands and artificial obstacles with complicated boundaries. Second, although we have employed as simple models as possible to abstract the essential nonlinear characteristics in aquatic ecosystems, an alternative mathematical approach due to the theory of complex systems may reveal more profound dynamics in real systems. That is, a reduction theory for a many-components model may prove more than two stable states (attractors), exhibiting multi-stable states dynamics. Such an approach may reveal why bistable states become dominant in many model aquatic ecosystems, or possibly in real aquatic ecosystems. Third, we should reconsider stable states in reaction-diffusion-advection (PDE) models, since it is not evident that stable states for the PDE models exist in the neighborhood of those in ODE models mainly analyzed in the present study.

# REFERENCES

Abraham ER (1998) The generation of plankton patchiness by turbulent stirring. *Nature* 391:577-580.

Amemiya T, Enomoto T, Rossberg AG, Takamura N, Itoh K (2005) Lake restoration in terms of ecological resilience: a numerical study of biomanipulations under bistable conditions. *Ecol. Soc.* 10:3.

Amemiya, T, Enomoto T, Rossberg A G, Yamamoto T, Inamori Y, Itoh K (2007) Stability and dynamical behavior in a lake-model and implications for regime shifts in real lakes. *Ecol. Modell.* 206: 54-62.

Aref H (2002) The development of chaotic advection. *Phys. Fluids* 40: 1315-1325.

Bohannan BJM, Lenski RE (1997) Effect of resource enrichment on a chemostat community of bacteria and bacteriophage. *Ecology* 78:2303-2315.

Brookes JD, Ganf GG, Oliver RL (2000) Heterogeneity of cyanobacterial gas-vesicle volume and metabolic activity. *J. Plankton. Res*. 22:1579-1589.

Brookes JD, Ganf GG (2001) Variations in the buoyancy response of *Microcystis aeruginosa* to nitrogen, phosphorus and light. *J. Plankton. Res*. 23:1399-1411.

Burkert U, Hyenstrand P, Drakare S, Blomqvist P (2001) Effects of the mixotrophic flagellate *Ochromonas* sp. on colony formation in *Microcystis aeruginosa. Aquat. Ecol*. 35:9-217.

Crowdy D, Marshall J (2005) Analytical formulae for the Kirchhoff-Routh path function in multiply connected domains. *Proc. Roy. Soc*. A, 461:.2477–2501.

DeAngelis DL, Bartell SM, Brenkert AL (1989) Effects of nutrient recycling and food-chain length on resilience. *Am. Nat.* 134:778-805.

Dokulil MT, Teubner K (2000) Cyanobacterial dominance in lakes. *Hydrobiologia* 438:1-12.

Drenner RW, Hambright KD (1999) Review: biomanipulation of fish assemblages as a lake restoration technique. *Archiv für. Hydrobiolgie.* 146:129-165.

Dyke PPG, Robertson T (1985) The simulation of offshore turbulent dispersion using seeded eddies. *Appl. Math. Model.* 9:429-433.

Fennel K, Boss E (2003) Subsurface maxima of phytoplankton and chlorophyll: steady-state solutions from a simple model. *Limnol. Oceanogr*. 48:1521-1534.

Fulton III RS, Paerl HW (1987) Effects of colonial morphology on zooplankton utilization of algal resources during blue-green algal (*Microcystis aeruginosa*) blooms. Limnol Oceanogr 32:634-644.

Gentleman W, Leising A, Frost B, Strom S, Murray J (2003) Functional responses for zooplankton feeding on multiple resources: a review of assumptions and biological dynamics. *Deep-Sea Res. II* 50:2847-2875.

Hansson LA, Annadotter H, Bergman E, Hamrin SF, Jeppesen E, Kairesalo T, Luokkanen E, Nilsson, PÅ, Søndergaard M, Strand J (1998) Biomanipulation as an application of food-chain theory: constraints, synthesis, and recommendations for temperate lakes. *Ecosystems* 1:558-574.

Holling CS (1973) Resilience and stability of ecological systems. *Annu. Rev. Ecol. Syst.* 4:1-23.

Huisman J, Thi NNP, Karl DM, Sommeijer B (2006) Reduced mixing generates oscillations and chaos in the oceanic deep chlorophyll maximum. *Nature* 439:322-325.

Ibelings BW, Mur LR, Walsby AE (1991) Diurnal changes in buoyancy and vertical distribution in populations of *Microcystis* in two shallow lakes. *J. Plankton. Res.* 13:419-436.

Jakobsen HH, Tang KW (2002) Effects of protozoan grazing on colony formation in *Phaeocystis globosa* (Prymnesiophyceae) and the potential costs and benefits. *Aquat. Microb. Ecol.* 27:261-273.

Lampert W, Rothhaupt KO, von Elert E (1994) Chemical induction of colony formation in a green alga (*Scenedesmus acutus*) by grazers (*Daphnia*). *Limnol. Oceanogr.* 39:1543-1550.

Long BM, Jones GJ, Orr PT (2001) Cellular microcystin content in N-limited *Microcystis aeruginosa* can be predicted from growth rate. *Appl. Environ. Microbiol.* 67:278-283.

Luckinbill LS (1974) The effects of space and enrichment on a predator-prey system. *Ecology* 55:1142-1147.

Lürling M (2003a) Phenotypic plasticity in the green algae *Desmodesmus* and *Scenedesmus* with special reference to the induction of defensive morphology. *Ann. Limnol-Int. J. Lim.* 39:85-101.

Lürling M (2003b) The effect of substances from different zooplankton species and fish on the induction of defensive morphology in the green algae *Scenedesmus obliquus*. *J. Plankton. Res.* 25:979-989.

Martin AP (2003) Phytoplankton patchiness: the role of lateral stirring and mixing. *Prog. Oceanogr.* 57:125-174.

McCauley E, Murdoch WW (1990) Predator-prey dynamics in environments rich and poor in nutrients. *Nature* 343:455-457.

Medvinsky AB, Petrovskii SV, Tikhonova IA, Malchow H, Li B-L (2002) Spatiotemporal complexity of plankton and fish dynamics. *SIAM Rev.* 44:311-370.

Mullin MM, Stewart EF, Fuglister FJ (1975) Ingestion by planktonic grazers as a function of food concentration. *Limnol. Oceanogr.* 20:259-262.

Neufeld Z, Haynes PH, Garçon V, Sudre J (2002) Ocean fertilization experiments may initiate a large scale phytoplankton bloom. *Geophys. Res. Lett.* 29:10.1029/2001GL013677.

Oberholster PJ, Botha A-M, Grobbelaar JU (2004) *Microcystis aeruginosa*: source of toxic microcystins in drinking water. *Afr. J. Biotechnol.* 3:159-168.

Ozawa K, Fujioka H, Muranaka M, Yokoyama A, Katagami Y, Homma T, Ishikawa K, Tsujimura S, Kumagai M, Watanabe MF, Park H-D (2005) Spatial distribution and temporal variation of *Microcystis* spesies composition and Microcystin concentration in Lake Biwa. *Wiley InterScience* DOI 10.1002/tox.20117.

Petrovskii SV, Malchow H (1999) A minimal model of pattern formation in a prey-predator sysyem. *Math. Comput. Model* 29:49-63.

Petrovskii SV, Li B-L, Malchow H (2004) Transition to spatiotemporal chaos can resolve the paradox of enrichment. *Ecol. Complex* 1:37-47.

Reynolds CS, Oliver RL, Walsby AE (1987) Cyanobacterial dominance: the role of buoyancy regulation in dynamic lake environments. *NZ J. Mar. Freshwater. Res.* 21:379-390.

Rietkerk M, Dekker SC, de Ruiter PC, van de Koppel J (2004) Self-organized patchiness and catastrophic shifts in ecosystems. *Science* 305:1926-1929.

Rosenzweig ML (1971) Paradox of enrichment: destabilization of exploitation ecosystem in ecological time. *Science* 171:385-387.

Scheffer M (1991) Fish and nutrients interplay determines algal biomass: a minimal model. *Oikos* 62:271-282.

Scheffer M (1998) Ecology of shallow lakes. Kluwer Academic Publishers, Dordrecht, Netherlands.

Scheffer M, Carpenter SR, Foley JA, Folke C, Walker B (2001) Catastrophic shifts in ecosystems. *Nature* 413:591-596.

Scheffer M, Carpenter SR (2003) Catastrophic regime shifts in ecosystems: linking theory to observation. *Trends Ecol. Evol.* 18:648-656.

Serizawa H, Amemiya T, Itoh K (2008a) Patchiness in a minimal nutrient-phytoplankton model. *J. Biosci.* (Submitted).

Serizawa H, Amemiya T, Rossberg AG, Itoh K (2008b) Computer simulations of seasonal outbreak and diurnal vertical migration of cyanobacteria. *Limnology DOI* 10.1007/s10201-008-0245-5.

Serizawa H, Amemiya T, Enomoto T, Rossberg AG, Itoh K (2008c) Mathematical modeling of colony formation in algal blooms: phenotypic plasticity in cyanobacteria. *Ecol. Res.* DOI 10.1007/s11284-007-0447-z.

Shapiro J, Lamarra V, Lynch M (1975) Biomanipulation: an ecosystem approach to lake restoration. Pages 85-96 *in* P. L. Brezonik, and J. L. Fox, editors. *Proceedings of a symposium on water quality management through biological control*. University of Florida, Gainesville.

Sigee DC (2005) Freshwater microbiology. John Wiley and Sons Ltd, West Sussex, England.

Takamura N, Yasuno M (1984) Diurnal changes in the vertical distribution of phytoplankton in hypertrophic Lake Kasumigaura, Japan. *Hydrobiologia* 112:53-60.

Tang KW (2003) Grazing and colony size development in *Phaeocystis globosa* (Prymnesiophyceae): the role of chemical signal. *J. Plankton. Res.* 25:831-842.

Turing AM (1952) On the chemical basis of morphogenesis. *Philos. Trans. R. Soc. Ser. B.* 237:37-72.

Tzella A, Haynes PH (2007) Small-scale spatial structure in plankton distributions. *Biogeosci.* 4:173-179.

van Holthoon FL, van Beek TA, Lürling M, van Donk E, de Groot A (2003) Colony formation in *Scenedesmus*: a literature overview and further steps towards the chemical characterisation of the *Daphnia* kairomone. *Hydrobiologia* 491:241-254.

Visser PM, Passarge J, Mur LR (1997) Modelling vertical migration of the cyanobacterium *Microcystis*. *Hydrobiologia* 349:99-109.

Vos M, Kooi BW, DeAngwlis DL, Mooij WM (2004a) Inducible defences and the paradox of enrichment. *Oikos* 105:471-480.

Vos M, Verschoor AM, Kooi BW, Wäckers FL, DeAngwlis DL, Mooij WM (2004b) Inducible defenses and trophic structure. *Ecology* 85:2783-2794.

Wallace BB, Hamilton DP (2000) Simulation of water-bloom formation in the cyanobacterium *Microcystis aeruginosa*. *J. Plankton. Res.* 22:1127-1138.

Walsby AE (1994) Gas vesicles. *Microbiol. Rev.* 58:94-144.

Watanabe MF, Harada K, Carmichael WW, Fujiki H (1996) Toxic *Microcystis*. CRC Press Inc, Boca Raton, Florida, U.S.A.

Yang Z, Kong F, Shi X, Cao H (2006) Morphological response of *Microcystis aeruginosa* to grazing by different sorts of zooplankton. *Hydrobiologia* 563:225-230.

Yoshinaga I, Hitomi T, Miura A, Shiratani E, Miyazaki T (2006) Cyanobacterium *Microcystis* bloom in a eutrophicated regulating reservoir. *JARQ* 40:283-289.

Yoshiyama K, Nakajima H (2002) Catastrophic transition in vertical distributions of phytoplankton: alternative equilibria in a water column. *J. Theor. Biol.* 216:397-408.

In: Aquatic Ecosystem Research Trends
Editor: George H. Nairne

ISBN 978-1-60692-772-4
© 2009 Nova Science Publishers, Inc.

*Chapter 4*

# SOURCES, PRESENCE, ANALYSIS, AND FATE OF STEROID SEX HORMONES IN FRESHWATER ECOSYSTEMS – A REVIEW

### *Robert B. Young and Thomas Borch**

Colorado State University,
Departments of Chemistry and Soil and Crop Sciences,
1170 Campus Delivery, Fort Collins, Colorado 80523-1170

## ABSTRACT

Natural and synthetic steroid sex hormones, including those administered to humans and livestock as pharmaceuticals, constitute an important class of environmental endocrine disruptors. Steroid sex hormones generally have high biological activities at low concentrations, and in river systems steroid sex hormones have been linked to various adverse effects on fish, including altered sex ratios, intersex fish, and diminished reproduction. In a national reconnaissance study conducted by the U.S. Geological Survey (USGS) from 1999 to 2000, steroid sex hormones were observed at various concentrations and frequencies in water samples from 139 stream sites across the United States. Steroid sex hormones also have been observed in rivers and river sediments in other countries throughout the world. Generally, the reported concentrations are in the low ng $L^{-1}$ (parts per trillion) range. However, in a 7-year, whole-lake experiment in northwestern Ontario, Canada, chronic exposure of fathead minnows to 17α-ethinylestradiol (the active ingredient in most oral contraceptive pills) at concentrations ranging from 5 to 6 ng $L^{-1}$ adversely affected gonadal development in males and egg production in females, and led to a near extinction of fathead minnows from the lake. Because steroid sex hormones and other endocrine disruptors have been detected in locations around the world at concentrations that could have adverse biological and ecological effects, it is important to understand their sources, the processes that transform them, and their ultimate fate in the environment. Therefore, the objective of this article is to survey and critically review existing literature regarding the presence of steroid sex hormones in freshwater ecosystems, their sources, and potential fate. In particular, the

* Corresponding author: thomas.borch @ colostate.edu

article will examine discharges from wastewater treatment plants, and transport from agricultural operations. The article also will consider how biodegradation, photodegradation, and sorption to sediments influence the fate of steroid sex hormones in freshwater ecosystems. Finally, the article will compile relevant information about the physical and chemical properties of selected steroid sex hormones, and conclude with suggestions for future research.

# INTRODUCTION

The endocrine system produces hormones, which travel through the bloodstream in extremely small concentrations ($\sim 10^{-9}$ g L$^{-1}$ to $10^{-12}$ g L$^{-1}$) to specialized receptors in target organs and tissues, including mammary glands, bone, muscle, the nervous system, and male and female reproductive organs [1]. Hormones bind to hormone receptors, and the resulting complexes help to regulate gene expression, cell differentiation, hormone secretion, and other bodily processes [2]. Broadly speaking, the endocrine system uses hormones as chemical signals to regulate various important biological functions, including homeostasis (the body's ability to maintain a state of balance), growth, development, sexual differentiation, and reproduction [1, 3].

In recent years, scientists have become concerned about the exposure of humans and wildlife to chemicals in the environment that can disrupt the normal function of their endocrine systems, especially during critical stages of growth and development [4, 5]. Suspected endocrine disruptors in aquatic environments include natural hormones, synthetic hormones, plant sterols, phytoestrogens (plant compounds that are structurally similar to estrogens), and organic chemicals used in pesticides, detergents, plastics, and other products [6, 7].

The mechanisms of endocrine disruption are complex. Endocrine disruptors operate by mimicking, enhancing, or inhibiting the actions of endogenous (i.e., self-produced) hormones, interfering with hormone synthesis or metabolism, disrupting hormone transport, or altering hormone receptor populations [5, 6, 8]. An endocrine disruptor's potency appears to be related primarily to its affinity for binding to hormone receptors, and to the shape of the resulting complex, but its potency can be affected by subsequent interactions and rate-limiting events [9]. The relationship between endocrine disruptor potency and concentration is often nonlinear (e.g., U-shaped), which could reflect different mechanisms of action at different concentrations [10-12]. In addition, mixtures of endocrine disruptors can have additive or even synergistic effects [13-18]. When attempting to assess the environmental risks of endocrine disruption, it is difficult to generalize across species, because the basic mechanisms of sexual differentiation, metabolism, and receptor structure and function differ across species [19-21].

The health effects of endocrine disruption have been extensively reviewed [1, 4-6, 8, 22-24]. Most of the evidence for endocrine disruption in wildlife has come from studies on species living in, or closely associated with, aquatic environments [20]. Many of the observed effects appear to result from disruption of the functions of steroid sex hormones, and particularly those of estrogens [20]. Adverse effects of endocrine disruption include abnormal blood hormone levels, masculinization of females, feminization of males, altered sex ratios, intersexuality, and reduced fertility and fecundity [20, 24, 25]. In fish, for example, intersexuality is characterized by the presence of oocytes (female gametophytes) within testicular tissue, or disruption of reproductive duct development [25, 26]. One study from 1995 to 1996 examined wild populations of freshwater fish (roach; *Rutilus rutilus*), and reported a high incidence of intersexuality across the United Kingdom [26]. Other studies also have reported evidence of endocrine disruption in freshwater ecosystems [25, 27-29].

Among suspected endocrine disruptors, exogenous steroid sex hormones (i.e., not self-produced) generally have the highest affinities for binding to steroid sex hormone receptors, and the highest potencies for disrupting steroid sex hormone functions [13, 19, 21, 30, 31]. In laboratory experiments with some fish species, steroid sex hormones have been linked to endocrine disruption after three weeks of exposure to concentrations of 17α-ethinylestradiol as low as 1 ng $L^{-1}$ (fathead minnows; *Pimephales promelas*), 17β-estradiol as low as 1-10 ng $L^{-1}$ (rainbow trout; *Oncorhynchus mykiss*), and estrone as low as 25-50 ng $L^{-1}$ (rainbow trout; *Oncorhynchus mykiss*) [32, 33]. In a 7-year, whole-lake experiment in northwestern Ontario, Canada, chronic exposure of fathead minnows to 17α-ethinylestradiol at concentrations ranging from 5 to 6 ng $L^{-1}$ adversely affected gonadal development in males and egg production in females, and led to a near extinction of fathead minnows from the lake [34]. After a review of more than 100 studies on the effects of 17α-ethinylestradiol on aquatic organisms, 0.35 ng $L^{-1}$ has been recommended as the predicted no-effect concentration (PNEC) for 17α-ethinylestradiol in surface water [35]. Because steroid sex hormones and other endocrine disruptors have been detected in locations around the world at concentrations that could have adverse biological and ecological effects, it is important to understand their sources, the processes that transform them, and their ultimate fate in the environment.

Therefore, the objective of this article is to survey and critically review existing literature regarding the presence of steroid sex hormones in freshwater ecosystems, their sources, and potential fate. In particular, the article will examine discharges from wastewater treatment plants, and transport from agricultural operations. The article also will consider how biodegradation, photodegradation, and sorption to sediments influence the fate of steroid sex hormones in freshwater ecosystems. Finally, the article will compile relevant information about the physical and chemical properties of selected steroid sex hormones, and conclude with suggestions for future research.

## ABBREVIATIONS

A list of the abbreviations used in this article, and their respective meanings, is set forth in Table 1.

**Table 1. List of Abbreviations and their Meanings**

| HPLC | High performance liquid chromatography | mg kg$^{-1}$ | Milligrams per kilogram (parts per million) |
|------|----------------------------------------|--------------|---------------------------------------------|
| UPLC | Ultra performance liquid chromatography | µg kg$^{-1}$ | Micrograms per kilogram (parts per billion) |
| GC | Gas chromatography | µg L$^{-1}$ | Micrograms per liter (parts per billion) |

**Table 1. (Continued)**

| | | | |
|---|---|---|---|
| MS | Mass spectrometry | ng L$^{-1}$ | Nanograms per liter (parts per trillion) |
| MS/MS | Tandem mass spectrometry | TOF MS | Time-of flight mass spectrometry |
| DAD | Diode array detection | pK$_a$ | Acid dissociation constant |
| FID | Flame ionization detection | RBA | Relative binding affinity |
| SPE | Solid phase extraction | CLLE | Continuous liquid-liquid extraction |
| SIM | Selected ion monitoring | BET | Brunauer/ Emmett/ Teller |
| MRM | Multiple reaction monitoring | pK$_a$ | Acid dissociation constant |
| m/z | Mass-to-charge ratio | EEq | 17β-estradiol equivalents |
| K$_{ow}$ | Octanol-water partition coefficient | S$_w$ | Water solubility |
| K$_{oc}$ | Carbon normalized partition coefficient | $f_{oc}$ | Organic carbon fraction |
| K$_d$ | Sorption coefficient | S | Sorbed chemical amount |
| K$_f$ | Freundlich sorption coefficient | C$_r$ | Concentration in retentate |
| PNEC | Predicted no effect concentration | C$_p$ | Concentration in permeate |
| SAR | Structure-activity relationship | WWTP | Wastewater treatment plant |
| C-18 | Octadecyl stationary phase | APS | Aminopropyl silica stationary phase |
| SVB | Polystyrene divinylbenzene stationary phase | HLB | Hydrophilic-lipophilic balance stationary phase (n-vinyl pyrrolidone + divinylbenzene) |
| EVB | Ethinylbenzene-divinylbenzene stationary phase | SE | Solubility enhancement |
| CFF | Cross-flow ultrafiltration | FQ | Fluorescence quenching |
| MDL | Method detection limit | LOD | Limits of detection |
| UV | Ultraviolet | λ | Wavelength |

## STEROID SEX HORMONES – IN GENERAL

Steroid sex hormones and their receptors are found in a range of vertebrate and invertebrate species [5]. Steroid sex hormones are hydrophobic in nature, and commonly act by diffusing through cell membranes and binding to nuclear hormone receptors, although interactions with transmembrane receptors also occur [6, 9, 11].

There are three classes of steroid sex hormones: androgens, estrogens, and progestagens [3]. In vertebrates, androgens play a key role in the development of male traits, spermatogenesis, mating and breeding behaviors, reproduction, and muscle growth [3, 31]. The most common androgens among vertebrates are testosterone and 5α-dihydrotestosterone, although 11-ketotestosterone is common among fish [3, 6]. In vertebrates, estrogens are crucial for the development of female traits, ovulation, reproduction, mating and breeding behaviors, and somatic cell function [3, 20, 23]. In egg-laying vertebrates, estrogens also stimulate the liver to produce vitellogenin, a precursor of egg yolk constituents and eggshell proteins [6]. The most common estrogens among vertebrates are 17β-estradiol, estrone, and estriol [6]. In vertebrates, progestagens influence water and salt metabolism, and help to maintain pregnancy through various anti-estrogenic and anti-androgenic effects [36]. The most common progestagen among vertebrates is progesterone, although 17α, 20β-dihydroxyprogesterone is important among fish [3]. Like vertebrates, the endocrine systems of invertebrates regulate growth, development, and reproduction, but the endocrine systems of invertebrates are more diverse, and less well-documented, than vertebrates [20, 37]. Testosterone, 17β-estradiol, estrone, and progesterone have been reported in many invertebrate groups, but their role is not well understood [20, 37]. In addition, progesterone has been detected in the dry mature wood, pine bark, and pine needles of loblolly pine (*Pinus taeda* L.) [38].

Humans and animals excrete steroid sex hormones primarily in the form of sulfate or glucuronide conjugates, which are biologically inactive and more water soluble than unconjugated hormones [39-42]. Studies have suggested that glucuronide conjugates are deconjugated by sewage bacteria (e.g., *Escherichia coli*) before they reach WWTPs [40, 41, 43]. Sulfate conjugates are more recalcitrant, and have been detected in WWTP influent and effluent [40, 44, 45]. The types of natural steroid sex hormones that are excreted, and the degree of conjugation, varies with species, gender, and stage of reproduction, as reviewed previously [41, 42, 46, 47].

Exogenous natural and synthetic steroid hormones are administered to humans and livestock for a variety of pharmaceutical purposes. In humans, 17β-estradiol, equine-derived estrogens (e.g., equilin and equilenin), synthetic estrogens (e.g., 17α-ethinylestradiol and mestranol), natural and synthetic progestagens (e.g., progesterone and norethindrone), and testosterone are used for contraception, palliative care during cancer treatment, and hormone replacement therapy for menopause and osteoporosis [22, 48, 49]. In livestock, testosterone, trenbolone (synthetic androgen), 17β-estradiol, zeranol (non-steroidal estrogen), progesterone, and melengestrol (synthetic progestagen) are used as growth promoters [22, 46-49]. Synthetic steroid sex hormones (e.g., 17α-ethinylestradiol) are specifically designed for increased potency, bioavailability, and degradation resistance, and might be persistent if discharged to the environment [49].

# STEROID SEX HORMONES – PHYSICAL AND CHEMICAL PROPERTIES

The physical and chemical properties of steroid sex hormones can influence their ability to bind with steroid sex hormone receptors, and their distribution, bioavailability, and persistence in freshwater ecosystems [50].

The molecular structure of a representative steroid (a 27 carbon cholestane) is set forth in Figure 1. In general, steroids are characterized by a carbon skeleton consisting of four fused rings (a cyclopentan-o-perhydrophenanthrene ring) [51]. Differences among steroids arise from variations in the number and location of double bonds, and the type and stereochemical arrangements of functional groups along the carbon skeleton [52]. The steroid skeleton causes steroids to be rigid and hydrophobic, and variations in double bonds and functional groups along the steroid skeleton determine the intermolecular interactions that are possible (e.g., hydrogen bonding, dipole-induced dipole interactions, etc.) [31]. For example, the phenolic A-ring common to estrogens is polarizable, and can accept or donate hydrogen bonds.

Selected physical and chemical properties of a representative group of steroid sex hormones are set forth in Table 2. The representative group includes all three classes of steroid sex hormones (androgens, estrogens, and progestagens), and many of the steroid sex hormones that are common among vertebrates. The physical and chemical properties include molecular structures, molecular weights, water solubilities ($S_w$), octanol-water partition coefficients ($K_{ow}$), and relative binding affinities (RBAs) for androgen and estrogen receptors. Acid dissociation constants ($pK_a$) are not given, because the reported $pK_a$ values for steroid sex hormones substantially exceed expected pH values in freshwater ecosystems (e.g., 17β-estradiol has a $pK_a$ of 10.23) [53, 54]. Likewise, vapor pressures are not given, because steroid sex hormones generally are not volatile, and their vapor pressures are very low (e.g., the vapor pressure of 17β-estradiol is approximately $3 \times 10^{-8}$ Pa) [51, 55].

Water solubility data in the scientific literature can be highly variable. For example, some reported estrogen water solubilities vary by up to a factor of approximately 15 (17β-estradiol = 3.1 to 12.96 mg $L^{-1}$; 17α-ethynylestradiol = 3.1 to 19.1 mg $L^{-1}$; and estrone = 0.8 to 12.4 mg $L^{-1}$) [50]. Experimental conditions such as temperature, pH, and ionic strength can affect water solubilities, and probably represent the source of variability [50, 53, 56].

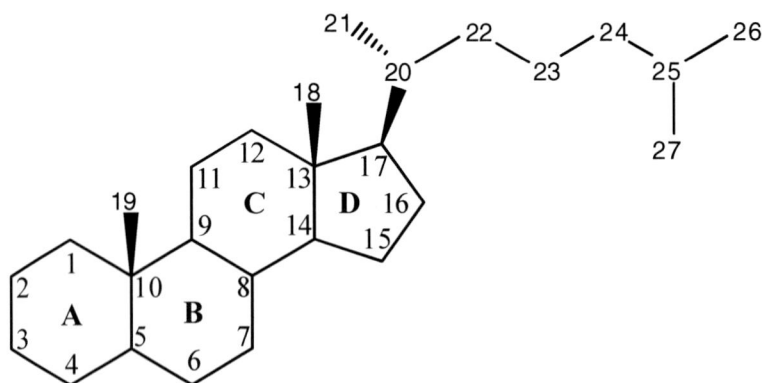

Figure 1. Basic steroid hormone structure (27 carbon cholestane). The steroid skeleton is characterized by four fused rings, labeled from A to D. Each carbon is labeled from 1 to 27.

**Table 2. Steroid hormone physical and chemical properties, including molecular structures, molecular weights (MW), water solubilities ($S_w$), octanol-water partition coefficients ($K_{ow}$), and relative binding affinities (RBA) for androgen and estrogen receptors (determined using in vitro competitive binding assays). NA = not available; SB = slight binder (< 50% inhibition); NB = non-binder**

| | Chemical Data | Function | $S_w$ (mg $L^{-1}$) | log $K_{ow}$ | Androgen log RBA | Estrogen log RBA | Source | Ref. |
|---|---|---|---|---|---|---|---|---|
| **ANDROGENS** | | | | | | | | |
| androstenedione | $C_{19}H_{26}O_2$ MW: 286.42 CAS: 63-05-8 | reproductive hormone | 37-41 | | | | Literature values (37 °C) | [56] |
| | | | 50.5 ± 2.1 | | | | Experimental (pH 6.8; 23 °C; n=6) | |
| | | | 57.8 | 2.75 | | | Literature value (25 °C) | [59] |
| | | | | 2.76 | | | Literature value | |
| | | | | | -0.62 | NA | Experimental (competitive binding assay) | [31] |
| 17α-methyltestosterone | $C_{20}H_{30}O_2$ MW: 302.46 CAS: 58-18-4 | androgen replacement (human use) | 33.9 | 3.72 | | | Literature value (25 °C) | [59] |
| | | | | 3.36 | | | KOWWIN™ computer model, v. 1.67 | |
| | | | | | | | Literature value | |
| | | | | | 1.28 | NA | Experimental (competitive binding assay) | [31] |
| cis-androsterone | $C_{19}H_{30}O_2$ MW: 290.45 CAS: 53-41-8 | reproductive hormone | 12 | 3.07 | | | Literature value (23 °C) | [59] |
| | | | 20.2 | | | | Literature value (23 °C) | |
| | | | | 3.69 | | | KOWWIN™ computer model, v. 1.67 | |
| | | | | | | | Literature value | |
| | | | | | −2.12 | NA | Experimental (competitive binding assay) | [31] |
| testosterone | $C_{19}H_{28}O_2$ MW: 288.43 CAS: 58-22-0 | reproductive hormone | 18-25 | | | | Literature values | [56] |
| | | | 23.2 ± 1.6 | | | | Experimental (pH 6.8; 23 °C; n=6) | |
| | | | 23.4 | 3.27 | | | Literature value (25 °C) | [59] |
| | | | | 3.32 | | | KOWWIN™ computer model, v. 1.67 | |
| | | | | | | | Literature value | |
| | | | | | 1.28 | | Experimental (competitive binding assay) | [31] |

| Chemical Data | Function | Sw (mg L⁻¹) | log Kow | Androgen log RBA | Estrogen log RBA | Source | Ref. |
|---|---|---|---|---|---|---|---|
| 17β-trenbolone — C$_{18}$H$_{22}$O$_2$; MW: 270.37; CAS: 10161-33-8 | growth promoter (animal use) | | | | NB | Experimental (competitive binding assay) | [30] |
| | | | 2.65 | | | KOWWIN™ computer model, v. 1.67 | [59] |
| | | | | 2.05 | NA | Experimental (competitive binding assay) | [31] |
| **ESTROGENS** | | | | | | | |
| 17α-estradiol — C$_{18}$H$_{24}$O$_2$; MW: 272.39; CAS: 57-91-0 | reproductive hormone | 3.6 | | | | Literature value (25 °C) | |
| | | 3.9 | | | | Literature value (27 °C) | [59] |
| | | | 3.94 | | | KOWWIN™ computer model, v. 1.67 | |
| | | | 4.01 | | | Literature value | |
| | | | | -2.40 | | Experimental (competitive binding assay) | [31] |
| | | | | | 0.49 | Experimental (competitive binding assay) | [30] |
| 17β-estradiol — C$_{18}$H$_{24}$O$_2$; MW: 272.39; CAS: 50-28-2 | reproductive hormone | 3.1-12.96 | | | | Literature values | [50] |
| | | 1.51 ± 0.04 | | | | Experimental (pH 7; 25 ± 0.5 °C; n=6) | |
| | | 3.1 ± 0.02 | | | | Experimental (pH 6.8; 23 °C; n=6) | [53] |
| | | 3.6 | | | | Literature value (25 °C) | |
| | | 3.9 | | | | Literature value (27 °C) | [59] |
| | | | 3.94 | | | KOWWIN™ computer model, v. 1.67 | |
| | | | 4.01 | | | Literature value | |
| | | | | -0.12 | | Experimental (competitive binding assay) | [31] |
| | | | | | 2.00 | Experimental (competitive binding assay) | [30] |
| 17α-ethinylestradiol — C$_{20}$H$_{24}$O$_2$; MW: 296.41; CAS: 57-63-6 | ovulation inhibitor (human use) | 3.1-19.1 | | | | Literature values | [50] |
| | | 9.20 ± 0.09 | | | | Experimental (pH 7; 25 ± 0.5 °C; n=6) | |
| | | 3.1 ± 0.03 | | | | Experimental (pH 6.8; 23 °C; n=6) | [53] |
| | | 11.3 | | | | Literature value (27 °C) | [59] |
| | | | 4.12 | | | KOWWIN™ computer model, v. 1.67 | |

| | Chemical Data | Function | Sw (mg L$^{-1}$) | log Kow | Androgen log RBA | Estrogen log RBA | Source | Ref. |
|---|---|---|---|---|---|---|---|---|
| estrone | C$_{18}$H$_{22}$O$_2$ MW: 270.37 CAS: 53-16-7 | reproductive hormone; (17β-estradiol metabolite) | | 3.67 | | | Literature value | [31] |
| | | | | | -1.42 | | Experimental (competitive binding assay) | [30] |
| | | | | | | 2.28 | Experimental (competitive binding assay) | [30] |
| | | | 0.8-12.4 | | | | Literature values | [50] |
| | | | 1.30 ± 0.08 | | | | Experimental (pH 7; 25 ± 0.5 °C; n=6) | [53] |
| | | | 2.1 ± 0.03 | | | | Experimental (pH 6.8; 23 °C; n=6) | [59] |
| | | | | 3.43 | | | KOWWIN™ computer model, v. 1.67 | |
| | | | | 3.13 | | | Literature value | |
| | | | | | SB | | Experimental (competitive binding assay) | [31] |
| | | | | | | 0.86 | Experimental (competitive binding assay) | [30] |
| estriol | C$_{18}$H$_{24}$O$_3$ MW: 288.39 CAS: 50-27-1 | reproductive hormone (17β-estradiol metabolite) | 13 | | | | Literature value | [55] |
| | | | | 3.67 | | | Literature value | |
| | | | | 2.81 | | | KOWWIN™ computer model, v. 1.67 | [59] |
| | | | | 2.45 | | | Literature value | |
| | | | | | -3.15 | | Experimental (competitive binding assay) | [31] |
| | | | | | | 0.99 | Experimental (competitive binding assay) | [30] |
| mestranol | C$_{21}$H$_{26}$O$_2$ MW: 310.44 CAS: 72-33-3 | ovulation inhibitor (human use) | 0.3 | | | | Literature value | [55] |
| | | | | 4.10 | | | Literature value | [55] |
| | | | | 3.68 | | | KOWWIN™ computer model, v. 1.67 | [59] |
| | | | | | NA | 0.35 | Experimental: competitive binding assay | [30] |
| PROGESTAGENS norethindrone | C$_{20}$H$_{26}$O$_2$ MW: 298.43 CAS: 68-22-4 | ovulation inhibitor (human use) | 7.04 | | | | Literature value (25 °C) | |
| | | | | 2.99 | | | KOWWIN™ computer model, v. 1.67 | [59] |
| | | | | 2.97 | | | Literature value | |
| | | | | | 0.41 | NA | Experimental (competitive binding assay) | [31] |

| Chemical Data | Function | Sw (mg L$^{-1}$) | log Kow | Androgen log RBA | Estrogen log RBA | Source | Ref. |
|---|---|---|---|---|---|---|---|
| medroxyprogesterone<br>C$_{22}$H$_{32}$O$_3$<br>MW: 344.50<br>CAS: 520-85-4 | ovulation inhibitor (human use); estrus regulator (animal use) |  | 3.50 |  |  | KOWWIN™ computer model, v. 1.67 | [59] |
|  |  |  |  | −0.41 | NA | Experimental (competitive binding assay) | [31] |
| progesterone<br>C$_{21}$H$_{30}$O$_2$<br>MW: 314.47<br>CAS: 57-83-0 | reproductive hormone | 8.81 |  |  |  | Literature value (25 °C) |  |
|  |  |  | 3.67 |  |  | KOWWIN™ computer model, v. 1.67 | [59] |
|  |  |  | 3.87 |  |  | Literature value |  |
|  |  |  |  | −0.70 |  | Experimental (competitive binding assay) | [31] |
|  |  |  |  |  | NB | Experimental (competitive binding assay) | [30] |
| **OTHER** |  |  |  |  |  |  |  |
| bisphenol-A<br>C$_{15}$H$_{16}$O$_2$<br>MW: 228.29<br>CAS: 80-05-7 | plasticizer | 120 |  |  |  | Literature value (25 °C) |  |
|  |  |  | 3.64 |  |  | KOWWIN™ computer model, v. 1.67 | [59] |
|  |  |  | 3.32 |  |  | Literature value |  |
|  |  |  |  | −2.39 |  | Experimental (competitive binding assay) | [31] |
|  |  |  |  |  | −2.11 | Experimental (competitive binding assay) | [30] |
| diethylstilbestrol<br>C$_{18}$H$_{20}$O$_2$<br>MW: 268.36<br>CAS: 56-53-1 | synthetic non-steroidal estrogen (use banned) | 12 |  |  |  | Literature value (25 °C) |  |
|  |  |  | 5.64 |  |  | KOWWIN™ computer model, v. 1.67 | [59] |
|  |  |  | 5.07 |  |  | Literature value |  |
|  |  |  |  | −1.66 |  | Experimental (competitive binding assay) | [31] |
|  |  |  |  |  | 2.60 | Experimental (competitive binding assay) | [30] |

**Table 3. Selected steroid hormone occurrence data, organized by steroid hormone and country, including median or mean (bold) concentrations (conc.), maximum concentrations (max.), numbers of samples (n), limits of detection (LOD), recovery percentages (recov.), sample types, forms of analysis, and references (refs.). NA = not available**

| | Conc. (ng L$^{-1}$) | Max. (ng L$^{-1}$) | n | LOD (ng L$^{-1}$) | Recov. (%) | Sample Type | Form of Analysis | Refs. |
|---|---|---|---|---|---|---|---|---|
| **ANDROGENS** | | | | | | | | |
| androstenedione | | | | | | JAPAN | | [194, 195] |
| | 5.2 | NA | 2 | 1.2 | 91 ± 8.1 | WWTP effluent | SPE (HLB, silica); UPLC-MS/MS (MRM) | |
| | 0.38 | NA | 4 | 0.06 | 85 ± 3.6 | Surface water | | [194] |
| cis-androsterone | | | | | | JAPAN | | [194, 195] |
| | <LOD | NA | 2 | 10 | 86 ± 7.2 | WWTP effluent | SPE (HLB, silica); UPLC-MS/MS (MRM) | [194] |
| | <LOD | NA | 4 | 5.0 | 82 ± 5.8 | Surface water | | |
| | | | | | | USA | | [77, 119, 120] |
| | 17 | 214 | 70 | 5 | 148.5 (d3-testosterone) | Surface water | CLLE; GC-MS (SIM) | [119, 120] |
| testosterone | | | | | | JAPAN | | [194, 195] |
| | <LOD | NA | 2 | 0.12 | 87 ± 5.6 | WWTP effluent | SPE (HLB, silica); UPLC-MS/MS (MRM) | [194] |
| | <LOD | NA | 4 | 0.06 | 81 ± 4.1 | Surface water | | |
| | | | | | | USA | | [77, 119, 120] |
| | 116 | 214 | 70 | 5 | 148.5 (d3-testosterone) | Surface water | CLLE; GC-MS (SIM) | [119, 120] |
| 17β-trenbolone | | | | | | JAPAN | | |
| | <LOD | NA | 2 | 0.30 | 88 ± 2.4 | WWTP effluent | SPE (HLB, silica); UPLC-MS/MS (MRM) | [194] |
| | <LOD | NA | 4 | 0.10 | 84 ± 5.4 | Surface water | | |

| | Conc. (ng L⁻¹) | Max. (ng L⁻¹) | n | LOD (ng L⁻¹) | Recov. (%) | Sample Type | Form of Analysis | Refs. |
|---|---|---|---|---|---|---|---|---|
| **ESTROGENS** | | | | | | | | |
| | | | | | | NETHERLANDS | | |
| | <LOD | 5.0 | 4 | 0.1-1.2 | 88 ± 12 | WWTP effluent | SPE (SVB, C-18, amino); HPLC fractionation; GC-MS/MS | [43, 196] |
| | <LOD | 3.0 | 6 | 0.1-0.3 | | Surface water | | [43] |
| | | | | | | GERMANY | | |
| | 0.5 | 4.5 | 16 | 0.15 | NA | WWTP effluent | SPE (EVB); HRGC-MS (SIM) | [197] |
| | 0.4 | 2.0 | 31 | 0.15 | NA | Surface water | | [77, 119, 120] |
| | | | | | | USA | | |
| 17α-estradiol | 30 | 74 | 70 | 5 | 128.8 (d4-estradiol) | Surface water | CLLE; GC-MS (SIM) | [119, 120] |
| | | | | | | CHINA | | |
| | 1 | 2 | 3 | 0.5 | NA | Surface water (low-flow) | SPE (C-18); GC-MS (SIM) | [198] |
| | | | | | | GERMANY | | |
| | <LOD | 4 | 16 | 1 | 74 ± 1 | WWTP effluent | SPE (C-18; silica gel cleanup; GC-MS/MS (MRM) | [116] |
| | <LOD | <LOD | 15 | 0.5 | 90 ± 1 | Surface water | | [77, 119, 120] |
| | | | | | | USA | | |
| mestranol | 17 | 407 | 70 | 5 | 128.8 (d4-estradiol) | Surface water | CLLE; GC-MS (SIM) | [119, 120] |
| | | | | | | CHINA | | |
| | <LOD | <LOD | 1 | 1 | 105 ± 19 | Surface water (low-flow) | SPE (C-18); GC-MS (SIM) | [198] |
| | | | | | | NETHERLANDS | | |
| | 0.9 | 12 | 6 | 0.5-2.4 | 88 ± 9 | WWTP effluent | SPE (SVB, C-18, amino); HPLC fractionation; GC-MS/MS | [43, 196] |
| 17β-estradiol | <LOD | 5.5 | | 0.3-0.6 | | Surface water | | [43] |

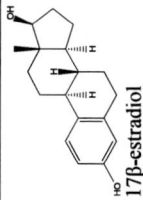

17β-estradiol

| Conc. (ng L⁻¹) | Max. (ng L⁻¹) | n | LOD (ng L⁻¹) | Recov. (%) | Sample Type | Form of Analysis | Refs. |
|---|---|---|---|---|---|---|---|
| | | | | | GERMANY | | [116, 197, 199] |
| <LOD | 3 | 16 | 1 | 76 ± 14 | WWTP effluent | SPE (C-18); silica gel cleanup; GC-MS/MS (MRM) | |
| <LOD | <LOD | 15 | 0.5 | 77 ± 14 | Surface water | | [116] |
| | | | | | FRANCE | | [200, 201] |
| 2.1 ± 0.6 | | | | | Upstream (1 km) | | |
| 6.6 ± 1.4 | | | | | WWTP effluent (Colombes) | | |
| 3.2 ± 0.3 | NA | 6 | NA | 84-98% (all estrogens) | Downstream | SPE (C-18); GC-MS (SIM) | [200] |
| 1.4 ± 0.6 | | | | | Upstream (1 km) | | |
| 8.6 ± 0.9 | | | | | WWTP effluent (Achères) | | |
| 3.0 ± 0.6 | | | | | Downstream (1 km) | | |
| | | | | | SPAIN | | [29, 202, 203] |
| <LOD | | | | | WWTP effluent (Calaf) | | |
| <LOD – 7.6 | | | | | WWTP effluent (Igualada) | | |
| <LOD | NA | 6 | 5.0 | 93 | WWTP effluent (Piera) | SPE (C-18); LC-DAD-MS (SIM) | [29, 202] |
| <LOD | | | | | WWTP effluent (Manresa) | | |
| <LOD | | | | | Surface water | | |
| | | | | | USA | | [77, 119, 120] |
| 9 | 93 | 70 | 5 | 128.8 (d4-estradiol) | Surface water | CLLE; GC-MS (SIM) | [119, 120] |
| | | | | | CHINA | | |
| 1 | 2 | 2 | 1 | 113 ± 5 | Surface water (low-flow) | SPE (C-18); GC-MS (SIM) | [198] |

| Conc. (ng L$^{-1}$) | Max. (ng L$^{-1}$) | n | LOD (ng L$^{-1}$) | Recov. (%) | Sample Type | Form of Analysis | Refs. |
|---|---|---|---|---|---|---|---|
| | | | | | NETHERLANDS | | [43, 196] |
| <LOD | 7.5 | 6 | 0.3-1.8 | 96 ± 8 | WWTP effluent | SPE (SVB, C-18, amino); HPLC fractionation; GC-MS/MS | [43] |
| <LOD | 4.3 | | 0.1-0.3 | | Surface water | | |
| | | | | | GERMANY | | [116, 197, 199] |
| 1 | 15 | 16 | 1 | 76 ± 0 | WWTP effluent | SPE (C-18); silica gel cleanup; GC-MS/MS (MRM) | [116] |
| <LOD | <LOD | 15 | 0.5 | 85 ± 0 | Surface water | | |
| | | | | | FRANCE | | [200, 201] |
| 1.1 ± 0.1 | NA | 6 | NA | 84-98 (all estrogens) | Upstream (1 km) | SPE (C-18); GC-MS (SIM) | [200] |
| 2.7 ± 0.8 | | | | | WWTP effluent (Colombes) | | |
| 2.3 ± 0.7 | | | | | Downstream | | |
| 1.5 ± 0.5 | | | | | Upstream (1 km) | | |
| 4.5 ± 0.8 | | | | | WWTP effluent (Achères) | | |
| 2.9 ± 0.6 | | | | | Downstream (1 km) | | |
| | | | | | SPAIN | | |
| <LOD | NA | 6 | 5.0 | 94 | WWTP effluent (Calaf) WWTP effluent (Igualada) WWTP effluent (Piera) WWTP effluent (Manresa) | SPE (C-18); LC-DAD-MS (SIM) | [29, 202] |
| | | | | | USA | | [77, 119, 120] |
| 94 | 273 | 70 | 5 | 128.8 (d4-estradiol) | Surface water | CLLE; GC-MS (SIM) | [119, 120] |
| | | | | | CHINA | | |
| 0 | 1 | 1 | 1 | 98 ± 13 | Surface water (low flow) | SPE (C-18); GC-MS (SIM) | [198] |

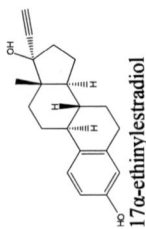

17α-ethinylestradiol

| Conc. (ng L⁻¹) | Max. (ng L⁻¹) | n | LOD (ng L⁻¹) | Recov. (%) | Sample Type | Form of Analysis | Refs. |
|---|---|---|---|---|---|---|---|
| | | | | | NETHERLANDS | | |
| 4.5 | 47 | 6 | 0.3-1 | 98 ± 14 | WWTP effluent | SPE (SVB, C-18, amino); HPLC fractionation; GC-MS/MS | [43, 196] |
| 0.3 | 3.4 | | 0.2-0.3 | | Surface water | | [43] |
| | | | | | GERMANY | | [116, 197, 199] |
| 9 | 70 | 16 | 1 | 82 ± 2 | WWTP effluent | SPE (C-18); silica gel cleanup; GC-MS/MS (MRM) | |
| <LOD | 1.6 | 15 | 0.5 | 90 ± 2 | Surface water | | [116] |
| | | | | | FRANCE | | [200, 201] |
| 1.2 ± 0.2 | NA | 6 | NA | 84-98% (all estrogens) | Upstream (1 km) | SPE (C-18); GC-MS (SIM) | [200] |
| 4.3 ± 0.6 | | | | | WWTP effluent (Colombes) | | |
| 2.2 ± 0.3 | | | | | Downstream | | |
| 1.1 ± 0.3 | | | | | Upstream (1 km) | | |
| 6.2 ± 0.8 | | | | | WWTP effluent (Achères) | | |
| 3.0 ± 0.9 | | | | | Downstream (1 km) | | |
| | | | | | SPAIN | | |
| <LOD – 8.1 | NA | 6 | 2.5 | 93 | WWTP effluent (Calaf) | SPE (C-18); LC-DAD-MS (SIM) | [29, 202] |
| <LOD – 2.7 | | | | | WWTP effluent (Igualada) | | |
| <LOD | | | | | WWTP effluent (Piera) | | |
| <LOD – 7.2 | | | | | WWTP effluent (Manresa) | | |
| 8.0 | | | | | Surface water | | |
| | | | | | USA | | [77, 119, 120] |
| 27 | 112 | 70 | 5 | 128.8 (d4-estradiol) | Surface water | CLLE; GC-MS (SIM) | [119, 120] |
| | | | | | CHINA | | |
| 34 | 65 | 8 | 0.5 | 118 ± 15 | Surface water (low-flow) | SPE (C-18); GC-MS (SIM) | [198] |

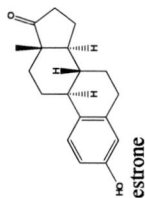

estrone

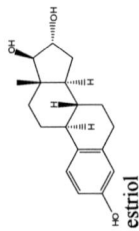
estriol

| Conc. (ng L⁻¹) | Max. (ng L⁻¹) | n | LOD (ng L⁻¹) | Recov. (%) | Sample Type | Form of Analysis | Refs. |
|---|---|---|---|---|---|---|---|
| | | | | | FRANCE | | [200, 201] |
| 1.0 ± 0.4 | | | | | Upstream (1 km) | | |
| 5.7 ± 1.6 | | | | | WWTP effluent (Colombes) | | |
| 2.1 ± 0.7 | NA | 6 | NA | 84-98% (all estrogens) | Downstream | SPE (C-18); GC-MS (SIM) | [200] |
| 1.5 ± 0.5 | | | | | Upstream (1 km) | | |
| 6.8 ± 0.6 | | | | | WWTP effluent (Achères) | | |
| 2.5 ± 0.6 | | | | | Downstream (1 km) | | |
| | | | | | SPAIN | | |
| 4.8 – 18.9 | | | | | WWTP effluent (Calaf) | | |
| <LOD – | | | | | WWTP effluent (Igualada) | | |
| 4.1 | NA | 6 | 0.25 | 85 | WWTP effluent (Piera) | SPE (C-18); LC-DAD-MS (SIM) | [29, 202] |
| 1.7 – 5.8 | | | | | WWTP effluent (Manresa) | | |
| 10.3 – 21.5 | | | | | | | |
| 6.3 | | | | | Surface water | | |
| | | | | | USA | | [77, 119, 120] |
| 19 | 51 | 70 | 5 | 128.8 (d4-estradiol) | Surface water | CLLE; GC-MS (SIM) | [119, 120] |
| | | | | | CHINA | | |
| 0 | 1 | 1 | 1 | 68 ± 6 | Surface water (low-flow) | SPE (C-18); GC-MS (SIM) | [198] |

# PROGESTAGENS

| | Conc. (ng L⁻¹) | Max. (ng L⁻¹) | n | LOD (ng L⁻¹) | Recov. (%) | Sample Type | Form of Analysis | Refs. |
|---|---|---|---|---|---|---|---|---|
| **norethindrone** | | | | | | **JAPAN** | SPE (HLB, silica); UPLC-MS/MS (MRM) | [194, 195] |
| | <LOD | NA | 2 | 0.60 | 82 ± 3.6 | WWTP effluent | | [194] |
| | <LOD | NA | 4 | 0.30 | 79 ± 7.6 | Surface water | | |
| | | | | | | **USA** | | [77, 119, 120] |
| | 48 | 872 | 70 | 5 | 148.5 (d3-testosterone) 116.9 (d7-cholesterol) | Surface water | CLLE; GC-MS (SIM) | [119, 120] |
| **progesterone** | | | | | | **JAPAN** | SPE (HLB, silica); UPLC-MS/MS (MRM) | [194, 195] |
| | 0.34 | NA | 2 | 0.26 | 100 ± 12 | WWTP effluent | | [194] |
| | 0.07 | NA | 4 | 0.02 | 83 ± 10 | Surface water | | |
| | | | | | | **SPAIN** | | |
| | <LOD | NA | 6 | 0.20 | 113 | WWTP effluent (Calaf) | SPE (C-18); LC-DAD-MS (SIM) | [29, 202] |
| | <LOD – 1.1 | | | | | WWTP effluent (Igualada) | | |
| | <LOD | | | | | WWTP effluent (Piera) | | |
| | 0.3 – 1.5 | | | | | WWTP effluent (Manresa) | | |
| | 4.3 | | | | | Surface water | | |
| | | | | | | **USA** | | [77, 119, 120] |
| | 110 | 199 | 70 | 5 | 148.5 (d3-testosterone) 116.9 (d7-cholesterol) | Surface water | CLLE; GC-MS (SIM) | [119, 120] |

$K_{ow}$ measures a chemical's distribution at equilibrium between water and 1-octanol (an organic solvent), in order to determine the chemical's relative hydrophobicity [57]. Hydrophobicity is important for many biological endpoints, including a chemical's ability to bind to steroid sex hormone receptors [31, 58]. Because measured values range over 12 orders of magnitude ($10^{-4}$ to $10^8$), $K_{ow}$ values are usually expressed as logarithms (log $K_{ow}$ or log P). Experimental log $K_{ow}$ values, and modeled log $K_{ow}$ values calculated by KOWWIN$^{TM}$ computer software, are set forth in Table 2 [59]. A prior version of the same program, which uses an atom/fragment contribution method to estimate log $K_{ow}$ values, predicted log $K_{ow}$ values within ±0.8 log units for over 96% of an experimental dataset of 8,406 compounds [57].

The steroid sex hormones' RBAs for androgen and estrogen receptors were determined using *in vitro* competitive binding assays [31, 58]. *In vitro* assays generally fall into three categories: (a) competitive binding assays that measure a chemical's relative binding affinity for steroid sex hormone receptors; (b) reporter gene assays that measure receptor binding-dependent biological activities; and (c) cell proliferation assays that measure cell proliferation upon exposure to steroid sex hormones (e.g., human breast cancer MCF-7 cells exposed to estrogens) [2, 60]. Each assay type measures different end points at different levels of biological complexity, but relative receptor activities tend to be consistent across different assay methods and species, at least with respect to estrogen receptors [2]. Estrogen receptor binding appears to be the major determinant across all three levels of biological activity, but structural and functional differences among receptors do exist across species [2, 21].

To calculate a test chemical's RBA, the test chemical is combined with hormone receptor proteins and a radiolabeled hormone standard to determine the test chemical amount that is required to prevent 50% of the hormone standard from binding to hormone receptors (inhibitory concentration or $IC_{50}$) [61]. The test chemical's RBA is then calculated according to the following formula [31, 58]:

$$RBA = \frac{IC_{50} \; steroid \; hormone \; standard}{IC_{50} \; test \; chemical} \times 100 \qquad (1)$$

17β-estradiol is commonly used as the hormone standard for estrogen receptor studies, and 17α-methyltrienolone (a potent androgen) has been used as the hormone standard for androgen receptor studies [31, 58]. Because measured values can range over several orders of magnitude, RBAs are commonly expressed as logarithms (log RBAs). By definition, a hormone standard has an RBA of 100, and a log RBA of 2. Greater numbers indicate that the test chemical has a hormone receptor binding affinity that is stronger than the hormone standard, and lesser numbers indicate a weaker binding affinity. For purposes of comparison, test chemicals can be divided into five categories: (a) strong binders (log RBA > 0); (b) moderate binders (−2 < log RBA < 0); (c) weak binders (log RBA < −2); (d) nonbinders that fail to compete with the steroid sex hormone standard (NBs); and (e) slight binders that fail to prevent 50% of the steroid sex hormone standard from binding (SBs) [31, 58].

Because chemicals with similar biological activities commonly share structural features, structure-activity relationships (SARs) have been studied to identify the molecular features that determine a chemical's biological activity. In an SAR analysis of estrogen activity, several structural features were found to be important for estrogen activity, using 17β-estradiol as a template: (a) hydrophobicity, because significant portions of the estrogen receptor are hydrophobic; (b) the presence of a hydrogen bond donor to mimic the phenolic A-ring common to estrogens; (c) the presence of a second hydrogen bond donor to mimic the alcohol group in the β position on carbon 17; (d) a distance between the hydrogen bond donors of approximately 9.7-12.3 Å (1 Å = $10^{-10}$ m); (e) a ring structure to increase rigidity; and (f) the presence of steric hydrophobic centers at the 7α and 11β positions, which must be able to fit in available cavities on the estrogen receptor [58]. For example, the molecular structure of estrone is identical to the molecular structure of 17β-estradiol, except that estrone has a ketone at carbon 17, instead of an alcohol group (Table 2). Ketones accept, but do not donate, hydrogen bonds. As a result, the RBA of estrone (7.3%) is approximately 14 times weaker than the RBA of 17β-estradiol (100%) [58]. Similar structural features, including the ability to form hydrogen bonds at carbons 3 and 17, also are important for androgen activity [31]. Among other things, information from SAR studies can be used to predict endocrine disruption risks posed by steroid sex hormone degradation products.

## ANALYSIS OF STEROID HORMONES IN ENVIRONMENTAL MATRICES

Analyzing environmental samples for trace levels of target compounds (e.g., ng $L^{-1}$) is challenging, because small amounts of target compounds must be extracted from large volumes or masses of sample material, and because environmental samples contain numerous potential interferences that can hide the presence of target compounds, mimic the presence of target compounds (generating "false positive" results), or cause overestimation or underestimation of target compound amounts. When target compounds are extracted and concentrated for analysis, potential interferences (e.g., natural organic matter) can be extracted and concentrated with them [40, 62]. For example, natural organic substances often remain after solid phase extraction (SPE) of natural waters, and give strong spectra during mass spectrometry (MS) across the 100-500 m/z range, where steroid sex hormones typically appear [63]. Therefore, considerable time is often spent developing extraction and cleanup methods to maximize the yields of target compounds, and eliminate the presence of interfering compounds [46].

After the sample has been prepared for analysis, some analytical instruments can target specific compounds to improve measurement sensitivity and selectivity, and thereby improve detection limits. For example, when mass spectrometry (MS) is used for analysis, selected ion monitoring (SIM) can improve the instrument's sensitivity and selectivity by focusing the instrument's resources on a narrow m/z range. When tandem mass spectrometry (MS/MS) is used, multiple reaction monitoring (MRM) can further improve selectivity by fragmenting m/z selected ions into characteristic fragments, which can be used to distinguish target compounds from interferences in the same m/z range [64-66]. In fact, many analytical instruments and methods are available to improve detection limits, but potentially useful information about non-target compounds is often lost when selective methods are employed to lower detection limits [63, 67].

Time-of-flight mass spectrometry (TOF-MS) and quadrupole orthogonal acceleration time-of-flight mass spectrometry (QTOF-MS/MS) provide full-scan data over a broad m/z range, and accurate mass measurements that can distinguish isobaric ions with the same nominal m/z value (e.g., a nominal m/z value of 156, but different exact mass m/z values of 156.0119 and 156.0773) [63, 68]. As a result, TOF-MS and QTOF-MS provide distinctive information about both target and non-target compounds, and help to identify new and unknown compounds [63, 69].

Any study on the presence of steroid sex hormones in the environment should be considered in the context of the analytical techniques that were used, and the compounds that were targeted.

# STEROID SEX HORMONES –
## SOURCES AND PRESENCE IN THE ENVIRONMENT

## 1. Sources and "Hotspots"

Steroid sex hormones have many natural sources. They are excreted continuously by vertebrates, and have been reported in many invertebrate groups and loblolly pine [37]. In addition, many microbial species can transform cholesterol and plant sterols into steroid sex hormones (e.g., plant sterols → androstenedione) [70, 71]. Other potential sources of steroid sex hormones include WWTPs, septic systems, concentrated animal feeding operations (CAFOs), agricultural operations, rangeland grazing, paper mills, and aquaculture [38, 42, 46, 47·52, 72-78]. Once steroid sex hormones enter the environment, they are subject to a variety of transport and removal processes (e.g., sorption, dilution, biodegradation, and photodegradation). Therefore, detectable and biologically relevant concentrations are likely to diminish with time and space, unless continuous inputs act to create steady-state conditions.

For these reasons, it is important to identify the "hotspots" where steroid sex hormones occur at biologically relevant concentrations, and to understand what happens over time as transport and removal processes take place [28, 76, 77, 79-83]. It is also important to understand the factors that may contribute to hormone hotspots, including such diverse factors as manure management and pharmaceutical disposal practices.

## 2. WWTPs

In a series of studies conducted in the U.K. from 1986 to 1989, after anglers casually observed the occurrence of intersex fish in WWTP lagoons, caged rainbow trout were placed in WWTP effluent, and subsequent measurements revealed increased vitellogenin concentrations in their plasma (500 to 100,000 times higher than fish maintained in spring or tap water) [84]. Increased vitellogenin concentrations were also observed in carp, but to a lesser extent. A separate study of seven U.K. WWTPs was conducted to identify and quantify the causes of estrogenic activity in treated wastewater [85]. Various analytical techniques (including a reporter gene assay, SPE, and HPLC) were used to isolate the estrogenic compounds from the effluent, and the resulting fractions were analyzed by GC-MS. The most active fraction (> 80% of the total effluent activity) was found to contain 17β-estradiol (1-50 ng $L^{-1}$), estrone (1-80 ng $L^{-1}$), and 17α-ethinylestradiol (up to 7 ng $L^{-1}$). Because the effluent came from urban sewage treatment plants, the estrogens were assumed to originate from humans.

Additional examples of steroid sex hormone concentrations in WWTP effluents are set forth in Table 3. The concentrations of steroid sex hormones in WWTP effluent are influenced by several factors, including the composition of WWTP influents, and the treatment processes used [29, 41, 82, 86-89].

## 3. Steroid Sex Hormones – CAFOs

Endocrine disruption has been observed in fathead minnows (*Pimephales promelas*) exposed to feedlot effluent, and female painted turtles (*Chrysemys picta*) in ponds near livestock pastures [90, 91]. In concentrated animal feeding operations (CAFOs), solid wastes are commonly separated, dewatered, and collected for application as fertilizers, and liquid wastes are collected in lagoons, diluted with irrigation water, and applied as fertilizers [74, 92, 93]. Lagoons function as holding reservoirs or anaerobic digesters, and livestock wastes are typically applied to land without additional treatment [93, 94].

In one study, whole lagoon effluents from swine, cattle, and poultry CAFOs were analyzed to determine concentrations of free estrogens and estrogen conjugates [94]. Lagoon samples were centrifuged, and separated into liquid and solid components. The liquid components were filtered through 1.2 µm glass fiber filters, and split into two components for separate analyses of four free estrogens (17α-estradiol, 17β-estradiol, estrone, and estriol) and 13-16 estrogen sulfate and glucuronide conjugates. The conjugate samples were treated by enzyme hydrolysis, and all liquid samples were preserved with formaldehyde. The free-estrogen samples were extracted with SPE, derivatized, and analyzed by GC-MS/MS, and MDLs ranged from 4 ng $L^{-1}$ (17α-estradiol) to 20 ng $L^{-1}$ (17β-estradiol). The conjugate samples were extracted with SPE, and analyzed by LC-MS/MS, and the LOD was determined from the lowest quantitation standard (1 ng $L^{-1}$). The solid components were freeze-dried, extracted (liquid extraction, sonication, and SPE), derivatized, and analyzed by GC-MS/MS.

In the swine and poultry primary lagoons, free estrogens were distributed as follows: estrone > estriol > 17α-estradiol > 17β-estradiol. By comparison, in the dairy operation (secondary lagoon, 10,000 cows), free estrogens were distributed as follows: 17α-estradiol > 17β-estradiol > estrone > estriol. Generally, swine and poultry excrete more 17β-estradiol than 17α-estradiol, but fecal bacteria can oxidize 17β-estradiol to estrone, and reduce estrone back to 17α-estradiol and 17β-estradiol (17β-estradiol ↔ estrone ↔ 17α-estradiol) [95]. Microbial interconversion also has been observed with 17β-trenbolone (17β-trenbolone ↔ trendione ↔ 17α-trenbolone) [96, 97]. Estrogen sulfate conjugates were detected in the swine sow, poultry, and dairy lagoons, but not in the swine finisher, swine nursery, and beef feedlot lagoons.

Primary lagoon estrogen concentrations appear to be directly related to the number of animals housed in the CAFO. In the swine sow primary lagoon, which housed 662 swine sows, mean estrogen concentrations were 9,940 ng $L^{-1}$ (estrone), 1,200 ng $L^{-1}$ (17α-estradiol), 194 ng $L^{-1}$ (17β-estradiol), and 6,290 ng $L^{-1}$ (estriol). In a large swine sow facility that housed more animals (19,920 gestating sows, 4,980 farrowing sows, and 100 boars), a related study observed mean estrogen concentrations equal to 17,400 ng $L^{-1}$ (estrone), 2,460 ng $L^{-1}$ (17β-estradiol), and 7,830 ng $L^{-1}$ (estriol) [98]. In the primary lagoon of a swine farrowing facility that housed fewer animals (100 sows and 15 boars), a separate study found mean estrogen concentrations equal to 81 ng $L^{-1}$ (estrone), 3 ng $L^{-1}$ (17β-estradiol), and 9.2 ng $L^{-1}$ (estriol) [99].

Another study examined the transport of steroid sex hormones from dairy CAFOs. Groundwater samples were collected from two dairy farms immediately prior to the irrigation season (June – August), and one month after the end of the irrigation system [74]. The groundwater samples were collected from 13 shallow groundwater monitoring wells (7-10 m), including 3 wells located downgradient from dairy waste lagoons, 4 wells located within a feedlot, 5 wells located downgradient from fields receiving regular manure applications, and 1 well located outside a dairy's influence (to serve as a control). Sandy to loamy sand soils caused high percolation rates, and groundwater ages ranged from a few days to 1-2 years, based on the depth where the water entered the well screen. Additional samples were collected from a deep aquifer well (> 25 m), a dairy waste lagoon, and surface water sites likely to be affected by agricultural operations, including sites upstream and downstream from dairy farms and irrigation canal discharge points, near tile drain pump discharge points, and in irrigation canals. The samples were prepared by various analytical techniques (including filtration, centrifugation, and SPE), derivatized, and analyzed by GC-MS/MS for 7 steroid sex hormones: 17β-estradiol, estrone, estriol, progesterone, medroxyprogesterone, testosterone, and androstenedione. For every 10 samples, one sample was spiked with 10 ng $L^{-1}$ of the steroid sex hormones, and recoveries of the steroid sex hormones were correlated with recoveries of 100 ng $L^{-1}$ mesterolone (used as a surrogate standard). Recoveries of the spiked analytes ranged from 56-85%, and method detection limits ranged from 0.1-0.2 ng $L^{-1}$.

Steroid sex hormone concentrations in the lagoon samples varied considerably between the two sampling dates. Estrone was the only steroid sex hormone detected on both sampling dates, and estrone concentrations varied by an order of magnitude (~65 ng $L^{-1}$ to 650 ng $L^{-1}$). Steroid hormones were detected in 7 of the 26 shallow groundwater samples, at concentrations significantly below levels detected in the lagoons (6 detections < 10 ng $L^{-1}$; all 7 detections < 20 ng $L^{-1}$). Estriol, androstenedione, and progesterone were not detected in any groundwater samples. Among the groundwater samples, there was no consistency in the hormones detected, or the wells where they were detected. In addition, no steroids were detected in the deep aquifer well, or in samples from the fields' tile drainage system, which are believed to represent a composite average of shallow groundwater throughout the dairy. In the surface water samples, estrone was most frequently detected (12 of 26 samples), with a maximum concentration of 17 ng $L^{-1}$. Testosterone was second most frequently detected (6 of 26 samples), with a maximum concentration of 1.9 ng $L^{-1}$. Estriol, androstenedione, and progesterone were not detected in any surface water samples.

It was suggested that steroid sex hormones are generally adsorbed or degraded over distances of 10-100 m when dairy wastewater infiltrates groundwater, because detections were sporadic and hormones were not detected in the tile drain samples. The sporadic detections were believed to result from preferential flow paths in the subsurface. Finally, runoff was suggested to be more important than leaching, because concentrations of steroid sex hormones in irrigation canal and river samples exceeded tile drain concentrations (e.g., estrone was detected in 13 of 26 surface water samples, and no tile drain samples) [74].

In another study, the transport of steroid sex hormones from a dairy waste lagoon to groundwater was studied at an Israeli dairy farm with approximately 60 dairy cows and 30 calves [100]. The facility used an unlined, earthen waste lagoon with an average depth of

0.5 m, and excess wastewater flowed directly into a dry creek. Two boreholes were drilled for sediment sampling and groundwater monitoring. The first borehole was drilled directly under the waste lagoon, after a portion of the lagoon was dried. The second borehole was drilled upgradient, in an agricultural field used for growing barley. Manure was not used to fertilize the agricultural field. Groundwater levels ranged from 42 m (agricultural field) to 47 m (waste lagoon) below the surface. Three types of sediment were found below the waste lagoon: (i) a layer from 0-6 m characterized by a high content of clay materials and organic matter; (ii) a transition layer of sandy loam from 6-8 m; and (iii) a layer > 8 m of sand with high calcareous content and low organic matter. Groundwater and sediment samples were prepared by various analytical techniques (including filtration, centrifugation, liquid extraction and SPE), and analyzed by radioimmunoassay for androgens (testosterone, and possibly dihydrotestosterone) and estrogens (17β-estradiol and estrone). The reported detection limit was 0.3 ng $L^{-1}$, and the reported hormone recoveries were 90% if sample concentrations were above 1.0 ng $L^{-1}$, and < 50% if sample concentrations were below 0.5 ng $L^{-1}$. Radioimmunoassays have been criticized for lack of specificity, but the estrogen antibody was reported to exhibit negligible cross-reactions with other steroids [101, 102]. Testosterone and estrogens were found deep in the sediment profile under the waste lagoon, and were generally absent from the agricultural field. Testosterone was detected down to the groundwater surface (47 m), and estrogen was detectable to a depth of 32 m. The results suggested that steroid sex hormones from the dairy farm could be transported throughout the vadose zone, and that a clay lagoon lining may be insufficient to protect groundwater from dairy farm leachates.

Another study examined the transport of 17β-estradiol and estrone from swine manure to tile drainage systems at two Danish field sites on structured loamy soil, in part to account for effects from preferential transport in structured soils [103]. At both sites, the uppermost meter of soil was heavily fractured and bioturbated, and the water table was located approximately 1-3 m below ground surface. The tile drains were located at an average depth of 1 m, and laterally spaced 12-18 m apart. The average soil temperature of the plow layer ranged from 7 °C in April to 17-18 °C in July. The swine manure slurry was applied in accordance with Danish regulations on dose and application methods. Drainage water samples were collected during storm flow events over the 12 month period following manure application. Water samples were immediately adjusted to pH 3, and frozen until analysis. The water samples were prepared for analysis by various techniques (including filtration, SPE, and silica gel cleanup), derivatized, and analyzed by GC-MS/MS. Manure samples were freeze-dried and frozen until analysis. The manure samples were prepared for analysis by various techniques (including mechanical homogenization, pressurized liquid extraction, centrifugation, SPE, and silica gel cleanup), derivatized, and analyzed by GC-MS/MS.

Estrone and 17β-estradiol leached from the root zone into tile drains at both field sites. Approximately one-half of the 27 samples had detectable estrone concentrations (maximum = 68.1 ng $L^{-1}$). At one site (Estrup), rapid leaching of estrone and 17β-estradiol occurred during the first storm event (2 days at 23 and 16 mm $d^{-1}$) at concentrations as high as 0.9 ng $L^{-1}$ (estrone) and 0.2 ng $L^{-1}$ (17β-estradiol). Thereafter, 17β-estradiol leached on a few occasions, and estrone was detected up to 11 months after the manure applications. At the other site (Silstrup), a storm event (26 mm $d^{-1}$) induced estrone leaching at a concentration of 68.1 ng $L^{-1}$, and subsequent events did not cause major estrone leaching. 17β-estradiol leached on one occasion at a concentration of 1.8 ng $L^{-1}$. Among other things, the study demonstrated that preferential transport and soil temperatures may significantly influence steroid sex hormone leaching to groundwater.

## 4. Steroid Sex Hormones – Biosolids

Biosolids (treated sewage sludge) are also applied to agricultural fields as fertilizers and soil amendments. Biosolids improve soil physical properties, and add important nutrients including nitrogen and phosphorous [104]. Estrogen sorption to activated and inactivated sewage sludge has been observed, and both androgenic and estrogenic activities have been detected in municipal biosolids [105-107]. In general, hormone activities were substantially higher after anaerobic digestion (mean estrogen: 1,233 ng $g^{-1}$ dry weight; mean androgen: 543 ng $g^{-1}$ dry weight) than after aerobic digestion (mean estrogen: 11.3 ng $g^{-1}$ dry weight; androgen: < LOD) [107].

During a survey of organic wastewater contaminants in nine biosolids products destined for land application, estrone was detected in one product at a concentration of 150 µg $kg^{-1}$ [104] (GC-MS; full scan). No MDL was reported for estrone, but the reported MDLs (n = 4) for other steroids ranged from 168 µg $L^{-1}$ (cholesterol) to 367 µg $L^{-1}$ (stigmastanol), while the reported MDLs (n = 19) for non-steroidal pharmaceuticals ranged from 0.76 µg $L^{-1}$ (acetaminophen) to 5.5 µg $L^{-1}$ (gemfibrozil). Estrone might have been detected more frequently, and other steroid sex hormones might have been detected, if the method had been targeted to steroid sex hormones.

It is possible that steroid sex hormones will be transported like non-steroidal pharmaceuticals, which have been detected in drainage and runoff after land applications of biosolids (no steroids were tested) [108, 109].

## 5. Runoff and Leaching from Agricultural Operations – Other Studies

Other studies have determined that runoff and leaching from agricultural operations can be an important source of steroid sex hormones in ground and surface waters [76, 110-112]. Many factors appear to influence steroid sex hormone runoff and leaching from agricultural operations, including soil type, soil structure, precipitation amounts, soil temperature, sorption to colloids, and irrigation with wastewater [113-115]. Studies that fail to take these factors into account are unlikely to accurately describe steroid sex hormone transport.

## 6. Presence

Studies throughout the world have examined receiving waters for the presence of steroid sex hormones [29, 44, 45, 86, 89, 116-118]. Data from selected studies are set forth in Table 3 to illustrate the ubiquitous presence of steroid sex hormones around the world.

In one such study, water samples were collected from 5 WWTPs and 11 coastal and freshwater locations throughout the Netherlands [43]. The surface water locations were part of a water quality monitoring program, including river sample locations downstream from densely populated and heavily industrialized areas, and coastal sample locations in areas dominated by agriculture. Various analytical procedures (including SPE, HPLC fractionation, and GC-MS/MS) were used to determine the presence of 17β-estradiol, 17α-estradiol, estrone, 17α-ethinylestradiol, and their glucuronide conjugates. Glucuronide conjugates are not amenable to GC because of their low volatility and thermal instability [45]. Therefore, duplicate samples were taken at the WWTPs and three of the surface water locations, and deconjugated overnight at 37 °C with β-glucuronidase enzyme. The glucuronide conjugates were determined indirectly by comparing untreated samples with enzyme-treated duplicates. Data from the study are set forth in Table 3. Generally, steroid sex hormone recoveries ranged from 88-98%. Separate tests indicated that the enzyme reaction was complete, but the recovery of estradiol-17-glucuronide was still only 59%. Limits of detection ranged from 0.1-2.4 ng $L^{-1}$. 17β-estradiol (median 0.9 ng $L^{-1}$) and estrone (median 4.5 ng $L^{-1}$) were detected in most effluent samples, and the highest concentration detected in any sample was 47 ng $L^{-1}$ (estrone). Hormone levels in the untreated samples and the enzyme-treated duplicates generally matched, suggesting that no glucuronide conjugates were present, even though humans excrete steroid sex hormones primarily as conjugates. In the surface water samples, estrone (median 0.3 ng $L^{-1}$) was detected most frequently, and the highest concentration detected in any sample was 5.5 ng $L^{-1}$ (17β-estradiol). In general, the study demonstrated that steroid sex hormones are ubiquitous in Dutch surface waters at low ng $L^{-1}$ concentrations.

In a national reconnaissance study conducted by the U.S. Geological Survey (USGS) from 1999 to 2000, water samples from 139 stream sites across the United States were analyzed to determine the concentrations of 95 selected organic wastewater contaminants, including 14 steroid compounds [119, 120]. Each stream site was sampled once, using standard width and depth integrating techniques to obtain representative samples of the stream waters. Five different analytical methods were used, based on the type of compound being analyzed (i.e., antibiotics, hormones, etc.). In the steroid method, steroid compounds were extracted from the water samples by CLLE, and the extracts were concentrated, derivatized, and analyzed by GC-MS. Data from the national reconnaissance study are set forth in Table 3. Cholesterol (84.3%), coprostanol (85.7%), and estriol (21.4%) were among the thirty most frequently detected compounds. In addition, cis-androsterone (14.3%), 17β-estradiol (10.0%), and norethindrone (12.8%) were detected in at least ten percent of the stream sites. Among the steroid compounds analyzed with the steroid method, cholesterol (a lipid commonly associated with animals, but synthesized by all eukaryotes) and coprostanol (a metabolite formed from the hydrogenation of cholesterol) were detected at the highest median concentrations (cholesterol: 0.83 µg $L^{-1}$; coprostanol: 0.088 µg $L^{-1}$). Stigmastanol (a plant steroid) experienced poor average recoveries (< 60%) under a different method, but its estimated median concentration was 2 µg $L^{-1}$. The median concentrations of all other steroid compounds ranged from 9 ng $L^{-1}$ (17β-estradiol) to 147 ng $L^{-1}$ (equilin). In general, the reconnaissance study demonstrated that, like the Netherlands, steroid compounds are ubiquitous in U.S. surface waters at low ng $L^{-1}$ concentrations.

# STEROID SEX HORMONES –
## SORPTION AND TRANSFORMATION PROCESSES

Once steroid sex hormones enter the environment, their fate is influenced by a variety of physical and transformation processes, including sorption to soils and sediments, microbial degradation, and abiotic transformation processes including photodegradation [55, 121]. Individually, these processes are complex, and not well understood. Collectively, these processes influence each other, complicating attempts to understand the environmental fate of steroid sex hormones.

## 1. Sorption

### a) In General

Sorption has been defined as "the accumulation of a substance or material at an interface between the solid surface and the bathing solution" [122]. Sorption can occur through a variety of mechanisms, including hydrophobic partitioning, hydrogen bonding, and nonspecific van der Waals interactions [123]. The actual mechanism is influenced by physical and chemical properties of the sorbate (e.g., a steroid sex hormone) and the solid-phase sorbent. For example, a finely divided solid, such as a clay particle (diameter < 2 μm), will have a high sorption capacity because its surface area is large relative to its volume [124]. Often, sorption to sand (50 μm < diameter < 2,000 μm) is relatively insignificant, and the sand fraction can be treated as diluting the sorptive capacity of clay (diameter < 2 μm) and silt (2 μM < diameter < 50 μm) [125]. In addition to particle size, the sorption of neutral hydrophobic contaminants to soil has been positively correlated with soil organic matter content [125, 126].

Sediments have been characterized as largely eroded soils that have been subjected to redispersion and particle-size fractionation by runoff and other water processes [125]. Because the properties of the parent soil and the dynamics of the specific stream, river, lake, or pond differ, the sediment within a given water compartment may contain a narrow range of particle sizes [125]. For example, suspended sediment in a river might be mostly clay, sediment from the middle of the river might be mostly sand, and sediment from the edge of the river might be mostly silt [125]. As a result, sorption will vary according to water compartment.

Sorption has the potential to affect the fate and transport of steroid sex hormones in the environment in various ways. Sorption to immobile soil components can inhibit leaching, and reduce bioavailability to microorganisms. On the other hand, sorption to mobile soil particles, such as clay or dissolved organic matter, can enhance steroid transport via runoff or leaching, and enhance bioavailability to solid phase bacteria (e.g., bacteria in biofilms) [124, 127-130]. Also, photodegradation can be inhibited if steroids diffuse into unreactive micropores and other microenvironments in soil particles or organic matter [131].

Methods for determining chemical sorption to soils and sediments include batch equilibrium experiments and column displacement studies [132]. In batch equilibrium experiments, multiple chemical concentrations are used to measure the effect of concentration on sorption at apparent equilibrium [132]. In column displacement studies, sorption is studied by introducing a chemical into a glass column that is packed with soil, sediment, or another solid-phase sorbent, flowing water through the column, and measuring the chemical's concentration at the column exit as a function of time or relative pore volume (the amount of water in the packed column when saturated) [133, 134].

Frequently, sorption to soils and sediments can be described by the Freundlich equation, which describes a nonlinear relationship between the sorbed chemical amount and the dissolved chemical amount at apparent equilibrium:

$$S = K_f C^{1/n} \qquad (2)$$

where: $S$ = amount of chemical sorbed to soil (mg kg$^{-1}$)

$K_f$ = Freundlich sorption coefficient

$C$ = equilibrium solution concentration (mg L$^{-1}$)

$n$ = index of nonlinearity

When n = 1, the Freundlich equation simplifies to a linear function, which indicates that the sorbed chemical amount is proportional to its solution concentration at apparent equilibrium:

$$S = K_d C \qquad (3)$$

where: $S$ = amount of chemical sorbed to soil (mg kg$^{-1}$)

$K_d$ = sorption coefficient

$C$ = equilibrium solution concentration (mg L$^{-1}$)

Generally, the sorption coefficient ($K_f$ or $K_d$) for a chemical will vary with the soil's physical and chemical properties. However, because soil organic matter acts as a quasi-solvent for many hydrophobic contaminants, the carbon normalized partition coefficient ($K_{oc}$) can be reasonably constant across soil and sediment types [126, 132, 135]:

$$K_{oc} = \frac{K_d}{f_{oc}} \qquad (4)$$

where: $K_{oc}$ = carbon normalized sorption coefficient

$K_d$ = sorption coefficient

$f_{oc}$ = fraction of total mass attributable to organic carbon

The $K_{oc}$ values of hydrophobic contaminants often correlate well with their octanol-water distribution coefficients ($K_{ow}$) [57, 125]. Although water solubility is a good estimator of $K_{oc}$ values for slightly soluble organic compounds, the $K_{oc}$ values of hydrophobic contaminants do not always correlate well with their water solubility values ($S_w$) [125, 136]. This result might be explained by differences between the partitioning and dissolution processes, but also might be explained by differences in the experimental conditions used to determine water solubility [50, 125].

As the organic carbon content of a soil or sediment decreases, other mechanisms, besides partitioning into soil organic matter, can have increasing influence on sorption. Such other mechanisms include sorption to mineral surfaces, and diffusion or intercalation into clay mineral structures [137].

Sorption often occurs in stages. Generally, rapid sorption occurs first, followed by slower sorption that might last days, weeks, or even months before equilibrium is reached [137, 138]. The different rates of sorption have been attributed to separate sorption mechanisms. For example, rapid sorption has been attributed to hydrophobic partitioning and sorption to external mineral surfaces, and slower sorption has been attributed to diffusion into clay mineral structures and condensed, organic matter [137-139]. The fraction that sorbs slowly tends to be inversely dependent on initial concentration, which suggests that slow sorption is more important at lower concentrations [138]. As a result, sorption experiments that presume rapid sorption may underestimate actual sorption [138].

### b) Sorption to Minerals

Some studies have examined the sorption of estrogens to selected minerals. One study used batch and continuous mode experiments to examine the sorption of 17β-estradiol to the iron oxyhydroxide goethite (BET surface area: $49.6 \pm 0.1$ m$^2$ g$^{-1}$), and the clay minerals kaolinite (BET surface area: $14.7 \pm 0.1$ m$^2$ g$^{-1}$), illite (BET surface area: $123.3 \pm 0.3$ m$^2$ g$^{-1}$), K-montmorillonite (BET surface area: $31.8 \pm 0.1$ m$^2$ g$^{-1}$) and Ca-montmorillonite (BET surface area: $31.8 \pm 0.1$ m$^2$ g$^{-1}$) [137]. To conduct the experiments, suspensions were created in glass reaction vessels for each mineral by adding the mineral to a 10 mM aqueous KNO$_3$ solution to obtain a surface area equivalent of 100 m$^2$ L$^{-1}$ of suspension, and by pre-equilibrating the suspension under nitrogen for 18-25 h at $25.0 \pm 0.5$ °C. The suspensions were spiked with 17β-estradiol in methanol to an initial concentration of 4.17 μM (~ 1.14 mg L$^{-1}$). Batch experiments were conducted by transferring aliquots of the suspension to 50 mL polyethylene centrifuge tubes, purging with nitrogen, and mixing at $25 \pm 2$ °C for the required time. Continuous experiments were conducted in the original reaction vessel by maintaining a temperature of $25.0 \pm 0.5$ °C, purging continuously with nitrogen, and sampling the suspension periodically. Kinetic experiments were performed in both modes at pH 4.5, 6.5, and 8.0 by sampling periodically for 72 h. Desorption experiments were conducted by equilibrating for 1 week, centrifuging the suspension, and resuspending the mineral paste in 10 mM aqueous KNO$_3$ solution or methanol. For each sample, aqueous 17β-estradiol was determined by centrifuging the sample and analyzing the supernatant by HPLC.

Initially, 17β-estradiol sorbed rapidly to all substrates. Sorption to goethite ceased after 30 minutes, but illite and kaolinite continued to sorb 17β-estradiol slowly over a three-day period, until 10-15% of the 17β-estradiol was sorbed. The montmorillonite samples sorbed 60% of the 17β-estradiol over the same period. Degradation at the mineral surface was ruled out because there was no evidence of breakdown products, so the slow sorption period was attributed to diffusion-controlled sorption into clay mineral structures. The montmorillonite samples were believed to have higher sorption capacities because their interlayers are more accessible, and X-ray diffraction studies of the montmorillonite samples confirmed that interlayer spacing (c-axis spacing) changed significantly when 17β-estradiol was present. The rapid sorption period was attributed to sorption on external surfaces. In methanol, 17β-estradiol desorbed completely from goethite, and desorption from kaolinite was greater than illite. In 10 mM aqueous $KNO_3$, the 17β-estradiol fraction desorbed from kaolinite was greater than illite and goethite. There was no evidence of desorption from the montmorillonite samples after 3 weeks. Sorption to goethite was believed to occur through weak interactions with uncharged surface hydroxyls, because sorption to goethite was greatest at mid pH (7.0), and desorption was complete in methanol. Sorption to the minerals' external surfaces was pH independent, and attributed to hydrophobic interactions (e.g., Van der Waals forces).

Another study examined the sorption of bisphenol A, 17α-ethinylestradiol, and estrone to goethite, kaolinite, and montmorillonite [140]. Like the prior study, suspensions were prepared by adding minerals to 10 mM aqueous $KNO_3$ solution to obtain a surface area equivalent of 100 $m^2 L^{-1}$ of suspension, and the suspensions were spiked with estrogens in methanol to an initial concentration of 3μM (817 µg $L^{-1}$). Continuous mode experiments were conducted in a nitrogen-purged glass reactor vessel pre-equilibrated for 20 h at 25.0 ± 0.5 °C, and batch experiments were conducted in nitrogen-purged glass centrifuge tubes mixed for the appropriate time at 25.0 ± 0.5 °C. Desorption experiments were conducted with 17α-ethinylestradiol, and kinetic experiments were performed in both modes at pH 4, 7, and 10. Separate experiments were also conducted to determine the effect of the flocculation state of suspended montmorillonite particles on sorption, because montmorillonite particles are believed to flocculate face-to-face at high pH, and edge-to-face at low pH.

Like 17β-estradiol in the prior study, 17α-esthinylestradiol and estrone sorbed rapidly to all substrates. Sorption to montmorillonite was slower, and occurred in two stages over a 48-hour period. The estrogens had a much higher affinity for montmorillonite than for goethite or kaolinite. In addition, sorption to montmorillonite increased steadily above pH 7, while sorption to goethite and kaolinite was pH independent. When pre-loaded goethite and kaolinite were resuspended in 10 mM aqueous $KNO_3$, approximately 80% of the 17α-esthinylestradiol desorbed within 30 minutes. A smaller fraction of the 17α-esthinylestradiol desorbed from montmorillonite at pH 4, and virtually none desorbed at pH 10 after 96 h. Like the prior study, the slow sorption period was attributed to diffusion-controlled sorption into clay mineral structures. Because sorption to goethite and kaolinite was pH independent and substantial desorption occurred in 10 mM aqueous $KNO_3$ solution, sorption to goethite and kaolinite was attributed to weak interactions with their external surfaces, and it was suggested that surface charge plays no significant role in the sorption process. Edge-to-face interactions of montmorillonite particles at lower pH were believed to reduce access to interlayer sites, inhibit sorption, and enhance desorption relative to montmorillonite at higher pH.

The influence of particle size (i.e., sand, silt, or clay) and mineral type (e.g., clay mineral or goethite) on estrogen sorption has been confirmed elsewhere [54]. It is clear that organic matter is not necessary for estrogen sorption, and that expanding clay minerals (e.g., montmorillonite) have a significant effect on the extent of sorption [54]. However, 17β-estradiol, estrone and, to a lesser extent, 17α-ethinylestradiol sorption has been shown to have a linear relationship with soil organic matter content [54]. Presumably, as organic matter content increases, the mineral influence on estrogen sorption and desorption will decrease. According to a study on pesticide sorption, clay and silt are primarily responsible for pesticide sorption when the soil organic matter content is less than 0.01%, organic matter is primarily responsible for pesticide sorption when the soil organic matter content is greater than 6-8%, and both are involved in the middle range [141]. The same idea may apply to other steroid sex hormones, even if the actual percentage amounts are different due to the hormone's individual physical and chemical properties.

Together, the preceding studies suggest that estrogen sorption is substantially influenced by particle size and mineral type, and that sorption to mineral surfaces is generally rapid, weak, and reversible. The studies also suggest that diffusion-controlled sorption into clay mineral structures is slow, and difficult to reverse. The same principles will likely apply to androgens and progestagens.

### c) Sorption to Colloids

Colloids are commonly defined as 1 nm to 1 μm sized particles, but larger particles also move with water through worm holes, plant root channels, and other soil macropores and fractures [124, 142]. To distinguish it from organic colloids, *dissolved* organic carbon can be defined as organic carbon that can pass through a molecular weight cutoff ultrafilter of a specified size (e.g., 1 kDa) [114]. Colloids are produced through microbiological processes and weathering, and commonly consist of clay particles, metal oxides, and organic material from manure, plant matter decomposition, and other biological sources [115, 124]. Relatively insoluble contaminants can precipitate as colloids, but sorption to organic or mineral colloids appears to be the most important method of colloidal transport [124]. Colloids have a high surface area in relation to their volume, and sorb contaminants through various mechanisms including hydrophobic partitioning, hydrogen bonding, London dispersion and other intermolecular forces [115, 123]. Physical and chemical properties, including size, determine whether colloids remain in suspension, or coagulate and settle as sediment [143].

Humic substances represent a significant part of organic colloids [144]. Humic substances are heterogeneous mixtures of various organic compounds, and are ubiquitous components of soils, sediments and surface waters [145]. Ordinarily, humic substances are divided into three groups based on their solubility in acids and bases [146]. Fulvic acids are smaller molecular weight molecules (~2 kDa), which are soluble in acids and bases. Humic acids are larger molecular weight molecules (~5-100 kDa), which are soluble only in bases. Humin describes the largest molecular weight molecules (~300 kDa), which are insoluble in acids and bases.

Several studies have examined steroid hormone sorption to colloidal material. However, the process of separating dissolved, colloidal, and particulate phases is challenging. In fact, the "conventional" dissolved phase, after filtration, often includes a mixture of dissolved and colloidal phases. Various methods have been used to determine the distribution of steroid hormones between dissolved and colloidal phases, and they vary in terms of accuracy, time efficiency, cost, and ease of use.

In one study, batch equilibrium experiments were used to investigate the sorption of 17β-estradiol, 17α-ethinylestradiol, estriol, and other suspected endocrine disruptors to seven representative organic colloids [127]. Aldrich humic acid, Suwannee River humic acid, Suwannee River fulvic acid, and Nordic fulvic acid were selected because humic acid and fulvic acid have been estimated to comprise 6-8% and 54-72%, respectively, of the total organic carbon in typical rivers and streams. To model aquatic humic acids better, the peat-based Aldrich humic acid was filtered with a 10 kDa dialysis membrane to eliminate the high molecular fraction, which is believed to cause excess sorption relative to aquatic humic acids. Alginic acid and dextran were selected as representative polysaccharides (a major component of microbial biofilms), which have been estimated to comprise 6-12% of the total organic carbon in typical rivers and streams. Finally, tannic acid (gallotannin) was selected to model plant residues. The batch equilibrium experiments were conducted in glass centrifuge tubes using four concentrations of the organic colloids, which ranged from 2-10 mg C L$^{-1}$ in accordance with natural organic carbon levels. The initial estrogen concentration in each centrifuge tube was 700 μg L$^{-1}$, a concentration significantly below each estrogen's reported water solubility limit (Table 2). Each solution was adjusted to a pH of 7 and an ionic strength of 0.2 M, and mixed for approximately 24 hours at room temperature until equilibrium was reached. Fluorescence quenching was used to measure estrogen sorption to the organic colloids, but only for estrogens that would not fluoresce when sorbed or that would fail to fluoresce in the presence of dissolved oxygen (a potential "quencher"). When the fluorescence quenching technique was used, organic carbon normalized partition coefficients (K$_{oc}$) were calculated according to the following formula:

$$\frac{F_o}{F} = 1 + K_{oc} \cdot [colloids]$$

(5)

where: $F_o$ = fluorescence without organic colloids

$F$ = fluorescence with organic colloids

[colloids] = the colloid concentration

Because 17β-estradiol and p-nonylphenol did fluoresce when sorbed to alginic acid and dextran, a more complex solubility enhancement technique was used to measure their aqueous concentrations. After analyzing selected compounds in Suwannee River humic acid and tannic acid under both techniques, the fluorescence quenching technique was believed to overestimate K$_{oc}$ values (Table 4), but deemed acceptable for providing easy K$_{oc}$ estimates.

Log K$_{oc}$ values from the batch equilibrium experiments are set forth in Table 4. According to the experimental data, estrogens sorbed to the organic colloids in the following order: tannic acid > humic acids > fulvic acids > polysaccharides. The experimental log K$_{oc}$ values for alginic acid were moderately correlated with the estrogens' reported octanol-water partition coefficients (log K$_{ow}$), which suggested that hydrophobic partitioning is important for estrogen sorption to polysaccharides. Otherwise, no trend was evident between the experimental log K$_{oc}$ values and reported log K$_{ow}$ values, suggesting that other mechanisms were also involved. No trends were evident between the experimental log K$_{oc}$ values and the organic colloids' molecular weight averages, carboxylic group concentrations, or elemental ratios (e.g., H/O or (O+N)/C). However, the experimental log K$_{oc}$ values did correlate with the phenolic group concentrations in the organic colloids, and with the estrogens' ability to absorb ultraviolet light at 272 nm, a wavelength associated with π electrons and aromaticity [147]. Together, these data suggest that π electron interactions play an important role in estrogen sorption to organic colloids. According to the experimental data, estriol sorbed to organic colloids more than any other estrogen, even though estriol was least hydrophobic estrogen tested (smallest K$_{ow}$ value). No mechanism was suggested, but estriol's sorption might be explained by its ability to donate more hydrogen bonds to acceptor sites in the organic colloids, based on the number of hydroxyl functional groups (−OH) in estriol, and the fact that carbonyl groups (C=O) account for a significant fraction of the carbon atoms in organic colloids [148]. In two additional, related studies, cross-flow ultrafiltration and batch equilibrium experiments were used to investigate the distribution of 17β-estradiol, 17α-ethinylestradiol, estrone, 16α-hydroxyestrone, and other compounds between dissolved and colloidal phases in river water, treated effluent, and seawater [115, 142].

**Table 4. Organic carbon normalized sorption coefficients ($K_{oc}$) for the sorption of selected steroid sex hormones to various colloids, the forms of analysis used (FQ = fluorescence quenching; SE = solubility enhancement; CFF = cross-flow ultrafiltration), and octanol-water partition coefficients ($K_{ow}$) for comparison**

| | log $K_{ow}$ | log $K_{oc}$ | Colloid Type (Form of Analysis) | Organic Carbon | Ref. |
|---|---|---|---|---|---|
| **ESTROGENS** | | | | | |
| 17β-estradiol | 4.01 (3.94) | 4.94 4.92 4.56 4.57 4.61 3.75 2.76 5.28 4.94 | Aldrich humic acid (FQ) Suwannee River humic acid (FQ) Suwannee River humic acid (SE) Suwannee River fulvic acid (FQ) Nordic fulvic acid (FQ) Alginic acid (SE) Dextran (SE) Tannic acid (FQ) Tannic acid (SE) | 100% | [127] |
| | 3.94, 4.01 | 3.98 3.94 4.04 3.85 4.86 | Longford stream colloids (CFF) River Ouse colloids (CFF) Horsham STW effluent colloids (CFF) River L'Aa colloids (CFF) Seawater colloids (CFF) | 2.3 mg L$^{-1}$ 3.0 mg L$^{-1}$ 9.2 mg L$^{-1}$ 2.9 mg L$^{-1}$ 0.4 mg L$^{-1}$ | [115] |
| 17α-ethinylestradiol | 3.67 (4.12) | 4.78 4.80 4.55 4.63 3.23 3.04 5.22 | Aldrich humic acid (FQ) Suwannee River humic acid (FQ) Suwannee River fulvic acid (FQ) Nordic fulvic acid (FQ) Alginic acid (SE) Dextran (SE) Tannic acid (FQ) | 100% | [127] |
| | 4.15, 3.67 | 4.58 4.81 4.73 4.85 5.47 | Longford stream colloids (CFF) River Ouse colloids (CFF) Horsham STW effluent colloids (CFF) River L'Aa colloids (CFF) Seawater colloids (CFF) | 2.3 mg L$^{-1}$ 3.0 mg L$^{-1}$ 9.2 mg L$^{-1}$ 2.9 mg L$^{-1}$ 0.4 mg L$^{-1}$ | [115] |

**Table 4. (Continued)**

| | log $K_{ow}$ | log $K_{oc}$ | Colloid Type (Form of Analysis) | Organic Carbon | Ref. |
|---|---|---|---|---|---|
| estriol | 2.45 (2.81) | 4.99 4.96 4.64 NA NA NA 5.32 | Aldrich humic acid (FQ) Suwannee River humic acid (FQ) Suwannee River fulvic acid (FQ) Nordic fulvic acid (FQ) Alginic acid (SE) Dextran (SE) Tannic acid (FQ) | 100% | [127] |
| estrone | 3.43, 3.13 | 4.85 4.83 4.30 4.67 5.04 | Longford stream colloids (CFF) River Ouse colloids (CFF) Horsham STW effluent colloids (CFF) River L'Aa colloids (CFF) Seawater colloids (CFF) | 2.3 mg L$^{-1}$ 3.0 mg L$^{-1}$ 9.2 mg L$^{-1}$ 2.9 mg L$^{-1}$ 0.4 mg L$^{-1}$ | 115 |
| 16α-hydroxyestrone | 3.23 | 4.20 3.75 3.88 4.41 4.67 | Longford stream colloids (CFF) River Ouse colloids (CFF) Horsham STW effluent colloids (CFF) River L'Aa colloids (CFF) Seawater colloids (CFF) | 2.3 mg L$^{-1}$ 3.0 mg L$^{-1}$ 9.2 mg L$^{-1}$ 2.9 mg L$^{-1}$ 0.4 mg L$^{-1}$ | [115] |

During cross-flow ultrafiltration, samples are passed through an ultrafilter membrane, and suspended colloids with molecular weights exceeding the pore size are concentrated in the retentate (i.e., the sample portion that fails to pass through the ultrafilter membrane). For the batch equilibrium experiments, water samples from five locations were filtered through 0.7 μm glass fiber filters, and then passed through a 1 kDa ultrafilter membrane during cross-flow ultrafiltration. The resulting retentate contained colloids ranging in size from 1 kDa to 0.7 μm, which were concentrated 15-18 times relative to the original samples. Four different colloid concentrations were prepared from each sample by diluting the retentate with the < 1 kDa permeate (i.e., the sample portion that did pass through the ultrafilter membrane), and 5 L test samples were prepared by spiking the colloid solutions with estrogens to an initial concentration of 600 ng L$^{-1}$. Various mixing times of up to 1 month were used to equilibrate the solutions, and 0.05% sodium azide was added to samples mixed for more than 5 days to inhibit biodegradation. After mixing, cross-flow ultrafiltration was used to separate estrogens in the < 1 kDa permeate from estrogens sorbed to organic colloids in the retentate. Then, SPE and GC-MS were used to isolate and quantify the estrogens in both phases. Partition constants ($K_p$) were calculated with the following formula, and divided by the colloidal organic carbon fraction ($f_{OC}$) to determine the organic carbon normalized partition constants ($K_{oc}$):

$$\frac{C_r}{C_f} = 1 + K_{oc} \text{ colloids} \tag{6}$$

where:  $C_r$ =  steroid hormone concentration in the retentate

$C_f$ =  steroid hormone concentration in the permeate

colloids  =  the colloid concentration

Log $K_{oc}$ values from the batch equilibrium experiments are set forth in Table 4. In general, the ratio of estrogens in the retentate to estrogens in the permeate increased as colloid concentrations increased, suggesting sorption to the colloids. According to the experimental data, 17α-ethinylestradiol sorbed to the colloids more than any other estrogen. The experimental data suggested that hydrophobic partitioning was not the only sorption mechanism involved, both because the experimental log $K_{oc}$ values for the effluent and river colloids were similar, despite differences in organic carbon content, and because no trend was evident between the experimental log $K_{oc}$ values and reported log $K_{ow}$ values. The experimental log $K_{oc}$ values did correlate with the colloids' ability to absorb ultraviolet light at 280 nm, suggesting again that π electron interactions play a substantial role in estrogen sorption to colloids.

In the same studies, additional experiments were conducted with water samples collected upstream, downstream, and near the outfall of a WWTP in West Sussex, UK, to determine the distribution of 17β-estradiol, 17α-ethinylestradiol, estrone, 16α-hydroxyestrone, and other compounds between dissolved and colloidal phases in field samples. Water samples were collected in 50 mL stainless steel barrels for cross-flow ultrafiltration, and separate samples were collected in 2.5 L glass bottles to analyze the "conventional" dissolved phase (a combination of dissolved and colloidal phases after filtration). Various analytical procedures (including SPE and GC-MS) were used to determine the estrogens' presence in the dissolved, colloidal, and conventional dissolved phases. Upstream, the mean conventional dissolved concentration of 17β-estradiol was approximately 7 ng $L^{-1}$, and the mean conventional dissolved concentrations of the other estrogens were significantly lower. At the outfall, mean conventional dissolved concentrations of the estrogens were distributed as follows: estrone (26.5 ng $L^{-1}$) > 17β-estradiol (22.5 ng $L^{-1}$) > 17α-ethinylestradiol (< 1 ng $L^{-1}$) ≈ 16α-hydroxyestrone (< 1 ng $L^{-1}$). Downstream, the mean conventional dissolved concentrations were substantially reduced (e.g., estrone = 3.9 ng $L^{-1}$). After the dissolved and colloidal phases were separated by cross-flow ultrafiltration, the mean percentages of colloid-bound estrogens (i.e., estrogens sorbed to colloids, but present in the conventional dissolved phase) were distributed as follows: 16α-hydroxyestrone (20-33%); 17β-estradiol (15-30%); 17α-ethinylestradiol (20-29%); and estrone (4-26%).

Other studies also have investigated steroid hormone sorption to colloids [114, 145]. In general, log $K_{oc}$ values and log $K_{ow}$ values often are not well-correlated in studies of estrogen sorption to colloids, which suggests that estrogen sorption to colloids is influenced by factors in addition to hydrophobic partitioning with organic matter. The same principles also may apply to androgens and progestagens.

### d) Sorption to Soils and Sediments

One study characterized sediments as major sinks of steroidal estrogens in river systems [149]. In that study, a yeast-based reporter gene assay was used to measure the estrogenic activities of effluents from two WWTPs in the U.K., and to measure the estrogenic activities of water and sediments from up to 1.5 km upstream and downstream of the WWTPs, along the Rivers Arun and Ouse (England). Various analytical methods (including SPE and liquid extraction) were used to extract estrogens from the waters and sediments. To identify the sources of estrogenic activity, sample extracts were separated into fractions with HPLC and analyzed by GC-MS. The estrogenic activities of the WWTP effluents ranged from 1.4-2.9 ng EEq $L^{-1}$, and the estrogenic activities of the river waters upstream and downstream from the WWTPs were below the limits of detection (0.04 ng $L^{-1}$). The estrogenic activities of the sediments were substantially higher than the overlying water, ranging from 21-30 ng EEq $kg^{-1}$. Estrone and 17β-estradiol were identified as the major active chemicals in the effluents and the sediments. No significant differences in estrogenic activity were observed between the sediments collected upstream and downstream from the WWTPs. Long distance transport was suggested as the source of estrogenic activity in the upstream sediments, because both upstream sites were located approximately 5 km downstream from other WWTPs. A related study reported 1-7 day biodegradation half-lives for 17β-estradiol and estrone at 20 °C (a common summer water temperature for the area), and direct photodegradation half-lives of not less than 10 days for 17β-estradiol and 17α-ethinylestradiol (using a polychromatic lamp equipped to simulate natural sunlight) [150]. Under average flow conditions, 17β-estradiol and estrone were estimated to travel 10 km in a few hours, so significant concentrations were expected to remain after transport for 5 km.

Other studies have sought to determine the rate and extent of steroid sex hormone sorption to soils and river sediments. One study examined the sorption of 17β-estradiol, estrone, estriol, 17α-ethinylestradiol, and mestranol to surficial bed sediments from one river (Thames) and one estuary (Blackwater) in the U.K. [55]. The total organic carbon content of the sediments ranged from 0.3-3.3%, and particle sizes were distributed as follows: sand (0-1.6%); silt (61-93%); and clay (7-39%). Batch equilibrium experiments were conducted in 250 mL Teflon bottles containing 200 mL of deionized water, 3 g of sediment, and a mixed estrogen standard containing 100 μg $L^{-1}$ of each estrogen. The bottles were mixed until apparent equilibrium, and centrifuged. Then, 1 mL samples were extracted, derivatized, and analyzed by GC-MS. Separate studies were conducted to examine (i) sorption rates, (ii) the effects of changing estrogen concentrations (10-1,000 μg $L^{-1}$) and sediment amounts (0.6-15 g), and (iii) competition with 100 μg $L^{-1}$ estradiol valerate (log $K_{ow}$ = 6.41).

Data from preliminary rate experiments with one estuarine suggested that estrogen sorption occurred in three stages. Rapid sorption occurred during the first half hour, followed by slower sorption for up to 1 hour, and then desorption. Based on these experiments, apparent equilibrium was believed to occur after 1 hour. The apparent decrease in sorption was attributed to estrogen sorption to dissolved organic matter released into solution. In general, slow sorption into clay mineral structures and condensed organic matter occurs by diffusion, and requires more than 1 h to occur [137, 140]. Therefore, the batch equilibrium experiments probably fail to take diffusion-controlled slow sorption into account.

According to the data, at apparent equilibrium, no linear relationship existed between estrogens sorbed to the sediments and estrogens remaining in solution. Because estrogen sorption increased when larger amounts of sediment were added, saturation of the available sediment might explain the nonlinear relationship. In the presence of 100 μg L$^{-1}$ estradiol valerate, the sorption of other estrogens was suppressed, and the magnitude of suppression grew with decreasing estrogen hydrophobicity (estriol > estrone > 17β-estradiol > 17α-ethinylestradiol > mestranol), suggesting that competitive sorption does occur among estrogens. Finally, the relationship between sorption and organic matter content was shown to be linear, which suggested that specific interactions were unimportant during the 1 hr equilibration period, despite nonlinear isotherms and relatively low organic matter content (0.3-3.3%).

Another study examined 17β-estradiol and 17α-ethinylestradiol sorption to suspended and bed sediments from three rivers (Aire, Calder, and Thames) and two estuaries (Tyne and Tees) in the U.K. [121]. The Thames River sampling sites were located in a rural area, and the other sampling sites were located near urban and industrial areas. Bulk water samples were collected in 1 L bottles at all sites for use in experiments with sediments from the same sites. Midstream bed sediments were collected from the two estuaries and one Thames River site with a mechanical grab, and all other bed sediment samples were collected near river banks by removing the top 2-5 cm of bed material. A continuous flow centrifuge was used to collect suspended sediment samples from the Aire, Calder, and Thames Rivers. Batch equilibrium experiments were conducted to determine 17β-estradiol and 17α-ethinylestradiol sorption to bed and suspended sediments at apparent equilibrium, and continuous mode experiments were conducted to study the rate of 17β-estradiol sorption to bed sediments. Suspensions were prepared in PTFE centrifuge tubes by mixing filtered river water with air-dried sediment and [$^{14}$C]-labeled estrogens to reach desired concentrations. For the continuous mode experiments, 15 mL of filtered river water was mixed with 1.0 g of sediment, and spiked with [$^{14}$C]-labeled 17β-estradiol for an initial concentration of 5 μg L$^{-1}$. The samples were placed in a 2.5 L anaerobic jar, mixed at room temperature, and sampled after 1, 2, and 6 days. For batch equilibrium experiments with bed sediments, 15 mL of filtered river water was mixed with 1-5 g of sediment (based on expected sorption), and spiked with [$^{14}$C]-labeled estrogens to initial concentrations of 0.5-10 μg L$^{-1}$. After 20 h, samples were collected and centrifuged, and the supernatant was prepared for liquid scintillation counter analysis. In some samples, the original river water was replaced with fresh river water, equilibrated for another 20 h, and examined for desorption. For batch equilibrium experiments with suspended sediments, 5 mL suspensions were prepared at 0.9-54 g L$^{-1}$ sediment concentrations (concentration factor of 60-1,550), and spiked with [$^{14}$C]-labeled estrogens to initial concentrations of 1.5-10 μg L$^{-1}$. After 1 h, the suspended sediment samples were analyzed like the bed sediment samples. In all cases, sorbed amounts were determined by subtracting the radioactivity remaining in solution from the initial radioactivity.

According to continuous mode experiments with bed sediments from the Thames and Aires Rivers, 80-90% of estradiol sorption occurred within the first day, but apparent equilibrium did not occur within two days. Because no steps were taken to inhibit biodegradation, it was acknowledged that no "true equilibrium" could be reached, and that any sorption coefficient calculated with reference to [$^{14}$C]-labeled 17β-estradiol (before biodegradation) would represent the sorption of 17β-estradiol and its metabolites (e.g., estrone).

An apparent equilibrium period of 20 h was chosen to represent a compromise between incomplete sorption and biodegradation. The resulting sorption isotherms were essentially linear (mean $1/n$ = 0.97), and no significant difference was found between sorption and desorption coefficients. However, the chosen 20 h equilibrium period may have eliminated the effects of slow sorption and desorption, which generally give rise to nonlinear isotherms. The sorption coefficients for 17α-ethinylestradiol ranged from 1.6 to 3.1 times higher than 17β-estradiol, indicating that 17α-ethinylestradiol sorbed more than 17β-estradiol. Desorption coefficients were 1.5 to 3.2 times higher than the original sorption coefficients, suggesting either that desorption occurs more slowly than sorption (i.e., due to the requirement for activation), or that some degree of hysteresis was occurring. A weak correlation was observed between sorption coefficients and the bed sediments' organic carbon content (0.1-10% organic carbon), suggesting that estrogen sorption to the bed sediments was influenced by factors in addition to hydrophobic partitioning with organic matter. In addition, higher sorption coefficients were associated with smaller particle sizes (i.e., sorption coefficients were negatively correlated with sand content). As a consequence, sorption would be strongest in water compartments with the lowest flows (e.g., along the banks), where smaller particle sizes are found. In general, the suspended sediments had higher sorption coefficients than the bed sediments, but they were determined to be much less important for steroid sex hormone removal because of their low concentrations (7.6-52 mg $L^{-1}$).

Additional studies have used batch and continuous mode experiments to determine the distribution of steroid sex hormones between water and various soils and sediments, and to examine sorption rates [53, 54, 56, 133, 134, 151-157]. Differences among sorption coefficients ($K_f$), indexes of linearity ($1/n$), and rates of sorption often can be attributed to differences in organic matter content, mineral types, and experimental conditions (e.g., the steroid and sediment concentrations used). For example, one study found that apparent equilibrium occurred in 1-2 weeks when initial hormone concentrations were 10,000 µg $L^{-1}$, and 2-3 weeks when initial concentrations were 300 µg $L^{-1}$ [56]. These concentrations are extremely high relative to environmental concentrations, but they do suggest that equilibrium times may vary with concentration. The same study also reported that sorption coefficients increased as steroid sex hormone concentrations decreased (i.e., smaller concentrations sorbed more), and that the order of androgen sorption changed with soil type (e.g., testosterone sorbed to agricultural soils more than androstenedione, and androstenedione sorbed to pond and creek sediments more than testosterone) [56].

## 2. Biodegradation

### a) In General

Bacteria can use steroids in redox (reduction/oxidation) reactions to gain energy, or metabolize the steroids completely as a carbon source for cell growth [158, 159]. In redox reactions, bacteria need an electron donor as the energy source, and an external electron acceptor to complete the respiration process. Organic carbon sources commonly serve as electron donors, and the available electron acceptor that will yield the most energy to the bacteria commonly serves as the electron acceptor. Under aerobic conditions, where sufficient oxygen is available, oxygen serves as the electron acceptor. Under anaerobic conditions, where sufficient oxygen is not available, bacteria generally use available electron acceptors in the following order: $NO_3^-$; Mn(IV); Fe(III); and $SO_4^{2-}$ [158]. As an example, during activated sludge treatment in WWTPs, 17β-estradiol is commonly oxidized to estrone under aerobic and anaerobic conditions [41, 160]. In nitrifying activated sludge, ammonia ($NH_4^+$) is oxidized to form nitrites and nitrates, and steroids appear to be oxidized co-metabolically [161-163].

Microorganisms are capable of transforming steroid compounds in various ways. For this reason, since the 1950's, the pharmaceutical industry has engaged in research to identify microorganisms and mechanisms that can assist in the production of steroidal drugs and hormones, as previously reviewed [159, 164-167]. Typically, natural plant and animal compounds, including cholesterol and plant sterols (e.g., stigmasterol, β-sitosterol, and campesterol), are used as the starting materials for steroidal drugs and hormones [164]. For example, several microbial species (including species of *Mycobacterium*, *Arthrobacter*, *Rhodococcus*, *Escherichia*, *Nocardia*, *Pseudomonas*, and *Micrococcus*) can be used to produce androstenedione, androstadienedione, and their derivatives from cholesterol and plant sterols [71, 164, 166]. In addition, species of *Bacillus* and *Nectria* have been used to transform progesterone into androstenedione and androstadienedione, and species of *Mycobacterium* and *Lactobacillus* have been used to transform cholesterol into testosterone [70, 164, 166]. Because several of these species have been isolated from soils (e.g., *Mycobacterium*, *Bacillus*, and *Micrococcus* species), it is reasonable to assume that the same reactions can happen in natural systems.

Actinobacteria are Gram-positive bacteria that are common to soil, and include genera such as *Arthrobacter*, *Micrococcus*, *Mycobacterium*, and *Nocardia*. Actinobacteria can introduce hydroxyl groups at numerous positions along the steroid skeleton, and sequential hydroxylations along carbons 22-27 of the sterol side chain have been identified as stages of degradation by actinobacteria [159]. In addition to hydroxylation, actinobacteria have the ability to transform steroids through double bond hydrogenation, single bond dehydrogenation, steroid alcohol oxidation (e.g., 17β-OH → 17C=O), steroid ketone reduction (e.g., 17C=O → 17β-OH), and double bond isomerization [159]. Many actinobacteria can metabolize steroids completely as a carbon source for cell growth, but the mechanism of degradation will depend on the steroid type and the genera of actinobacteria involved [159]. Estranes (18 carbons) are often degraded by cleavage of the A-ring, and androstanes (19 carbons) and pregnanes (21 carbons) are often degraded by cleavage of the B-ring [159]. Cholestane (27 carbons), on the other hand, are commonly degraded by cleavage both of the B-ring and the sterol side chain [159].

Actinobacteria (e.g., species of *Nocardia* and *Arthrobacter*) often degrade 3-keto-4-ene steroids such as androstenedione, testosterone and progesterone through cleavage of the B-ring, beginning with the introduction of a double bond between carbons 1 and 2, and hydroxylation at carbon 9, which causes aromatization of the A-ring, cleavage of the B-ring, and the formation of a labile metabolite (3-hydroxy-9,10-secoandrosta-1,3,5(10)-triene-9,17-dione) [159, 168]. The labile metabolite is further degraded by hydroxylation at carbon 4, and cleavage of the A-ring, prior to mineralization. The entire degradation pathway is illustrated in Figure 2.

Proteobacteria and bacteriodetes are Gram-negative bacteria. Proteobacteria (e.g., species of *Comamonas* and *Pseudomonas*) also use the pathway in Figure 2 to degrade 3-keto-4-ene steroids, and bacteriodetes (e.g., species of *Sphingobacterium*) use the same pathway to degrade 17α-ethinylestradiol [168-170]. As a result, the pathway in Figure 2 has been described as a general pathway of steroid degradation [171]. However, microbes that use this pathway generally seem unable to metabolize estrone [171].

A separate degradation pathway was proposed for estrone, after incubation for 5 days with a *Nocardia* sp. isolated from soil [171]. Three degradation products were obtained after hydroxylation at carbon 4 and the opening of ring A. One involved a reaction with ammonia, and the other two involved the opening of ring B, in addition to ring A. Estrone was not mineralized during the 5 day incubation period.

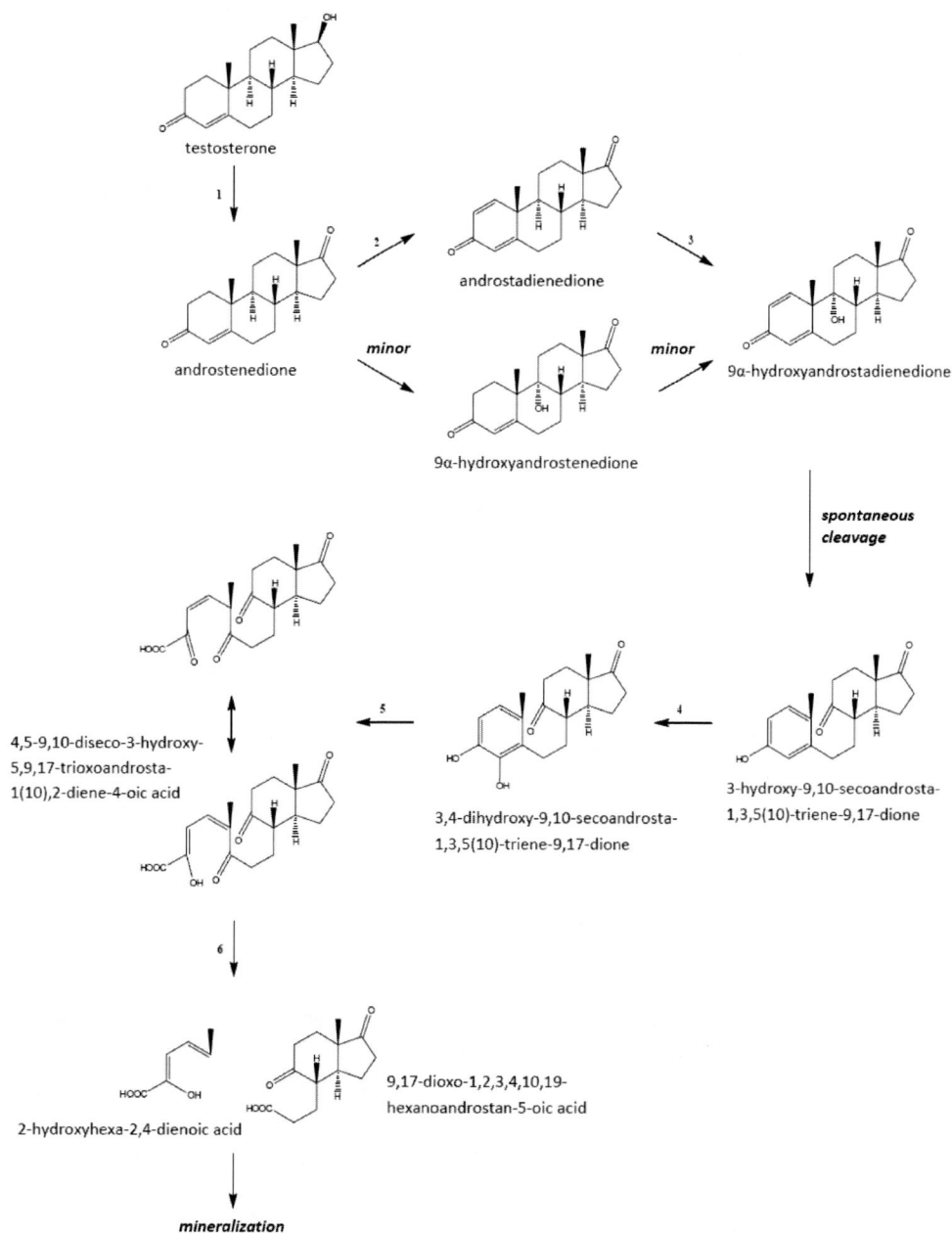

Figure 2. General pathway of steroid hormone biodegradation (illustrated with testosterone). Reaction steps are numbered (modified from ref. [169]).

Another 17β-estradiol metabolite was identified after incubation with a 17β-estradiol degrading culture [160]. Two-thirds of the 17β-estradiol (200 μg L$^{-1}$) was quantitatively oxidized to estrone, with little or no other metabolites, after contact with the culture for 22 h. Almost all of the 17β-estradiol was removed after approximately 3 days, and almost all of the estrone was removed after approximately 14 days. Very early in the process (1-5 h), a labile metabolite was observed, and tentatively identified as a gamma lactone (5 member ring) produced in the D-ring by estrone oxidation. Although some species of *Cylindrocarpon*, *Penicillium*, and *Streptomyces* can introduce a lactone into the D-ring to produce testolactone (a delta lactone with a 6 member ring), they generally do not metabolize the testolactone further [171]. As a result, if the tentative identification is correct, then either the gamma lactone is less stable, or different microbes participated in the gamma lactone degradation.

From the foregoing, it is clear numerous metabolites can form from steroid sex hormone biodegradation by a variety of bacterial species. Microalgae and fungi also have the ability to transform steroids through hydroxylation, steroid ketone reduction, double bond hydrogenation, and single bond dehydrogenation [167, 172-174]. In addition, some microalgae also have the ability to metabolize steroids, including androstenedione and 5α-androstanedione, completely [167, 175].

### b) Biodegradation in Agricultural Soils

Biodegradation in agricultural soils is relevant to aquatic ecosystems because runoff and leaching from agricultural soils represent potential sources of steroid sex hormones in aquatic ecosystems, and because many of the same microorganisms and degradation pathways may be involved with biodegradation in aquatic ecosystems.

In a series of related studies, the dissipation of testosterone, 17β-estradiol, estrone, and 17α-ethinylestradiol was examined in three agricultural soils: (i) loam (3.2% organic matter; particle sizes: 40% sand, 45% silt, 15% clay); (ii) sandy loam (0.8% organic matter; particles sizes: > 90% sand); and (iii) silt loam (2.9% organic matter; particle sizes: 32% sand, 52% silt, 16% clay) [176-178]. Microcosms were prepared by adding 25-100 g of soil (moist weight) to baby-food jars. Then, each jar was placed in a sealable 1 L mason jar with a vial containing 10 mL of water to maintain moisture, and a second vial containing 5 mL of 1 M NaOH solution to trap $^{14}CO_2$. The microcosms were spiked with $^3$H-labeled, $^{14}$C-labeled, or unlabeled steroid sex hormones to an initial concentration of 1-10 mg kg$^{-1}$ (moist weight), and incubated at 30 °C. Periodically, 5 g soil samples were collected from the microcosms, extracted by liquid extraction (ethyl acetate and acetone), and analyzed by liquid scintillation counter, HPLC-RD (radioactivity detection), HPLC-UV, HPLC-MS, or a yeast-based reporter gene assay. Preliminary studies indicated that hormones were extracted from the soils with the following efficiencies: testosterone (84 ± 2%; n = 9); 17β-estradiol (98.9 ± 10.4%; n = 9); estrone (72.4 ± 4.4%; n = 6); and 17α-ethinylestradiol (80.5 ± 3.3%; n = 5). When necessary, sterile soil was prepared by autoclaving twice (45 min. at 120 °C).

Steroid sex hormone dissipation was determined by the decrease in extractable [$^{14}$C] material over time. Non-extractable hormones were assumed to be dissipated, and extractable metabolites were distinguished from parent hormones by HPLC-RD. The mineralization of [$^{14}$C]-labeled hormones was determined by measuring the trapped $^{14}CO_2$ with a liquid scintillation counter. In the testosterone experiments, the mineralization of 1,2,6,7-[$^3$H]-testosterone was determined by collecting two extracts, and evaporating one to dryness under nitrogen. Then, the evaporated extract was redissolved in ethyl acetate, a liquid scintillation counter was used to measure the [$^3$H] material in both extracts, and the difference was assumed to be evaporated $^3H_2O$ attributable to testosterone mineralization.

During the 17β-estradiol experiments, most of the [$^{14}$C] material was not extractable after 3 days of incubation (loam: 90.7% non-extractable; sandy loam: 70.3% non-extractable; silt loam: 56.0% non-extractable). In all cases, the beginning 17β-estradiol concentration was 50% dissipated in less than one-half day, from 5.3 h (sandy loam) to 11.5 h (silt loam). As 17β-estradiol dissipated, estrone accumulated. Throughout the experiment, 17β-estradiol and estrone accounted for approximately 100% of the extractable radioactivity, suggesting that 17β-estradiol was being oxidized to estrone, and that no other transformation product was being formed from 17β-estradiol. In the loam soil, estrone diminished after 6 h, and was undetectable by the end of the experiment. In the sandy loam and silt loam soils, estrone remained detectable at the end of the experiment, and represented 100% of the extractable [$^{14}$C] material. Declines in extractable radioactivity closely paralleled declines in estrogenicity, taking into account the relative potencies of 17β-estradiol and estrone. 17β-estradiol was mineralized in all three soils, but only very slowly. After 3 months, only 11.5-17.1% of the 4-[$^{14}$C]-labeled 17β-estradiol was recovered as $^{14}CO_2$. In the autoclaved soils, 17β-estradiol was oxidized to estrone within 72 h, suggesting that 17β-estradiol was being oxidized abiotically to form estrone. Because extractable radioactivity remained constant in the autoclaved soils as 17β-estradiol was being oxidized to estrone, decreases in extractable radioactivity were associated with the formation of non-extractable residues during estrone degradation. In a similar experiment (described below), 50-67% of the initial radioactivity was not extractable after 5 days, and approximately 90% of the extractable radioactivity remained 17β-estradiol [146]. Possible explanations for the difference include different soil conditions (e.g., different oxidants), and different experimental conditions.

During the estrone experiments, most of the [$^{14}$C] material was not extractable after 3 days of incubation (loam: 88.2% non-extractable; sandy loam: 59.4% non-extractable; silt loam: 71.4% non-extractable). In all cases, the beginning estrone concentration was 50% dissipated in 14.7 h (silt loam) to 40.6 h (sandy loam).

17α-ethinylestradiol was 50% dissipated in 59.4 h (sandy loam soil; 12 % moisture) to 97.9 h (silt loam; 15% moisture). The decline in 17α-ethinylestradiol was accompanied by a corresponding decline in total estrogenicity, suggesting that no estrogenic metabolites were formed. 17α-ethinylestradiol appeared to be stable in the autoclaved soil. However, when 17α-ethinylestradiol was oxidized by Mn(III) in another study, a double bond was introduced between carbons 9 and 11 [179]. If this degradation product had been formed, it is unlikely that 17α-ethinylestradiol dissipation would have been detected.

Testosterone was 50% dissipated in 8.5 h (loam soil) to 21 h (silt loam soil), but dissipated more slowly at lower temperatures (4-12 °C). After approximately 120 h, testosterone was 50% mineralized in all soils. Androgenic activity declined faster than extractable [$^{14}$C] radioactivity, suggesting that the testosterone was transformed into metabolites with less androgenic activity. Three testosterone metabolites were identified: androstenedione; 5α-androstan-3,17-dione; and androstadienedione. None of the [$^3$H] material was recovered as $^3H_2O$, but the extractable [$^3$H] radioactivity dissipated rapidly (within 48 h) in moist or saturated soil. In autoclaved soil, extractable [$^3$H] radioactivity dissipated more slowly, and androstenedione formed, suggesting that testosterone was oxidized abiotically to form androstenedione.

Another study examined the biodegradation of 17β-estradiol and testosterone in an agricultural soil under aerobic and anaerobic conditions [146]. Each steroid sex hormone was investigated in four soil microcosms: (i) agricultural soil under aerobic conditions; (ii) agricultural soil under anaerobic conditions; (iii) autoclaved soil under aerobic conditions; and (iv) autoclaved soil under anaerobic conditions. The organic matter content of the agricultural soil was 2.23%, and the particle size distribution follows: 14% clay; 19% silt; and 67% sand. The autoclaved soils were sterilized for 40 min. at 122 °C. The incubation experiments were prepared by mixing radiolabeled hormones with approximately 210 g of soil (not sieved or air-dried) in 250 mL glass flasks. The initial concentrations of 1,500 μg L$^{-1}$ 17β-estradiol and 52 μg L$^{-1}$ testosterone were chosen to represent concentrations found in animal manures applied to agricultural fields. The flasks were covered with aluminum to eliminate photodegradation, and supplied with moist air to maintain aerobic conditions, or humified helium gas to maintain anaerobic conditions. The autoclaved soils were also spiked with 500 mg kg$^{-1}$ HgCl$_2$ to inhibit contamination by airborne bacteria. 200 mL glass flasks containing 3M NaOH were set up to trap $^{14}CO_2$ under all conditions, and a 200 mL flask containing Bray's solution was up to trap $^{14}CH_4$ under anaerobic conditions. Periodic samples were collected over a 132-hour period, and analyzed by liquid scintillation counter. At the end of the experiments, extracts from the soils were examined for radioactivity, thin layer chromatography was used to determine metabolites, and the distribution of non-extractable [$^{14}C$] among organic fractions (i.e., humic acid, fulvic acid, and humus) was determined with a multi-step fractionation process.

During the testosterone experiments in natural soil under *aerobic* conditions, 85.4% of the total [$^{14}C$] was recovered. Of the recovered amount, 63% was trapped after mineralization to $^{14}CO_2$, 3.4% was extractable from the soil with water or acetone, and 19% was associated with natural organic matter (3% humic acids, 9% fulvic acids, 7% humin) and non-extractable. Of the extractable [$^{14}C$] material (3.4%), 17% was testosterone, and 83% was testosterone metabolites. During the 17β-estradiol experiments in natural soil under aerobic conditions, 91% of the total [$^{14}C$] was recovered. Of the recovered amount, 6% was trapped after mineralization to $^{14}CO_2$, 12% was extractable from the soil with water or acetone, and 73% was associated with natural organic matter (37% humic acids, 17% fulvic acids, 19% humin) and non-extractable. All of the extractable [$^{14}C$] material (12%) was 17β-estradiol metabolites.

During the testosterone experiments in natural soil under *anaerobic* conditions, 89% of the total [$^{14}C$] was recovered. Of the recovered amount, 46% was trapped after mineralization to $^{14}CO_2$, 2% was trapped after mineralization to $^{14}CH_4$, 16% was extractable from the soil with water or acetone, and 25% was associated with natural organic matter (5% humic acids, 0% fulvic acids, 20% humin) and non-extractable. Of the extractable [$^{14}C$] material (16%), 13% was testosterone, and 87% was testosterone metabolites. During the 17β-estradiol experiments in natural soil under anaerobic conditions, 89.9% of the total [$^{14}C$] was recovered. Of the recovered amount, 0.9% was trapped after mineralization to $^{14}CO_2$, 19% was extractable from the soil with water or acetone, and 70% was associated with natural organic matter (37% humic acids, 22% fulvic acids, 11% humin) and non-extractable. Of the extractable [$^{14}C$] material (19%), 11% was 17β-estradiol, and 89% was 17β-estradiol metabolites.

During the testosterone and 17β-estradiol experiments in the *autoclaved* soils under all conditions (aerobic and anaerobic), no $^{14}CO_2$ was recovered in excess of 0.2%, and no $^{14}CH_4$

was recovered. Like the experiments in the natural soils, much of the [$^{14}$C] material in the autoclaved soil experiments was not extractable with water or acetone (49-67%).

In natural soils, testosterone was mineralized more than 17β-estradiol under all conditions (aerobic and anaerobic). This result was explained by testosterone's greater presence in the dissolved phase, and by noting that more energy is required to cleave the aromatic structure of 17β-estradiol where the 4-[$^{14}$C]-label was located [113]. The aerobic degradation rates of testosterone ($t_{1/2}$ = 58 h) and 17β-estradiol ($t_{1/2}$ = 1,150 h) were faster than their anaerobic degradation rates (testosterone: $t_{1/2}$ = 173 h; 17β-estradiol: $t_{1/2}$ = 6,930 h). This difference was explained by the ready availability of electron acceptors under aerobic conditions, and the possible involvement of different microbial species or different degradation pathways.

### c) Biodegradation in Rivers, Lakes and their Sediments

One environmental study examined whether the source of androstenedione in the Fenholloway River (Florida, U.S.A.) might be the microbial transformation of progesterone [71]. In prior studies, androstenedione and progesterone were detected in the water column and sediments of the Fenholloway River, which contains paper mill effluent and masculinized mosquitofish [72, 73]. The study hypothesized that progesterone was being produced by microbial transformation of plant sterols in the mill pulp, and that androstenedione was being produced by microbial transformation of progesterone. To test the hypothesis, progesterone (1 mM) was incubated with *Mycobacterium smegmatis* (a common soil bacterium). Samples of the slurried media were collected periodically over 36 days, and various analytical methods (including liquid extraction and SPE) were used to prepare the samples for HPLC-DAD analysis. 17α-hydroxyestrone, androstenedione, and androstadienedione were identified as transformation products by comparisons with HPLC standards [71]. The concentration of 17α-hydroxyestrone increased from day 0 through day 12, and the highest mean concentration occurred on day 20 (23.0 ± 0.7 µM). The concentration of androstenedione increased from day 0 to day 6 (7.2 ± 0.8 µM), remained constant until day 12, and decreased thereafter. The concentration of androstadienedione mirrored androstenedione, and reached a maximum on day 8 (5.2 ± 0.5 µM). Based on the rates of accumulation of the intermediates during the experiment, it was suggested that the steroid hormones were produced in the following order: progesterone → 17α-hydroxyprogesterone → androstenedione → androstadienedione.

In another study, water and sediment samples were collected from rivers influenced by rural (Thames) and urban or industrial activities (Aire and Calder) to examine the potential for biodegradation of 17β-estradiol and 17α-ethinylestradiol in freshwater ecosystems [150]. Bulk water samples were collected in 1 L glass bottles from the top 0.5 m of the water column during low, medium, and high flow periods between 1997 and 2000. In addition, bed sediments were collected 2-3 m from the banks of the Thames and Calder Rivers by skimming off the top few centimeters with a bucket. River water samples were prepared by adding 50 mL of water to autoclaved 125 mL PTFE flasks, and estrogens were spiked in to a nominal concentration of 100 µg L$^{-1}$. The water samples were incubated in darkness at 10 or 20 °C, and autoclaved samples of the same water were used as sterile controls. Periodically, water samples were combined for analysis by HPLC-MS or HPLC- DAD. Wet sediment and water from the same site were collected in 100 mL conical flasks, fitted with gas flushing heads, and placed under nitrogen. Autoclaved water and bed sediments were used as sterile controls. Another set of experiments compared aerobic to anaerobic conditions in the bed sediments. To study the potential for mineralization, radiolabeled hormones (4-[$^{14}$C]-labeled) were added to water samples, and a 50 mM solution of NaOH was used to trap $^{14}CO_2$ as it evolved. Periodically, samples were collected and analyzed by liquid scintillation counter.

In all the non-sterile river water samples, 17β-estradiol dissipated and estrone accumulated. No significant losses were observed in the sterile controls. Faster degradation rates were associated summer water samples, possibly because nutrient concentrations and temperatures are higher during summer. 17α-ethinylestradiol (t½ = 17 d) dissipated much more slowly than the 17β-estradiol (t½ = 1.2 d) under aerobic conditions. In the anaerobic bed sediments, 17β-estradiol was converted to estrone over the 2-day incubation period. The potential for 17α-ethinylestradiol to dissipate in the anaerobic bed sediments was not tested. After approximately 25 days, 24-45% of the [$^{14}C$] material had evolved as $^{14}CO_2$, suggesting that microorganisms in river waters and sediments can cleave the aromatic A-ring to release the 4-[$^{14}C$]-radiolabels. Mineralization rates slowed after 25 days, either because nutrient supplies diminished or, if the degradation process is cometabolic, because the cometabolic substrate was exhausted. At the end of the experiment, 18-32% of the original 17β-estradiol concentration was 17β-estradiol, 10-23% was hydrophilic by-products (18-32%); and 5% was sorbed to glassware or charcoal. After 8 days, 17β-estradiol and estrone became undetectable, but sample estrogenicity remained detectable, suggesting that estrone by-products could be slightly estrogenic. After 2 weeks, more than 99% of the initial estrogenicity was lost. Loss of estrogenicity is expected with cleavage of the A-ring, because the A-ring is essential for estrogen receptor binding.

Another study examined the anaerobic biodegradation of 17β-estradiol, estrone, and 17α-ethinylestradiol under methanogenic, sulfate-, iron-, and nitrate-reducing conditions [95]. Cultures were established in 160 mL serum bottles by combining 10 mL of lake sediment, 90 mL of freshwater mineral medium, resazurin (a redox indicator), and estrogens at initial concentrations of 5 mg L$^{-1}$. Strict anaerobic technique was used, and the following electron acceptor solutions were added to create reducing conditions: 20 mM NaNO$_3$ (nitrate-reducing conditions); 20 mM Fe$^{3+}$−nitrilotriacetic acid (iron-reducing conditions); 20 mM Na$_2$SO$_4$ (sulfate-reducing conditions); or water (methanogenic conditions). Iron-reducing culture samples were amended with 80 mg L$^{-1}$ ethylenediaminetetraacetic acid (EDTA) to chelate the iron. Aqueous samples were collected with glass syringes, transferred to microcentrifuge tubes containing methanol, and microcentrifuged. The supernatant was analyzed by HPLC-DAD, methane was analyzed by GC-FID, reduced iron was analyzed by ferrozine colorimetric assay, and sulfate and nitrate were analyzed by ion chromatography. Spiked water sample recoveries exceeded 95%, and quantitation limits were 20 μg L$^{-1}$. Metabolites were identified in separate culture experiments after liquid extraction, derivatization, and GC-MS/MS analysis.

In experiments combining 17α-ethinylestradiol, 17β-estradiol, and lake sediment, 17α-ethinylestradiol was not significantly degraded under anaerobic conditions, even after 35-38 months. By comparison, 17β-estradiol was partially removed under all four anaerobic conditions. 17β-estradiol was oxidized to estrone in each case, but the oxidation rate was slightly faster under nitrate-reducing conditions. HPLC-DAD appeared to detect estriol, but estriol was not confirmed by GC-MS/MS analysis. 17α-estradiol was detected under methanogenic, sulfate-reducing, and iron-reducing conditions, but not nitrate-reducing conditions. Despite the differences in potential energy (e.g., redox potential) available from different electron acceptors, there was no clear correlation between the use of electron acceptors and the rate or extent of 17β-estradiol transformation, suggesting that 17β-estradiol was not oxidized to gain energy. Instead, the oxidation of 17β-estradiol was believed to be used for co-factor regeneration.

The preceding studies demonstrate that steroid sex hormones biodegrade, but may not be mineralized, in rivers, lakes, and their sediments under aerobic or anaerobic conditions. Therefore, in some cases, biodegradation may produce metabolites that reduce, but do not eliminate, the biological activities of steroid sex hormones.


## 3. Photodegradation

### a) Introduction

Photodegradation is an important abiotic degradation pathway that takes place in natural waters [180]. Direct photodegradation occurs when an organic contaminant absorbs light, becomes excited, and undergoes chemical changes [49, 180, 181]. Indirect (or sensitized) photodegradation occurs when another compound absorbs light, becomes excited, and reacts with the organic contaminant [49, 180, 181]. Saturated organic compounds (i.e., molecules with single bonds) do not absorb light in the 200-700 nm range, which includes all visible light and much of the low energy UV spectrum (UV-A, UV-B, and some UV-C). Unsaturated molecules (e.g., molecules with C=C, C=O, and aromatic groups) can absorb light in this range [181]. The solar spectrum generally ranges from 290-800 nm, and natural waters contain many unsaturated molecules (e.g., dissolved organic matter) and other reactive intermediates (e.g., nitrate and nitrite) that absorb light in this range [182, 183]. As a result, natural water bodies have been described as "large photochemical reactor systems" [182]. Because steroid sex hormones are often unsaturated molecules, they have the potential to undergo direct and indirect photodegradation.

### b) Direct Photodegradation

In laboratory experiments with simulated sunlight, direct photodegradation of 100 µg $L^{-1}$ solutions of 17α-ethinylestradiol and 17β-estradiol was observed with half-lives of approximately 10 days [150]. Over a 60 minute period, another laboratory experiment observed direct photodegradation of 17β-estradiol and estrone in UV-C light ($\lambda_{max}$ = 254 nm), and direct photodegradation of estrone, but not 17β-estradiol, in UV-A plus visible light ($\lambda \geq$ 365 nm) [184]. Over a 90 minute period, a related experiment observed direct photodegradation of 17α-ethinylestradiol in UV-C light ($\lambda_{max}$ = 254 nm), and no direct photodegradation of 17α-ethinylestradiol in UV-A plus visible light ($\lambda \geq$ 365 nm) [185]. Because UV-C light generally fails to reach the Earth's surface, estrone is expected to undergo direct photodegradation more easily than 17α-ethinylestradiol, 17β-estradiol, and estriol [183].

In a separate study using UV-A light ($\lambda$ > 315 nm), direct photodegradation of testosterone and progesterone were observed [118].

### c) Indirect Photodegradation

In laboratory experiments with simulated sunlight (290nm < $\lambda$ < 700 nm), estrogens (17α-ethinylestradiol, 17β-estradiol, estrone, and estriol) experienced increased degradation rates in filtered and autoclaved Santa Ana river water (Southern California, USA; 4.6 mg L$^{-1}$ dissolved organic carbon; 22.3 mg L$^{-1}$ nitrate), suggesting indirect photodegradation [183].

In a separate study using UV-A light ($\lambda > 315$ nm) and 10 mg L$^{-1}$ nitrate and 5 mg L$^{-1}$ humic acid as potential reactive intermediates, no indirect photodegradation of progesterone was observed, but indirect photodegradation of testosterone was observed in the presence of 5 mg L$^{-1}$ humic acid [118].

To date, little is known about the mechanisms and products of steroid sex hormone photodegradation.


## STEROID SEX HORMONES – FATE AND TRANSPORT

Column studies have been used to study the fate and transport of steroid sex hormones in agricultural soils [128, 133, 134, 156]. Batch equilibrium experiments expose steroid sex hormones to water and soils and sediments, and measure their distribution at apparent equilibrium. Similarly, biodegradation studies incubate microbial cultures in controlled environments, and examine the rates and effects of biodegradation. Column studies, on the other hand, examine the sorption and degradation of steroid sex hormones under non-equilibrium conditions, as water flows through the soils and sediments. Packed columns eliminate preferential flow through soil macropores to simplify the transport analysis, but fail to consider the effects of natural soil structure and macropores on fate and transport [151]. Undisturbed soil columns attempt to preserve these features, in order to better reflect transport through soils under field conditions [151].

In a study of testosterone fate and transport, a 300 mL pulse of 4-[$^{14}$C]-testosterone (~5.9 µCi) was added to an undisturbed soil column (30 cm height × 15 cm inner diameter), and eluted with approximately 6 pore volumes of a weak salt solution (0.01 M CaCl$_2$) [152]. The soil had a bulk density of 1.54 g cm$^{-3}$, 0.42 porosity, 2.23% organic matter, 14% clay, 19% silt, and 67% sand. The column was modified to collect [$^{14}$C]-labeled volatile compounds, and 1 pore volume of 0.75 mg L HgCl$_2$ solution was added after the experiment to stop biological activity. Using the modified column, the pore water velocity was approximately 5 cm h$^{-1}$. The column effluent was collected in 3.5 minute fractions, and analyzed by liquid scintillation counter. Thin-layer chromatography was used to identify metabolites in the column effluent, and combustion analysis was used to determine the distribution of $^{14}$C inside the soil column after the experiment.

Using the modified column, approximately 80% of the [$^{14}$C] was recovered, and the remainder was deemed lost due to incomplete combustion (i.e., present in the column, but not recovered). 13.25% of the [$^{14}$C]-material was recovered in the column effluent, and 23.4% of the [$^{14}$C]-material was recovered as $^{14}$CO$_2$ following testosterone mineralization. Of the [$^{14}$C]-material recovered from the column, 70-74% was found in the top 1-5 cm of the column, possibly because more organic matter was present there. [$^{14}$C]-concentrations in the column effluent tailed off, and the tailing was attributed to chemical processes (e.g., rate-limited sorption), rather than physical processes (e.g., preferential transport), because no tailing was observed when a chloride ion tracer passed through the column. However, preferential transport is generally expected in undisturbed soil, and likely contributes to the leaching that has been observed in natural systems.

Another undisturbed soil column study was conducted to examine the fate and transport of testosterone and 17β-estradiol [151]. The soil samples were taken from no-till and conventionally tilled agricultural plots. In both plots, total organic carbon was higher in the first 10 cm (0.85-0.88%) than between 20-30 cm (0.38-0.43%). Particle sizes in both plots ranged from 60.7-65.3% sand, 15.3-19.3% silt, and 16.7-24% clay. A 1.1 L pulse containing 0.533 µg 17β-estradiol (17,700 Bq 6,7-[$^3$H]-17β-estradiol plus 0.525 µg unlabeled 17β-estradiol) and 1.21 µg 4-[$^{14}$C]-testosterone (9,430 Bq) was added to each undisturbed soil column (32 cm height × 15 cm inner diameter), and eluted with 21 L (~10 pore volumes) of weak salt solution (0.01 M CaNO$_3$). The column effluent was collected in ~67 mL fractions. After the experiment, soil cores were collected from the columns, divided into increments by depth, and oxidized with a biological oxidizer to release and capture the $^3$H and $^{14}$C in liquid scintillation cocktails. A liquid scintillation counter was used to measure the soil core cocktails and the column effluent.

17β-estradiol sorbed more strongly than testosterone, as an average of 27% of the 17β-estradiol and 42% of the testosterone leached from the columns. The sorption coefficients of both hormones decreased with depth, possibly because total organic carbon decreased with depth. Oxidation of the column samples recovered 29% of the 17β-estradiol and 17% of the testosterone, and the majority of sorbed hormones occurred in the top 10 cm of soil. The low recovery rates of 56% 17β-estradiol and 59% testosterone were believed to result from underestimation of sorption by the biological oxidizer method or sorption to column materials. $^{14}$C losses also might be attributable to biodegradation. Generally, hormone peak concentrations occurred simultaneously with the chloride tracer, indicating the influence of preferential transport.

# FUTURE RESEARCH

To date, considerable research has been done to understand the sources, presence, and fate of selected environmental endocrine disruptors, including selected steroid sex hormones, in the environment. In addition, considerable research has been done with certain species of organisms to understand the biological and ecological effects of environmental endocrine disruptors at concentrations already found in the environment. Additional research has addressed the development of treatment technologies, including advanced oxidation processes, and the development of best management practices to limit endocrine disruptor contributions to the environment [93, 99, 186-193].

Thus far, research has shown that the subject of endocrine disruption is complex, and difficult to generalize. Additional research must be done to understand the effects of suspected endocrine disruptors on untested species of organisms, and to understand the influence of untested mechanisms and conditions on the environmental fate of suspected endocrine disruptors. Specific subjects for further research include:

- Biosolids leaching and runoff
- Mechanisms and pathways of biodegradation and photodegradation, including the development of mass balance techniques to identify degradation products that might

be missed after the application of selective extraction and detection techniques (e.g., MS/MS in the MRM mode)
- Other mechanisms and pathways of abiotic transformation, including redox reactions involving metal oxides
- Presence and environmental effects of degradation products that have not been analyzed before
- Mixtures of endocrine disruptors, their biological effects, and their influence on environmental fate
- Alternate transport pathways, including atmospheric transport
- Specific sorption mechanisms
- Plant and animal sterols as sources of steroid sex hormones in the environment
- Plant uptake, and the influence of plants on environmental fate
- Isolation and identification of microbial species able to mineralize or transform steroid sex hormones, in order to optimize wastewater treatment, composting, and lagoon processes for steroid sex hormone treatment

As additional progress is made, improved treatment technologies and management practices will help to minimize the risks of endocrine disruption to freshwater ecosystems and the broader environment, even as human populations, agricultural operations, and other possible sources of endocrine disruptors continue to grow.

## ACKNOWLEDGMENTS

We gratefully acknowledge financial support from the Colorado Water Resources Research Institute (CWRRI) and The SeaCrest Group. We also thank Dr. Jessica G. Davis, Alex Makedonski and Jens Blotevogel for their technical support.

## REFERENCES

[1] Crisp, T. M.; Clegg, E. D.; Cooper, R. L.; Wood, W. P.; Anderson, D. G.; Baetcke, K. P.; Hoffmann, J. L.; Morrow, M. S.; Rodier, D. J.; Schaeffer, J. E.; Touart, L. W.; Zeeman, M. G.; Patel, Y. M., Environmental endocrine disruption: An effects assessment and analysis. *Environmental Health Perspectives* 1998, 106, 11-56.

[2] Fang, H.; Tong, W. D.; Perkins, R.; Soto, A. M.; Prechtl, N. V.; Sheehan, D. M., Quantitative comparisons of in vitro assays for estrogenic activities. *Environmental Health Perspectives* 2000, 108, (8), 723-729.

[3] Norris, D. O.; Carr, J. A., *Endocrine Disruption*. Oxford University Press: New York, 2006.

[4] Colborn, T.; Saal, F. S. V.; Soto, A. M., Developmental effects of endocrine-disrupting chemicals in wildlife and humans. *Environmental Health Perspectives* 1993, 101, (5), 378-384.

[5] Vos, J. G.; Dybing, E.; Greim, H. A.; Ladefoged, O.; Lambre, C.; Tarazona, J. V.; Brandt, I.; Vethaak, A. D., Health effects of endocrine-disrupting chemicals on wildlife,

with special reference to the European situation. *Critical Reviews in Toxicology* 2000, 30, (1), 71-133.

[6]   Tyler, C. R.; Jobling, S.; Sumpter, J. P., Endocrine disruption in wildlife: A critical review of the evidence. *Critical Reviews in Toxicology* 1998, 28, (4), 319-361.

[7]   Verger, P.; Leblanc, J. C., Concentration of phytohormones in food and feed and their impact on the human exposure. *Pure and Applied Chemistry* 2003, 75, (11), 1873-1880.

[8]   Sonnenschein, C.; Soto, A. M., An updated review of environmental estrogen and androgen mimics and antagonists. *Journal of Steroid Biochemistry and Molecular Biology* 1998, 65, (1-6), 143-150.

[9]   Katzenellenbogen, J. A.; Muthyala, R., Interactions of exogenous endocrine active substances with nuclear receptors. *Pure and Applied Chemistry* 2003, 75, (11), 1797-1817.

[10]  vom Saal, F. S.; Timms, B. G.; Montano, M. M.; Palanza, P.; Thayer, K. A.; Nagel, S. C.; Dhar, M. D.; Ganjam, V. K.; Parmigiani, S.; Welshons, W. V., Prostate enlargement in mice due to fetal exposure to low doses of estradiol or diethylstilbestrol and opposite effects at high doses. *Proceedings of the National Academy of Sciences of the United States of America* 1997, 94, (5), 2056-2061.

[11]  Welshons, W. V.; Thayer, K. A.; Judy, B. M.; Taylor, J. A.; Curran, E. M.; vom Saal, F. S., Large effects from small exposures. I. Mechanisms for endocrine-disrupting chemicals with estrogenic activity. *Environmental Health Perspectives* 2003, 111, (8), 994-1006.

[12]  Ankley, G. T.; Jensen, K. M.; Makynen, E. A.; Kahl, M. D.; Korte, J. J.; Hornung, M. W.; Henry, T. R.; Denny, J. S.; Leino, R. L.; Wilson, V. S.; Cardon, M. C.; Hartig, P. C.; Gray, L. E., Effects of the androgenic growth promoter 17β-trenbolone on fecundity and reproductive endocrinology of the fathead minnow. *Environmental Toxicology and Chemistry* 2003, 22, (6), 1350-1360.

[13]  Arnold, S. F.; Klotz, D. M.; Collins, B. M.; Vonier, P. M.; Guillette, L. J.; McLachlan, J. A., Synergistic activation of estrogen receptor with combinations of environmental chemicals. *Science* 1996, 272, (5267), 1489-1492.

[14]  Crews, D.; Putz, O.; Thomas, P.; Hayes, T.; Howdeshell, K., Wildlife as models for the study of how mixtures, low doses, and the embryonic environment modulate the action of endocrine-disrupting chemicals. *Pure and Applied Chemistry* 2003, 75, (11), 2305-2320.

[15]  Thorpe, K. L.; Hutchinson, T. H.; Hetheridge, M. J.; Scholze, M.; Sumpter, J. P.; Tyler, C. R., Assessing the biological potency of binary mixtures of environmental estrogens using vitellogenin induction in juvenile rainbow trout (Oncorhynchus mykiss). *Environmental Science and Technology* 2001, 35, (12), 2476-2481.

[16]  Lin, L. L.; Janz, D. M., Effects of binary mixtures of xenoestrogens on gonadal development and reproduction in zebrafish. *Aquatic Toxicology* 2006, 80, (4), 382-395.

[17]  Brian, J. V.; Harris, C. A.; Scholze, M.; Kortenkamp, A.; Booy, P.; Lamoree, M.; Pojana, G.; Jonkers, N.; Marcomini, A.; Sumpter, J. P., Evidence of estrogenic mixture effects on the reproductive performance of fish. *Environmental Science and Technology* 2007, 41, (1), 337-344.

[18]  Thorpe, K. L.; Cummings, R. I.; Hutchinson, T. H.; Scholze, M.; Brighty, G.; Sumpter, J. P.; Tyler, C. R., Relative potencies and combination effects of steroidal estrogens in fish. *Environmental Science and Technology* 2003, 37, (6), 1142-1149.

[19] Miyamoto, J.; Klein, W., Environmental exposure, species differences and risk assessment. *Pure and Applied Chemistry* 1998, 70, (9), 1829-1845.

[20] Jobling, S.; Casey, D.; Rodgers-Gray, T.; Oehlmann, J.; Schulte-Oehlmann, U.; Pawlowski, S.; Baunbeck, T.; Turner, A. P.; Tyler, C. R., Comparative responses of molluscs and fish to environmental estrogens and an estrogenic effluent. *Aquatic Toxicology* 2003, 65, (2), 205-220.

[21] Denny, J. S.; Tapper, M. A.; Schmieder, P. K.; Hornung, M. W.; Jensen, K. M.; Ankley, G. T.; Henry, T. R., Comparison of relative binding affinities of endocrine active compounds to fathead minnow and rainbow trout estrogen receptors. *Environmental Toxicology and Chemistry* 2005, 24, (11), 2948-2953.

[22] Arcand-Hoy, L. D.; Nimrod, A. C.; Benson, W. H., Endocrine-modulating substances in the environment: Estrogenic effects of pharmaceutical products. *International Journal of Toxicology* 1998, 17, (2), 139-158.

[23] Lai, K. M.; Scrimshaw, M. D.; Lester, J. N., The effects of natural and synthetic steroid estrogens in relation to their environmental occurrence. *Critical Reviews in Toxicology* 2002, 32, (2), 113-132.

[24] Jobling, S.; Tyler, C. R., Endocrine disruption in wild freshwater fish. *Pure and Applied Chemistry* 2003, 75, (11), 2219-2234.

[25] Vajda, A. M.; Barber, L. B.; Gray, J. L.; Lopez, E. M.; Woodling, J. D.; Norris, D. O., Reproductive disruption in fish downstream from an estrogenic wastewater effluent. *Environmental Science and Technology* 2008, 42, (9), 3407-3414.

[26] Jobling, S.; Nolan, M.; Tyler, C. R.; Brighty, G.; Sumpter, J. P., Widespread sexual disruption in wild fish. *Environmental Science and Technology* 1998, 32, (17), 2498-2506.

[27] Folmar, L. C.; Denslow, N. D.; Kroll, K.; Orlando, E. F.; Enblom, J.; Marcino, J.; Metcalfe, C.; Guillette, L. J., Altered serum sex steroids and vitellogenin induction in walleye (Stizostedion vitreum) collected near a metropolitan sewage treatment plant. *Archives of Environmental Contamination and Toxicology* 2001, 40, (3), 392-398.

[28] Harries, J. E.; Sheahan, D. A.; Jobling, S.; Matthiessen, P.; Neall, M.; Sumpter, J. P.; Taylor, T.; Zaman, N., Estrogenic activity in five United Kingdom rivers detected by measurement of vitellogenesis in caged male trout. *Environmental Toxicology and Chemistry* 1997, 16, (3), 534-542.

[29] Petrovic, M.; Sole, M.; de Alda, M. J. L.; Barcelo, D., Endocrine disruptors in sewage treatment plants, receiving river waters, and sediments: Integration of chemical analysis and biological effects on feral carp. *Environmental Toxicology and Chemistry* 2002, 21, (10), 2146-2156.

[30] Blair, R. M.; Fang, H.; Branham, W. S.; Hass, B. S.; Dial, S. L.; Moland, C. L.; Tong, W. D.; Shi, L. M.; Perkins, R.; Sheehan, D. M., The estrogen receptor relative binding affinities of 188 natural and xenochemicals: Structural diversity of ligands. *Toxicological Sciences* 2000, 54, (1), 138-153.

[31] Fang, H.; Tong, W. D.; Branham, W. S.; Moland, C. L.; Dial, S. L.; Hong, H. X.; Xie, Q.; Perkins, R.; Owens, W.; Sheehan, D. M., Study of 202 natural, synthetic, and environmental chemicals for binding to the androgen receptor. *Chemical Research in Toxicology* 2003, 16, (10), 1338-1358.

[32] Pawlowski, S.; van Aerle, R.; Tyler, C. R.; Braunbeck, T., Effects of 17α-ethinylestradiol in a fathead minnow (Pimephales promelas) gonadal recrudescence assay. *Ecotoxicology and Environmental Safety* 2004, 57, (3), 330-345.

[33] Routledge, E. J.; Sheahan, D.; Desbrow, C.; Brighty, G. C.; Waldock, M.; Sumpter, J. P., Identification of estrogenic chemicals in STW effluent. 2. In vivo responses in trout and roach. *Environmental Science and Technology* 1998, 32, (11), 1559-1565.

[34] Kidd, K. A.; Blanchfield, P. J.; Mills, K. H.; Palace, V. P.; Evans, R. E.; Lazorchak, J. M.; Flick, R. W., Collapse of a fish population after exposure to a synthetic estrogen. *Proceedings of the National Academy of Sciences of the United States of America* 2007, 104, (21), 8897-8901.

[35] Caldwell, D. J.; Mastrocco, F.; Hutchinson, T. H.; Lange, R.; Heijerick, D.; Janssen, C.; Anderson, P. D.; Sumpter, J. P., Derivation of an aquatic predicted no-effect concentration for the synthetic hormone, 17α-ethinyl estradiol. *Environ. Sci. Technol.* 2008.

[36] Sitruk-Ware, R., New progestagens for contraceptive use. *Human Reproduction Update* 2006, 12, (2), 169-178.

[37] Oehlmann, J.; Schulte-Oehlmann, U., Endocrine disruption in invertebrates. *Pure and Applied Chemistry* 2003, 75, (11), 2207-2218.

[38] Carson, J. D.; Jenkins, R. L.; Wilson, E. M.; Howell, W. M.; Moore, R., Naturally occurring progesterone in loblolly pine (Pinus taeda L.): A major steroid precursor of environmental androgens. *Environmental Toxicology and Chemistry* 2008, 27, (6), 1273-1278.

[39] Andreolini, F.; Borra, C.; Caccamo, F.; Dicorcia, A.; Samperi, R., Estrogen conjugates in late pregnancy fluids - Extraction and group separation by a graphitized carbon-black cartridge and quantification by high performance liquid chromatography. *Analytical Chemistry* 1987, 59, (13), 1720-1725.

[40] Gomes, R. L.; Birkett, J. W.; Scrimshaw, M. D.; Lester, J. N., Simultaneous determination of natural and synthetic steroid estrogens and their conjugates in aqueous matrices by liquid chromatography/mass spectrometry. *International Journal of Environmental Analytical Chemistry* 2005, 85, (1), 1-14.

[41] Ternes, T. A.; Kreckel, P.; Mueller, J., Behaviour and occurrence of estrogens in municipal sewage treatment plants - II. Aerobic batch experiments with activated sludge. *Science of the Total Environment* 1999, 225, (1-2), 91-99.

[42] Lange, I. G.; Daxenberger, A.; Schiffer, B.; Witters, H.; Ibarreta, D.; Meyer, H. H. D., Sex hormones originating from different livestock production systems: Fate and potential disrupting activity in the environment. *Analytica Chimica Acta* 2002, 473, (1-2), 27-37.

[43] Belfroid, A. C.; Van der Horst, A.; Vethaak, A. D.; Schafer, A. J.; Rijs, G. B. J.; Wegener, J.; Cofino, W. P., Analysis and occurrence of estrogenic hormones and their glucuronides in surface water and waste water in The Netherlands. *Science of the Total Environment* 1999, 225, (1-2), 101-108.

[44] Koh, Y. K. K.; Chiu, T. Y.; Boobis, A.; Cartmell, E.; Lester, J. N.; Scrimshaw, M. D., Determination of steroid estrogens in wastewater by high performance liquid chromatography - tandem mass spectrometry. *Journal of Chromatography A* 2007, 1173, (1-2), 81-87.

[45]  Gentili, A.; Perret, D.; Marchese, S.; Mastropasqua, R.; Curini, R.; Di Corcia, A., Analysis of free estrogens and their conjugates in sewage and river waters by solid-phase extraction then liquid chromatography-electrospray-tandem mass spectrometry. *Chromatographia* 2002, 56, (1-2), 25-32.

[46]  Lee, L. S.; Carmosini, N.; Sassman, S. A.; Dion, H. M.; Sepulveda, M. S., Agricultural contributions of antimicrobials and hormones on soil and water quality. In *Advances in Agronomy, Vol 93*, 2007; Vol. 93, pp 1-68.

[47]  Kolok, A. S.; Sellin, M. K., The environmental impact of growth-promoting compounds employed by the United States beef cattle industry: History, current knowledge, and future directions. In *Reviews of Environmental Contamination and Toxicology*, 2008; Vol. 195, pp 1-30.

[48]  Ingerslev, F.; Vaclavik, E.; Halling-Sorensen, B., Pharmaceuticals and personal care products: A source of endocrine disruption in the environment? *Pure and Applied Chemistry* 2003, 75, (11), 1881-1893.

[49]  Khetan, S. K.; Collins, T. J., Human pharmaceuticals in the aquatic environment: A challenge to green chemistry. *Chemical Reviews* 2007, 107, (6), 2319-2364.

[50]  Shareef, A.; Angove, M. J.; Wells, J. D.; Johnson, B. B., Aqueous solubilities of estrone, 17β-estradiol, 17α-ethynylestradiol, and bisphenol A. *Journal of Chemical and Engineering Data* 2006, 51, (3), 879-881.

[51]  Ying, G. G.; Kookana, R. S.; Ru, Y. J., Occurrence and fate of hormone steroids in the environment. *Environment International* 2002, 28, (6), 545-551.

[52]  Hanselman, T. A.; Graetz, D. A.; Wilkie, A. C., Manure-borne estrogens as potential environmental contaminants: A review. *Environmental Science and Technology* 2003, 37, (24), 5471-5478.

[53]  Yu, Z. Q.; Xiao, B. H.; Huang, W. L.; Peng, P., Sorption of steroid estrogens to soils and sediments. *Environmental Toxicology and Chemistry* 2004, 23, (3), 531-539.

[54]  Bonin, J. L.; Simpson, M. J., Sorption of steroid estrogens to soil and soil constituents in single- and multi-sorbate systems. *Environmental Toxicology and Chemistry* 2007, 26, (12), 2604-2610.

[55]  Lai, K. M.; Johnson, K. L.; Scrimshaw, M. D.; Lester, J. N., Binding of waterborne steroid estrogens to solid phases in river and estuarine systems. *Environmental Science and Technology* 2000, 34, (18), 3890-3894.

[56]  Kim, I.; Yu, Z. Q.; Xiao, B. H.; Huang, W. L., Sorption of male hormones by soils and sediments. *Environmental Toxicology and Chemistry* 2007, 26, (2), 264-270.

[57]  Meylan, W. M.; Howard, P. H., Atom fragment contribution method for estimating octanol-water partition coefficients. *Journal of Pharmaceutical Sciences* 1995, 84, (1), 83-92.

[58]  Fang, H.; Tong, W. D.; Shi, L. M.; Blair, R.; Perkins, R.; Branham, W.; Hass, B. S.; Xie, Q.; Dial, S. L.; Moland, C. L.; Sheehan, D. M., Structure-activity relationships for a large diverse set of natural, synthetic, and environmental estrogens. *Chemical Research in Toxicology* 2001, 14, (3), 280-294.

[59]  *EPI Suite (KOWWIN$^{TM}$ model)*, v3.20; U.S. Environmental Protection Agency: 2007.

[60]  Campbell, C. G.; Borglin, S. E.; Green, F. B.; Grayson, A.; Wozei, E.; Stringfellow, W. T., Biologically directed environmental monitoring, fate, and transport of estrogenic endocrine disrupting compounds in water: A review. *Chemosphere* 2006, 65, (8), 1265-1280.

[61] Soto, A. M.; Maffini, M. V.; Schaeberle, C. M.; Sonnenschein, C., Strengths and weaknesses of in vitro assays for estrogenic and androgenic activity. *Best Practice and Research Clinical Endocrinology and Metabolism* 2006, 20, (1), 15-33.

[62] Huang, C. H.; Sedlak, D. L., Analysis of estrogenic hormones in municipal wastewater effluent and surface water using enzyme-linked immunosorbent assay and gas chromatography/tandem mass spectrometry. *Environmental Toxicology and Chemistry* 2001, 20, (1), 133-139.

[63] Thurman, E. M.; Ferrer, I.; Benotti, M.; Heine, C. E., Intramolecular isobaric fragmentation: A curiosity of accurate mass analysis of sulfadimethoxine in pond water. *Analytical Chemistry* 2004, 76, (5), 1228-1235.

[64] Gomes, R. L.; Avcioglu, E.; Scrimshaw, M. D.; Lester, J. N., Steroid estrogen determination in sediment and sewage sludge: a critique of sample preparation and chromatographic/mass spectrometry considerations, incorporating a case study in method development. *Trac-Trends in Analytical Chemistry* 2004, 23, (10-11), 737-744.

[65] de Alda, M. J. L.; Diaz-Cruz, S.; Petrovic, M.; Barcelo, D., Liquid chromatography-(tandem) mass spectrometry of selected emerging pollutants (steroid sex hormones, drugs and alkylphenolic surfactants) in the aquatic environment. *Journal of Chromatography A* 2003, 1000, (1-2), 503-526.

[66] Diaz-Cruz, M. S.; de Alda, M. J. L.; Lopez, R.; Barcelo, D., Determination of estrogens and progestogens by mass spectrometric techniques (GC/MS, LC/MS and LC/MS/MS). *Journal of Mass Spectrometry* 2003, 38, (9), 917-923.

[67] Gabet, V.; Miege, C.; Bados, P.; Coquery, M., Analysis of estrogens in environmental matrices. *Trac-Trends in Analytical Chemistry* 2007, 26, (11), 1113-1131.

[68] Clauwaert, K. M.; Van Bocxlaer, J. F.; Major, H. J.; Claereboudt, J. A.; Lambert, W. E.; Van den Eeckhout, E. M.; Van Peteghem, C. H.; De Leenheer, A. P., Investigation of the quantitative properties of the quadrupole orthogonal acceleration time-of-flight mass spectrometer with electrospray ionisation using 3,4-methylenedioxymethamphetamine. *Rapid Communications in Mass Spectrometry* 1999, 13, (14), 1540-1545.

[69] Ferrer, I.; Thurman, E. M., Multi-residue method for the analysis of 101 pesticides and their 14 degradates in food and water samples by liquid chromatography/time-of-flight mass spectrometry. *Journal of Chromatography A* 2007, 1175, (1), 24-37.

[70] Gopal, S. K. V.; Naik, S.; Somal, P.; Sharma, P.; Arjuna, A.; Hassan, R. U.; Khajuria, R. K.; Qazi, G. N., Production of 17-keto androstene steroids by the side chain cleavage of progesterone with Bacillus sphaericus. *Biocatalysis and Biotransformation* 2008, 26, (4), 272-279.

[71] Jenkins, R. L.; Wilson, E. M.; Angus, R. A.; Howell, W. M.; Kirk, M.; Moore, R.; Nance, M.; Brown, A., Production of androgens by microbial transformation of progesterone in vitro: A model for androgen production in rivers receiving paper mill effluent. *Environmental Health Perspectives* 2004, 112, (15), 1508-1511.

[72] Jenkins, R.; Angus, R. A.; McNatt, H.; Howell, W. M.; Kemppainen, J. A.; Kirk, M.; Wilson, E. M., Identification of androstenedione in a river containing paper mill effluent. *Environmental Toxicology and Chemistry* 2001, 20, (6), 1325-1331.

[73] Jenkins, R. L.; Wilson, E. M.; Angus, R. A.; Howell, W. M.; Kirk, M., Androstenedione and progesterone in the sediment of a river receiving paper mill effluent. *Toxicological Sciences* 2003, 73, (1), 53-59.

[74]  Kolodziej, E. P.; Harter, T.; Sedlak, D. L., Dairy wastewater, aquaculture, and spawning fish as sources of steroid hormones in the aquatic environment. *Environmental Science and Technology* 2004, 38, (23), 6377-6384.

[75]  Kolodziej, E. P.; Sedlak, D. L., Rangeland grazing as a source of steroid hormones to surface waters. *Environmental Science and Technology* 2007, 41, (10), 3514-3520.

[76]  Soto, A. M.; Calabro, J. M.; Prechtl, N. V.; Yau, A. Y.; Orlando, E. F.; Daxenberger, A.; Kolok, A. S.; Guillette, L. J.; le Bizec, B.; Lange, I. G.; Sonnenschein, C., Androgenic and estrogenic activity in water bodies receiving cattle feedlot effluent in eastern Nebraska, USA. *Environmental Health Perspectives* 2004, 112, (3), 346-352.

[77]  Kolok, A. S.; Snow, D. D.; Kohno, S.; Sellin, M. K.; Guillette, L. J., Occurrence and biological effect of exogenous steroids in the Elkhorn River, Nebraska, USA. *Science of the Total Environment* 2007, 388, (1-3), 104-115.

[78]  Swartz, C. H.; Reddy, S.; Benotti, M. J.; Yin, H. F.; Barber, L. B.; Brownawell, B. J.; Rudel, R. A., Steroid estrogens, nonylphenol ethoxylate metabolites, and other wastewater contaminants in groundwater affected by a residential septic system on Cape Cod, MA. *Environmental Science and Technology* 2006, 40, (16), 4894-4902.

[79]  Martinovic, D.; Denny, J. S.; Schmieder, P. K.; Ankley, G. T.; Sorensen, P. W., Temporal variation in the estrogenicity of a sewage treatment plant effluent and its biological significance. *Environmental Science and Technology* 2008, 42, (9), 3421-3427.

[80]  Panter, G. H.; Thompson, R. S.; Sumpter, J. P., Intermittent exposure of fish to estradiol. *Environmental Science and Technology* 2000, 34, (13), 2756-2760.

[81]  Palace, V. P.; Wautier, K. G.; Evans, R. E.; Blanchfield, P. J.; Mills, K. H.; Chalanchuk, S. M.; Godard, D.; McMaster, M. E.; Tetreault, G. R.; Peters, L. E.; Vandenbyllaardt, L.; Kidd, K. A., Biochemical and histopathological effects in pearl dace (Margariscus margarita) chronically exposed to a synthetic estrogen in a whole lake experiment. *Environmental Toxicology and Chemistry* 2006, 25, (4), 1114-1125.

[82]  Jin, S.; Yang, F.; Liao, T.; Hui, Y.; Xu, Y., Seasonal variations of estrogenic compounds and their estrogenicities in influent and effluent from a municipal sewage treatment plant in China. *Environmental Toxicology and Chemistry* 2008, 27, (1), 146-153.

[83]  Williams, R. J.; Johnson, A. C.; Smith, J. J. L.; Kanda, R., Steroid estrogens profiles along river stretches arising from sewage treatment works discharges. *Environmental Science and Technology* 2003, 37, (9), 1744-1750.

[84]  Purdom, C. E.; Hardiman, P. A.; Bye, V. V. J.; Eno, N. C.; Tyler, C. R.; Sumpter, J. P., Estrogenic effects of effluents from sewage treatment works. *Chemistry and Ecology* 1994, 8, (4), 275 - 285.

[85]  Desbrow, C.; Routledge, E. J.; Brighty, G. C.; Sumpter, J. P.; Waldock, M., Identification of estrogenic chemicals in STW effluent. 1. Chemical fractionation and in vitro biological screening. *Environmental Science and Technology* 1998, 32, (11), 1549-1558.

[86]  Baronti, C.; Curini, R.; D'Ascenzo, G.; Di Corcia, A.; Gentili, A.; Samperi, R., Monitoring natural and synthetic estrogens at activated sludge sewage treatment plants and in a receiving river water. *Environmental Science and Technology* 2000, 34, (24), 5059-5066.

[87]  Layton, A. C.; Gregory, B. W.; Seward, J. R.; Schultz, T. W.; Sayler, G. S., Mineralization of steroidal hormones by biosolids in wastewater treatment systems in Tennessee USA. *Environmental Science and Technology* 2000, 34, (18), 3925-3931.

[88]  Kirk, L. A.; Tyler, C. R.; Lye, C. M.; Sumpter, J. P., Changes in estrogenic and androgenic activities at different stages of treatment in wastewater treatment works. *Environmental Toxicology and Chemistry* 2002, 21, (5), 972-979.

[89]  Johnson, A. C.; Sumpter, J. P., Removal of endocrine-disrupting chemicals in activated sludge treatment works. *Environmental Science and Technology* 2001, 35, (24), 4697-4703.

[90]  Orlando, E. F.; Kolok, A. S.; Binzcik, G. A.; Gates, J. L.; Horton, M. K.; Lambright, C. S.; Gray, L. E.; Soto, A. M.; Guillette, L. J., Endocrine-disrupting effects of cattle feedlot effluent on an aquatic sentinel species, the fathead minnow. *Environmental Health Perspectives* 2004, 112, (3), 353-358.

[91]  Irwin, L. K.; Gray, S.; Oberdorster, E., Vitellogenin induction in painted turtle, Chrysemys picta, as a biomarker of exposure to environmental levels of estradiol. *Aquatic Toxicology* 2001, 55, (1-2), 49-60.

[92]  Hanselman, T. A.; Graetz, D. A.; Wilkie, A. C.; Szabo, N. J.; Diaz, C. S., Determination of steroidal estrogens in flushed dairy manure wastewater by gas chromatography-mass spectrometry. *Journal of Environmental Quality* 2006, 35, (3), 695-700.

[93]  Zheng, W.; Yates, S. R.; Bradford, S. A., Analysis of steroid hormones in a typical dairy waste disposal system. *Environmental Science and Technology* 2008, 42, (2), 530-535.

[94]  Hutchins, S. R.; White, M. V.; Hudson, F. M.; Fine, D. D., Analysis of lagoon samples from different concentrated animal feeding operations for estrogens and estrogen conjugates. *Environmental Science and Technology* 2007, 41, (3), 738-744.

[95]  Czajka, C. P.; Londry, K. L., Anaerobic biotransformation of estrogens. *Science of the Total Environment* 2006, 367, (2-3), 932-941.

[96]  Schiffer, B.; Daxenberger, A.; Meyer, K.; Meyer, H. H. D., The fate of trenbolone acetate and melengestrol acetate after application as growth promoters in cattle: Environmental studies. *Environmental Health Perspectives* 2001, 109, (11), 1145-1151.

[97]  Khan, B.; Lee, L. S.; Sassman, S. A., Degradation of synthetic androgens 17α- and 17β-trenbolone and trendione in agricultural soils. *Environ. Sci. Technol.* 2008, 42, (10), 3570-3574.

[98]  Hutchins, S. R.; White, M. V.; Fine, D. D.; Breidenbach, G. P., *Analysis of swine lagoons and ground water for environmental estrogens.* Battelle Press: Columbus, OH, 2003.

[99]  Shappell, N. W.; Billey, L. O.; Forbes, D.; Matheny, T. A.; Poach, M. E.; Reddy, G. B.; Hunt, P. G., Estrogenic activity and steroid hormones in swine wastewater through a lagoon constructed-wetland system. *Environmental Science and Technology* 2007, 41, (2), 444-450.

[100] Arnon, S.; Dahan, O.; Elhanany, S.; Cohen, K.; Pankratov, I.; Gross, A.; Ronen, Z.; Baram, S.; Shore, L. S., Transport of testosterone and estrogen from dairy-farm waste lagoons to groundwater. *Environmental Science and Technology* 2008, 42, (15), 5521-5526.

[101] Snyder, S. A.; Keith, T. L.; Verbrugge, D. A.; Snyder, E. M.; Gross, T. S.; Kannan, K.; Giesy, J. P., Analytical methods for detection of selected estrogenic compounds in aqueous mixtures. *Environmental Science and Technology* 1999, 33, (16), 2814-2820.

[102] Guo, T. D.; Chan, M.; Soldin, S. J., Steroid profiles using liquid chromatography - Tandem mass spectrometry with atmospheric pressure photoionization source. *Archives of Pathology and Laboratory Medicine* 2004, 128, (4), 469-+.

[103] Kjær, J.; Olsen, P.; Bach, K.; Barlebo, H. C.; Ingerslev, F.; Hansen, M.; Sorensen, B. H., Leaching of estrogenic hormones from manure-treated structured soils. *Environmental Science and Technology* 2007, 41, (11), 3911-3917.

[104] Kinney, C. A.; Furlong, E. T.; Zaugg, S. D.; Burkhardt, M. R.; Werner, S. L.; Cahill, J. D.; Jorgensen, G. R., Survey of organic wastewater contaminants in biosolids destined for land application. *Environmental Science and Technology* 2006, 40, (23), 7207-7215.

[105] Clara, M.; Strenn, B.; Saracevic, E.; Kreuzinger, N., Adsorption of bisphenol-A, 17 beta-estradiole and 17 alpha-ethinylestradiole to sewage sludge. *Chemosphere* 2004, 56, (9), 843-851.

[106] Andersen, H. R.; Hansen, M.; Kjolholt, J.; Stuer-Lauridsen, F.; Ternes, T.; Halling-Sorensen, B., Assessment of the importance of sorption for steroid estrogens removal during activated sludge treatment. *Chemosphere* 2005, 61, (1), 139-146.

[107] Lorenzen, A.; Hendel, J. G.; Conn, K. L.; Bittman, S.; Kwabiah, A. B.; Lazarovitz, G.; Masse, D.; McAllister, T. A.; Topp, E., Survey of hormone activities in municipal biosolids and animal manures. *Environmental Toxicology* 2004, 19, (3), 216-225.

[108] Lapen, D. R.; Topp, E.; Metcalfe, C. D.; Li, H.; Edwards, M.; Gottschall, N.; Bolton, P.; Curnoe, W.; Payne, M.; Beck, A., Pharmaceutical and personal care products in tile drainage following land application of municipal biosolids. *Science of the Total Environment* 2008, 399, (1-3), 50-65.

[109] Topp, E.; Monteiro, S. C.; Beck, A.; Coelho, B. B.; Boxall, A. B. A.; Duenk, P. W.; Kleywegt, S.; Lapen, D. R.; Payne, M.; Sabourin, L.; Li, H.; Metcalfe, C. D., Runoff of pharmaceuticals and personal care products following application of biosolids to an agricultural field. *Science of the Total Environment* 2008, 396, (1), 52-59.

[110] Nichols, D. J.; Daniel, T. C.; Moore, P. A.; Edwards, D. R.; Pote, D. H., Runoff of estrogen hormone 17β-estradiol from poultry litter applied to pasture. *Journal of Environmental Quality* 1997, 26, (4), 1002-1006.

[111] Finlay-Moore, O.; Hartel, P. G.; Cabrera, M. L., 17β-estradiol and testosterone in soil and runoff from grasslands amended with broiler litter. *Journal of Environmental Quality* 2000, 29, (5), 1604-1611.

[112] Peterson, E. W.; Davis, R. K.; Orndorff, H. A., 17 beta-estradiol as an indicator of animal waste contamination in mantled karst aquifers. *Journal of Environmental Quality* 2000, 29, (3), 826-834.

[113] Stumpe, B.; Marschner, B., Long-term sewage sludge application and wastewater irrigation on the mineralization and sorption of 17β-estradiol and testosterone in soils. *Science of the Total Environment* 2007, 374, (2-3), 282-291.

[114] Holbrook, R. D.; Love, N. G.; Novak, J. T., Sorption of 17β-estradiol and 17α-ethinylestradiol by colloidal organic carbon derived from biological wastewater treatment systems. *Environmental Science and Technology* 2004, 38, (12), 3322-3329.

[115] Zhou, J. L.; Liu, R.; Wilding, A.; Hibberd, A., Sorption of selected endocrine disrupting chemicals to different aquatic colloids. *Environmental Science and Technology* 2007, 41, (1), 206-213.

[116] Ternes, T. A.; Stumpf, M.; Mueller, J.; Haberer, K.; Wilken, R. D.; Servos, M., Behavior and occurrence of estrogens in municipal sewage treatment plants - I. Investigations in Germany, Canada and Brazil. *Science of the Total Environment* 1999, 225, (1-2), 81-90.

[117] Halling-Sorensen, B.; Nielsen, S. N.; Lanzky, P. F.; Ingerslev, F.; Lutzhoft, H. C. H.; Jorgensen, S. E., Occurrence, fate and effects of pharmaceutical substances in the environment - A review. *Chemosphere* 1998, 36, (2), 357-394.

[118] Borch, T.; Young, R. B.; Gray, J. L.; Foreman, W. T.; Yang, Y., Presence and fate of steroid hormones in a Colorado river. *Abstracts of Papers of the American Chemical Society (preprint)* 2008, 8, (2), 689-694.

[119] Kolpin, D. W.; Furlong, E. T.; Meyer, M. T.; Thurman, E. M.; Zaugg, S. D.; Barber, L. B.; Buxton, H. T., Pharmaceuticals, hormones, and other organic wastewater contaminants in US streams, 1999-2000: A national reconnaissance. *Environmental Science and Technology* 2002, 36, (6), 1202-1211.

[120] Kolpin, D. W.; Furlong, E. T.; Meyer, M. T.; Thurman, E. M.; Zaugg, S. D.; Barber, L. B.; Buxton, H. T., Response to Comment on,"Pharmaceuticals, hormones, and other organic wastewater contaminants in US streams, 1999-2000: A national reconnaissance". *Environmental Science and Technology* 2002, 36, (18), 4007-4008.

[121] Holthaus, K. I. E.; Johnson, A. C.; Jurgens, M. D.; Williams, R. J.; Smith, J. J. L.; Carter, J. E., The potential for estradiol and ethinylestradiol to sorb to suspended and bed sediments in some English rivers. *Environmental Toxicology and Chemistry* 2002, 21, (12), 2526-2535.

[122] Sparks, D. L., *Environmental Soil Chemistry.* 2nd ed.; Academic Press: San Diego, 2002; p 352.

[123] Nguyen, T. H.; Goss, K. U.; Ball, W. P., Polyparameter linear free energy relationships for estimating the equilibrium partition of organic compounds between water and the natural organic matter in soils and sediments. *Environmental Science and Technology* 2005, 39, (4), 913-924.

[124] McGechan, M. B.; Lewis, D. R., Transport of particulate and colloid-sorbed contaminants through soil, part 1: General principles. *Biosystems Engineering* 2002, 83, (3), 255-273.

[125] Karickhoff, S. W.; Brown, D. S.; Scott, T. A., Sorption of hydrophobic pollutants on natural sediments. *Water Research* 1979, 13, (3), 241-248.

[126] Chiou, C. T.; Peters, L. J.; Freed, V. H., Physical concept of soil-water equilibria for non-ionic organic compounds. *Science* 1979, 206, (4420), 831-832.

[127] Yamamoto, H.; Liljestrand, H. M.; Shimizu, Y.; Morita, M., Effects of physical-chemical characteristics on the sorption of selected endocrine disruptors by dissolved organic matter surrogates. *Environmental Science and Technology* 2003, 37, (12), 2646-2657.

[128] Das, B. S.; Lee, L. S.; Rao, P. S. C.; Hultgren, R. P., Sorption and degradation of steroid hormones in soils during transport: Column studies and model evaluation. *Environmental Science and Technology* 2004, 38, (5), 1460-1470.

[129] Bosma, T. N. P.; Middeldorp, P. J. M.; Schraa, G.; Zehnder, A. J. B., Mass transfer limitation of biotransformation: Quantifying bioavailability. *Environmental Science and Technology* 1997, 31, (1), 248-252.

[130] Xia, X. H.; Yu, H.; Yang, Z. F.; Huang, G. H., Biodegradation of polycyclic aromatic hydrocarbons in the natural waters of the Yellow River: Effects of high sediment content on biodegradation. *Chemosphere* 2006, 65, (3), 457-466.

[131] Zepp, R. G.; Schlotzhauer, P. F., Effects of equilibration time on photoreactivity of the pollutant DDE sorbed on nautral sediments. *Chemosphere* 1981, 10, (5), 453-460.

[132] Tolls, J., Sorption of veterinary pharmaceuticals in soils: A review. *Environmental Science and Technology* 2001, 35, (17), 3397-3406.

[133] Casey, F. X. M.; Larsen, G. L.; Hakk, H.; Simunek, J., Fate and transport of 17β-estradiol in soil-water systems. *Environmental Science and Technology* 2003, 37, (11), 2400-2409.

[134] Casey, F. X. M.; Hakk, H.; Simunek, J.; Larsen, G. L., Fate and transport of testosterone in agricultural soils. *Environmental Science and Technology* 2004, 38, (3), 790-798.

[135] Grathwohl, P., Influence of organic matter from soils and sediments from various origins on the sorption of some chlorinated aliphatic hydrocarbons - Implications on $K_{oc}$ correlations. *Environmental Science and Technology* 1990, 24, (11), 1687-1693.

[136] Chiou, C. T.; Freed, V. H.; Schmedding, D. W.; Kohnert, R. L., Partition coefficient and bioaccumulation of selected organic chemicals. *Environmental Science and Technology* 1977, 11, (5), 475-478.

[137] Van Emmerik, T.; Angove, M. J.; Johnson, B. B.; Wells, J. D.; Fernandes, M. B., Sorption of 17β-estradiol onto selected soil minerals. *Journal of Colloid and Interface Science* 2003, 266, (1), 33-39.

[138] Pignatello, J. J.; Xing, B. S., Mechanisms of slow sorption of organic chemicals to natural particles. *Environmental Science and Technology* 1996, 30, (1), 1-11.

[139] Xing, B. S.; Pignatello, J. J., Dual-mode sorption of low-polarity compounds in glassy poly(vinyl chloride) and soil organic matter. *Environmental Science and Technology* 1997, 31, (3), 792-799.

[140] Shareef, A.; Angove, M. J.; Wells, J. D.; Johnson, B. B., Sorption of bisphenol A, 17α-ethynylestradiol and estrone to mineral surfaces. *Journal of Colloid and Interface Science* 2006, 297, (1), 62-69.

[141] Gao, J. P.; Maguhn, J.; Spitzauer, P.; Kettrup, A., Sorption of pesticides in the sediment of the Teufelsweiher Pond (Southern Germany). I: Equilibrium assessments, effect of organic carbon content and pH. *Water Research* 1998, 32, (5), 1662-1672.

[142] Liu, R. X.; Wilding, A.; Hibberd, A.; Zhou, J. L., Partition of endocrine-disrupting chemicals between colloids and dissolved phase as determined by cross-flow ultrafiltration. *Environmental Science and Technology* 2005, 39, (8), 2753-2761.

[143] Newman, M. E.; Filella, M.; Chen, Y.; Nègre, J.-C.; Perret, D.; Buffle, J., Submicron particles in the Rhine River- II. Comparison of field observations and model predictions. *Water Research* 1994, 28, (1), 107-118.

[144] Chin, Y. P.; Gschwend, P. M., Partitioning of polycyclic aromatic hydrocarbons to marine porewater organic colloids. *Environmental Science and Technology* 1992, 26, (8), 1621-1626.

[145] Sun, W. L.; Ni, J. R.; Xu, N.; Sun, L. Y., Fluorescence of sediment humic substance and its effect on the sorption of selected endocrine disruptors. *Chemosphere* 2007, 66, (4), 700-707.

[146] Fan, Z. S.; Casey, F. X. M.; Hakk, H.; Larsen, G. L., Persistence and fate of 17β-estradiol and testosterone in agricultural soils. *Chemosphere* 2007, 67, (5), 886-895.

[147] Chin, Y. P.; Aiken, G.; Oloughlin, E., Molecular weight, polydispersity, and spectroscopic properties of aquatic humic substances. *Environmental Science and Technology* 1994, 28, (11), 1853-1858.

[148] Canonica, S.; Jans, U.; Stemmler, K.; Hoigne, J., Transformation kinetics of phenols in water - Photosensitization by dissolved natural organic material and aromatic ketones. *Environmental Science and Technology* 1995, 29, (7), 1822-1831.

[149] Peck, M.; Gibson, R. W.; Kortenkamp, A.; Hill, E. M., Sediments are major sinks of steroidal estrogens in two United Kingdom rivers. *Environmental Toxicology and Chemistry* 2004, 23, (4), 945-952.

[150] Jurgens, M. D.; Holthaus, K. I. E.; Johnson, A. C.; Smith, J. J. L.; Hetheridge, M.; Williams, R. J., The potential for estradiol and ethinylestradiol degradation in English rivers. *Environmental Toxicology and Chemistry* 2002, 21, (3), 480-488.

[151] Sangsupan, H. A.; Radcliffe, D. E.; Hartel, P. G.; Jenkins, M. B.; Vencill, W. K.; Cabrera, M. L., Sorption and transport of 17β-estradiol and testosterone in undisturbed soil columns. *Journal of Environmental Quality* 2006, 35, (6), 2261-2272.

[152] Fan, Z. S.; Casey, F. X. M.; Hakk, H.; Larsen, G. L., Discerning and modeling the fate and transport of testosterone in undisturbed soil. *Journal of Environmental Quality* 2007, 36, (3), 864-873.

[153] Lee, L. S.; Strock, T. J.; Sarmah, A. K.; Rao, P. S. C., Sorption and dissipation of testosterone, estrogens, and their primary transformation products in soils and sediment. *Environmental Science and Technology* 2003, 37, (18), 4098-4105.

[154] Ying, G. G.; Kookana, R. S., Sorption and degradation of estrogen-like-endocrine disrupting chemicals in soil. *Environmental Toxicology and Chemistry* 2005, 24, (10), 2640-2645.

[155] Ying, G. G.; Kookana, R. S.; Dillon, P., Sorption and degradation of selected five endocrine disrupting chemicals in aquifer material. *Water Research* 2003, 37, (15), 3785-3791.

[156] Casey, F. X. M.; Simunek, J.; Lee, J.; Larsen, G. L.; Hakk, H., Sorption, mobility, and transformation of estrogenic hormones in natural soil. *Journal of Environmental Quality* 2005, 34, (4), 1372-1379.

[157] Hildebrand, C.; Londry, K. L.; Farenhorst, A., Sorption and desorption of three endocrine disrupters in soils. *Journal of Environmental Science and Health Part B-Pesticides Food Contaminants and Agricultural Wastes* 2006, 41, (6), 907-921.

[158] Korom, S. F., Natural denitrification in the saturated zone - A review. *Water Resources Research* 1992, 28, (6), 1657-1668.

[159] Donova, M. V., Transformation of steroids by actinobacteria: A review. *Applied Biochemistry and Microbiology* 2007, 43, (1), 1-14.

[160] Lee, H. B.; Liu, D., Degradation of 17β-estradiol and its metabolites by sewage bacteria. *Water Air and Soil Pollution* 2002, 134, (1-4), 353-368.

[161] Shi, J.; Fujisawa, S.; Nakai, S.; Hosomi, M., Biodegradation of natural and synthetic estrogens by nitrifying activated sludge and ammonia-oxidizing bacterium Nitrosomonas europaea. *Water Research* 2004, 38, (9), 2323-2330.

[162] Ren, Y. X.; Nakano, K.; Nomura, M.; Chiba, N.; Nishimura, O., Effects of bacterial activity on estrogen removal in nitrifying activated sludge. *Water Research* 2007, 41, (14), 3089-3096.

[163] Yi, T.; Harper, W. F., The link between nitrification and biotransformation of 17 alpha-ethinylestradiol. *Environmental Science and Technology* 2007, 41, (12), 4311-4316.

[164] Fernandes, P.; Cruz, A.; Angelova, B.; Pinheiro, H. M.; Cabral, J. M. S., Microbial conversion of steroid compounds: recent developments. *Enzyme and Microbial Technology* 2003, 32, (6), 688-705.

[165] Donova, M. V.; Egorova, O. V.; Nikolayeva, V. M., Steroid 17β-reduction by microorganisms - A review. *Process Biochemistry* 2005, 40, (7), 2253-2262.

[166] Malaviya, A.; Gomes, J., Androstenedione production by biotransformation of phytosterols. *Bioresource Technology* 2008, 99, (15), 6725-6737.

[167] Faramarzi, M. A.; Adrangi, S.; Yazdi, M. T., Microalgal biotransformation of steroids. *Journal of Phycology* 2008, 44, (1), 27-37.

[168] Dodson, R. M.; Muir, R. D., Microbiological transformations. 6. Microbiological aromatization of steroids. *Journal of the American Chemical Society* 1961, 83, (22), 4627-and.

[169] Horinouchi, M.; Hayashi, T.; Yamamoto, T.; Kudo, T., A new bacterial steroid degradation gene cluster in Comamonas testosteroni TA441 which consists of aromatic-compound degradation genes for seco-steroids and 3-ketosteroid dehydrogenase genes. *Applied and Environmental Microbiology* 2003, 69, (8), 4421-4430.

[170] Ren, H. Y.; Ji, S. L.; Ahmad, N. U. D.; Dao, W.; Cui, C. W., Degradation characteristics and metabolic pathway of 17α-ethynylestradiol by Sphingobacterium sp. JCR5. *Chemosphere* 2007, 66, (2), 340-346.

[171] Coombe, R. G.; Tsong, Y. Y.; Hamilton, P. B.; Sih, C. J., Mechanisms of steroid oxidation by microorganisms. X. Oxidative cleavage of estrone. *Journal of Biological Chemistry* 1966, 241, (7), 1587-and.

[172] Porter, R. B. R.; Gallimore, W. A.; Reese, P. B., Steroid transformations with Exophiala jeanselmei var. lecanii-corni and Ceratocystis paradoxa. *Steroids* 1999, 64, (11), 770-779.

[173] Wilson, M. R.; Gallimore, W. A.; Reese, P. B., Steroid transformations with Fusarium oxysporum var. cubense and Colletotrichum musae. *Steroids* 1999, 64, (12), 834-843.

[174] Lamm, A. S.; Chen, A. R. M.; Reynolds, W. F.; Reese, P. B., Steroid hydroxylation by Whetzelinia sclerotiorum, Phanerochaete chrysosporium and Mucor plumbeus. *Steroids* 2007, 72, (9-10), 713-722.

[175] Fiorentino, A.; Pinto, G.; Pollio, A.; Previtera, L., Biotransformation of 5α-androstane-3,17-dione by microalgal cultures. *Bioorganic and Medicinal Chemistry Letters* 1991, 1, (12), 673-674.

[176] Lorenzen, A.; Chapman, R.; Hendel, J. G.; Topp, E., Persistence and pathways of testosterone dissipation in agricultural soil. *Journal of Environmental Quality* 2005, 34, (3), 854-860.

[177] Colucci, M. S.; Bork, H.; Topp, E., Persistence of estrogenic hormones in agricultural soils: I. 17β-estradiol and estrone. *Journal of Environmental Quality* 2001, 30, (6), 2070-2076.

[178] Colucci, M. S.; Topp, E., Persistence of estrogenic hormones in agricultural soils: II. 17α-ethynylestradiol. *Journal of Environmental Quality* 2001, 30, (6), 2077-2080.

[179] Hwang, S.; Lee, D. I.; Lee, C. H.; Ahn, I. S., Oxidation of 17 alpha-ethynylestradiol with Mn(III) and product identification. *Journal of Hazardous Materials* 2008, 155, (1-2), 334-341.

[180] Vialaton, D.; Richard, C., Phototransformation of aromatic pollutants in solar light: Photolysis versus photosensitized reactions under natural water conditions. *Aquatic Sciences* 2002, 64, (2), 207-215.

[181] Turro, N. J., *Modern Molecular Photochemistry*. University Science Books: Sausalito, 1991.

[182] Leifer, A., *The kinetics of environmental aquatic photochemistry: Theory and practice*. American Chemical Society: Washington, DC, 1988; p 304.

[183] Lin, A. Y. C.; Reinhard, M., Photodegradation of common environmental pharmaceuticals and estrogens in river water. *Environmental Toxicology and Chemistry* 2005, 24, (6), 1303-1309.

[184] Liu, B.; Liu, X. L., Direct photolysis of estrogens in aqueous solutions. *Science of the Total Environment* 2004, 320, (2-3), 269-274.

[185] Liu, B.; Wu, F.; Deng, N. S., UV-light induced photodegradation of 17α-ethynylestradiol in aqueous solutions. *Journal of Hazardous Materials* 2003, 98, (1-3), 311-316.

[186] Khanal, S. K.; Xie, B.; Thompson, M. L.; Sung, S. W.; Ong, S. K.; Van Leeuwen, J., Fate, transport, and biodegradation of natural estrogens in the environment and engineered systems. *Environmental Science and Technology* 2006, 40, (21), 6537-6546.

[187] Rosenfeldt, E. J.; Chen, P. J.; Kullman, S.; Linden, K. G., Destruction of estrogenic activity in water using UV advanced oxidation. *Science of the Total Environment* 2007, 377, (1), 105-113.

[188] Agustina, T. E.; Ang, H. M.; Vareek, V. K., A review of synergistic effect of photocatalysis and ozonation on wastewater treatment. *Journal of Photochemistry and Photobiology C-Photochemistry Reviews* 2005, 6, (4), 264-273.

[189] Belgiorno, V.; Rizzo, L.; Fatta, D.; Della Rocca, C.; Lofrano, G.; Nikolaou, A.; Naddeo, V.; Meric, S., Review on endocrine disrupting-emerging compounds in urban wastewater: occurrence and removal by photocatalysis and ultrasonic irradiation for wastewater reuse. *Desalination* 2007, 215, (1-3), 166-176.

[190] Zhao, Y.; Hu, J.; Jin, W., Transformation of oxidation products and reduction of estrogenic activity of 17β-estradiol by a heterogeneous photo-fenton reaction. *Environ. Sci. Technol.* 2008, 42, (14), 5277-5284.

[191] Hakk, H.; Millner, P.; Larsen, G., Decrease in water-soluble 17β-estradiol and testosterone in composted poultry manure with time. *Journal of Environmental Quality* 2005, 34, (3), 943-950.

[192] Shappell, N. W.; Vrabel, M. A.; Madsen, P. J.; Harrington, G.; Billey, L. O.; Hakk, H.; Larsen, G. L.; Beach, E. S.; Horwitz, C. P.; Ro, K.; Hunt, P. G.; Collins, T. J., Destruction of estrogens using Fe-TAML/peroxide catalysis. *Environmental Science and Technology* 2008, 42, (4), 1296-1300.

[193] Nichols, D. J.; Daniel, T. C.; Edwards, D. R.; Mooe, P. A.; Pote, D. H., Use of grass filter strips to reduce 17β-estradiol in runoff from fescue-applied poultry litter. *Journal of Soil and Water Conservation* 1998, 53, (1), 74-77.

[194] Chang, H.; Wu, S.; Hu, J.; Asami, M.; Kunikane, S., Trace analysis of androgens and progestogens in environmental waters by ultra-performance liquid chromatography-electrospray tandem mass spectrometry. *Journal of Chromatography A* 2008, 1195, (1-2), 44-51.

[195] Yamamoto, A.; Kakutani, N.; Yamamoto, K.; Kamiura, T.; Miyakoda, H., Steroid hormone profiles of urban and tidal rivers using LC/MS/MS equipped with electrospray ionization and atmospheric pressure photoionization sources. *Environmental Science and Technology* 2006, 40, (13), 4132-4137.

[196] Vethaak, A. D.; Lahr, J.; Schrap, S. M.; Belfroid, A. C.; Rijs, G. B. J.; Gerritsen, A.; de Boer, J.; Bulder, A. S.; Grinwis, G. C. M.; Kuiper, R. V.; Legler, J.; Murk, T. A. J.; Peijnenburg, W.; Verhaar, H. J. M.; de Voogt, P., An integrated assessment of estrogenic contamination and biological effects in the aquatic environment of The Netherlands. *Chemosphere* 2005, 59, (4), 511-524.

[197] Kuch, H. M.; Ballschmiter, K., Determination of endocrine-disrupting phenolic compounds and estrogens in surface and drinking water by HRGC-(NCI)-MS in the picogram per liter range. *Environmental Science and Technology* 2001, 35, (15), 3201-3206.

[198] Peng, X.; Yu, Y.; Tang, C.; Tan, J.; Huang, Q.; Wang, Z., Occurrence of steroid estrogens, endocrine-disrupting phenols, and acid pharmaceutical residues in urban riverine water of the Pearl River Delta, South China. *Science of the Total Environment* 2008, 397, (1-3), 158-166.

[199] Zuehlke, S.; Duennbier, U.; Heberer, T., Determination of estrogenic steroids in surface water and wastewater by liquid chromatography-electrospray tandem mass spectrometry. *Journal of Separation Science* 2005, 28, (1), 52-58.

[200] Cargouet, M.; Perdiz, D.; Mouatassim-Souali, A.; Tamisier-Karolak, S.; Levi, Y., Assessment of river contamination by estrogenic compounds in Paris area (France). *Science of the Total Environment* 2004, 324, (1-3), 55-66.

[201] Labadie, P.; Budzinski, H., Determination of steroidal hormone profiles along the Jalle d'Eysines River (near Bordeaux, France). *Environmental Science and Technology* 2005, 39, (14), 5113-5120.

[202] de Alda, M. J. L.; Barcelo, D., Determination of steroid sex hormones and related synthetic compounds considered as endocrine disrupters in water by liquid chromatography-diode array detection-mass spectrometry. *Journal of Chromatography A* 2000, 892, (1-2), 391-406.

[203] Sole, M.; de Alda, M. J. L.; Castillo, M.; Porte, C.; Ladegaard-Pedersen, K.; Barcelo, D., Estrogenicity determination in sewage treatment plants and surface waters from the Catalonian area (NE Spain). *Environmental Science and Technology* 2000, 34, (24), 5076-5083.

In: Aquatic Ecosystem Research Trends
Editor: George H. Nairne

ISBN 978-1-60692-772-4
© 2009 Nova Science Publishers, Inc.

*Chapter 5*

# CHANGES IN MACROINVERTEBRATE ELEMENTAL STOICHIOMETRY BELOW A POINT SOURCE IN A MEDITERRANEAN STREAM

*Jesús D. Ortiz[1],\* , M. Ángeles Puig[1], Eugènia Martí[1], Francesc Sabater[2] and Gora Merseburger[2],\**

[1]Center of Advanced Studies of Blanes (CEAB-CSIC),
Camí d'accés a la cala St. Francesc, 14, E-17300 Blanes,
Girona (Catalonia, Spain)
[2]Department of Ecology, University of Barcelona, Av. Diagonal 645,
E-08028 Barcelona (Catalonia, Spain)
\*current adress: Rheos ecology, Camí de Valls 81-87, E-43204 Reus, Tarragona
(Catalonia, Spain) www.rheosecology.com

## ABSTRACT

Patterns regarding elemental stoichiometry in stream ecosystems remain incompletely understood. We analyzed C, N, and P contents in water, benthic macroinvertebrates and their potential food resources in two reaches located upstream and downstream of a point source input in La Tordera stream (Catalonia, NE Spain). The nutrient contents in periphyton and mosses did not vary between the two reaches. The percentage of C and N in filamentous algae was also similar in the two reaches, but percentage of P was two times higher downstream than upstream of the point source. Stoichiometric ratios (C:nutrient) for CPOM, FPOM, and SPOM decreased considerably below the point source. Elemental contents and ratios were highly variable among macroinvertebrate taxa but did not differ significantly between the two reaches. Dipterans, caddisflies, and mayflies had similar elemental contents and stoichiometry, whereas C and N were lower in mollusks and P in beetles. Elemental contents and ratios of functional feeding groups were not significantly different between the two reaches. Predators had the higher C, N and P contents, and P was also high in filterers. Scrapers harbored the lowest elemental contents. Elemental imbalances between consumers and resources were reduced at the downstream reach relative to the upstream reach, reflecting lower nutrient retention efficiency by consumers at the altered reach. The understanding of the effects of stoichiometric constraints derived from human disturbances on in-stream processes provide valuable information that could pay for future research and the development of future management plans.

# INTRODUCTION

Ecological stoichiometry is a conceptual framework that provides an integrative approach for the analysis of the balance among chemical elements in ecological interactions and links their cycling in ecosystems [Cross *et al*. 2005, Elser *et al*. 1996, Moe *et al*. 2005, Reiners 1986, Sterner and Elser 2002]. Elements cannot be synthesized or interconverted by organisms, and must therefore be acquired in quantities sufficient to attain their requirements for maintenance, growth, and reproduction [Anderson *et al*. 2004, Frost *et al*. 2005]. In opposition to autotrophic organisms, metazoans are often realized as being stoichiometrically homeostatic, i.e. having a relatively constant body nutrient proportions regardless of the chemical composition in their food [Elser *et al*. 1996]. This implies that differences in nutrient proportions between consumers and their food resources may lead to stoichiometric constraints. Recent research have shown that elemental imbalances between consumers and their food resources can impinge on population dynamics [Burkhardt and Lehman 1994, Loladze *et al*. 2000], trophic interactions [Elser *et al*. 1998, Sterner 1997], and community structure [DeMott and Gulati 1999, Grover 2002].

After the works of Redfield [1958], who demonstrated that biogeochemical cycles of C, N, and P are strongly coupled in pelagic oceanic ecosystems, ecological stoichiometry has been extensively studied in marine and lake planktonic ecosystems [e.g., Andersen and Hessen 1991, Elser and Urabe 1999, Stauffer 1985, Sterner *et al*. 1997, Urabe *et al*. 2002]. These works provided valuable outcomes that supposed the starting point for further development of the stoichiometric theory. The establishment of a more comprehensive framework encouraged researchers to set up the study of biological systems with increasing degrees of complexity, from benthic and terrestrial autotrophs [Enríquez *et al*. 1993, Kahlert 1998] to terrestrial insects [Schade *et al*. 2003] and vertebrates [Schindler and Eby 1997, Sterner and George 2000, Vanni *et al*. 2002]. Simultaneously, the implementation of laboratory and field experiments [e.g., Elser *et al*. 1998, Frost and Elser 2002a, Stelzer and Lamberti 2002, Sterner 1997, Urabe and Watanabe 1992] increased the understanding of processes and patterns involving nutrient cycling in ecosystems and generated valuable hypotheses for future research addressing elemental stoichiometry. However, there is yet little empirical support for stoichiometric theory in benthic ecosystems, because the elemental content of many biomass compartments (i.e., autotrophs, organic matter, macroinvertebrates, etc.) remain incompletely understood.

Frost *et al*. [2003] presented a pioneer research examining the elemental composition of lake benthic macroinvertebrates across a wide range of dissolved nutrient concentrations that lent support to the hypothesis of homeostatic regulation. Two years after, Evans-White *et al*. [2005] supplied similar results for stream macroinvertebrates. However, in a likewise interesting study, Cross *et al*. [2003] found that some stream insects did not exhibit strict homeostasis in a detritus-based stream. This finding was supported by posterior research in a Swedish lake [Liess and Hillebrand 2005] and in mountain rivers [Bowman *et al*. 2005]. Homeostatic regulation has been established for lake zooplankton [Hessen and Lyche 1991] and *Elimia* snails [Stelzer and Lamberti 2001] through manipulations of food elemental composition. On the other hand, some experiments where mayfly nymphs and snails were grew under different quantities and qualities of food supply demonstrated that body stoichiometry is susceptible to change under high elemental mismatches between consumers and their food resources [Frost and Elser 2002a, Liess and Hillebrand 2006]. Similar results have been also obtained in lake planktonic ecosystems for *Daphnia* [DeMott *et al.* 1998].

Within this framework, we determined the C, N, and P contents in coarse, fine, and suspended particulate organic matter (CPOM, FPOM, and SPOM), periphyton, filamentous algae, mosses, and benthic macroinvertebrates in two reaches located upstream and downstream of a point source input in La Tordera stream. Our objectives were to examine the effects of the point source on stoichiometric ratios of macroinvertebrates and their potential food resources with a special insight on the degree of homeostatic regulation in stream macroinvertebrates and the variability of elemental imbalances between consumers and food resources.

## METHODS

### Study Site

The study was conducted in La Tordera stream, located in Catalonia (NE Spain; Figure 1).

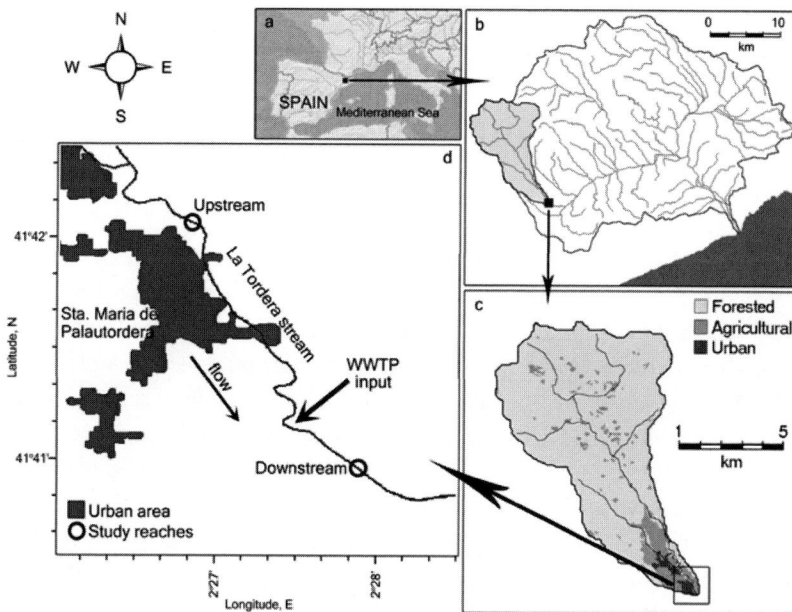

Figure 1. Location of La Tordera stream in Catalonia, NE Spain (a), La Tordera catchment and the subcatchment affecting the sampling site, highlighted in grey (b), land uses of the subcatchment affecting the study reaches (c), and location of the study reaches in relation to the wastewater treatment plant (WWTP) input (d). Data from the Catalan Cartographic institute (ICC).

We selected one reach upstream and one reach downstream of the input of the wastewater treatment plant (WWTP) effluent of Sta. Maria de Palautordera. At the sampling site (41°41' N, 2°27' E, 200 m a. s. l.), La Tordera is a 3rd-order stream draining a catchment of 80 km$^2$ dominated by a sclerophyllous forest of several species of *Quercus* (Figure 1). Small patches of irrigated crops are present in the lower part of the catchment, surrounding the urban area. The geology of the catchment is mainly siliceous, dominated by slates and phyllites. The climate is Mediterranean, with mean air temperatures ranging from 5 °C (January) to 23 °C (August). Mean annual precipitation is 575 mm, which mostly occurs in spring and fall.

The population of the catchment in 2001 was 8564 inhabitants, 93% of which were concentrated in the lower part of the catchment, in the villages of Sta. Maria de Palautordera and St. Esteve de Palautordera. The mean outflow of the WWTP effluent to La Tordera stream is 1300 m$^3$/day. The WWTP perform a biological treatment with activated sludge, but lack the technology to actively remove nitrogen or phosphorus. The point source considerably increased nutrient concentrations in La Tordera stream and persisted several hundred m below.

## Experimental Setting

We selected two 100-m long run-riffle reaches with similar substrata type and canopy cover upstream and downstream of the WWTP effluent input. The upstream reach was located three km above the WWTP effluent input and served as a reference reach. The downstream reach was located 500 m below the WWTP effluent input. We collected samples on six dates over the hydrologic year 2001-2002: November of 2001 and January, March, April, June, and September of 2002.

## Physical and Chemical Parameters

Discharge was calculated according to the velocity-area method described in Gordon *et al.* [1992] using a Neurtek Instruments® Miniair 2 flow meter. Water temperature and dissolved oxygen (DO) concentration were measured in each reach over 24 h cycles on each sampling date with a WTW® Oxi 340-A oxygen meter. We measured conductivity using a WTW® LF 340 conductivity meter. Water samples for nutrient analysis and dissolved organic carbon (DOC) were taken in each reach, filtered on site through preashed Whatman® GF/F glass fiber filters and stored on ice. Ammonium ($NH_4^+$-N) concentration was analyzed on a Bran-Luebbe® Technicon Autoanalyzer II. Nitrate ($NO_3^-$-N), and soluble reactive phosphorus (SRP) concentrations were analyzed on a Bran-Luebbe® TRAACS 2000 Autoanalyzer. $NO_3^-$-N was analyzed using the cadmium-copper reduction method, and SRP was done using the molybdenum blue colorimetric method. DOC concentration was analyzed using high-temperature catalytic oxidation (Shimadzu® TOC 5000 analyzer).

## Periphyton

We measured periphyton biomass by randomly collecting six stones from each reach. We broke each collected stone and selected flat pieces from the upper surface of approximately four cm$^2$. Samples were frozen until analysis. Samples were dried at 60 °C until constant weight, weighted, ashed at 450 °C for 4-5 h, and reweighted to obtain ash free dry mass (AFDM) and estimate dry mass. We measured the area of each rock piece using the computer program Scion Image (for Windows release Beta 4.0.2, Scion Corp., Frederick, Maryland) through high-resolution digital photographs of the stone pieces to determine AFDM per unit area.

## Benthic Macroinvertebrates

Six modified Surber samples (625 cm$^2$, 250 μm mesh size) were taken in each reach and sampling date. Samples were preserved in the field with 4% formaldehyde solution or frozen. In the laboratory, heavier inorganic substrates were removed by elutriation. All large invertebrates (> 5 mm) were hand-picked from the samples. If necessary, the smallest invertebrates (250 μm-5 mm) were subsampled on an area basis [Moulton II et al. 2000]. Invertebrates were counted by handpicking with the aid of a dissecting microscope at 15x magnification, identified to the lowest practical taxonomic level, enumerated, dried (60 °C until constant weight) and weighted to obtain dry mass. Taxa were assigned to their relative contribution to each functional feeding group (FFG) according to Moog [2002] and Tachet et al. [2000].

## Benthic Organic Matter and Primary Producers

We quantified standing stocks of benthic organic matter (BOM), filamentous algae, and mosses from the macroinvertebrate samples. After macroinvertebrates were removed, remaining organic matter was sorted into BOM, filamentous algae, and mosses. BOM was separated into coarse particulate organic matter (CPOM; > 1mm) and fine particulate organic matter (FPOM; 250 μm-1 mm) using nested sieves. Each fraction was dried at 60 °C for one week and weighted to obtain dry mass. Suspended particulate organic matter (SPOM) concentrations were obtained by filtering known volumes of stream water. SPOM samples were obtained through a peristaltic pump from the nearest four cm to the streambed and filtered with ashed and weighed Whatman$^{®}$ GF/F glass fiber filters (pore size = 0.7 μm). Filters were dried at 60 °C until constant weight and weighted to obtain dry mass.

## Elemental Analysis

We collected samples of macroinvertebrates, CPOM, FPOM, SPOM, periphyton, filamentous algae, and mosses in each reach and sampling date. Periphyton was brushed from the upper part of cobbles from the streambed. Samples of SPOM, periphyton, filamentous algae, and mosses were dried at 60 °C until constant weight and homogenized. Samples of macroinvertebrates, CPOM, and FPOM were immediately frozen. When samples were thawed, macroinvertebrates were sorted, identified, dried and homogenized. All live stages were pooled to obtain enough weight. Remaining BOM was sorted into CPOM and FPOM using nested sieves as described above, dried, and homogenized. All samples were weighed on a microbalance to the nearest μg. For C and N analysis, samples were weighed in tin capsules and analyzed with a Carlo Erba NA 2100 CHN analyzer (Carlo Erba Instruments, Milan, Italy). For P analysis, samples were oxidized with potassium persulphate in a microwave and were analyzed using the malachite green colorimetric technique [Fernández *et al.* 1985]. When sample biomass for elemental analysis was limited, we combined macroinvertebrate taxa from different sampling dates. Mollusks were analyzed including the shell, because the shell is an intrinsic part of mollusks and it was virtually impossible to separate the shell from living tissue of certain taxa and small individuals. Elemental contents are expressed as percent of dry mass and all stoichiometric ratios are molar.

## Data Analysis

We contrasted physical and chemical variables, between the two study reaches over the sampling period through paired *T*-test. We compared %C, %N, and %P and C:N, C:P, and N:P ratios in taxa from the two reaches by using paired *T*-tests to examine differences in macroinvertebrate elemental contents and elemental ratios between the two reaches. We compared elemental contents and elemental ratios between the two reaches and among functional feeding groups using two-way analysis of variance (ANOVA) procedures. Pairwise comparisons among group means were made using Tukey's studentized range test (HSD). Data were either arcsin-square root or $\log_{10}(x+1)$ transformed prior to analysis. All analysis were done by using the statistical package SPSS (for Windows, version 11.0.1, SPSS Inc., Chicago, Illinois).

Bowman *et al.* [2005] proposed a method to calculate elemental imbalances between consumers and their resources that is applicable to non-homeostatic conditions. This calculation was developed in opposition to the formulae assuming strict homeostasis of Elser and Hassett [1994]. Between these two methods, the former become more pragmatic but it means a lamentable loss of information. As each component is multiplied by itself, it losses the sign and, therefore, the direction of the difference (i.e. above or below the 1:1 line). With the objective of solving this difficulty, we multiplied the equation of Bowman *et al.* [2005] by the arithmetic difference between the *X:Y* ratio of the resource and the consumer divided by the absolute value of this difference. We designated elemental imbalances with the Greek symbol "*ι*" (*iota*) that we calculated using the following equation:

$$\iota_{X:Y} = a \cdot \sqrt{(X:Y_{resource} - b)^2 + (X:Y_{consumer} - b)^2},$$

where:

$$a = \frac{X:Y_{resource} - X:Y_{consumer}}{\left|X:Y_{resource} - X:Y_{consumer}\right|}, \text{ if } X:Y_{resource} \neq X:Y_{consumer} \text{ or } a = 0, \text{ if } X:Y_{resource} =$$

$X:Y_{consumer}$

and:

$$b = (X:Y_{resource} + X:Y_{consumer})/2$$

This formula, however, does not solve the trend of overestimating elemental imbalances of arithmetic methods described in [Frost et al. 2006].

# RESULTS

## Physical and Chemical Parameters

During the sampling period, discharge was 1.4 times higher on average at the downstream reach than at the upstream reach (paired $T$-test, $P = 0.014$; Table 1). Water temperature was slightly lower in winter and higher in summer at the downstream reach than at the upstream reach. The downstream reach also had a higher diel temperature range, especially during the summer, but mean values were not significantly different (paired $T$-test, $P = 0.260$). Mean DO concentration at the downstream reach was consistently lower (1.1 mg/L lower on average) than at the upstream reach (paired $T$-test, $P = 0.013$).

**Table 1. Mean ± SE, minimum (min.) and maximum (max.) values of physical and chemical parameters measured in the two reaches over the sampling period in La Tordera stream ($n = 6$)**

| variable | Upstream | | | Downstream | | |
|---|---|---|---|---|---|---|
| | mean ± SE | min. | max. | mean ± SE | min. | max. |
| Discharge (L/s) | 174.8 ± 56.8 | 50.1 | 414.9 | 249.3 ± 65.1 | 83.1 | 508.4 |
| Temperature[1] (°C) | 12.9 ± 1.5 | 8.0 | 20.0 | 13.6 ± 1.9 | 7.5 | 22.8 |
| DO[1] (mg/L) | 10.0 ± 1.9 | 7.9 | 12.1 | 8.9 ± 0.5 | 6.2 | 11.2 |
| Conductivity (µS/cm) | 152 ± 16.0 | 92.0 | 205.2 | 262.4 ± 28.8 | 173.8 | 327.7 |
| DOC (mg/L) | 1.23 ± 0.32 | 0.55 | 2.39 | 1.91 ± 0.25 | 1.01 | 2.71 |
| $NH_4^+$-N (mg N/L) | 0.04 ± 0.01 | 0.01 | 0.07 | 0.65 ± 0.31 | 0.15 | 2.14 |
| $NO_3^-$-N (mg N/L) | 1.22 ± 0.23 | 0.74 | 2.20 | 2.65 ± 0.38 | 1.72 | 4.11 |
| SRP (mg P/L) | 0.010 ± 0.002 | 0.006 | 0.015 | 0.262 ± 0.106 | 0.052 | 0.628 |

[1] Values of temperature and DO correspond to values registered over 24 h cycles performed for each reach and sampling date.

Conductivity was moderate in the two reaches, but it was almost two times higher at the downstream reach than at the upstream reach (paired $T$-test, $P = 0.001$). Nutrient concentrations were higher at the downstream reach than at the upstream reach on all dates, but the effect differed among nutrients (Table 1).

On average, $NO_3^--N$ concentration increased twofold below the point source (paired $T$-test, $P = 0.012$), while $NH_4^+-N$ concentrations increased 20-fold (paired $T$-test, $P = 0.037$). Although mean SRP concentration was much higher at the downstream reach than at the upstream reach, differences were only marginally significant (paired $T$-test, $P = 0.055$). The effect of the point source on DOC concentration was less noticeable but statistically significant (paired $T$-test, $P = 0.032$).

## Food Resources

Elemental contents for periphyton and mosses were similar in the two reaches (Figure 2). Among primary producers, periphyton had the lower C content, filamentous algae had the highest N content, and mosses had the highest P content. Filamentous algae had similar %C and %N in the two reaches but at the downstream reach %P was, on average, two times higher than at the upstream reach. The differences in elemental content of autotrophs between the two reaches were reflected in the respective elemental ratios (Figure 2). CPOM and FPOM contained less C and more P below the point source than in the upstream reach. In consequence, C:N, C:P, and N:P ratios of CPOM and FPOM were substantially higher in the upstream reach than in the downstream reach.

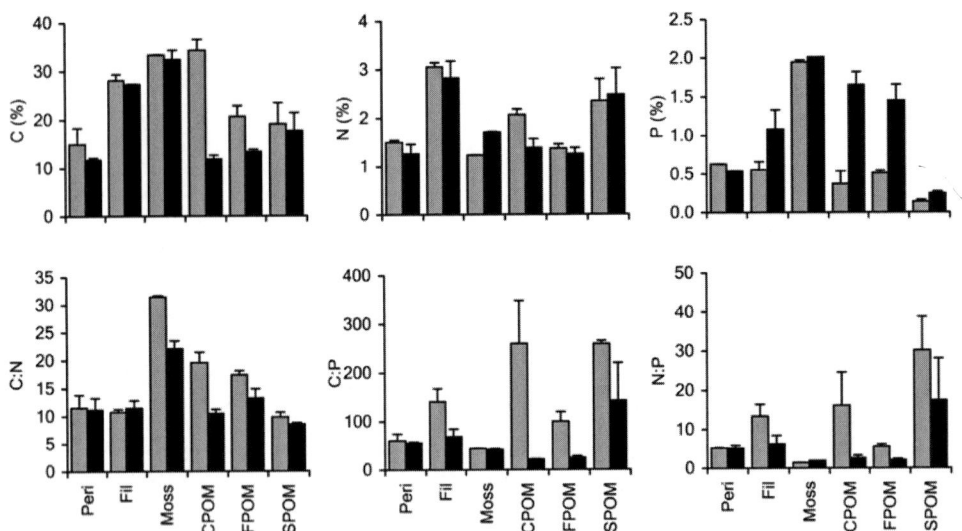

Figure 2. Mean elemental contents of C, N, and P (% of dry mass) and their molar elemental ratios C:N, C:P, and N:P of primary producers and organic matter in the upstream reach (grey bars) and the downstream reach (black bars) in La Tordera stream. for %C, %N, and C:N in the upstream reach and the downstream reach, respectively: periphyton ($n = 2, 2$), filamentous algae ($n = 3, 2$), mosses ($n = 2, 2$), CPOM ($n = 4, 6$), FPOM ($n = 6, 6$), SPOM ($n = 6, 6$). For %P, C:P, and N:P in the upstream reach and the downstream reach, respectively: periphyton ($n = 2, 2$), filamentous algae ($n = 3, 2$), mosses ($n = 2, 2$), CPOM ($n = 2, 2$), FPOM ($n = 2, 2$), SPOM ($n = 2, 2$). See also Table 2.

**Table 2. Mean, median, standard deviation (SD), and coefficient of variation (CV) of macroinvertebrate C, N, P, C : N, C : P, and N : P at the upstream reach (up, n = 33) and the downstream reach (dw, n = 23) in La Tordera stream and previous studies (See also Figure 2)**

| | Mean | | Median | | SD | | CV (%) | |
|---|---|---|---|---|---|---|---|---|
| | up | dw | up | dw | up | dw | up | dw |
| Stream macroinvertebrates in this study | | | | | | | | |
| C (%) | 45.90 | 38.97 | 49.24 | 41.14 | 11.79 | 9.96 | 26 | 26 |
| N (%) | 8.85 | 7.75 | 9.63 | 8.62 | 3.13 | 2.90 | 35 | 37 |
| P (%) | 1.42 | 1.32 | 1.57 | 1.26 | 0.65 | 0.58 | 46 | 44 |
| C : N | 6.6 | 6.7 | 6.1 | 5.7 | 1.8 | 2.9 | 28 | 43 |
| C : P | 102 | 95 | 78 | 78 | 61 | 56 | 60 | 59 |
| N : P | 16.20 | 15.10 | 13.21 | 13.85 | 9.96 | 7.91 | 61 | 52 |
| Stream macroinvertebrates (Cross *et al.* 2003) | | | | | | | | |
| C : N | 5.99 | | - | | 0.88 | | 15 | |
| C : P | 268.25 | | - | | 158.33 | | 60 | |
| N : P | 44.25 | | - | | 22.29 | | 51 | |
| Lake macroinvertebrates (Frost *et al.* 2003) | | | | | | | | |
| C : N | 5.6 | | 5.5 | | 0.75 | | 13 | |
| C : P | 148 | | 141 | | 51.2 | | 34 | |
| N : P | 27.3 | | 25.8 | | 9.92 | | 36 | |
| Mayflies (Bowman *et al.* 2005) | | | | | | | | |
| C : N | 5.0 | 5.6 | - | - | - | - | - | - |
| C : P | 210 | 250 | - | - | - | - | - | - |
| N : P | 40 | 44 | - | - | - | - | - | - |
| Stream insects (Evans-White *et al.* 2005) | | | | | | | | |
| C : N | 5.7 | | - | | 0.8 | | - | |
| C : P | 263 | | - | | 113 | | - | |
| N : P | 46 | | - | | 20 | | - | |
| Terrestrial invertebrates (Elser *et al.* 2000) | | | | | | | | |
| C : N | 6.5 | | 6.4 | | 1.9 | | 29 | |
| C : P | 116 | | 73.2 | | 72.4 | | 62 | |
| N : P | 26.4 | | 22.6 | | 10.1 | | 38 | |
| Lake zooplankton (Elser *et al.* 2000) | | | | | | | | |
| C : N | 6.3 | | 6.0 | | 1.3 | | 21 | |
| C : P | 124 | | 114 | | 48.0 | | 38 | |
| N : P | 22.3 | | 18.5 | | 10.5 | | 47 | |

We observed no important differences in elemental contents for SPOM between the two reaches, but C:P and N:P ratios for the upstream reach were almost twofold those for the downstream reach. The P content for SPOM was between three and seven times lower than that observed for CPOM or FPOM.

## Benthic Macroinvertebrates

We analyzed the nutrient content of 33 and 23 macroinvertebrate taxa from the upstream and the downstream reaches, respectively. All analyzed beetles were larvae. Elemental contents varied considerably among the macroinvertebrates sampled in the two reaches (Table 2 and Figure 3).

Macroinvertebrate C content varied more than 15-fold, from 16.43% in *Ancylus fluviatilis* at the downstream reach to 58.99% in Elmidae at the upstream reach. The macroinvertebrate %N was also highly variable and ranged from 1.06% in *Pisidium casertanum* at the downstream reach and 14.65% in *Erpobdella* sp. at the upstream reach. Macroinvertebrate P varied from 0.18% in *A. fluviatilis* at the downstream reach to 2.76% in *Polycentropus* sp. at the upstream reach. Mean elemental contents of macroinvertebrates were slightly higher at the upstream reach than at the downstream reach. Macroinvertebrate %C distribution was considerably different between the two reaches.

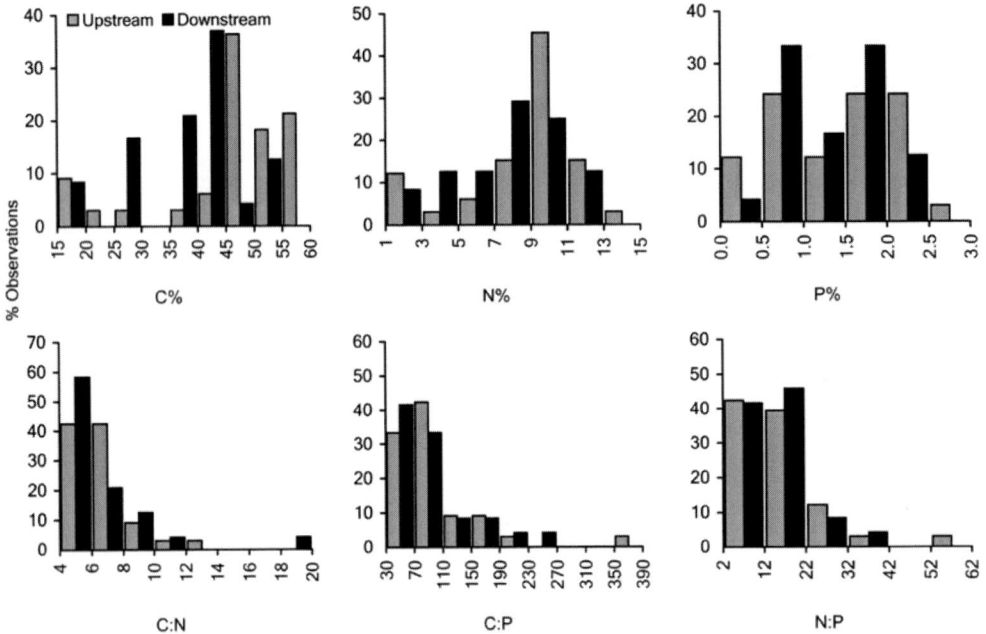

Figure 3. Frequency histograms of body C, N, and P content (% of dry mass) and their molar elemental ratios C:N, C:P, and N:P of macroinvertebrates in the upstream reach (grey bars) and the downstream reach (black bars) in La Tordera stream.

In contrast, the distribution of macroinvertebrate N and P contents and elemental ratios differed little between the two reaches. The coefficient of variation for %P was higher than those for %N and %C, but did not differ greatly between the two reaches. High variability of %P was reflected in elemental ratios, where the coefficient of variation was relatively high for C:P and N:P ratios while the C:N ratio was more constrained for the two reaches, but especially for the upstream reach. The mean ratios for macroinvertebrates found in our study were slightly higher than those previously reported for stream and lake macroinvertebrates and were more similar to those for terrestrial invertebrates or lake zooplankton (Table 2).

Variation in macroinvertebrate nutrient contents was relatively high among taxa phylogenetically related. However, mollusks generally had the lowest C, N, and P contents. Leeches were among the taxa with highest %N while elmid beetles had a very low P content. Paired comparisons between the two reaches revealed that most taxa had a similar elemental content in the two reaches (Figure 4). Macroinvertebrate C, N, and P contents were not significantly different between the two reaches (paired *T*-test, *P* = 0.079, *P* = 0.185, *P* = 0.807, respectively). However, the elemental content of some macroinvertebrates, such as *Serratella ignita*, *Calopteryx virgo*, *Bythiospeum* sp., and *Physella acuta* varied greatly between the two reaches. The C, N and P content of the snails *P. acuta* and *Radix* sp. were higher at the downstream than at the upstream reach, whereas the %P of Orthocladiinae, *C. virgo*, and *S. ignita* were much higher at the upstream than at the downstream reach.

Overall, mollusks showed the highest C:N ratios while midges generally had the lowest C:P and N:P ratios (Figure 5). Elmids, water mites, and *A. fluviatilis* showed the highest C:P and N:P ratios, while *P. acuta*, *Radix* sp., and *P. casertanum* had the lowest ratios. Stoichiometric ratios substantially decreased in the downstream reach relative to the upstream reach for some taxa (e.g., *P. acuta*, Elmidae) while increased for other (e.g., *C. virgo*, *S. ignita*) and remained similar for some taxa (e.g., *Baetis* spp., *Caenis* spp., *Hydropsyche instabilis*).

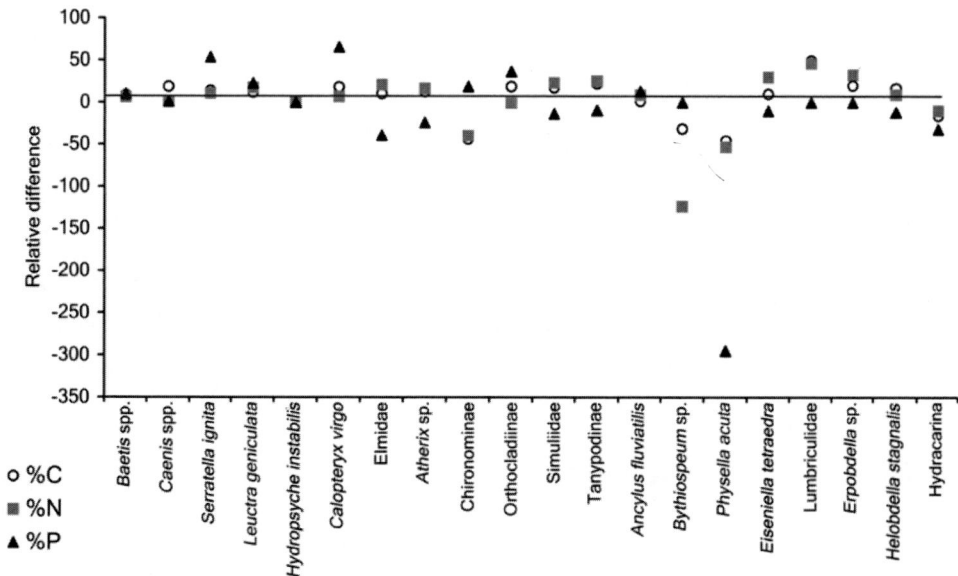

Figure 4. Relative difference (%) in mean %C, %N, and %P of paired macroinvertebrate taxa from the upstream reach and the downstream reach in La Tordera stream.

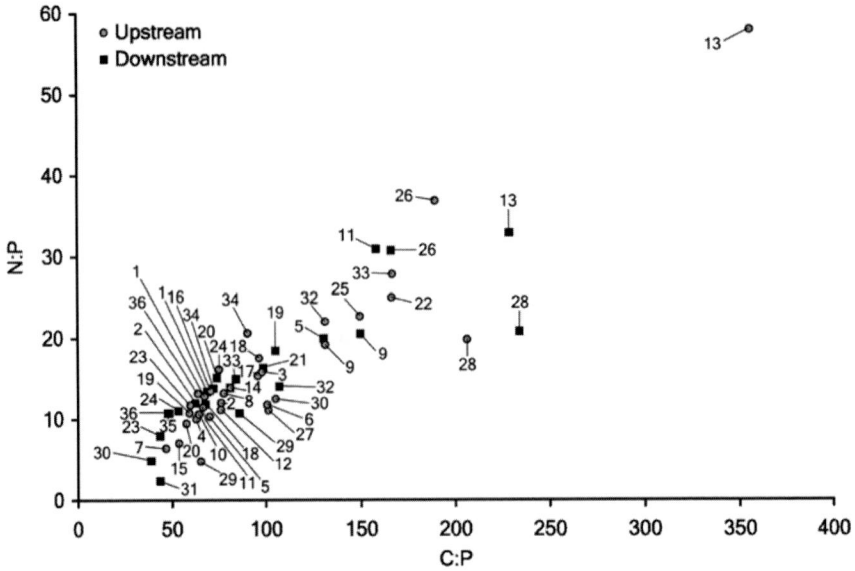

Figure 5. Stoichiometric variation among macroinvertebrate taxa in the upstream reach and the downstream reach in La Tordera stream. 1 = *Baetis* spp., 2 = *Caenis* spp., 3 = *Ecdyonurus angelieri*, 4 = *Epeorus torrentium*, 5 = *Serratella ignita*, 6 = *Amphinemura* sp., 7 = *Capnioneura mitis*, 8 = *Isoperla grammatica*, 9 = *Leuctra geniculata*, 10 = *Siphonoperla torrentium*, 11 = *Calopteryx virgo*, 12 = *Agabus* sp. L, 13 = *Elmidae*, 14 = *Hydropsyche instabilis*, 15 = *Polycentropus* sp., 16 = *Rhyacophila dorsalis*, 17 = *Sericostoma personatum*, 18 = *Atherix* sp., 19 = Chironominae, 20 = Orthocladinae, 21 = Psychodidae, 22 = Rhagionidae, 23 = Simuliidae, 24 = Tanypodinae, 25 = Tipulidae, 26 = Hydracarina, 27 = *Radix* sp., 28 = *Ancylus fluviatilis*, 29 = *Bythiospeum* sp., 30 = *Physella acuta*, 31 = *Pisidium casertanum*, 32 = *Eiseniella tetraedra*, 33 = Lumbriculidae, 34 = *Erpobdella* sp., 35 = *Glossiphonia* sp., 36 = *Helobdella stagnalis*.

Although several taxa greatly differed in C:N, C:P, and N:P ratios in the two reaches, differences were not significant among paired taxa between the two reaches (paired $T$-test, $P = 0.475$, $P = 0.301$, $P = 0.466$, respectively). We found no significant differences in any elemental content or elemental ratio between the two reaches with regard to the functional feeding groups (two-way ANOVA, reach and reach $x$ FFG interaction: $P > 0.05$; Figure 6). However, differences among FFGs were significant when data from the two reaches were combined (two-way ANOVA, FFG: $P < 0.05$). The elemental contents harbored by predators was significantly higher than those for scrapers (Tukey's HSD, C: $P < 0.0005$, N: $P < 0.0005$, P: $P = 0.045$). Filterers and predators had the highest %P, while shredders at the downstream reach (mainly Elmidae) had the lowest %P and, consequently, the highest C:P and N:P ratios.

## Consumer-Resource Stoichiometry

Overall, elemental imbalances between consumers and their presumed food resources were higher at the upstream reach (on average, $\iota_{C:N} = 4.6$, $\iota_{C:P} = 56.4$, $\iota_{N:P} = 0.5$) than at the downstream reach (on average, $\iota_{C:N} = 1.9$, $\iota_{C:P} = -27.8$, $\iota_{N:P} = -5.9$; Figure 7). All FFG tended to have a lower C:N ratio than their presumed food resources, indicating that consumers accumulated N.

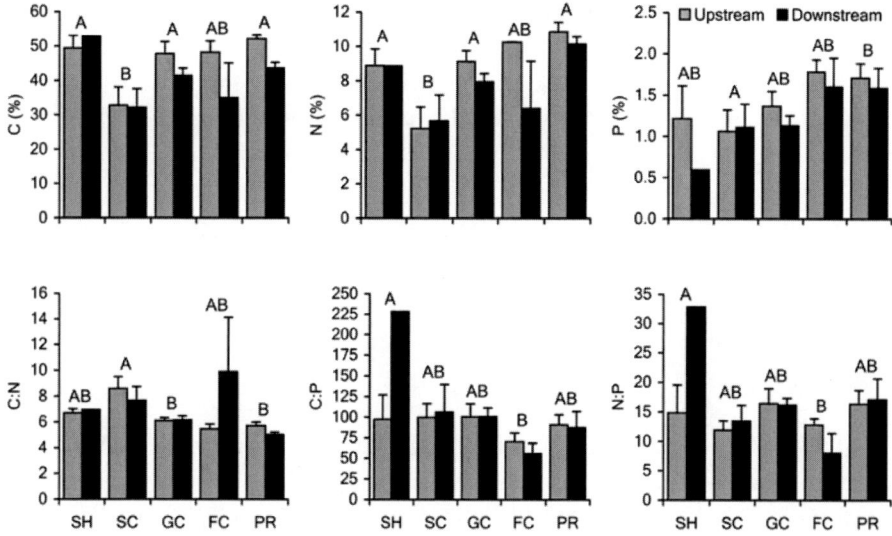

Figure 6. Mean %C, %N, and %P (± SE) of functional feeding groups in the upstream reach (grey bars) and the downstream reach (black bars) in La Tordera stream. SH = shredders, SC = scrapers, GC = gatherers, FC = filterers, PR = predators. Results of two-way ANOVA: FFG factor significant for all variables (%C, $p = 0.001$; %N, $p < 0.0005$; %P, $p = 0.036$; C:N, $p = 0.001$; C:P, $p = 0.050$; N:P, $p = 0.036$), reach factor not significant for any variable ($p > 0.05$), FFG x reach not significant for any variable ($p > 0.05$). Different capital letters indicate significant differences (Tuckey's HSD) based on data from the two reaches combined. For all elemental contents and ratios in the upstream reach and the downstream reach, respectively: SH ($n = 4, 1$), SC ($n = 8, 5$), GC ($n = 7, 8$), FC ($n = 2, 3$), and PR ($n = 12, 7$).

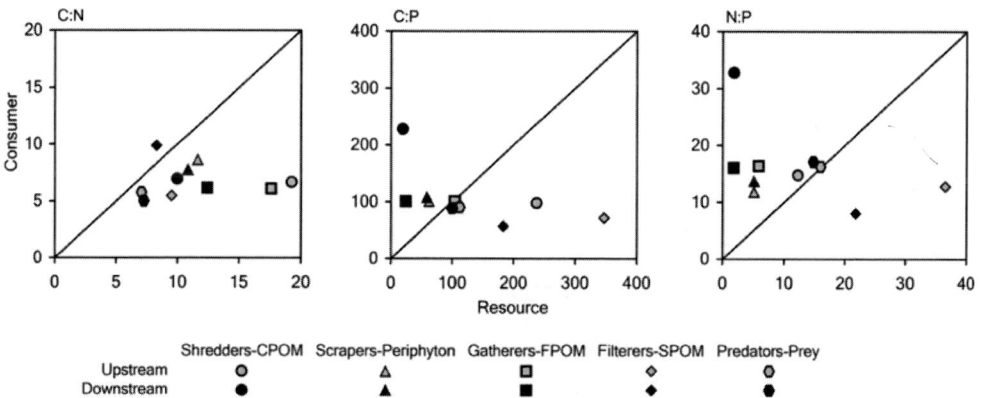

Figure 7. Stoichiometric relationships between consumers and their presumed food resources for the two reaches in La Tordera stream. Points on the 1:1 line (slope 1, intercept 0) represent identical stoichiometry in consumers and resources. All ratios are molar.

Filterers had a positive imbalance between their C:P (upstream $\iota_{C:P} = 195.4$, downstream $\iota_{C:P} = 89.4$) and N:P (upstream $\iota_{N:P} = 16.8$, downstream $\iota_{N:P} = 9.8$) ratios and those for their food resources (i.e., consumer ratio lower than the ratio for its food resource), indicating that their relative P content was higher than that in their food resources. At the upstream reach, shredders ($\iota_{C:N} = 8.9$) and gatherers ($\iota_{C:N} = 8.1$) were far out of balance from their food resources for C:N ratio. Predators were generally closer to be stoichiometrically balanced with their food, based on consumption of N and P-rich prey. At the downstream reach, most consumer-resource relationships for C:P and N:P ratios were located above the 1:1 line, indicating that the relative P contents in food resources were higher than in their respective consumers.

## DISCUSSION

### Differential Susceptibility in the Stoichiometry of Food Resources to the Point Source

Overall, food resources in La Tordera stream tended to be P-rich both upstream and downstream of the point source input. Based on studies from a wide range of freshwater ecosystems from worldwide, Kahlert [1998] proposed the $C_{158}:N_{18}:P_1$ ratio as the optimal median value and for freshwater algae in opposition, but close, to the Redfield theoretical ratio of $C_{106}:N_{16}:P_1$ based on oceanic systems [Redfield 1958] (Figure 8).

In addition, she concluded that the optimal ratio may largely vary among algal communities and provided optimal ranges for C:N (8-11), C:P (99-369), and N:P (11-32) ratios. In La Tordera stream, the C, N, and P contents for periphyton were similar in the two reaches, indicating a potential homeostatic regulation. This result disagrees with the findings of Cross et al. [2003], Stelzer and Lamberti [2001], and Liess and Hillebrand [2006] who found that %N and %P in stream periphyton increased after nutrient enrichment.

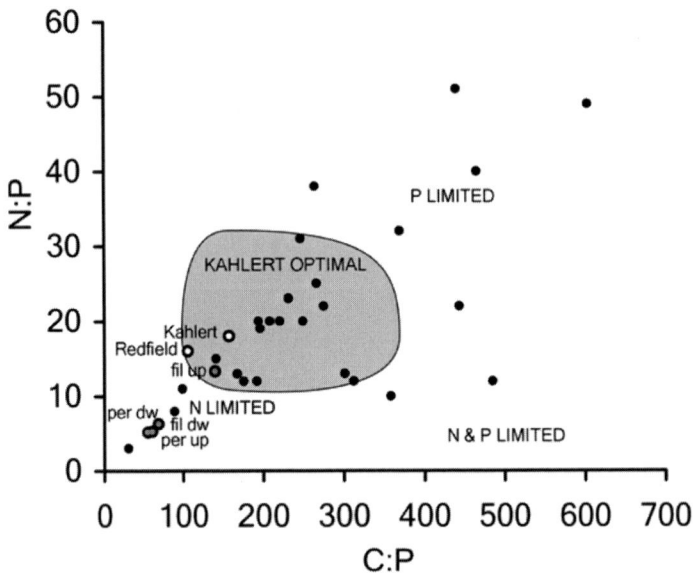

Figure 8. N:P and C:P stoichiometric ratios for algae. The spots correspond to the median ratio of Redfield [1958] and Kahlert [1998] (white spots), the optimal range of Kahlert (grey zone), data from Kahlert (black spots), and data from this study (grey spots) for periphyton (per) and filamentous algae in the upstream (up) and the downstream (dw) reaches. The zones where algae are limited by N, P or both are also indicated.

In the two reaches, periphyton C:N ratio was within the Kahlert optimal range for freshwater algae. However, C:P and N:P ratios in periphyton were extremely low compared to previous studies [Bowman et al. 2005, Cross et al. 2003, Kahlert 1998, Kahlert et al. 2002, Stelzer and Lamberti 2002], indicating P accumulation. At the downstream reach, elemental ratios for filamentous algae were similar to those for periphyton. At the upstream reach, in contrast, C:P and N:P ratios for filamentous algae were twofold those at the downstream reach and where located within the optimal range proposed by Kahlert [1998]. The higher P content relative to C and N indicates that filamentous algae were potentially N-limited at the downstream reach even though such differences may also be due to differences in species composition [Sterner and Elser 2002] or growth rates [Agren 2004]. Although nutrient concentrations were much higher at the downstream reach than at the upstream reach, the low DIN:SRP ratios in water at the downstream reach, lends support to the assumption that autotrophs were N-limited. The C:N ratio in mosses was much higher than in periphyton or filamentous algae, presumably because of a higher content in phenolic compounds among other structural differences [Stream Bryophyte Group 1999]. The C:N ratios for mosses were slightly lower than those previously reported in a forested stream in eastern Tennessee [Mulholland et al. 2000]. In our study, however, N and P contents for mosses were, respectively, two and 40 times higher than those previously reported for aquatic bryophytes in North America [Bedford et al. 1999]. We expected that N:P ratios should be lower at the downstream reach than at the upstream reach, since aquatic mosses are known to take up P rapidly [Meyer 1979, Steinman 1994]. However, elemental contents and ratios in mosses were similar above and below the point source in discrepancy with the findings of Christmas and Whitton [1998], probably because nutrient concentrations in the two reaches were much higher than those considered in previous studies.

CPOM and FPOM represent the major C pathway in most ecosystems [Wallace et al. 1997], and were the major food resources in La Tordera stream over the sampling period. Below the point source, C:N, C:P, and N:P ratios for CPOM and FPOM decreased relative to the upstream reach. Cross et al. [2003] and Kaushik and Hynes [1971] obtained similar results, apparently because microbes growing on BOM can uptake N and P from dissolved pools [Frost et al. 2002, Robinson and Gessner 2000, Suberkropp 1998]. In addition, CPOM and FPOM were higher in N content at the upstream than at the downstream reach, but the contrary occurred for P. This pattern was closely related with differences in DIN:SRP ratios between the two reaches. The C:N content for CPOM and FPOM in La Tordera stream were lower than those previously reported [Cross et al. 2003, Mulholland et al. 2000], but the C:P ratio was one order of magnitude lower than those provided by Cross et al. [2003], probably because of differences in stream water concentrations.

Research from a [15]N tracer addition in a forested stream [Mulholland et al. 2000] found that SPOM was not generated only by entrainment of FPOM but also from other biomass compartments such as periphyton or consumer feces. This may explain the observed differences in nutrient contents between FPOM and SPOM in La Tordera stream. Differences in nutrient contents of SPOM between the two reaches were not as high as the observed for FPOM or CPOM. However, N and P were slightly higher at the downstream reach and, combined with a lower C content, resulted in C:P and N:P ratios almost two times lower at the downstream than at the upstream reach. Similar to CPOM and FPOM, such differences may also be explained by differences in water chemistry. Few publications supplied elemental analysis for SPOM, among them, Mulholland *et al.* [2000] provided a C:N ratio that was twofold those found in our study, probably because of differences in material composition.


## Homeostatic Regulation in Macroinvertebrates

Although stoichiometric ratios may largely vary in autotrophic organisms [Bedford *et al.* 1999, Kahlert 1998], metazoans have been seen as having relatively invariant elemental ratios [Sterner and Elser 2002]. The number of published studies is scarce, but stoichiometric homeostasis has been demonstrated for freshwater and marine zooplankton [Andersen and Hessen 1991, Anderson *et al.* 2004] and also for macroinvertebrates [Evans-White *et al.* 2005, Frost *et al.* 2003, Stelzer and Lamberti 2002]. We found no significant differences in mean elemental content between the taxa collected above and below of the point source. However, certain taxa showed important differences in nutrient content between the two reaches. Cross *et al.* [2003] and Bowman *et al.* [2005] found that stream macroinvertebrates might not be strictly homeostatic in headwater flowing ecosystems. They suggested that their study streams were strongly limited by P and that their fauna were adapted to such nutritional constraints as seen before [Fagan *et al.* 2002]. Similarly, in Lake Erken (Sweden), Liess and Hillebrand [2005] found that although benthic invertebrates showed a taxon-specific stoichiometry, it may also vary seasonally and in space. These results may indicate that stream macroinvertebrates have varying degrees of homeostasis according to the stoichiometry of their food resources [Sterner and Elser 2002]. In natural ecosystems, macroinvertebrates may experience nutritional constraints that lead to suboptimal growth conditions [Elser *et al.* 1996].


## Elemental Imbalances between Consumers and Resources

The calculation of elemental imbalances provides a measure of the dissimilarity in relative supply of an element between consumers and their presumed resources [Sterner and Elser 2002] and has direct effects on nutrient cycling [Vanni 2002] and growth [Frost and Elser 2002a, Söderström 1988, Stelzer and Lamberti 2002]. The elemental imbalances in C:N between scrapers and periphyton were similar to those previously reported for lake zooplankton-phytoplankton [Elser *et al.* 2000, Sterner and Hessen 1994] and lake benthic invertebrates-algae [Frost and Elser 2002b], but stand out against those found between terrestrial insects and leaves [Elser *et al.* 2000]. Overall, our results lend support to the findings of Cross *et al.* [2003] in a detritus-based headwater stream that consumer-resource elemental imbalances were more relevant for shredders, less severe for scrapers and gatherers, while for predators were the least out of balance. However, they found that imbalances in elemental composition between consumers and resources were much higher than that for La Tordera stream, probably because of very low SRP concentrations in their study stream. Overall, nutrient enrichment below the point source tend to reduce elemental imbalances between the requirements of consumers and the consumed food in agreement with previous research [Bowman *et al.* 2005, Cross *et al.* 2003, Frost and Elser 2002a, Stelzer and Lamberti 2002]. Such changes may have severe consequences on stream processes [Cross *et al.* 2005, Fisher *et al.* 1998, Newbold *et al.* 1982] through changes on population dynamics [Burkhardt and Lehman 1994, Loladze *et al.* 2000], trophic interactions [Elser *et al.* 1998, Sterner *et al.* 1997], and community structure [DeMott and Gulati 1999, Grover 2002].

## Variability in Stoichiometric Relationships among Macroinvertebrate Taxa

Mean stoichiometric ratios of macroinvertebrates were slightly higher than those previously reported for stream invertebrates [Cross *et al.* 2003] and lake littoral invertebrates [Frost *et al.* 2003] (however see Evans-White *et al.* [2005]), and similar to those found for terrestrial invertebrates or zooplankton [Elser *et al.* 2000]. The differences in C, N, and P content among macroinvertebrate taxa were noticeable. The high content in crystalline calcium carbonate of mollusks shell translates in a lower C, N, and P content relative to insects. The high N content in leeches has been attributed to high concentrations of structural proteins [Frost *et al.* 2003]. Similarly, the low P content in beetles may result from their hard chitin-rich exoskeleton. On the other hand, the high P content in midges may be explained by their high growth rates [Benke 1998] and consequent high content in ribosomal RNA according to the growth rate hypothesis [Sterner and Elser 2002, Vrede *et al.* 2004]. Although differences in nutrient contents were considerable among certain insect taxa and species composition varied between the two reaches [Ortiz and Puig 2007], we found no substantial differences among insect orders. The number of studies considering stoichiometric differences among taxonomic groups is scarce and patterns are inconsistent. For example, Cross *et al.* [2003] found that dipterans harbored the highest amount of P, while Ephemeroptera was the most P-poor insect order. In contrast, Frost *et al.* [2003] found mayflies among the taxa with higher P content. Such incongruence may be merely due to the high intra-group variability explained by differences in species composition and ontogeny, lessen the sense of making comparisons among high taxonomic groups.

## ACKNOWLEDGEMENTS

We are indebted to Dr. Nicolás Ubero-Pascal for his assistance in the field and laboratory work. We also express our appreciation to all those students who helped us in the field. Jesús D. Ortiz benefited from a fellowship of the Department of Universities, Research and the Information Society of the Generalitat, Government of Catalonia (Spain). This study was supported by fundings of the STREAMES European project (EVK1-CT-2000-00081). The Serveis Cientifico Tècnics of the University of Barcelona provided their facilities and technical help in elemental analyses.

## REFERENCES

Agren, G. I. (2004). The C : N : P stoichiometry of autotrophs - theory and observations. *Ecology Letters*, 7, 185-191.

Andersen, T. and Hessen, D. O. (1991). Carbon, nitrogen, and phosphorus content of freshwater zooplankton. *Limnology and Oceanography*, 36, 807-814.

Anderson, T. R., Boersma, M., and Raubenheimer, D. (2004). Stoichiometry: linking elements to biochemicals. *Ecology*, *85*, 1193-1202.

Bedford, B. L., Walbridge, M. R., and Aldous, A. (1999). Patterns in nutrient availability and plant diversity of temperate north American wetlands. *Ecology*, *80*, 2151-2169.

Benke, A. C. (1998). Reproduction dynamics of riverine chironomids: extremely high biomass turnover rates of primary consumers. *Ecology*, *79*, 899-910.

Bowman, M. F., Chambers, P. A., and Schindler, D. W. (2005). Changes in stoichiometric constraints on epilithon and benthic macroinvertebrates in response to slight nutrient enrichment of mountain rivers. *Freshwater Biology*, *50*, 1836-1852.

Burkhardt, S. and Lehman, J. T. (1994). Prey consumption and predatory effects of an invertebrate predator (Bythotrephes: Cladocera, Cercopagidae) based on phosphorus budgets. *Limnology and Oceanography*, *39*, 1007-1019.

Christmas, M. and Whitton, B. A. (1998). Phosphorus and aquatic bryophytes in the Swale-Ouse river system, north-east England. 1. Relationship between ambient phosphate, internal N:P ratio and surface phosphate activity. *The Science of the Total Environment*, *210/211*, 389-399.

Cross, W. F., Benstead, J. P., Frost, P. C., and Thomas, S. A. (2005). Ecological stoichiometry in freshwater benthic systems: recent progress and perspectives. *Freshwater Biology*, *50*, 1895-1912.

Cross, W. F., Benstead, J. P., Rosemond, A. D., and Wallace, J. B. (2003). Consumer-resource stoichiometry in detritus-based streams. *Ecology Letters*, *6*, 721-732.

DeMott, W. R. and Gulati, R. D. (1999). Phosphorus limitation in *Daphnia*: Evidence from a long term study of three hypereutrophic Dutch lakes. *Limnology and Oceanography*, *44*, 1557-1564.

DeMott, W. R., Gulati, R. D., and Siewertsen, K. (1998). Effects of phosphorus-deficient diets on the carbon and phosphorus balance of *Daphnia magna*. *Limnology and Oceanography*, *43*, 1147-1161.

Elser, J. J., Chrzanowski, T. H., Sterner, R. W., and Mills, K. H. (1998). Stoichiometric constraints on food-web dynamics: a whole-lake experiment on the Canadian shield. *Ecosystems*, *1*, 120-136.

Elser, J. J., Dobberfuhl, D. R., MacKay, N. A., and Schampel, J. H. (1996). Organism size, life history, and N:P stoichiometry. Toward a unified view of cellular and ecosystem processes. *Bioscience*, *46*, 674-684.

Elser, J. J., Fagan, W. F., Denno, R. F., Dobberfuhl, D. R., Folarin, A., Huberty, A., Interlandi, S., Kilham, S. S., McCauley, E., Schulz, K., Siemann, E. H., and Sterner, R. W. (2000). Nutritional constrains in terrestrial and freshwater food webs. *Nature*, *408*, 578-580.

Elser, J. J. and Hassett, R. P. (1994). A stoichiometric analysis of the zooplankton-phytoplankton interaction in marine and freshwater ecosystems. *Nature*, *370*, 211-213.

Elser, J. J. and Urabe, J. (1999). The stoichiometry of consumer-driven nutrient recycling: theory, observations, and consequences. *Ecology*, *80*, 735-751.

Enríquez, S., Duarte, C. M., and Sand-Jensen, K. (1993). Patterns in decomposition rates among photosynthetic organisms: the importance of detritus C:N:P content. *Oecologia*, *94*, 457-471.

Evans-White, M. A., Stelzer, R. S., and Lamberti, G. A. (2005). Taxonomic and regional patterns in benthic macroinvertebrate elemental composition in streams. *Freshwater Biology*, *50*, 1786-1799.

Fagan, W. F., Siemann, E., Mitter, C., Denno, R. F., Huberty, A. F., Woods, H. A., and Elser, J. J. (2002). Nitrogen in insects: implications for trophic complexity and species diversification. *The American Naturalist*, *160*, 784-802.

Fernández, J. A., Niell, F. X., and Lucena, J. (1985). A rapid sensitive automated determination of phosphate in natural waters. *Limnology and Oceanography*, *30*, 227-230.

Fisher, S. G., Grimm, N. B., Martí, E., Holmes, R. M., and Jones, J. B. Jr. (1998). Material spiraling in stream corridors: A telescoping ecosystem model. *Ecosystems*, *1*, 19-34.

Frost, P. C., Benstead, J. P., Cross, W. F., Hillebrand, H., Larson, J. H., Xenopoulos, M. A., and Yoshida, T. (2006). Threshold elemental ratios of carbon and phosphorus in aquatic consumers. *Ecology Letters*, *9*, 774-779.

Frost, P. C. and Elser, J. J. (2002a). Growth responses of littoral mayflies to the phosphorus content of their food. *Ecology Letters*, *5*, 232-240.

Frost, P. C. and Elser, J. J. (2002b). Effects of light and nutrients on the net accumulation and elemental composition of epilithon in boreal lakes. *Freshwater Biology*, *47*, 173-183.

Frost, P. C., Evans-White, M. A., Finkel, Z. V., Jensen, T. C., and Matzek, V. (2005). Are you what you eat? Physiological constraints on organismal stoichiometry in an elementally imbalanced world. *Oikos*, *109*, 18-28.

Frost, P. C., Stelzer, R. S., Lamberti, G. A., and Elser, J. J. (2002). Ecological stoichiometry of trophic interactions in the benthos: understanding the role of C:N:P ratios in lentic and lotic habitats. *Journal of the North American Benthological Society*, *21*, 515-528.

Frost, P. C., Tank, S. E., Turner, M. A., and Elser, J. J. (2003). Elemental composition of littoral invertebrates from oligotrophic and eutrophic Canadian lakes. *Journal of the North American Benthological Society*, *22*, 51-62.

Gordon, N. D., McMahon, T. A., and Finlayson, B. L. (1992). *Stream hydrology: an introduction for ecologists*. Chichester, England: John Wiley and Sons Ltd.

Grover, J. P. (2002). Stoichiometry, herbivory and competition for nutrients: simple models based on planktonic ecosystems. *Journal of Theoretical Biology*, *214*, 599-618.

Hessen, D. O. and Lyche, A. (1991). Inter- and intraspecific variations in zooplankton element composition. *Archiv für. Hydrobiologie.*, *121*, 343-353.

Kahlert, M. (1998). C:N:P ratios of freshwater benthic algae. *Archiv für. Hydrobiologie., Special Issues: Advances in Limnology*, *51*, 105-114.

Kahlert, M., Hasselrot, A. T., Hillebrand, H., and Pettersson, K. (2002). Spatial and temporal variation in the biomass and nutrient status of epilithic algae in Lake Erken, Sweden. *Freshwater Biology*, *47*, 1191-1215.

Kaushik, N. K. and Hynes, H. B. N. (1971). The fate of the dead leaves that fall into streams. *Archiv für. Hydrobiologie.*, *68*, 465-515.

Liess, A. and Hillebrand, H. (2005). Stoichiometric variation in C:N, C:P, and N:P ratios of littoral benthic invertebrates. *Journal of the North American Benthological Society*, *24*, 256-269.

Liess, A. and Hillebrand, H. (2006). Role of nutrient supply in grazer-periphyton interactions: reciprocal influences of periphyton and grazer nutrient stoichiometry. *Journal of the North American Benthological Society*, *25*, 632-642.

Loladze, I., Kuang, Y., and Elser, J. J. (2000). Stoichiometry in producer-grazer systems: linking energy flow with element cycling. *Bulletin of Mathematical Biology*, *62*, 1137-1162.

Meyer, J. L. (1979). The role of sediments and bryophytes in phosphorus dynamics in a headwater stream ecosystem. *Limnology and Oceanography*, *24*, 365-375.

Moe, S. J., Stelzer, R. S., Forman, M. R., Harpole, W. S., Daufresne, T., and Yoshida, T. (2005). Recent advances in ecological stoichiometry: insights for population and community ecology. *Oikos*, *109*, 29-39.

Moog, O. (2002). *Fauna aquatica Austriaca: Katalog zur autökologischen einstufung a quatischer organismen Österreichs*. Vienna, Austria: Bundesministerium für Land - und Forstwirtschaft, Umwelt und Wasserwirtschaft.

Moulton II, S. R., Carter, J. L., Grotheer, S. A., Cuffney, T. F., and Short, T.M. (2000). Methods of analysis by the U.S. Geological Survey National Water Quality Laboratory – processing, taxonomy, and quality control of benthic macroinvertebrate samples. Report number 00-212. U.S. Geological Survey, Denver, Colorado.

Mulholland, P. J., Tank, J. L., Sanzone, D. M., Wollheim, W. M., Peterson, B. J., Webster, J. R., and Meyer, J. L. (2000). Nitrogen cycling in a forest stream determined by an [15]N tracer addition. *Ecological Monographs*, *70*, 471-493.

Newbold, J. D., O'Neill, R. V., Elwood, J. W., and Van Winkle, W. (1982). Nutrient spiralling in streams: implications for nutrient limitation and invertebrate activity. *The American Naturalist*, *120*, 628-652.

Ortiz, J. D. and Puig, M. A. (2007). Point source effects on density, biomass and diversity of benthic macroinvertebrates in a Mediterranean stream. *River Research and Applications*, *23*, 155-170.

Redfield, A. C. (1958). The biological control of chemical factors in the environment. *American Scientist*, *46*, 205-222.

Reiners, W. A. (1986). Complementary models for ecosystems. *The American Naturalist*, *127*, 59-73.

Robinson, C. T. and Gessner, M. O. (2000). Nutrient addition accelerates leaf breakdown in an alpine springbrook. *Oecologia*, *122*, 258-263.

Schade, J. D., Kyle, M., Hobbie, S. E., Fagan, W. F., and Elser, J. J. (2003). Stoichiometric tracking of soil nutrients by a desert insect herbivore. *Ecology Letters*, *6*, 96-101.

Schindler, D. E. and Eby, L. A. (1997). Stoichiometry of fishes and their prey: implications for nutrient recycling. *Ecology*, *78*, 1816-1831.

Söderström, O. (1988). Effects of temperature and food quality on life-history parameters in *Parameletus chelifer* and *P. minor* (Ephemeroptera): a laboratory study. *Freshwater Biology*, *20*, 295-303.

Stauffer, R. E. (1985). Nutrient internal cycling and the trophic regulation of Green Lake, Wisconsin. *Limnology and Oceanography*, *30*, 347-363.

Steinman, A. D. (1994). The influence of phosphorus enrichment on lotic bryophytes. *Freshwater Biology*, *31*, 53-63.

Stelzer, R. S. and Lamberti, G. A. (2001). Effects of N:P ratio and total nutrient concentration on stream periphyton community structure, biomass, and elemental composition. *Limnology and Oceanography*, *46*, 356-367.

Stelzer, R. S. and Lamberti, G. A. (2002). Ecological stoichiometry in running waters: periphyton chemical composition and snail growth. *Ecology*, *83*, 1039-1051.

Sterner, R. W. (1997). Modelling interactions of food quality and quantity in homeostatic consumers. *Freshwater Biology*, *38*, 473-481.

Sterner, R. W. and Elser, J. J. (2002). *Ecological stoichiometry: the biology of elements from molecules to the biosphere*. Princeton, New Jersey, USA: Princeton University Press.

Sterner, R. W., Elser, J. J., Fee, E. J., Guildford, S. J., and Chrzanowski, T. H. (1997). The light : nutrient ratio in lakes: the balance of energy and materials affects ecosystem structure and process. *The American Naturalist*, *150*, 663-684.

Sterner, R. W. and George, N. B. (2000). Carbon, nitrogen, and phosphorus stoichiometry of cyprinid fishes. *Ecology*, *81*, 127-140.

Sterner, R. W. and Hessen, D. O. (1994). Algal nutrient limitation and the nutrition of aquatic herbivores. *Annual Reviews of Ecology and Systematics*, *25*, 1-29.

Stream Bryophyte Group (1999). Roles of bryophytes in stream ecosystems. *Journal of the North American Benthological Society*, *18*, 151-184.

Suberkropp, K. (1998). Effect of dissolved nutrients on two aquatic hyphomycetes growing on leaf litter. *Mycological Research*, *102*, 998-1002.

Tachet, H., Richoux, P., Bournaud, M., and Usseglio-Polatera, P. (2000). *Invertébrés d'eau douce*. Paris: CNRS Éditions.

Urabe, J., Kyle, M., Makino, W., Yoshida, T., Andersen, T., and Elser, J. J. (2002). Reduced light increases herbivore production due to stoichiometric effects of light/nutrient balance. *Ecology*, *83*, 619-627.

Urabe, J. and Watanabe, Y. (1992). Possibility of N or P limitation for planktonic cladocerans: An experimental test. *Limnology and Oceanography*, *37*, 244-251.

Vanni, M. J. (2002). Nutrient cycling by animals in freshwater ecosystems. *Annual Reviews of Ecology and Systematics*, *33*, 341-370.

Vanni, M. J., Flecker, A. S., Hood, J. M., and Headworth, J. L. (2002). Stoichiometry of nutrient recycling by vertebrates in a tropical stream: linking species identity and ecosystem processes. *Ecology Letters*, *5*, 285-293.

Vrede, T., Dobberfuhl, D. R., Kooijman, S. A. L. M., and Elser, J. J. (2004). Fundamental connections among organism C:N:P stoichiometry, macromolecular composition, and growth. *Ecology*, *85*, 1217-1229.

Wallace, J. B., Eggert, S. L., Meyer, J. L., and Webster, J. R. (1997). Multiple trophic levels of a forest stream linked to terrestrial litter inputs. *Science*, *277*, 102-104.

In: Aquatic Ecosystem Research Trends
Editor: George H. Nairne

ISBN 978-1-60692-772-4
© 2009 Nova Science Publishers, Inc.

*Chapter 6*

# THE FUNCTIONAL ROLE OF *PHRAGMITES AUSTRALIS* IN THE INHIBITION OF CYANOBACTERIAL GROWTH AND THE TRANSITION OF PLANT SPECIES

*Sheng Zhou[1], Satoshi Nakai[2], Megumi Nishikawa[1] and Masaaki Hosomi[1]*

1. Faculty of Engineering, Tokyo University of Agriculture and Technology,
2-24-16 Naka, Koganei, Tokyo 184-8588, Japan
2. Graduate School of Engineering, Hiroshima University, 1-4-1 Kagamiyama, Higashi-
Hiroshima, Hiroshima 739-8527, Japan

## ABSTRACT

*Phragmites australis* (common reed) is a large perennial macrophyte found in aquatic ecosystems throughout the world. *P. australis* has many important roles in aquatic ecosystems. A functional allelopathic role, inhibiting cyanobacterial growth has been suggested for this species. To confirm this characteristic a series of assays using "rotting-reed solution (RRS)" against *Phormidium tenue* and *Microcystis aeruginosa* were performed. The results demonstrated the growth inhibition of both cyanobacteria. The degree of growth inhibition due to RRS was correlated to the dissolved organic carbon concentration of the RRS, which changed seasonally. *P. tenue* was more sensitive to the allelopathic effect of *P. australis* than *M. aeruginosa*. To identify anti-cyanobacterial allelochemicals released from the reed, extracts of RRS were analyzed using GC/MS. Eight phenols and four fatty acids were identified. Of these, four phenols and one fatty acid were found to inhibit the growth of both cyanobacteria. The remaining compounds were species-specific in their inhibition or else exhibited no effect.

Other roles of *P. australis* include river water treatment, accumulation of sediment and plant transition. These were surveyed within a series of 10 year old constructed wetlands. The average total sediment accumulation in constructed wetlands with *P. australis* was $3.61 - 4.43$ kg m$^{-2}$ yr$^{-1}$. This resulted in some land formation within the wetlands which may affect the plant community. The average community diversity index of four representative constructed wetland units was 0.88 in 2004, which was significantly higher than 10 years previously, indicating an increase in diversity. In addition, surveys during 2005 demonstrated that *P. australis* in these constructed wetlands had more gene types in comparison to *P. australis* populations in native wetlands.

**Keywords:** Allelopathic effect, Constructed wetland, Cyanobacteria, Genetic diversity, *Phragmites australis*, Plant transition, Sediment accumulation

# 1. INTRODUCTION

*Phragmites australis* (common reed) is a large perennial macrophyte found in natural and constructed wetlands throughout temperate and tropical regions of the world. Moreover, *P. australis* is one of the most important aquatic plant communities in Japan. The annual stems develop from a perennial rhizome system, which is responsible for the rapid vegetative expansion of the species in native wetlands. In Japan *P. australis* grows in a variety of wetland habitats such as the littoral zone of lakes (e.g. at Lake Biwa (670 km$^2$) Lake Kasumigaura (220 km$^2$, Figure 1), and Lake Teganuma), rivers (e.g. Tamagawa River (Figure 1)) and marshs (e.g. Kushiro-shitsugen). Within these habitats it dominates stands or co-dominates with other plants (*Typha latifolia, Typha angustifolia, Typha orientalis, Miscanthus sacchariflorus, Zizania latifolia*). It plays a multi-purpose role in nature conservation and environmental protection (Karunaratne et al., 2003). It is of ecological importance for shoreline protection (Coops et al., 1996), conservation of aquatic biodiversity (Kira, 1991), and nutrient retention (Kovacs et al., 1978; Brix, 1994; Yuji and Hosomi, 2002).

Organisms in aquatic ecosystems are influenced by changes of the state of eutrophication. Eutrophication is the process whereby water bodies, such as lakes, slow-moving streams, or estuaries receive excess nutrients. Natural eutrophication is the process by which lakes gradually age and become more productive. It normally takes thousands of years. However, humans have greatly accelerated this process in thousands of lakes around the globe through various cultural activities. Surveys have shown that in the Asia-Pacific Region, 54% of lakes are eutrophic; in Europe, 53%; in Africa, 28%; in North America, 48%; and in South America, 41% (UNEP, 1994). In Japan, 34% of all lakes with an area greater than 1ha are confirmed to be eutrophic (JWRC, 1995). Eutrophication stimulates the excessive growth of algae, with subsequent changes in water quality in many lakes. Algal blooms reduce dissolved oxygen in the water when the dead algae decompose, and this oxygen depletion can lead to the death of other organisms. Oxygen depletion can cause serious perturbation of aquatic resources such as fisheries and water-supply reservoirs. Watanabe (1989) reported that toxic substances, such as the microcystin formed by cyanobacteria can kill domestic animals. The World Health Organization (WHO) established a Tolerable Daily Intake (TDI) for microcystin in 1998. This allowed up to 0.04µg kg$^{-1}$ d$^{-1}$ microcystin in drinking water and set a tentative standard value for tap water quality of 1µg l$^{-1}$ microcystin. To control algal blooms, algaecides such as copper chelate have been used in ponds or lakes. However, there is concern that such chemicals could have adverse effects on the ecosystem and particularly on fishes (Hale, 1957). Hence, it is necessary to find feasible and less environmentally harmful methods for the control of algal blooms.

Allelopathy refers to the chemical inhibition of one species by another. Allelopathic chemicals released into the environment affect the development and growth of neighbouring plants. Allelopathic chemicals can be present in any part of a plant. They can be found in leaves, flowers, roots, fruits, or stems. These toxins affect target species in many different ways. Allelopathic chemicals may inhibit shoot or root growth, they may inhibit nutrient uptake, or they may attack a naturally occurring symbiotic relationship thereby destroying the plant's usable source of nutrients. Many examples of allelopathy between terrestrial plants have been studied (Lovett et al., 1989; David, 1996). However, some experiments have also been conducted in aquatic ecosystems. Macrophytes and algae have long been known to have an antagonistic relationship in both natural and experimental aquatic ecosystems. Some researchers have investigated the feasibility of controlling nuisance cyanobacterial growth using macrophytes (Pillinger et al., 1994; Barrett et al., 1996; Ridge and Pillinger, 1996; Everall and Lees, 1997). Either living or withered macrophytes can be used to control cyanobacterial growth. Some bioactive anti-algal compounds have been extracted from macrophytes. Saito et al. (1989) extracted hydrolysable tannins (quercetin, gallic acid) from *Myriophyllum brasilience* for inhibiting the growth of blue-green algae (*Anabaena flos-aquae* and *Microcystis aeruginosa*). In another study, *Pistia stratiotes* was shown to store a growth-inhibiting phenylpropanoid (α-asarone) and four fatty acids (Aliotta et al., 1991). In addition, some other compounds, e.g. ellagic acid, stigmast-4-ene-3,6-dione, 6z,9z,12z,15z-octadecatetraenoic acid, and 4-methylthio- 1,2-dithiolane have been identified (Anthoni et al., 1980; Planas et al., 1981; Kakisawa et al., 1988; Aliotta et al., 1992). The allelopathic effect of submerged macrophytes on algae growth has also been investigated (Nakai et al., 1999; Nakai et al., 2001; Nakai and Hosomi, 2002). Other studies, (for example, Everall and Lees, 1997) have demonstrated that the presence of rotting barley (*Hordeum vulgare*) straw significantly reduced cyanobacterial and general algal activity in a reservoir, whilst Ridge and Pillinger (1996) showed that autoclaving the straw before use enhanced its anti-algal activity.

*P. australis* has many important roles in aquatic ecosystems, including landscape creation, wildlife conservation, and water treatment. As a member of the poaceae (like barley), *P. australis* has been considered to inhibit algal growth in the same manner as rotting barley straw. In 2000, an experiment was performed to investigate the feasibility of controlling cyanobacterial growth using cut barley straw and reeds in isolated tanks set in Watarase reservoir in Japan (PWRC, 2002). Both reed and barley significantly inhibited the growth of cyanobacteria. However, no detailed information was recorded regarding the growth inhibition of specific cyanobacterial species or of the causes of the effects of the common reed. In addition, knowledge of seasonal variation and any difference in inhibitory effect between leaves and stems of reeds is required, since the carbon metabolism of reed leaves varies with the age of the plant (Antonielli et al., 2002). It is thus necessary to examine the allelopathic effect of *P. australis* on cyanobacterial growth with particular reference to variation among plants, leaves and stems and to investigate any seasonal variation in this effect.

On the other hand, to thorough resolve the problem of lake eutrophication and restore aquactic ecosystem, it is necessary to reduce nutrient load in lake, particularly load from non-point sources through rivers. Constructed wetlands have been used as nutrient sinks or removal systems for polluted river water (Mitsch et al., 2005a; Reilly et al., 2000; Jing et al., 2001; Zhou and Hosomi, 2008) and farm runoff (Comin et al., 1997; Fink and Mitsch, 2004). The effects of constructed wetland on water quality are described from several aspects, which include the settling of particulates, transferring oxygen into the rhizosphere, nitrification and denitrification, adsorption and absorption of phosphorus, and reduction of other pollutant concentrations (Kadlec and Knight, 1996). The most commonly used plants in constructed wetlands is *P. australis*. Nowadays, there are thousands of constructed wetlands operating as wastewater and river water purification systems in Europe and North America. In Europe, horizontal sub-surface flow wetland is the most popular type (Vymazal, 2005) while in the United States, most constructed wetlands are free water surface flow wetland (FWSF) for treating domestic wastewater, farmland runoff, and river water (Kadlec and Knight, 1996; Mitsch et al., 2002; Fink and Mitsch, 2004). For example, Bachand and Horne, 2000a, b) built six macrocosms (approximately 0.13 ha each) to study nitrogen transformation processes in the adjacent Prado Basin constructed wetland and attributed nitrate removal to denitrification. In addition, two created wetlands in the Mississippi River Basin were investigated periodically for 9 years to evaluate nitrogen retention capacity (Mitsch et al., 2005a). Furthermore, the National-Research-Council (1992) called for the creation and restoration of 4 million ha of wetlands in the United States by 2010. According to a survey made by Nakamura et al., 2007), constructed wetlands in Japan receive influent containing relatively low concentrations of nutrients (N, P) and rather high concentrations of suspended solids (SS) compared with wastewater from a secondary or tertiary treatment. SS would be deposited in constructed wetland planted with macrophytes that also contribute to sediment accumulation. It has been reported that dense *Typha* has elicited greater sediment deposition and reduced re-suspension in open water zones (Anderson and Mitsch, 2006). However, in constructed estuarine wetlands, sediment and phosphorus accumulation tends to occur rapidly in the first few years and then decline after 10 years depending on the conceptual model of the wetland (Craft, 1997).

Lake Kasumigaura (Figure 1), the second largest freshwater lake in Japan, is well known for its eutrophication, with mean concentrations of T-N and T-P of 1.10 mg $l^{-1}$ and 0.10 mg $l^{-1}$, respectively. N loads to Lake Kasumigaura from domestic wastewater and non-point sources account for 35% and 40%, respectively, of the lake's entire N load (Japan Society on Water Environment, 1999). There are 56 rivers flowing into this lake. Because most of the nutrient load is carried to the lake by rivers, to reduce the nutrient load to the lake, it is necessary to reduce the nutrient concentrations in the rivers. Several wetlands were constructed along Lake Kasumigaura to purify nutrient polluted rivers, most of which were based on the common reed. On the other hand, compared with the reed stand distribution around the entire littoral zone of Lake Kasumigaura about 100 years ago, most of the reed stands have decreased in size for various reasons relating to the past fifty years. For example, construction of bank and some facilities on the lakeside for fishery or yacht harbor. Besides these reasons, a rise in the water level of the lake and an increase in depth due to the removal of gravelwas also thought to eliminate reed stands (Onuma and Ikeda, 1999). It is drawing a lot of attention for reconstructing reed stand around this lake from nineties. Hence, besides purifying polluted river water, constructed wetlands with *P. australis* would contribute to the restoration of reed stands along the lake. However, different species of *P. australis* show enormous variations in morphology, ecology and cytogenetics (Haslam, 1972; Clevering and Lissner, 1999). Some studies have suggested that the high ecological plasticity of the species *P. australis* is often based on the high number of reed clones (Haslam, 1972; vanderToorn, 1972). Furthermore, Neuhaus et al., 1993) reported the stability of reed stands is related to genetic diversity, and monoclonal stands should have limited adaptive response to changing site conditions. In addition, the genetic variation and population genetic structure of aquatic plants are highly influenced by environmental conditions including water quality (Barrett et al., 1993; Laushman, 1993). Despite this, there are scarcely data for the genetic diversity evaluation of *P. australis* in constructed wetland. Hence, it is necessary to evaluate whether monoclonal reed stands form in constructed wetland and further compare the genetic diversity of *P. australis* in constructed wetland with that in the native wetland around the lake as well as evaluate the effect of constructed wetland on nutrient reduction, sediment accumulation, and plant transition during a relatively long term.

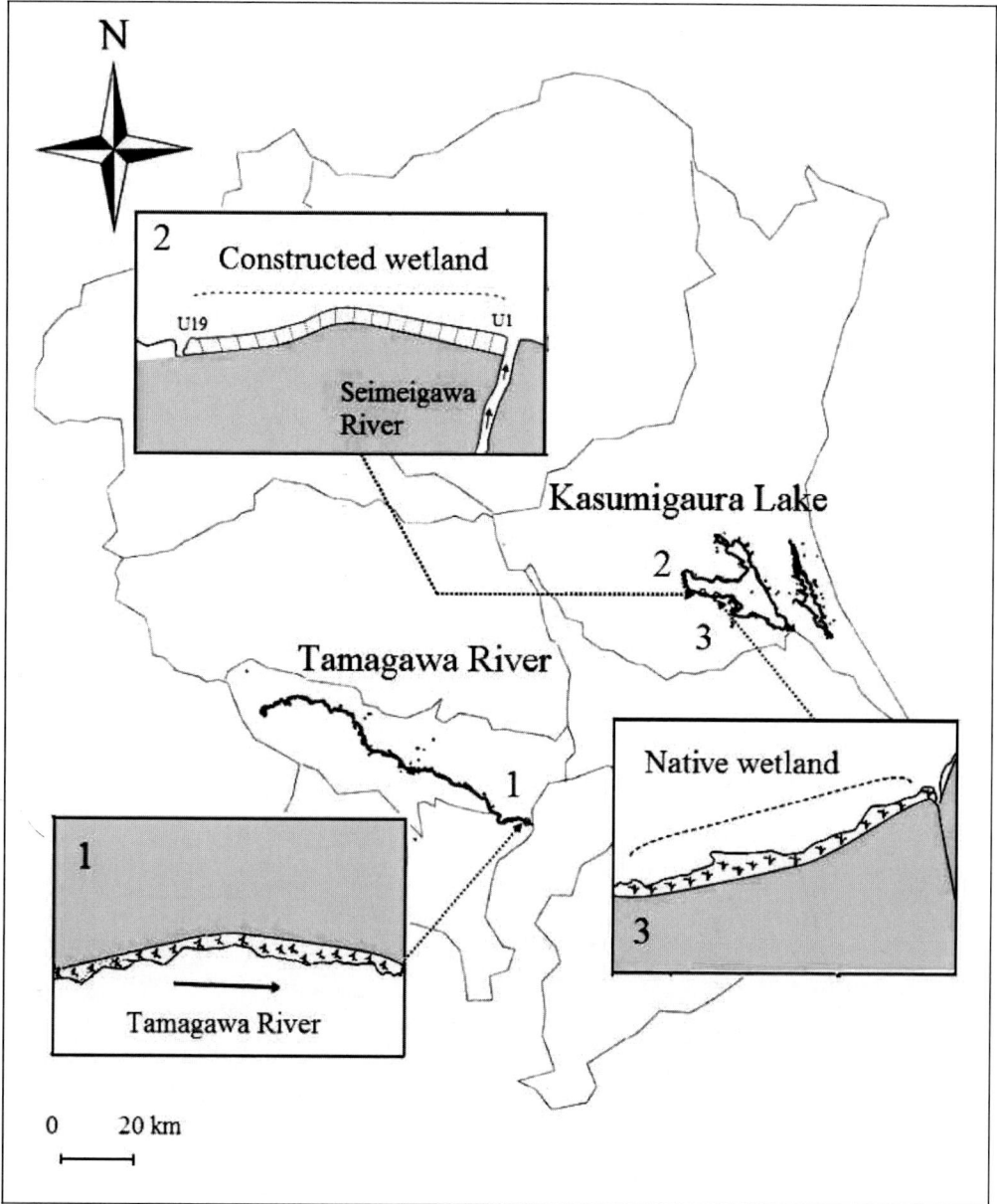

Figure 1. Map of Kanto region, Japan. Location of the Site 1 near Tamagawa River, Site 2 (constructed wetlands) and Site 3 (native wetlands) near Kasumigaura Lake are shown. Outlines of Site 1, 2, and 3 are shown in the insets.

## 2. METHOD AND MATERIALS

### 2.1. Inhibition of cyanobacterial growth by *P. australis*

#### 2.1.1. Cyanobacterial assays

Samples of *P. australis* were collected from a tidal flat at the mouth of the Tamagawa River (Figure 1, Site 1) from May to December. *Microcystis aeruginosa* and *Phormidium tenue*, obtained from the microbial collection of the National Institute for Environmental Studies (NIES), Japan, were used for the cyanobacterial assays.

To confirm the allelopathic effects of reed, a leachate of rotting reed was assayed against both cyanobacteria. At the time of collection the reed was cut into pieces 2–3 cm long and then added to pure water at concentrations of 50 g-wet $l^{-1}$ for *P. tenue* and 120 g-wet $l^{-1}$ for *M. aeruginosa*. The mixture was stored at 25 °C for 13 days with aeration to obtain "rotting-reed solution" (RRS). For the *P. tenue* inhibition experiments, the reed plants were further divided into leaves and stems. The RRS of leaves and stems were made following the same method. Two different culture media, CB (Table 1 for *M. aeruginosa*) and C (Table 2 for *P. tenue*) were prepared (Watanabe and Satake, 1991). 100 ml measures of the different types of RRS were filtered through autoclaved membrane filters (0.22 µm) and added to 500-ml Erlenmeyer flasks (Figure 2). The RRS was diluted to varying degrees by addition of one or other of the cyanobacterial media. *P. tenue* and *M. aeruginosa* were respectively inoculated at a range of turbidity between 0.1–1.0. Each experiment was replicated 3 times and all treatments were cultured under a light intensity of 5000 lux at 25 °C for 10–15 days at a rotation of 70 rpm.

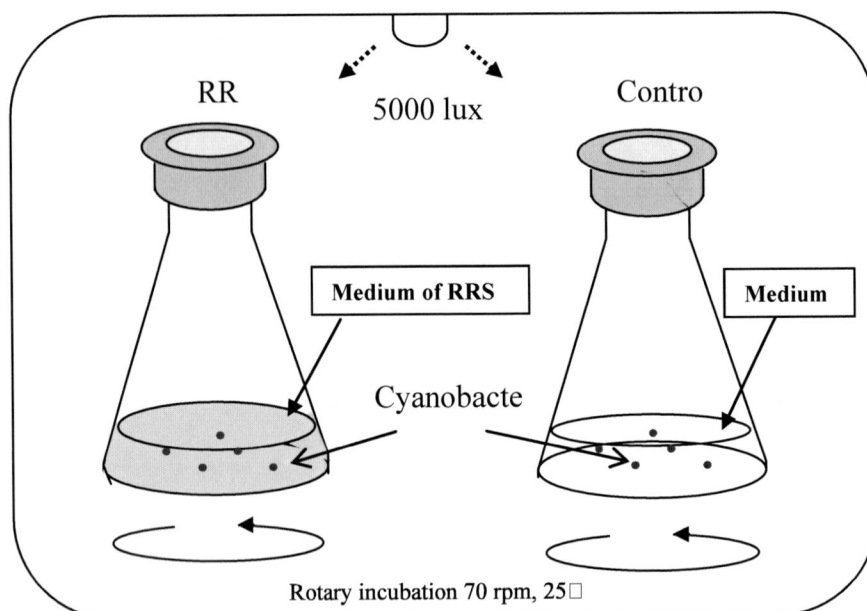

Figure 2 Schematic diagram of bioassay for RRS.

**Table 1. CB culture medium composition.**

| | |
|---|---|
| $Ca(NO_3)_2\epsilon 4H_2O$ | 15 mg |
| $KNO_3$ | 10 mg |
| $\beta$-$Na_2$glycerophosphate | 5 mg |
| $MgSO_4\epsilon 7H_2O$ | 4 mg |
| Vitamin $B_{12}$ | 0.01µg |
| Biotin | 0.01µg |
| Thiamine HCl | 1µg |
| PIV metals | 0.3 ml |
| Bicine | 50 mg |
| Distilled water | 99.8 ml |

**Table 2. C culture and PIV metals medium composition**

| C culture | | PIV metals | |
|---|---|---|---|
| $Ca(NO_3)_2\epsilon 4H_2O$ | 15 mg | $FeCl_3\epsilon 6H_2O$ | 19.6 mg |
| $KNO_3$ | 10 mg | $MnCl_2\epsilon 4H_2O$ | 3.6 mg |
| $\beta$-$Na_2$glycerophosphate | 5 mg | $ZnSO_4\epsilon 7H_2O$ | 2.2 mg |
| $MgSO_4\epsilon 7H_2O$ | 4 mg | $CoCl_2\epsilon 6H_2O$ | 0.4 mg |
| Vitamin $B_{12}$ | 0.01µg | $Na_2MoO_4\epsilon 2H_2O$ | 0.25 mg |
| Biotin | 0.01µg | $Na_2EDTA\epsilon 2H_2O$ | 100 mg |
| Thiamine HCl | 1µg | Distilled water | 100 ml |
| PIV metals | 0.3 ml | | |
| Hepes | 50 mg | | |
| Distilled water | 99.8 ml | | |

**Table 3. Operating conditions for GC/MS.**

| | |
|---|---|
| Gas chromatograph | Hewlett Packard 6890 series |
| Column | HP5-MS (30.0 m × 320 µm, 0.25 µm) |
| Carrier gas | He (99.9999%), 1.3 mL/min |
| Oven temp. | 50°C (1 min); 10°C/min to 280°C; 20°C/min to 310°C |
| Injection mode | Splitless |
| Injection volume | 1 µL |
| Injector temp. | 250°C |
| Mass spectrometer | Hewlett Packard 5973 series |
| Ionization mode | Electron ionization |
| Mass interface temp. | 280°C |
| Ion source temp. | 280°C |

### 2.1.2 Measurement of cyanobacterials growth

The growth of both cyanobacteria was monitored by measuring turbidity (T-2600 DA turbidity meter, Tokyo-Denshoku, Tokyo, Japan) and thus maximum growth was determined. Dissolved organic carbon (DOC) concentrations within the RRS were measured using a total organic carbon analyzer (TOC-5000, Shimadzu, Japan).

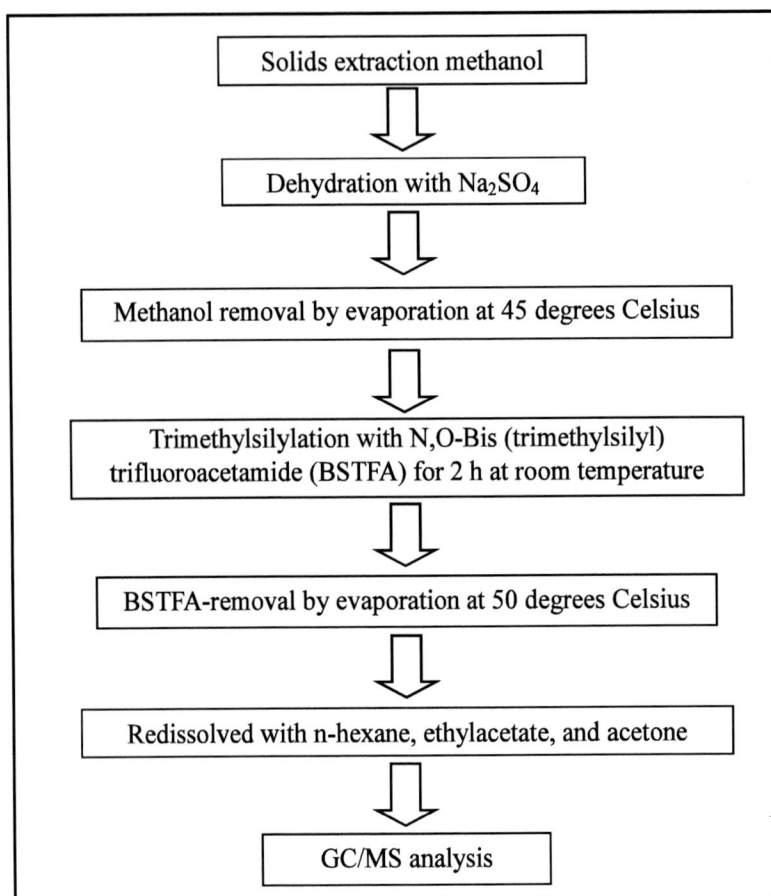

Figure 3. Scheme of GC/MS analyses

## 2.2. Identification of Allelopathic Chemicals

To identify the anti-cyanobacterial allelochemicals released by the reed, RRS extracts were analyzed using a gas chromatograph equipped with a mass-selective detector (GC/MS). RRS extract was prepared using a solid extraction technique using C18HD (3M, Japan) and OASIS HLB (Waters, Japan) solid extraction cartridges. In brief, 100 ml of RRS was filtered through an autoclaved membrane filter (0.22 μm) and then adjusted to pH 3 (for C18HD) or pH 1 (for OASIS HLB) with 0.1 N HCl. After passing the solution through the solid extraction cartridge it was eluted using methanol. All methanol solutions were treated with sodium sulphate anhydrate and the methanol fully evaporated at 45 °C. The resultant residues were treated with N,O-bis(trimethylsilyl) trifluoroacetamide (BSTFA: Tokyo Kasei, Tokyo, Japan) for 2 h at room temperature for trimethylsilyl derivatization (Figure 3). Subsequently, samples were dissolved in a small volume of ethylacetate (100 μl) and subjected to GC/MS analysis in electron ionization mode (Table 3).

## 2.3. Nutrient Retention by *P. australis* in Constructed Wetland

### 2.3.1. Site Description

The Seimeigawa River (36°01'52''N, 140°16'44''E; Figure 1) is one of several rivers that drain water from the nearby basin and transport domestic wastewater and agricultural land runoff to Lake Kasumigaura. The average flow rate of Seimeigawa River was 0.52 $m^3 s^{-1}$ and T-N and T-P concentrations were 3.05 mg $L^{-1}$ and 0.28 mg $L^{-1}$, respectively.

To reduce nutrient loading from the river to Lake Kasumigaura using constructed wetland, a series of surface-flow wetlands with *P. australis* that taken from littoral zones of Lake Kasumigaura were constructed near the estuary of the Seimeigawa River (Site 2 in Figure 1) by the Kasumigaura River Office, Japan. There are 19 units (U1 to U19, with one unit being 50 m × 40 m; the total area of the 19 units is 38,000 $m^2$) at the lakeside were constructed from 1991. By 1995-1996, all 19 units had been constructed and applied for purifying river water from the Seimeigawa River. The designed flow rate into the entire constructed wetland is 0.21 $m^3 s^{-1}$ (hydraulic loading rate: 0.48 $m^3 m^{-2} d^{-1}$), which accounted for about 40% of river flow. To separate river water from the channel into each unit equally, there are 6 slits of equal width for influent at the inlet of each unit. The sides facing the lake have outlets.

### 2.3.2. Sampling and Measurement

Most units were constructed by 1995. Water quality and sediment component were investigated during 1995-1996. Then, investigation of the water quality was conducted every year (U2, U5, U11, and U18) from 1999. Furthermore, measurements of sediment accumulation, hydraulic retention time, and vegetation were conducted for some years.

There are nine sampling points for sediment sampling for each unit, which consist of three points at three lines that located at 5 m, 15 m, and 25 m of the flow direction. Samples from each point were taken using cores and grouped by depth (0-5 cm; 5-10 cm; 10-20 cm; 20-30 cm) with other points. Sediment samples were analyzed for water content, organic content, nitrogen content, and phosphorus content as described by JSWA, 1984). To evaluate the sediment accumulation, ground levels of the sediment surface were measured in 1999, 2002, and 2005 based on a 5 m grid system. The sediment surface being higher than ground level was considered to be net accumulation during the experimental period. Nitrogen and phosphorus contents accumulated in the sediment were calculated by multiplying the net sediment accumulation and average nitrogen and phosphorus contents in the sediment.

Samples of influent and effluent were taken periodically. For effluent, six samples were taken along the outlet side of each unit and combined as one sample. Water samples were analyzed for SS, T-N, T-P, and particulate and soluble phosphorus. T-N and T-P were analyzed by absorption spectrophotometry after decomposition with potassium peroxodisulfate ($K_2S_2O_8$) (JSWA, 1984). The ion concentration was measured using an ion chromatograph (ICS-90, Dionex, USA).

### 2.3.3. Plant Community Transition in Constructed Wetland

Macrophytes coverage by the dominant community was estimated at the period of peak biomass (July or August) by ground surveys. A 2 m × 2 m grid system was used to identify the location of plant communities in U2, U5, U11, and U18. In some areas, different plant species mix and they are difficult to separate as different communities. In this chapter, these areas are referred to as a mixed community and half the area was used to calculate the community diversity index.

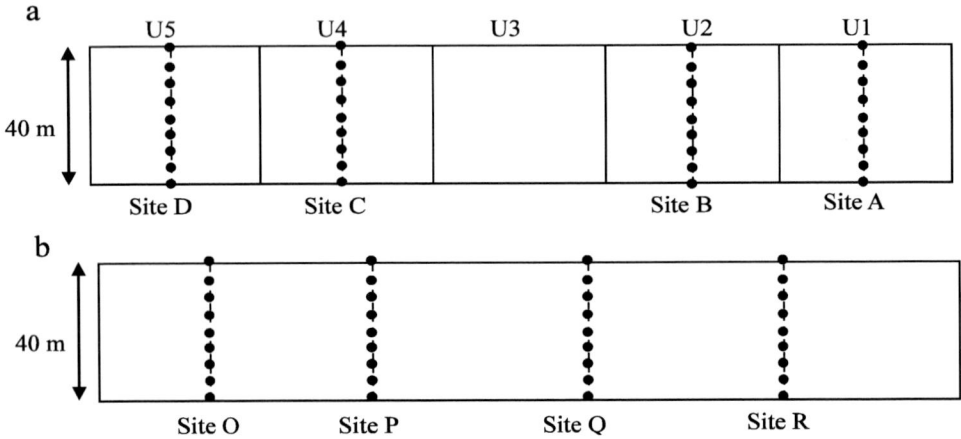

Figure 4 Details of *P. australis* sampling sites for genetic diversity investigation in constructed wetland (a: sites A-D) and native wetland (b: sites O-R). At each site, nine points (●) were sampled at 5 meter intervals.

## 2.4. Comparison of Genetic Diversity of *P. australis*

As a comparative analysis of the genetic diversity of *P. australis* between the constructed and native wetlands, another site of native wetland (Site 3 in Figure 1) was chosen for native *P. australis* sampling. The native wetland is one of the largest natural reed-beds around Lake Kasumigaura (36°01′46″N, 140°16′59″E). The native site is 2.0 ha in area, 20–50 m in width and 600 m in length, and situated to the east of the river mouth of the Seimeigawa River. The site comprises mainly reeds (*P. australis*), a *Carex–Phragmites* community, cattail (*Typha latifolia*) or wild rice (*Zizania latifolia*), and scattered willows (*Salix* spp.) or rose (*Rosa multifilia*) bushes from lakeside to bankside.

### 2.4.1. Plant material and Total DNA Extraction

Samples of *P. australis* used in this study were obtained from constructed wetland (Site 1: U1, U2, U4, U5) and native wetland (Site 2) around Lake Kasumigaura in 2005. About 5-10 g of fresh leaves per plant was collected. A total of 72 individuals from these populations were included in an inter-simple sequence repeat (ISSR) study. Frozen leaves were powdered in liquid nitrogen and genomic DNA was extracted. Approximately 0.8 g of leaf material was used for DNA extraction using the ISOPLANT II kit (NIPPON GENE CO., LTD, Tokyo, Japan) according to manual. The cell wall, cell membrane, and nucleus membrane of *P. australis* leaves are destroyed by benzyl chloride as the main ingredient in solution, and DNA melts to an aqueous phase. Extracted DNA was ethanol precipitated and then pelleted at 6,000 × g for 5 min. The pellet was washed in 300 μL of 70% ethanol. The dried pellet was resuspended in 100 μL of TE (Tris-EDTA buffer) at pH 8.0. The DNA quality and quantity were estimated by separating 1 μL on 1.5% agarose gels.

**Table 4. Name and sequence of primers used in the present study**

| Primer name | Sequence (5'-3') | Annealing temperature (°C) |
|---|---|---|
| AM-1 | $(GGC)_5AT$ | 58 |
| AM-2 | $(AAG)_5GC$ | 38 |
| AM-3 | $(AAG)_5TG$ | 38 |
| AM-4 | $(AAG)_5CC$ | 40 |
| AM-5 | $(AGC)_5CA$ | 57 |
| AM-6 | $(AGC)_5GG$ | 51 |
| AM-7 | $(GGC)_5TA$ | 60 |
| AM-8 | $(AGC)_5GA$ | 53 |
| AM-9 | $(AAG)_5CG$ | 40 |

## 2.4.2. ISSR Polymerase Chain Reaction (PCR) Amplification

Primers shown in Table 4 (Parsons et al., 1997) were selected for the preliminary ISSR analysis of *P. australis*. PCRs were carried out on a thermalcycler (Biometra, Inc., Japan) and programmed as shown in Table 5. Reactions were carried out in a volume of 25 μl, containing 0.2 mM of each dNTP (deoxyribonucleotide triphosphate), 2 mM $MgCl_2$, 0.2 mM primer, 0.625 U Taq DNA polymerase and 2 μl (approximately 120 ng) of DNA template. Amplification products were electrophoresed on 1.5% agarose gels stained with ethidium bromide, visualized under ultraviolet light, and photographed. Sizes of amplification products were estimated using a 100 bp DNA ladder (Takara Bio, Inc., Japan). Basing on the result of the preliminary experiment, the primers of AM-3, AM-7, and AM-8 are suitable and used for band generation of *P. australis* from constructed and native wetlands.

**Table 5. PCR program in thermocycler**

| Step | Temperature (°C) | Time (min) | |
|---|---|---|---|
| 1 | 95 | 3 | |
| 2 | 95 | 0.5 | |
| 3 | T(°C) in Table 2+ 2 | 1 | 2 cycles |
| 4 | 72 | 2 | |
| 5 | 95 | 0.5 | |
| 6 | T(°C) in Table 2 | 1 | 2 cycles |
| 7 | 72 | 2 | |
| 8 | 94 | 0.5 | |
| 9 | T(°C) in Table 2 | 1 | 40 cycles |
| 10 | 72 | 2 | |
| 11 | 72 | 5 | |

## 2.5. Data Analysis

PCR-fingerprints of all reed samples at each sample point were made. Band positions on the gels were determined visually and the fingerprint patterns were transformed into a binary character matrix with 1 indicating the presence and 0 the absence of a band at a particular position in a lane. The percentages of polymorphic bands in constructed and native wetlands for genetic diversity comparison were also calculated. The genetic similarity coefficients were used to cluster samples based on their degree of genetic similarity, which calculated by the unweighted pair group method using arithmetic averages (UPGMA), and finally for computing the dendrograms.

# 3. RESULTS AND DISCUSSION

## 3.1. The allelopathic effect of *P. australis*

Figure 5 shows growth curves of *P. tenue* in the RRS during June. The growth of *P. tenue* in RRS was inhibited in comparison to the rapid growth of *P. tenue* in controls. Thus the release of anti-cyanobacterial compounds from *P. australis* was confirmed.

As shown in figure 5, RRS also inhibited the growth of *M. aeruginosa*. Although the RRS for *M. aeruginosa* was prepared a higher concentration of rotting reed (120 g-wet $l^{-1}$) than that for *P. tenue*, *M. aeruginosa* still achieved 30% of the control's maximum growth. These results indicate that *P. tenue* may be more sensitive to the allelopathic effects of *P. australis* than *M. aeruginosa*. Therefore, *P. tenue* was selected for further tests using RRS.

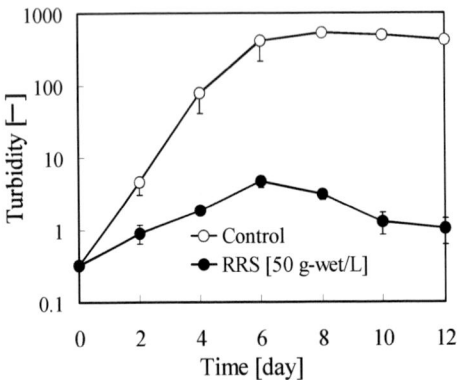

Figure 5. Growth of *P. tenue* as affected by RRS prepared from *P. australis* collected on June. Bars indicate standard deviations (n = 3).

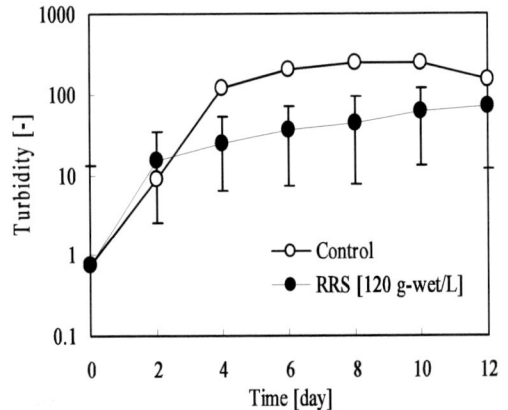

Figure 6. Growth of *M. aeruginosa* as affected by RRS prepared from *P. australis* collected in May. Bars indicate standard deviations (*n* = 3).

### 3.1.1. Comparison of cyanobacterial growth inhibition among RRS from whole plant, leaves, and stems.

To confirm variations in the growth inhibitory effect among whole plant RRS, leaf RRS, and stem RRS, each kind of RRS was manufactured using *P. australis* plants collected in August. The growth curves of *P. tenue* for each of the three kinds of RRS are shown in figure 7. The growth of *P. tenue* in whole plant RRS was inhibited completely. The stem RRS had no inhibitory effect on *P. tenue*, whilst the leaf RRS exhibited a significant inhibitory effect similar to that for the whole plant RRS. A similar pattern was observed for plants collected in September (figure 8). However, in this case the inhibitory effect of whole plant RRS was stronger than either that for leaf RRS or for stem RRS alone. This suggests an additive or multiplicative inhibitory effect on *P. tenue* among compounds released from leaves and stems. Furthermore, in November (figure 9) 78% growth inhibition was recorded for stem RRS. This was stronger than that for either August or September.

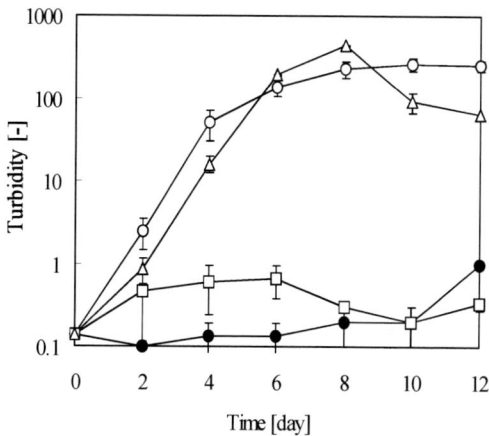

Figure 7 Growth curve of *P. tenue* in the RRS (Aug). Symbols: o, Control; •, Whole plant; □, Leaf; △, Stem.

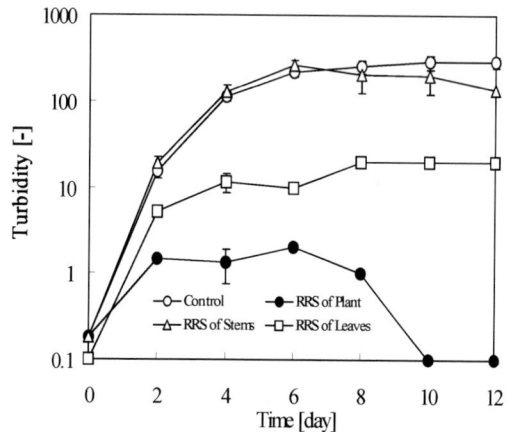

Figure 8 Growth curve of *P. tenue* in the RRS (Sep.). Symbols: o, Control; •, Whole plant □, Leaf; △, Stem.

Figure 10 shows the variation in inhibition of growth according to the time of reed collection. Here, the inhibitory effect is quantified as the proportion decrease in maximum growth (as measured without inhibition). As shown in figure 10, the whole plant RRS exhibited a stronger inhibitory effect after May and could inhibit more than 99% of the maximum growth of *P. tenue* from June to November. The inhibitory effect of stem RRS increased from September to November, indicating an increased stem allelopathic effect with the growth of reed. On the other hand, the inhibitory effect of leaf RRS was strong from August to November, suggesting that the leaves provide a major contribution to the whole plant allelopathic effect of *P. tenue*.

### 3.1.2. Seasonal changes in DOC and inhibition of cyanobacterial growth

Figure 11 shows the seasonal variation in DOC concentration of the RRS prepared using 50 g-wet l$^{-1}$rotting *P. australis*. Peak DOC occurred in October. The flowering time of *P. australis* in Japan is August to October (Kadono, 1996). Thus flowering may affect the leachability of organic compounds from *P. australis* plants. The relationship between DOC and the growth inhibition effect of the RRS is illustrated in figure 12. Growth inhibition of *P. tenue* appears at between 10 and 15 mg-DOC $l^{-1}$. It is notable that RRS containing 15 mg $l^{-1}$ of DOC suppressed the maximum growth of *P. tenue* to about 1% of the control. These results indicate that increased leachability of organic compounds from *P. australis* plants may enhance the inhibitory effects of the reed on cyanobacteria. *P. australis* collected from May to November inhibited the growth of *P. tenue*.

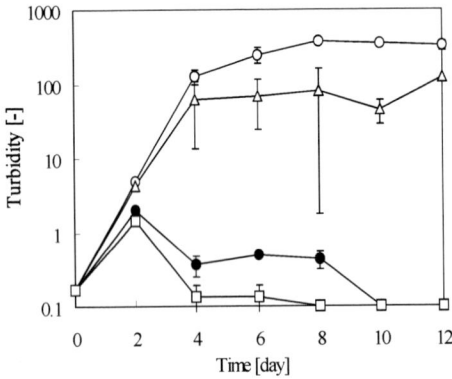

Figure 9 Growth curve of *P.tenue* in the RRS (Nov). Symbols: o, Control; ●, Whole plant; □, Leaf; △, Stem.

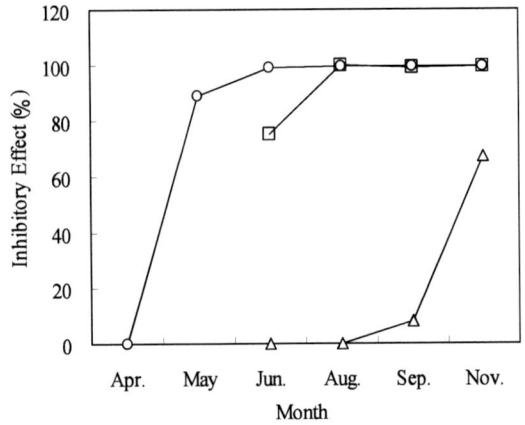

Figure 10.Comparison of inhibition effect among three kinds of RRS. Symbols: o, Whole plant; □, Leaf; △, Stem.

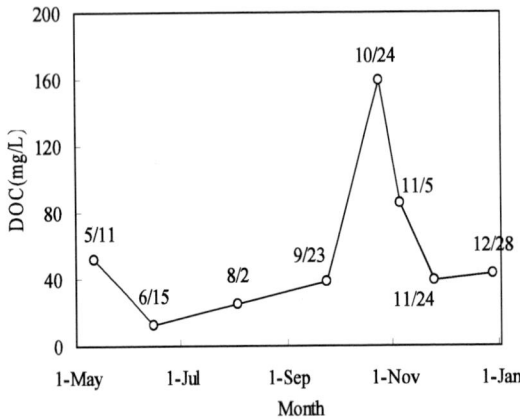

Figure 11 Seasonal changes in the DOC of RRS prepared by rotting *P. australis* at 50 g-wet/L.

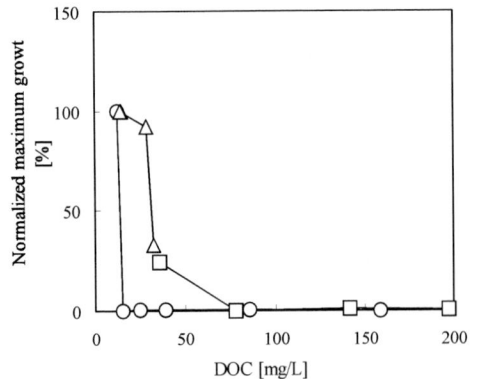

Figure 12. Relationship between the DOC of RRS and its inhibitory effect on the maximum growth of *P. tenue*. Symbols: o, Whole plant; □, Leaf; △, Stem.

Although data regarding leaf RRS and stem RRS is limited, Figure 12 suggests that the inhibitory effect of leaf RRS was stronger than that of stem RRS because the DOC produced by leaf RRS was greater than that of stem RRS. In addition, it also indicates an additive or multiplicative effect between the anti-cyanobacterial compounds released from leaves and stems.

### 3.1.3. Identification of allelopathic chemical candidates

Nishikawa (1999) reported that several kinds of bioactive compounds were released from *P. australis*, including: saturated and unsaturated fatty acids, such as palmitic and stearic acids, oleic and linoleic acids, phenolic compounds, flavonoids, and vitamins had been identified. GC/MS analysis showed the existence of many compounds in the extracts of RRS reflecting these prior studies.

C18HD and HLB were used to extract compounds from RSS. Before identification of the compounds, it was confirmed that the inhibition of cyanobacterial growth due to the solid extracts matched that of the RRS (data not shown). These analyses showed that the extracts did indeed contain allelopathic chemicals from the RRS.

Figure 13 (a, b) shows total ion chromatograms of the C18HD extract of RRS. The mass spectral patterns of the respective peaks were identified by comparing them with patterns stored in the mass spectral library (version 2) of the U.S. National Institute of Standards and Technology. Two phenols (protocatechuic and ferulic acids), and four fatty acids (myristic, palmitic, pelargonic and stearic acids) were identified. Figure 14 and figure 15 show the fragment pattern of pelargonic acid TMS ester and myristic acid TMS ester, respectively. The GC/MS analysis of the HLB extract resulted in the detection of another 6 phenols (p-coumaric, caffeic, sinapic, gallic, vanillic and syringic acids).

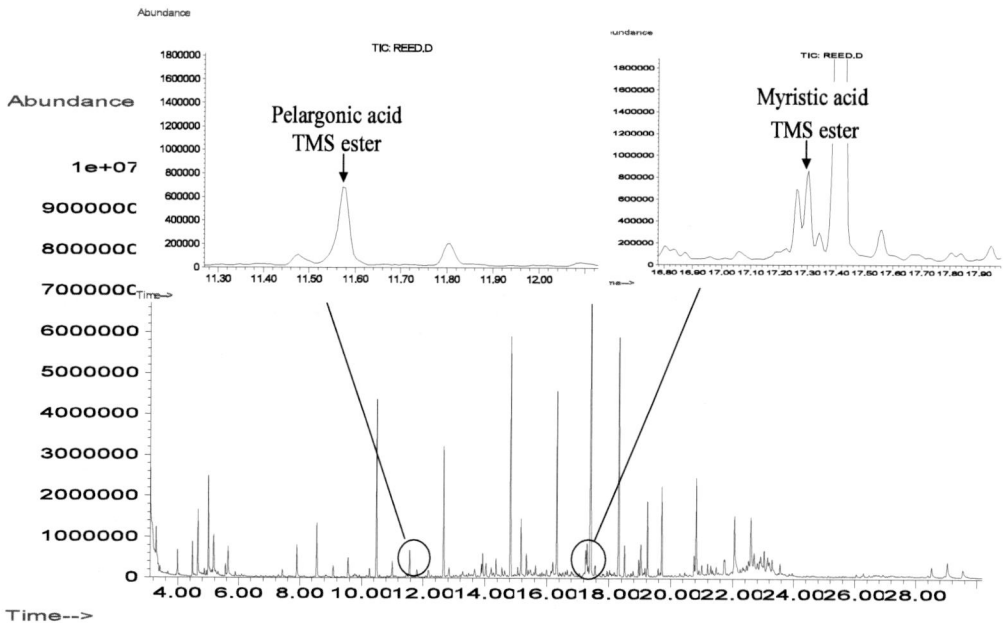

Figure 13a Total ion chromatogram of the C18HD extract of RRS.

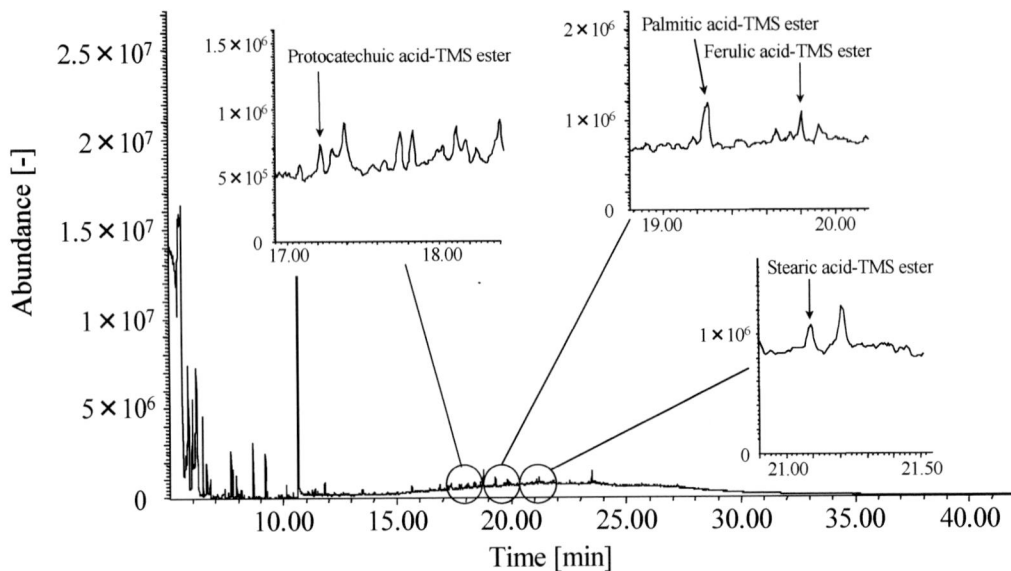

Figure 13b Total ion chromatogram of the C18HD extract of RRS.

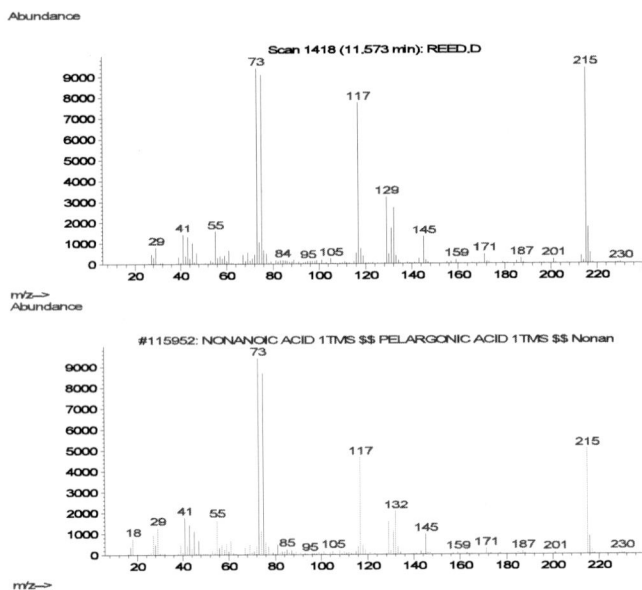

Figure 14 Comparison of a) fragment pattern of the peak at 11.573 with b) that of Nonanoic acid (Pelargonic acid) TMS ester.

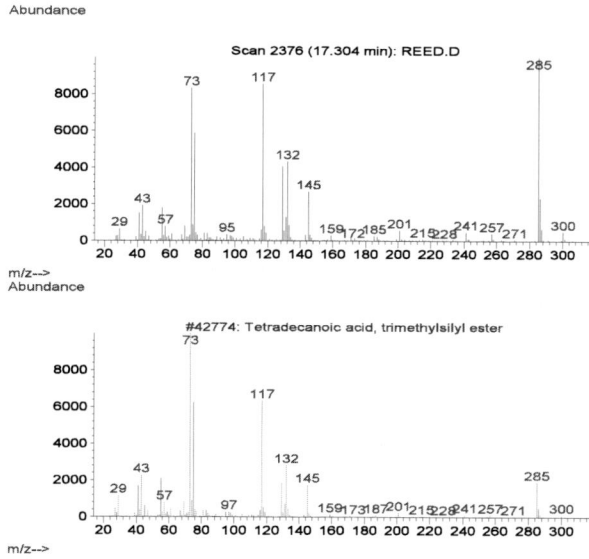

Figure 15 Comparison *of* a) fragment pattern of the peak at 17.304 with b) that of Tetradecanoic acid (Myristic acid) TMS ester.

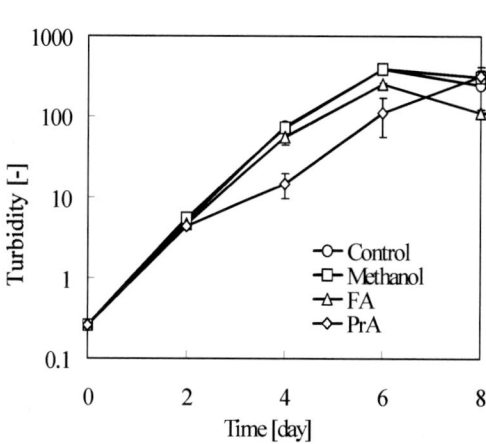

Figure 16. Growth curve of P. tenue as affected by Ferulic acid (FA) and Protocatechuic acid (PrA) at 10 mg l-1. Bars indicate standard deviations (n = 3).

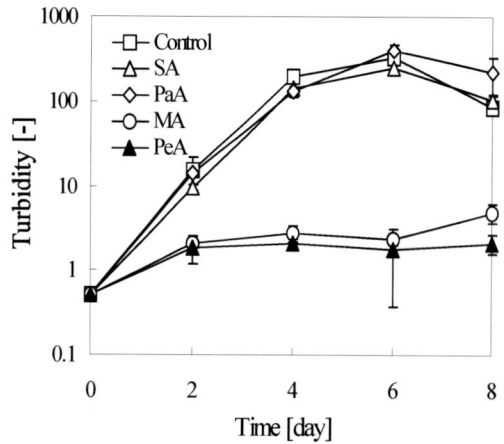

Figure 17. Growth of P. tenue as affected by Myristic (MA), Palmitic (PaA), Pelargonic (PeA), and Stearic acids (SA) at 10 mg l-1. Bars indicate standard deviations (n = 3).

It is interesting that pelargonic acid was detected in the RRS extract. Pelargonic acid was used as a herbicide component in Japan (Division, 1999). Although pelargonic acid sometimes has adverse effects on plants, some terrestrial plants are known to contain its derivatives (Okuno et al., 1993; Pelissier et al., 2001). However, it is also known that plants generally produce fatty acids with an even number of carbon. Since the *P. australis* samples used in this research were not axenic, it is possible that the pelargonic acid was produced by either microorganisms living within the reed and / or the reed itself. The other fatty acids identified: myristic, palmitic, and stearic acids; are all allelochemicals contributing to allelopathy among algae (Rice, 1984).

### *3.1.4. Growth inhibition effects of the identified compounds on* **P. tenue**

In order to confirm the possibility that these identified compounds contribute to the allelopathic effects of reed on *P. tenue* and *M. aeruginosa*, their inhibitory effects against both cyanobacteria were examined. Figure 16 indicates that two phenolic compounds (ferulic acid and protocatechuic acid: 10 mg $l^{-1}$) did not inhibit growth of *P. tenue*. Figure 17 demonstrates how growth of *P. tenue* was affected by 10 mg $l^{-1}$ of the four fatty acids, myristic, palmitic, pelargonic and stearic acids. Pelargonic and myristic acids inhibited growth of *P. tenue*. The dose–response relationships for pelargonic acid and myristic acid were measured. The EC50 of pelargonic acid, at which pelargonic acid inhibited normal growth by 50% of control, was 0.7 mg $l^{-1}$ while the EC50 of the myristic acid was 1.5 mg $l^{-1}$.

The inhibitory effects of all of the identified compounds on the maximum growth of *P. tenue* and *M. aeruginosa*, as indicated by the EC50s, were determined by the assays. These revealed species-specific inhibitory effects of the compounds. P-coumaric, vanillic, and myristic acids inhibited growth of *P. tenue* but not *M. aeruginosa*, whereas protocatechuic acid significantly inhibited growth of *M. aeruginosa* but not *P. tenue*. The phenols, sinapic, syringic, caffeic, and gallic acids and the fatty acid, pelargonic acid significantly inhibited the growth of both cyanobacteria. However, ferulic, palmitic, and stearic acids did not exhibit any effect. Among the twelve compounds identified, caffeic and pelargonic acids demonstrated the strongest growth inhibition on *P. tenue*; *M. aeruginosa* was most sensitive to pelargonic acid. These results suggest that p-coumaric, sinapic, syringic, vanillic caffeic, gallic acids, myristic, and pelargonic acids together and sinapic, syringic, caffeic, gallic, protocatechuic, and pelargonic acids together must respectively contribute to the allelopathic effect of *P. australis* on *P. tenue* and *M. aeruginosa*.

## 3.2. Nutrient Retention Effect of *P. australis* in Constructed Wetland

### *3.2.1. Nutrient Reduction*

The average flow rate was 0.25 m$^3$ s$^{-1}$. The SS concentration varied monthly with the average concentration being 20.2±11.5 mg L$^{-1}$, which resulted in about 160 ton SS entering the entire constructed wetland every year. The hydraulic retention time measured by the tracer was 84 minutes in 2005, which was significantly shorter than the designed value (5 hours) owing to the accumulated sediment reducing the volume of the unit and forming water ways.

The major form of nitrogen flowing into the wetland was NO$_3^-$, accounting for 52% of T-N, whereas NH$_4^+$ accounted for 5% of T-N, which is similar to what is found for other rivers around Lake Kasumigaura (Zhou and Hosomi, 2008). Furthermore, the concentrations of influent from spring to summer (April – September) are mostly in the low range, while those from autumn to winter (October – March) have higher values. The average concentrations of T-N over these two yearly periods from 1996 to 2005 were 1.45±0.40 mg L$^{-1}$ and 2.42±0.53 mg L$^{-1}$, respectively. The difference is probably owing to the average temperature decreasing from October to March, which would influence denitrification in the river water. On the other hand, the concentrations of T-P during different seasons were similar (April – September:

$0.15 \pm 0.04$ mg $L^{-1}$; October – March: $0.13 \pm 0.04$ mg $L^{-1}$). The T-P in influent was approximately 60% particulate phosphorus and 40% soluble phosphorus.

**Table 6. Comparison of SS, T-N, and T-P removal**
**in one unit between 1996 and 2005 (kg $yr^{-1}$)**

|      | U2 | | U5 | | U11 | | Average | |
|------|------|------|------|------|------|------|------|------|
|      | 1996 | 2005 | 1996 | 2005 | 1996 | 2005 | 1996 | 2005 |
| SS   | 1388 | 1222 | 1535 | 185  | 2632 | 720  | $1852 \pm 680$ | $709 \pm 519$ |
| T-N  | 116  | 85   | 133  | 43   | 117  | 73   | $122 \pm 10$   | $67 \pm 22$   |
| T-P  | 8.1  | 5.2  | 11.2 | 1.3  | 4.3  | 2.3  | $7.9 \pm 3.5$  | $2.9 \pm 2.0$ |

\* Data from the Kasumigaura River Office, Japan.

**Table 7. Amounts of nutrients accumulated in sediment by 2005 and total nutrient**
**loadings and removal during 10 years (kg per unit)**

| Unit | Sediment accumulation | | Accumulated removal | | Nutrient loading | |
|------|------|------|------|------|------|------|
|      | T-N  | T-P  | T-N  | T-P  | T-N  | T-P  |
| U2   | 207  | 67   | 825  | 62   | 9892 | 698  |
| U5   | 199  | 66   | 489  | 29   | 8390 | 542  |
| U11  | 137  | 40   | 556  | 14   | 5626 | 374  |
| U18  | 228  | 47   | 571  | 37   | 3186 | 184  |
| Average | 193 | 55  | 610  | 36   | 6774 | 450  |

\* Data from the Kasumigaura River Office, Japan.

To evaluate the SS and nutrient removal loads, data for U2, U5, and U11 collected every month in 1996 and 2005 were calculated and are shown in Table 6. The average removal of SS, T-N, and T-P decreased in 2005 compared with 1996 and accounted for 38%, 55%, and 37% of removal loads in 1996, which was probably due to the reduced HRT as described above. It is interesting that the decrease in SS and T-P removal loads were almost same, suggested again that sedimentation is the major mechanism for phosphorus removal in this study. In contrast, the constructed wetland retained a relatively high T-N removal capacity probably owing to the denitrification potential of mature wetland (Bachand and Horne, 2000b). However, despite the total amount of SS, T-N, and T-P removed decreasing to some degree compared with removal loads in 1996, the constructed wetland has proven to remove pollutants in river water purification even after 10 years running. In particular, the T-N removal rate in 2005 was calculated as 34 g-N $m^{-2}$ $yr^{-1}$, which is similar to the result (average $NO_3^-$ retention of 39 g-N $m^{-2}$ $yr^{-1}$) for the long term running of the created wetland (Mitsch et al., 2005a).

### 3.2.2. Sediment Accumulation

Figure 18 shows the average increase in sediment accumulation in U2, U5, U11 and U18. By 2005, the average sediment accumulation (180-296 $m^3$) already exceeded the designed volume (200 $m^3$) after 10 years running. Because the sediment surface in a unit is convexo-concave, some channelized flow formed even though the height at some places was higher than the designed levee crown (Figure 18, right). Sediment accumulation occurred mostly owing to the prevailing influence of SS sedimentation and dead plant accumulation.

Sedimentation can be estimated by calculating the total removed suspended solid from the influent. However, the amount of net solid removal ranged from 4.14 to 22.5 ton for each unit over the 10 years. The volume of removed solid modified by moisture content and bulk density ranged from 11 to 62 $m^3$, which was significantly lower than the accumulated sediment volume, suggesting that most sediment accumulated from dead plant. Supposing sediment accumulation only consisted of removed suspended solid and dead plant in each unit, the volume of dead plant accumulation can be calculated by subtracting the removed suspended solid volume from the accumulated sediment volume. The accumulation of dead plant in each unit was calculated as 3.4 kg $m^{-2}$ $yr^{-1}$, which is within the *P. australis* production range of 0.2 – 6.0 kg $m^{-2}$ $yr^{-1}$ (Shimatani et al., 2003). The average total sediment accumulation including removed SS and dead plant was 3.61 – 4.43 kg $m^{-2}$ $yr^{-1}$, which is similar to the sediment accumulation (3.9±0.3 kg $m^{-2}$ $yr^{-1}$) reported by Anderson and Mitsch, 2006) in investigating created riverine marshes over 10 years.

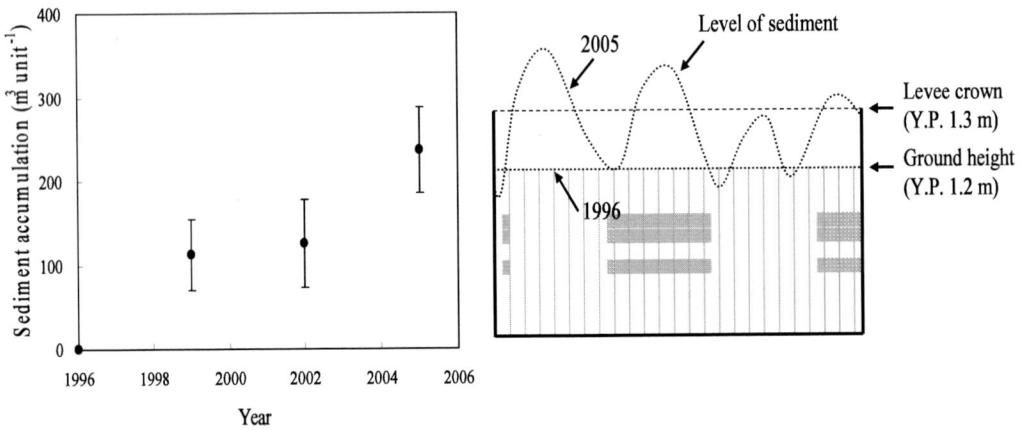

Figure 18. Average sediment accumulation from 1996 to 2005 (left) and a description of the shape of the unit sediment surface in 2005 (right) (data from the Kasumigaura River Office, Japan).

The changes in average organic matter, T-N, and T-P contents in soil at 0-5 cm, 5-10 cm, 10-20 cm, and 20-30 cm in U2, U5, U11, and U18 from 2000 to 2004 are shown in Figure 19. Compared with the stable values below 5 cm depth, contents of organic matter, T-N, and T-P in surface sediment increased from 2000 to 2004 by 1.6, 3.7, and 3.6 times, suggesting the nutrient sediment and residue accumulated with time. The amounts of nutrients accumulated in sediment by 2005 and total nutrient loadings and removal during the 10 years are shown in Table 5. The T-P content retained in sediment was calculated as having an average rate of 2.8 g P $m^{-2}$ $yr^{-1}$, which is within the range of 0.3–3.6 g P $m^{-2}$ $yr^{-1}$ for T-P deposition reported for natural wetlands (Mitsch et al., 1979; Mitsch and Reeder, 1991). However, compared with accumulated T-P removed from the inflow, there was more T-P accumulated in sediment, which was probably owing to macrophytes pumping phosphorus from deep sediments, simulated at a rate of 0.31–1.66 g P $m^{-2}$ $yr^{-1}$ (Wang and Mitsch, 2000). In contrast, comparing the amounts of nutrients accumulated in sediment with those removed during the 10 years, unlike the T-P content in sediment being similar or slight higher than the amount removed, the T-N content in sediment is significantly less than the amount removed. The average T-N removal rate was calculated as 31 g N m$^{-2}$ yr$^{-1}$ while only 9.6 g N m$^{-2}$ yr$^{-1}$ was stored in sediment. The loss of nitrogen was probably due to denitrification in the wetland (Bachand and Horne, 2000b; Spieles and Mitsch, 2000) since NO$_3^-$ is a major form of nitrogen in influent.

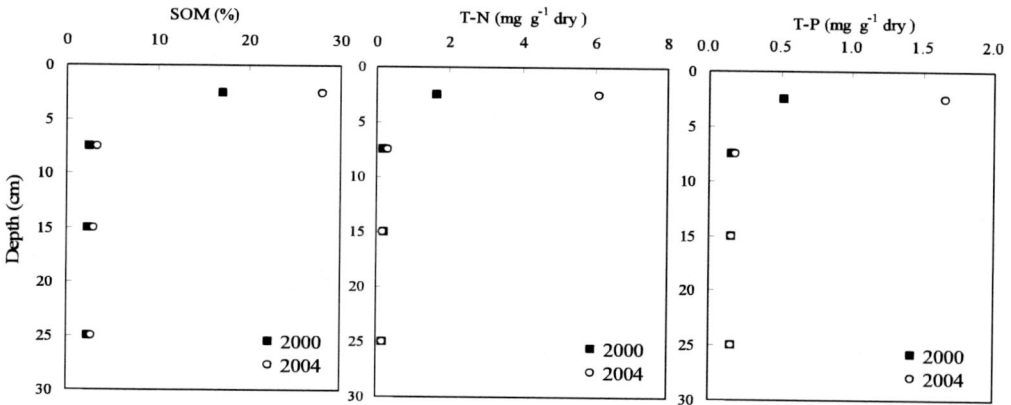

Figure 19. Organic matter, nitrogen, and phosphorus contents in sediment at different depths (data from the Kasumigaura River Office, Japan).

## 3.3. Plant Species Transition in Constructed Wetland with *P. australis*

*P. australis* was planted in each unit and covered 100% of the area of each unit at the beginning of the experiment after construction was completed in 1996. By 2000, as shown in Figure 20, the coverage of the uniform *P. australis* community decreased to 77% owing to other plant invasion (e.g. *Zizania latifolia, Carex dispalata, Lycopus lucidus*). Most of the invading species are indigenous plants around the basin (Kusumoto et al., 2007), which are probably in the soil or came from river water. Furthermore, although the *P. australis* community still dominated in each unit, different spatial community diversity developed in 2004. In particular, the mixing of *P. australis* and *Zizania latifolia*, and *P. australis* and *Carex dispalata* expanded. As a result, the average uniform *P. australis* community coverage was 53% in 2004, a further decrease from 2000. In contrast, although the distributions in each unit differed, the average coverage of *Zizania latifolia* and *Carex dispalata* increased significantly and finally reached 13% and 15%, respectively. On the other hand, not only macrophytes but also woody terrestrial plants such as *Salix* spp. were observed after several years running.

The macrophyte community diversity index (CDI) reported by Mitsch et al., 2005b) was used to quantify the spatial diversity in the constructed wetland. The index is expressed as

$$CDI = \sum_{i=1}^{N}(C_i \ln(C_i))$$

where $C_i$ is the percentage coverage of community "$i$" (0–1) and $N$ is the number of plant/aquatic communities. In 1996, the CDI was zero for U2, U5, U11, and U18 since all of units were covered only by *P. australis*. However, after 4 years and 8 years running, as shown in Figure 21, the CDI of U2, U5, U11, and U18 significantly increased. By 2004, the average CDI of these four units was 0.88±0.04, which was also significantly higher than that in 2000 (0.64±0.14).

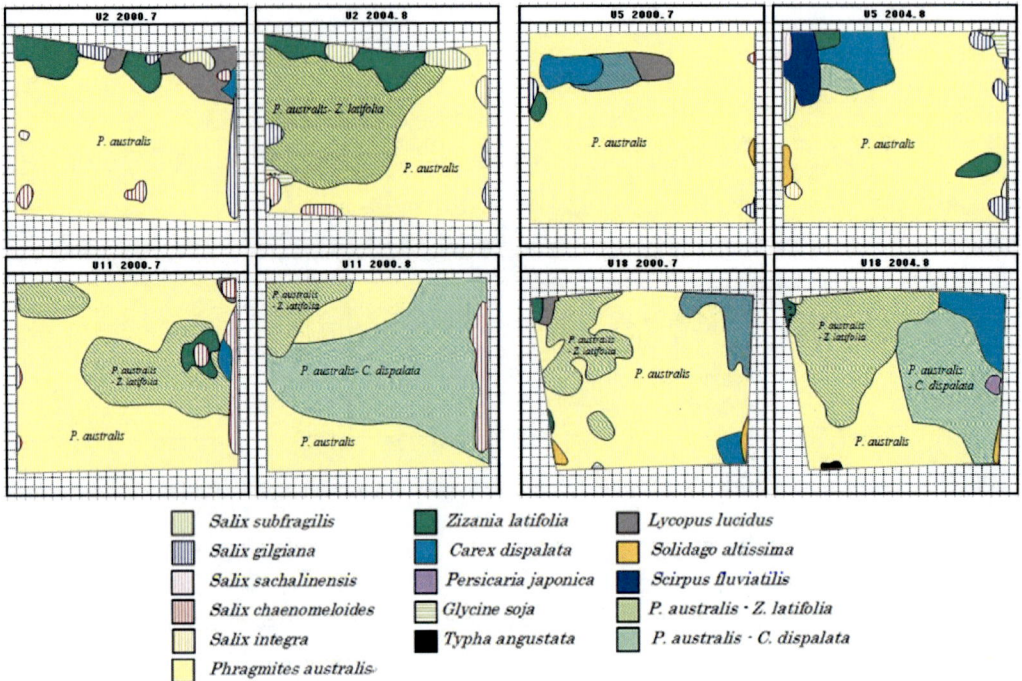

Figure 20. Vegetation community coverage in U2, U5, U11, and U18 in July 2000 and August 2004 (data from the Kasumigaura River Office, Japan).

The hydrology of wetland plays an important role in directing vegetation change by affecting the germination potential, growth capability and mortality of plants (Ellison and Bedford, 1995). Although macrophyte vegetation affects and is affected by water quality in riparian wetlands (Fink and Mitsch, 2007), the vegetation transition in this study probably occurred due to hydrology changes along with the ground level being elevated by sediment accumulation. Particularly after several years running, some areas rose above the flood water level to form land inside the constructed wetland (Figure 16, right), which is suitable for other vegetation invasion such as that by *Salix* spp. The phenomenon agrees with the result reported by Grace and Wetzel (1981) that less flood-tolerant species are superior competitors in drier locations, allowing invasion of occupied space when water tables are low, because more energy is needed for tolerance to flood conditions.

## 3.4. Comparison of Genetic Diversity of *P. australis* between Constructed and Native Wetland

Using AM-3, AM-7, and AM-8 primers, 39-40 bands were identified of which 64-73% were polymorphic at each site (A-D) of constructed wetland. In contrast, 28-38 bands were identified of which 43-67% were polymorphic at each site (O-R) of native wetland (Table 6).

Although the average percentages of polymorphic bands of constructed wetland (68±5%) were slightly higher than those of native wetland (57±11%), there was no significant difference ($p = 0.142$) between these two groups basing on statistic analysis.

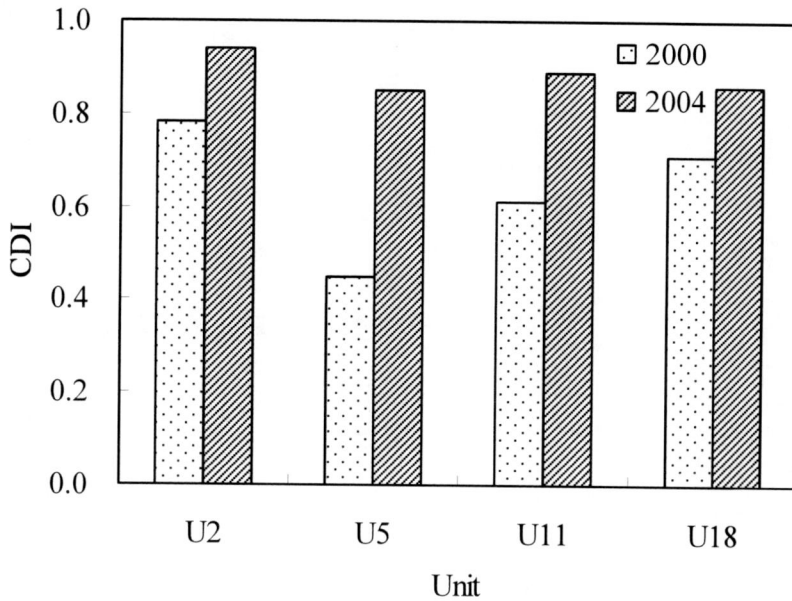

Figure 21. Vegetation diversities in U2, U5, U11, and U18 in July 2000 and August 2004. CDI is the community diversity index, which is defined in the text.

**Table 8. Numbers of polymorphic and monomorphic bands determined by AM-3, AM-7, and AM-8 ISSR-PCR primers. Total polymorphic bands, total monomorphic bands, and percentage of polymorphic bands of *P. australis* at each site are indicated**

| Site | AM-3 | | AM-7 | | AM-8 | | Total | | |
|------|----|----|----|----|----|----|----|----|------|
|      | PB | MB | PB | MB | PB | MB | PB | MB | PPB % |
| A | 3 | 5 | 9 | 6 | 13 | 3 | 25 | 14 | 64 |
| B | 6 | 3 | 9 | 5 | 13 | 3 | 28 | 11 | 72 |
| C | 3 | 6 | 10 | 5 | 12 | 3 | 25 | 14 | 64 |
| D | 6 | 4 | 9 | 5 | 14 | 2 | 29 | 11 | 73 |
| O | 6 | 5 | 10 | 4 | 8 | 3 | 24 | 12 | 67 |
| P | 2 | 6 | 9 | 6 | 10 | 5 | 21 | 17 | 55 |
| Q | 1 | 5 | 5 | 6 | 6 | 5 | 12 | 16 | 43 |
| R | 3 | 4 | 9 | 7 | 12 | 2 | 24 | 13 | 65 |

PB: polymorphic bands; MB: monomorphic bands; PPB: percentage of polymorphic bands.

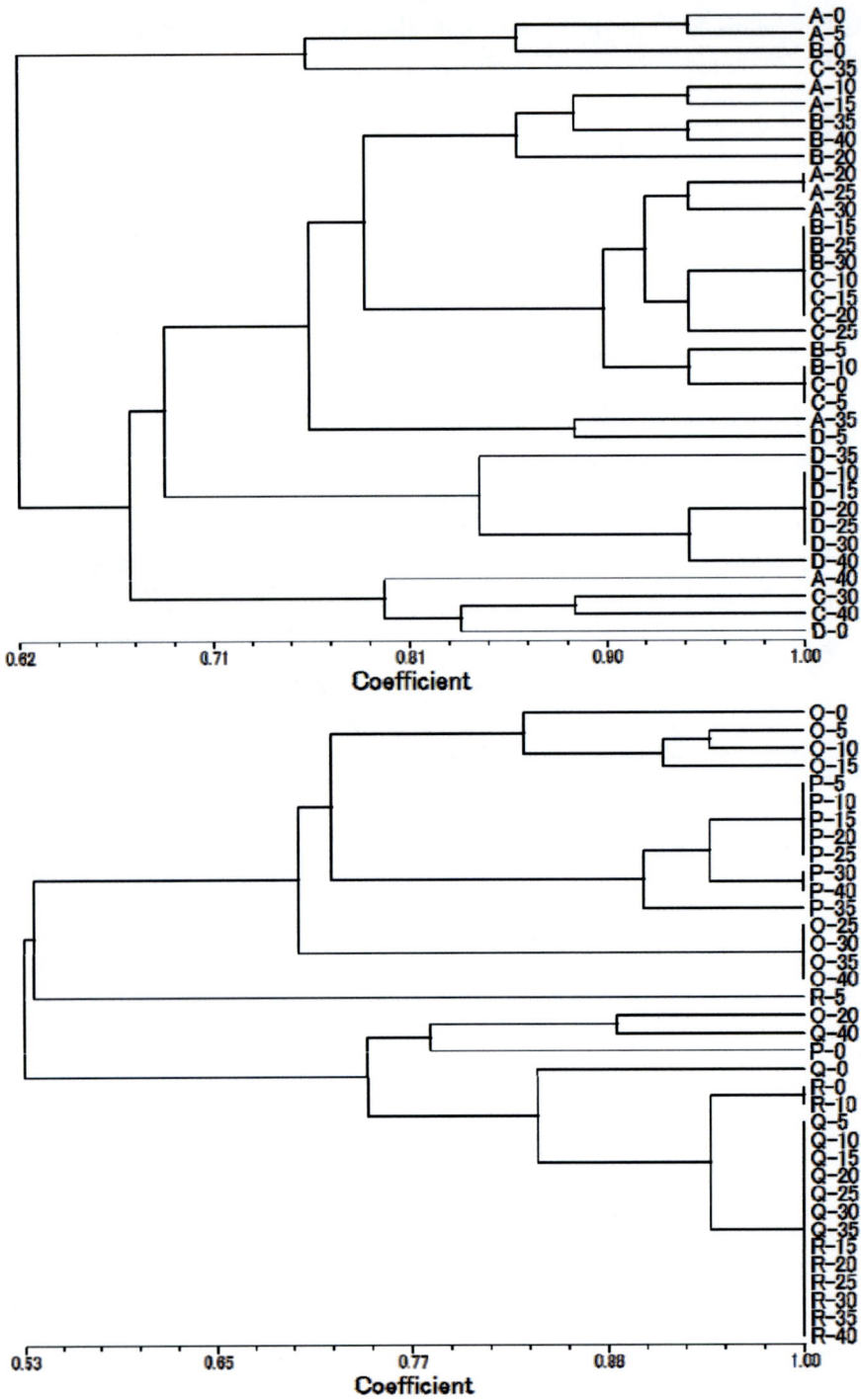

Figure 22. Dendrograms of *P. australis* samples from constructed wetland (a) and native wetland (b) revealed by UPGMA cluster analysis based on genetic similarity coefficients. Results are based on PCR fingerprinting analysis with primer AM-7.

Figure 22 shows the UPGMA-based dendrogram of reed samples from constructed and native wetlands basing on the AM-7 primer, which has the similar patterns with the results basing on the other two primers. The genetic similarity coefficients within the reed stand in the constructed wetland ranged from 0.62 to 1.00 while those in native wetland ranged from 0.53 to 1.00. However, as shown in Figure 14, more gene types appeared in the constructed wetland than in the native wetland. In the constructed wetland, samples had significant variations among sampling points even though the sampling intervals were only 5 m, except at site D. Compared with other sites of the constructed wetland, reed from most sampling points at site D had a relatively high similarity coefficient (>0.94). On the other hand, the reed from the native wetland can be divided into several main clusters according to sampling sites. One group consisted of site Q and site R, which had a similarity coefficient between them of approximately 0.95. Reed from all sampling points at site P also had a high similarity coefficient (>0.90) while reed from the two halves of site O can be divided into two groups.

Because seedlings of *P. australis* were taken from littoral zones around Lake Kasumigaura and planted in the constructed wetland at the beginning of construction, it can be considered that various genetic types of *P. australis* were introduced at the same time, which probably resulted in a relatively high number of gene types (Figure 20 (a)). Moreover, although the investigation was conducted 10 years after construction, the constructed wetland was considered to be in an establishment stage compared with the mature native wetland. At this stage, *P. australis* not only competes with itself to fill the spaces between plants, but competes with other invading species while coping with ground level elevation as mentioned above, which results in a higher genetic diversity owing to different clones intermingling. Furthermore, as shown in Figure 19, although the CDI was similar between U2 (Site B) and U5 (Site D) in 2004, the uniform distribution of *P. australis* in U5 (80%) was approximately twice that in U2 (42%), which probably results in a higher similarity coefficient at site D than at site B.

In contrast, the native wetland (Site 3) along Lake Kasumigaura has existed for 100 years as reported by Onuma and Ikeda (1999), and is thus considered well adapted to the local environmental conditions and to have entered a stationary stage. In addition, seedling establishment is not very frequent in mature *P. australis* populations (Barrett et al., 1993). Populations initiated by seeds are probably initially genetically diverse and over time become dominated by a few clones as a result of competition and selection. Genetic diversity would decrease during the stationary stage, in which a small number of clones well adapted to the local environmental conditions prevail. Relatively low genetic diversity and monoclonal populations at each site of the native wetland could be the result of such a selection process and be an indication that *P. australis* stands have grown under stable conditions for a long time (Watkinson and Powell, 1993). Lambertini et al., 2008) also reported that one large monoclonal stand was present in an old wetland with rather stable environmental conditions over a long time period whereas polyclonal stands were younger and characterized by disturbance.

# 4. CONCLUSION

*P. australis* has many important roles in aquatic ecosystems, including landscape creation, wildlife conservation, and water treatment. In this study, the allelopathic effect on cyanobacterial growth and potential for nutrient retention of *P. australis* in constructed wetlands were investigated. Assays of the RRS against *P. tenue* and *M. aeruginosa* revealed allelopathic growth inhibition of both cyanobacteria by *P. australis*. However, *P. tenue* was more sensitive to the allelopathic effect of *P. australis*. Whole plant RRS produced a stronger inhibitory effect after May and could inhibit over 99% of maximum growth of *P. tenue* from June to November. Leaf RRS and stem RRS exhibited different inhibitory effects on the growth of *P. tenue*. The leaves of *P. australis* in particular contribute a large part of the whole plant allelopathic effect. Results indicated a relationship between the inhibitory effect and the DOC concentration of the RRS. Growth inhibition of *P. tenue* appeared when DOC reached levels between 10 and 15 mg $l^{-1}$. GC/MS analysis indicated that *P australis* released eight phenols (p-coumaric, ferulic, vanillic, sinapic, syringic, caffeic, protocatechuic, and gallic acids) and four fatty acids (myristic, palmitic, pelargonic, and stearic acids).

On the other hand, considerable amounts of SS, nitrogen, and phosphorus have been retained in the constructed wetland planted with *P. australis* used in this study, during the 10 years since its inception. The average total sediment accumulation was 3.61 – 4.43 kg $m^{-2}$ $yr^{-1}$, mostly due to SS sedimentation and accumulation of dead plant matter. The average rate of T-P retained in sediment was calculated as 2.8 g P $m^{-2}$ $yr^{-1}$. Nevertheless the amount of T-P was neither higher nor even similar to that in the inflow river water. In fact the amount of T-N in sediment was significantly less than in the inflow water, suggesting that phosphorus was removed during particle sedimentation while most nitrogen was removed by denitrification. Furthermore, with the ground level being elevated by sediment deposition, woody terrestrial plants and other aquatic plants had invaded the constructed wetland over the course of its life time and had mixed with *P. australis*, developing population diversity in the constructed wetland. By 2004, the average CDI of the four units was 0.88±0.04, significantly more than during the initial stages of the constructed wetland. In addition, in comparison to the native wetland, more *P. australis* gene types occurred in the constructed wetland. This may reflect the constructed wetland being in an establishment stage whilst the native wetland had already entered a stationary stage.

# ACKNOWLEDGEMENT

We express our great thanks to the Kasumigaura River Office (Ministry of Land, Infrastructure, Transport and Tourism, Japan) for providing the data in section 3.2-3.3.

# REFERENCES

Aliotta, G., A. Molinaro, P. Monaco, G. Pinto and L. Previtera, 1992. Three biologically active phenylpropanoid glucosides from Myriophyllum verticillatum. *Phytochem.*, 31: 109-111.

Aliotta, G., P. Monaco, G. Pinto, A. Pollio and L. Previtera, 1991. Potential allelochemicals from Pistia stratiotes L. *J.Chem.Ecol.*, 17: 2223-2234.

Anderson, C.J. and W.J. Mitsch, 2006. Sediment, carbon, and nutrient accumulation at two 10-year-old created riverine marshes. *Wetlands*, 26: 779-792.

Anthoni, U., C. Christophersen, J.O. Madsen, S. Wium-Andersen and N. Jacobsen, 1980. Biologically active sulphur compounds from the green alga Chara globularis. *Phytochem.*, 19: 1228-1229.

Antonielli, M., S. Pasqualini, P. Batini, L. Ederli, A. Massacci and F. Loreto, 2002. Physiological and anatomical characterization of Phragmites australis leaves. *Aquat. Bot.*, 72: 55-66.

Bachand, P.A.M. and A.J. Horne, 2000a. Denitrification in constructed free-water surface wetlands: I. Very high nitrate removal rates in a macrocosm study. *Ecological Engineering*, 14: 9-15.

Bachand, P.A.M. and A.J. Horne, 2000b. Denitrification in constructed free-water surface wetlands: II. Effects of vegetation and temperature. *Ecological Engineering*, 14: 17-32.

Barrett, P.R.F., J.C. Curnow and J.W. Littlejohn, 1996. The control of diatom and cyanobacterial blooms in reservoirs using barley straw. *Hydrobiol.*, 340: 307-311.

Barrett, S.C.H., C.G. Eckert and B.C. Husband, 1993. Evolutionary Processes in Aquatic Plant-Populations. *Aquatic Botany*, 44: 105-145.

Brix, H., 1994. Functions of Macrophytes in Constructed Wetlands. *Water Science and Technology*, 29: 71-78.

Clevering, O.A. and J. Lissner, 1999. Taxonomy, chromosome numbers, clonal diversity and population dynamics of *Phragmites australis*. *Aquatic Botany*, 64: 185-208.

Comin, F.A., J.A. Romero, V. Astorga and C. Garcia, 1997. Nitrogen removal and cycling in restored wetlands used as filters of nutrients for agricultural runoff. *Water Science and Technology*, 35: 255-261.

Coops, H., N. Geilen, H.J. Verheij, R. Boeters and G. vanderVelde, 1996. Interactions between waves, bank erosion and emergent vegetation: An experimental study in a wave tank. *Aquatic Botany*, 53: 187-198.

Craft, C.B., 1997. Dynamics of nitrogen and phosphorus retention during wetland ecosystem succession. *Wetlands Ecology and Management*, 4: 177-187.

David, S.S., 1996. Chemistry and mechanism of allelopathic interactions. Agron. J., 88: 876-885.

Division, J.A., 1999. Summary of toxicity studies on pelargonic acid. *J. Pest. Sci.*, 24: 421-423.

Ellison, A.M. and B.L. Bedford, 1995. Response of a Wetland Vascular Plant Community to Disturbance - a Simulation Study. *Ecological Applications*, 5: 109-123.

Everall, N.C. and D.R. Lees, 1997. The identification and significance of chemicals released from decomposing barley straw during reservoir algal control. *Water Res.*, 31: 614-620.

Fink, D.F. and W.J. Mitsch, 2004. Seasonal and storm event nutrient removal by a created wetland in an agricultural watershed. *Ecological Engineering*, 23: 313-325.

Fink, D.F. and W.J. Mitsch, 2007. Hydrology and nutrient biogeochemistry in a created river diversion oxbow wetland. *Ecological Engineering*, 30: 93-102.

Grace, J.B. and R.G. Wetzel, 1981. Habitat Partitioning and Competitive Displacement in Cattails (Typha) - Experimental Field Studies. *American Naturalist*, 118: 463-474.

Hale, F.E., 1957. The use of copper sulfate in control of microscopic Organisms. Phelps Dodge Ref. Corp., 44 pp.

Haslam, S.M., 1972. Biological flora of the British Isles. *Phragmites communis Trin. J. Ecol.*, 60: 585-610.

Jing, S.R., Y.F. Lin, D.Y. Lee and T.W. Wang, 2001. Nutrient removal from polluted river water by using constructed wetlands. *Bioresource Technology*, 76: 131-135.

JSWA, J.S.W.A., 1984. Analysis Methods of Wastewater. Tokyo, 649 pp.

JWRC, 1995. Lakes and marshes environment of Japan II. Japan Wildlife Research Center, Tokyo.

Kadlec, R.H. and R.L. Knight, 1996. *Treatment wetlands*. Lewis Publishers, Florida.

Kadono, Y., 1996. *Aquatic plants of Japan*. Bun-ichi Sogo Shuppan, Tokyo.

Kakisawa, H., F. Asari, T. Kusumi, T. Toma, T. Sakurai, T. Oohusa, Y. Hara and M. Chihara, 1988. An allelopathic fatty acid from the brown alga Cladosiphon okamuranus. *Phytochem.*, 27: 731-735.

Karunaratne, S., T. Asaeda and K. Yutani, 2003. Growth performance of Phragmites australis in Japan: influence of geographic gradient. *Environmental and Experimental Botany*, 50: 51-66.

Kira, T., 1991. Brief review on ecology of reed. *Report of Lake Biwa Research Institute*, 9: 29-37.

Kovacs, M., I. Precsenyi and J. Podani, 1978. Accumulation of Elements in Reeds of Lake Balaton (Phragmites Communis). *Acta Botanica Academiae Scientiarum Hungaricae*, 24: 99-111.

Kusumoto, Y., S. Yamamoto, T. Ohkuro and M. Ide, 2007. The relationship between the paddy landscape structure and biodiversity of plant community in the Tone river basin. *Journal of The Japanses Institute of Landscape Architecture* 70: 445-448.

Lambertini, C., M.H.G. Gustafsson, J. Frydenberg, M. Speranza and H. Brix, 2008. Genetic diversity patterns in Phragmites australis at the population, regional and continental scales. *Aquatic Botany*, 88: 160-170.

Laushman, R.H., 1993. Population-Genetics of Hydrophilous Angiosperms. *Aquatic Botany*, 44: 147-158.

Lovett, J.V., M.Y. Ryuntyu and D.L. Liu, 1989. Allelopathy, chemical communication, and plant defense. *J. Chem. Ecol.* , 15: 1193-1202.

Mitsch, W.J., J.W. Day, L. Zhang and R.R. Lane, 2005a. Nitrate-nitrogen retention in wetlands in the Mississippi river basin. *Ecological Engineering*, 24: 267-278.

Mitsch, W.J., C.L. Dorge and J.R. Wiemhoff, 1979. Ecosystem Dynamics and a Phosphorus Budget of an Alluvial Cypress Swamp in Southern Illinois. *Ecology*, 60: 1116-1124.

Mitsch, W.J., J.C. Lefeuvre and V. Bouchard, 2002. Ecological engineering applied to river and wetland restoration. *Ecological Engineering*, 18: 529-541.

Mitsch, W.J. and B.C. Reeder, 1991. Modelling nutrient retention of a freshwater coastal wetland: estimating the roles of primary productivity, sedimentation, resuspension and hydrology. *Ecological Modelling* 54: 151-187.

Mitsch, W.J., L. Zhang, C.J. Anderson, A.E. Altor and M.E. Hernandez, 2005b. Creating riverine wetlands: Ecological succession, nutrient retention, and pulsing effects. *Ecological Engineering*, 25: 510-527.

Nakai, S. and M. Hosomi, 2002. Allelopathic inhibitory effects of polyphenols released by Myriophyllum spicatum on algal growth. *Allelopathy Journal* 10: 123-132.

Nakai, S., Y. INOUE and M. HOSOMI, 2001. Algal growth inhibition effects and inducement modes by plant-produced phenols. *Water Res.*, 35: 1855-1859.

Nakai, S., Y. Inoue, M. Hosomi and A. Murakami, 1999. Growth inhibition of blue-green algae by allelopathic effects of macrophytes. *Water Science & Technology*, 39: 47-53.

Nakamura, K., T. Chiba, K. Sato, Y. Morita, M. Hosomi and s. Tanaka, 2007. A survey of constructed wetlands in Japan. *Report of Research Institute of River & Watershed Environment Management (RIREM)*, 19: 41-46.

National-Research-Council, 1992. *Restoration of aquatic ecosystems.* National Academy Press, Washington, DC, 522 pp.

Neuhaus, D., H. Kuhl, J.G. Kohl, P. Dorfel and T. Borner, 1993. Investigation on the Genetic Diversity of Phragmites Stands Using Genomic Fingerprinting. *Aquatic Botany*, 45: 357-364.

Nishikawa, Y., 1999. Relationship between reed and human life. *Report of Kansai Organization for Nature Conservation*, 21: 301-338.

Okuno, M., H. Kameoka, M. Yamasita and M. Miyazawa, 1993. Components of volatile oil from plants of Polypodiaceae. *J. Jpn. Oil Chemists' Soc.*, 42: 44-48.

Onuma, H. and H. Ikeda, 1999. Reduction processes of reed community in the Lake Kasumigaura. *Bulletin of Environmental Research Center, the University of Tsukuba*, 24: 49-58.

Parsons, B.J., H.J. Newbury, M.T. Jackson and B.V. FordLloyd, 1997. Contrasting genetic diversity relationships are revealed in rice (Oryza sativa L) using different marker types. *Molecular Breeding,* 3: 115-125.

Pelissier, Y., C. Haddad, C. Marison, M. Miharu and J.-M. Bessier, 2001. Volatile constituents of fruit pulp of Dialium guineense Willd. *J. Essent. Oil Res.*, 13: 103-104.

Pillinger, J.M., J.A. Cooper and I. Ridge, 1994. Role of phenolic compounds in the anti-algal activity of barley straw. *J. Chem. Ecol.*, 20: 1557-1569.

Planas, D., F. Sarhan, L. Dube, H. Godmaire and C. Cadieux, 1981. Ecological significande of phenolic compounds of Myriophyllum spicatum. *Verh.Internat.Verein.Limnol.*, 21: 1492-1496.

PWRC, 2002. Report on control of phytoplankton growth. Public Works Research Center, Tokyo.

Reilly, J.F., A.J. Horne and C.D. Miller, 2000. Nitrate removal from a drinking water supply with large free-surface constructed wetlands prior to groundwater recharge. *Ecological Engineering,* 14: 33-47.

Rice, E.L., 1984. *Allelopathy.* 2nd Ed. Academic Press, Orlando, FL, USA.

Ridge, I. and J.M. Pillinger, 1996. Towards understanding the nature of algal inhibitors from barley straw. *Hydrobiol.*, 340: 301-305.

Saito, K., M. Matsumoto, T. Sekine and I. Murakoshi, 1989. Inhibitory substances from Myriophyllum brasilience on growth of blue-green algae. *J.Nat.Prod.*, 52: 1221-1226.

Shimatani, Y., M. Hosomi and K. Nakamura, 2003. *Water Quality Improvement by ecotechnology.* Soft Science, Inc., Tokyo.

Spieles, D.J. and W.J. Mitsch, 2000. The effects of season and hydrologic and chemical loading on nitrate retention in constructed wetlands: a comparison of low- and high-nutrient riverine systems. *Ecological Engineering*, 14: 77-91.

UNEP, 1994. Survey on the state of World Lakes.

vanderToorn, J., 1972. Variability of Phragmites australis (Cav.) Trin ex Steudel in relation to the environment. *Van Zee tot Land*, 48: 1-122.

Vymazal, J., 2005. Constructed wetlands for wastewater treatment. *Ecological Engineering*, 25: 475-477.

Wang, N.M. and W.J. Mitsch, 2000. A detailed ecosystem model of phosphorus dynamics in created riparian wetlands. *Ecological Modelling*, 126: 101-130.

Watanabe, M., 1989. Recent knowledge on the toxic blue-green Algal Blooms and Problems on its Study. *Japan Journal of Water Pollution Research,* 12: 750-756.

Watanabe, M.M. and N. Satake, 1991. NIES-Collection List of Strains. Microalgae and Protozoa. National Institute for Environmental Studies, Environmental Agency, Tokyo, 31-32 pp.

Watkinson, A.R. and J.C. Powell, 1993. Seedling Recruitment and the Maintenance of Clonal Diversity in Plant-Populations - a Computer-Simulation of Ranunculus-Repens. *Journal of Ecology*, 81: 707-717.

Yuji, S. and M. Hosomi, 2002. Killifish and Reed. IWANAMI SHOTEN, Tokyo, 186 pp.

Zhou, S. and M. Hosomi, 2008. Nitrogen transformations and balance in a constructed wetland for nutrient-polluted river water treatment using forage rice in Japan. *Ecological Engineering*, 32: 147-155.

In: Aquatic Ecosystem Research Trends
Editor: George H. Nairne

*Chapter 7*

# IS ALA-D ACTIVITY A USEFUL BIOMARKER OF LEAD EXPOSURE FOR FIELD MONITORING PROGRAMS?

## *Noemí R. Verrengia Guerrero\* and Paula E. Lombardi*
Toxicology and Legal Chemistry, Dept. of Biological Chemistry,
Faculty of Exact and Natural Sciences,
University of Buenos Aires. 4 Piso, Pabellón II, Ciudad Universitaria
Buenos Aires - Argentina. C1428EHA

## ABSTRACT

Lead has been used since ancient times and its toxic effects are well documented. Although the introduction of unleaded gasoline has contributed enormously to the decrease of lead emissions to the environment, the metal is still one of the most commonly used in many industries.

For a long time, the activity of the enzyme δ-aminolevulinic acid dehydratase (ALA-D) has been recognized as a valuable and sensitive tool to assess the extent of lead exposure. Regarded as a specific biomarker, its usefulness has been investigated in different aquatic organisms, including fish. A great number of laboratory studies have proved a strong correlation between the inhibition of the enzyme activity and the lead content, encouraging its use for field monitoring programs. Previously, we have found variable levels of lead in liver and gills samples of the fish *Prochilodus lineatus* collected from a coastal area in the La Plata River, Argentina, where sewage effluents are still discharged without any kind of treatment. The aim of this work was to analyze ALA-D activity in blood and liver samples of *P. lineatus* to determine the extent of enzyme inhibition. Correlations between the enzyme activity and the levels of lead bioaccumulated by the fish were also investigated. The analyses were performed in adult organisms, collected each two or three months during June/2002 and May/2004. According to the data, any significant temporal trend could be observed. In addition, and perhaps the most surprising result, poor correlations between the enzyme activity and the lead content were found. In view of these results we decided to optimize the methodology to reactivate the enzyme, a technique already used for mammalian samples but not for fish. Only the enzyme from blood samples could be effectively reactivated. In all these

\* Tel/Fax: 54 11 4576 3342, e-mail: noev@qb.fcen.uba.ar

samples, values of the reactivated enzyme were not significantly different among them, but significantly higher than those from the non-reactivated enzyme, demonstrating that ALA-D activity was actually inhibited. Good correlations were found between the percentages of enzyme inhibition and the blood lead levels. The results clearly show that for field studies, and even when no reference data are available, ALA-D activity may be regarded as a reliable biomarker of lead contamination.

# 1. INTRODUCTION

From ancient times mankind has taken advantage of the multiple benefits provided by aquatic systems. Most populations have been established along the coasts of both freshwater and marine environments. These systems have provided not only water and food supplies but also routes for transport and recreational areas. In parallel with the social and economical benefits, aquatic systems have been the principal vehicle for any kind of waste disposal. Paradoxically, anthropogenic activities have threatened the water quality and the biological community integrity of this vital natural resource (Monserrat et al., 2006). As an unwanted result, contamination has become the dark side of the industrial and population development.

It has long been recognized that estuaries and coastal areas have been the most severely affected by chemical pollution (GESAMP, 1991; Matthiessen and Law, 2002; Moore et al. 2004; Monserrat et al., 2006). On the other hand, these areas present a considerable high biological productivity, since they are the preferred sites for the development and growth of many fish, mollusks and crustacean species (Ketchum, 1983; Moore et al. 2004). In addition, a great number of these biological species have relatively short life cycles and frequent reproductive cycles that may be easily disturbed by changes in the environmental conditions (Ford, 1989). As a consequence, the most developed countries have undertaken different risk assessment monitoring programs, regulatory controls, and even remediation actions in order to protect and to recover the environmental health (Rand, 1995; Lehtonen et al., 2006; Pampanin et al., 2006).

Chemical analyses of environmental samples such as waters, sediments, and particulate matter, have been, and still are, conducted in order to identify ant to determine the concentration levels of the principal contaminant substances. Their disadvantages are that they are expensive and time consuming, since they usually require extreme care to avoid any accidental contamination of the samples, extensive sample cleanup and sophisticated analytical techniques. In addition, they do not provide evidence of the contaminant bioavailability (Nriagu et al., 1993; Kendall et al., 2001). Chemical analyses of biological samples (tissues, blood, urine, feces, etc.) give information about the internal concentration of contaminants, reflecting the internal dose that has been effectively incorporated by the organisms and reached a particular tissue regardless the route of uptake. However, these analyses alone do not provide any information about the toxic effects that may be induced on the biota (Lam and Grey, 2003).

On the other hand, biological studies give information about the functioning and structure of the biological communities. But when these effects are evident the chances for reversing them are scarce (Eason and O'Halloran, 2002; Hyne and Maher, 2003).

Among all the different alternatives the use of biomarkers offers an attractive and integrative approach.

## 1.2. Biomarkers

The term biomarker may be defined as any biochemical, physiological, histological, or behavioral response that is elicited by an organism as a consequence of the exposure to xenobiotics (NRC, 1987). Therefore, biomarkers are broadly defined as indicators signaling events in biological systems or samples (ATSDR, 2007). While some authors agree with this definition (Depledge et al., 1995; Handy and Depledge, 1999; van der Oost et al., 2003; Wu et al., 2005; Rodríguez-Castellanos and Sanchez-Hernández, 2007) others prefer limiting the term to those responses that are induced exclusively at the biochemical level (Engel and Vaughan, 1996; Hyne and Maher, 2003).

Under normalized and controlled laboratory bioassay conditions, it is possible to investigate the primary responses (that is: those that occur at the biochemical level) of the test organism, exposed to one more chemicals, in order to establish cause-effect relationships. This is a necessary step before validating the biomarker for field studies. It is assumed that most biochemical processes occur by similar mechanisms even among non-related species. However, it should be taken in mind that the extent of activity and the extent of conversion of these processes may exhibit a great variability among the different species (Koeman, 1991; Malins and Ostrander, 1991).

Table 1 lists some of the principal biomarkers routinely used in both laboratory and field studies.

**Table 1. List of biochemical biomarkers most frequently used**

| Biomarker | Modification | Affected by |
|---|---|---|
| Metallothioneins | Induction | Metals (Ag, Cd, Cu, Hg. Zn, etc.) |
| ALA-D | Inhibition | Pb |
| Porphyrins | Increase | Organochlorine compounds (HCB, PCBs) |
| Aceltylcholinesterase | Inhibition | Organophosphate and carbamate pesticides |
| MFO (EROD) activity | Induction | PAHs, PCBs, dioxins, organochlorine pesticides |
| Glutathione-S-transferase | Induction | Electrophilic substances |
| Stress proteins | Induction | Non-specific |
| Antioxidant enzymatic system | Induction/decrease | Non-specific |
| Glutathione levels | Induction/decrease | Non-specific |
| Lipid peroxidation (TBARS) | Induction | Non-specific |
| Total oxygen scavenging capacity (TOSC) | Decrease | Non-specific |
| ATP/ADP ratios | Decrease | Non-specific |
| Vitellogenin | Increase | Endocrine disrupting chemicals (EDCs) |
| DNA damage | Occurrence | Non-specific |
| Lysosomal system integrity and functioning | Decrease | Non-specific |

Metallothioneins are cytosolic non-enzymatic proteins, with low molecular weight, high cysteine content, no aromatic amino acids and heat stability that are present in almost all taxa, including mammals, fishes and invertebrates (Peakall, 1994; Amiard et al., 2006). These proteins are induced by several metals. Their principal physiological role is related to the homeostasis of some essential metals, in particular copper and zinc, but they are also involved in the detoxification of non-essential metals, such as silver, cadmium and mercury (Moffat and Denizeau, 1997; Vasseur and Cossu-Leguille, 2003; Amiard et al., 2006).

The enzyme δ-aminolevulinic acid dehydrase (ALA-D) is involved in the biosynthesis of the heme group (Figure 1). Its activity is inhibited by lead and since long has been recognized as a sensitive biomarker of the metal in Clinical Toxicology (Schäfer et al., 1999; Timbrell, 2000; ATSDR, 2007).

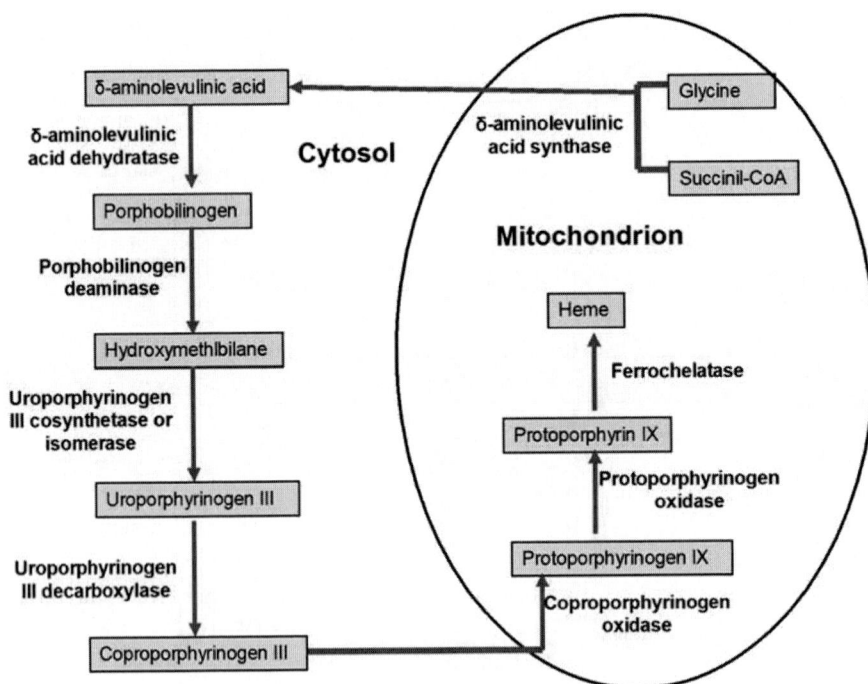

Figure 1. Schematic representation of the heme metabolic pathway.

Porphyrins are precursors of the heme group (Figure 1). At normal physiological conditions their levels are very low. However, several organochlorine compounds, such as hexachlorobenzene (HCB) and polychlorine biphenils (PCBs) induce an increase in porphyrin levels that are excreted in the urine (Timbrell, 2000; Walker et al., 2001). Although the mechanism is not completely elucidated, it is considered that the porphyrin induction is caused by the inhibition of the enzyme uroporphyrinogen decarboxylase (Walker et al., 2001).

The activity of the enzyme acetylcholinesterase (AChE) has been used as a sensitive biomarker of exposure to organophosphate and carbamate pesticides to monitor both workers and the general population (Timbrell, 2000; Walker et al., 2001). This enzyme catalyses the hydrolysis of the neurotransmitter acetylcholine. Its inhibition is linked with the pesticide mechanism of toxic action, irreversible or reversible binding to the esteratic site of the enzyme and potentiation of cholinergic effects in the nervous system.

A great number of the metabolic transformations involved in Phase I reactions are mediated by the mixed function oxidase system (MFO), located in microsomes. This system contains a cytochrome $P_{450}$, a flavoprotein, and b-nicotinamide adenine dinucleotide phosphate reduced form (NADPH) or b-nicotinamide adenine dinucleotide reduced form (NADH) as cofactor (Peakall, 1994). Many reactions of both endogenous and exogenous compounds are catalyzed by this system, therefore there are more than 500 cytochromes $P_{450}$ which are classified into several families (by a number) and then into subfamilies (by a letter) (Mansuy, 1998; Stegeman and Livingstone, 1998). Some isoenzymes are markedly induced by aromatic hydrocarbons, chlorinated dioxins and related halogenated aromatic hydrocarbons by linking to the aryl hydrocarbon receptor (AhR), a ligand-activated transcription factor involved in the regulation of the genes that codify those isoenzymes (Hahn, 1998). The most classical assay to determine this induction implies the measurement of the activity of the enzyme ethoxyresorufin-$O$-deethylase (EROD) (Förlin et al., 1994; Bonacci et al., 2003).

On the other hand, during the Phase II processes, the parent compound (or one of its metabolites of Phase I) is conjugated with a molecule of glucuronic acid, glutathione, sugar, sulphate or phosphate (Peakall, 1994). In particular, the conjugation of many electrophilic compounds with the tripeptide glutathione is mediated by the enzyme glutathione-S-transferase, also present as several isoenzymes (Hyne and Maher, 2003; Vasseur and Cossu-Leguille, 2003).

Stress proteins, also known as heat shock proteins, comprise a number of inducible proteins in response to both physical and chemical agents. Therefore, the synthesis of these proteins increases by changes in the temperature, light, levels of dissolved oxygen, salinity, and also by the exposure to both metal and organic contaminants (Stegeman et al., 1992; van der Oost et al., 2003; Kinder et al., 2007). Their induction is part of the cellular mechanism of defense against a potential harmful condition (Stegeman et al., 1992; van der Oost et al., 2003).

Many classes of contaminants, even those that are non-chemically related, exert their toxicity by generating reactive oxygen species (ROS) (Timbrell 2000, Walker et al. 2001; Valavanidis et al., 2006). ROS can react with vital macromolecules, such as membrane lipids, nucleic acids and other proteins, leading to peroxidation processes (Younes, 1999; Dalle-Donne et al., 2003; Hwang and Kim, 2007). Different antioxidant defenses are available to protect the cells from oxidative stress, both of enzymatic and non-enzymatic nature (Younes, 1999). The activities of the enzymes catalase, superoxide-dismutase and glutathione peroxidase are critically important in the detoxification of radicals to non-reactive molecules. Non-enzymatic antioxidant defenses comprise biomolecules of low molecular weight that act as free radical scavengers, such as glutathione, β- carotene (vitamin A), ascorbate (vitamin C), α-tocopherol (vitamin E) and ubiquinol$_{10}$. Almost all of them have been used as biomarkers of general stress (Stegeman et al., 1992; van der Oost et al., 2003).

The extent of lipoperoxidation processes may be estimated by measuring the amounts of primary peroxidation products formed by cleavage of fatty acids, in particular malondialdehyde (MDA). The most popular method is based on the reaction of MDA with the thiobarbituric acid (TBA) and known as the assay of TBA-reactive substances (TBARS) (Valavanidis et al., 2006; Hwang and Kim, 2007).

The induction of the different biomarkers of oxidative stress can be regarded as an adaptation response to an altered environment, while their inhibition may mean cell damage and toxicity (Falfushynska and Stolyar, *in press*).

The total oxidant scavenging capacity (TOSC) is a method that allows estimating the antioxidant defenses of a given biological sample (Regoli and Winston, 1998). It is an *in vitro* bioassay based on the production of ROS, induced by different compounds, in the presence of α-keto-γ-methyolbutiric acid (KMBA) which reacts yielding ethylene. The amount of ethylene formed will be highest when there is not any antioxidant capacity (control samples). On the contrary, if the biological sample has some antioxidant capacity to neutralize the oxidative stress induced by the exogenous substance, lower levels of ethylene will be formed. Different types of ROS may be produced depending on the compound selected, such as peroxil radicals by thermal homolysis of azocompounds, in particular 2,2'-azobis(2-methylpropionamidine: ABAP); hydroxyl radicals by iron plus ascorbate driven a Fenton reaction; and peroxinitriles by 3-morpholinosydnonomide: SIN-1 (Regoli and Winston, 1998, 1999).

The adenylate energy charge (AEC) is a parameter indicative of the metabolic energy available to an organism (Mayer et al., 1992). This parameter can be calculated as:

$$AEC = \frac{ATP + (1/2)ADP}{ATP + ADP + AMP}$$

where ATP = adenosine triphosphate concentration, ADP = adenosine diphosphate concentration, and AMP = adenosine monophosphate concentration.

High values for AEC are expected for the healthiest organisms (in the range 0.8 – 0.9) (Hemelraad et al., 1990; Mayer et al., 1992). On the contrary, low values are expected when the animals are under stress, due to a decrease in the concentration of ATP (Hemelraad et al., 1990; Mayer et al., 1992).

In oviparous organisms, vitellogenins (VGs) are the precursor of the egg-yolk proteins vitellins that provide the energy reserves for the embryonic development (Matozzo et al., 2007). Normally, these proteins are synthesized in the liver in response to endogenous estrogens, released to the blood and stored in developing oocytes in mature females (Matozzo et al., 2007). A male has also the vitellogenin gene, but is normally silent (Sumpter, 1995). Endocrine-disrupting chemicals (EDCs) are a wide variety of natural and anthropogenic substances that have the ability to interfere with the production, release, transport, binding, uptake, metabolism or action of endogenous hormones (Sumpter, 1995, 1998; Tolar et al., 2001; Matozzo et al., 2007). In consequence, both female and male organisms expose to them may increase the vitellogenin production.

The classical endpoints used to study DNA damage on aquatic organisms are focus on genotoxic effects that may lead to carcinogenesis rather than reproductive alterations. These endpoints are formation of DNA adducts, strand breakage and chromosomal aberrations such as sister chromatid exchange (Walker et al., 2001; van der Oost et al., 2003).

Lysosomes are membrane-bound subcellular organelles containing acid hydrolases that under normal conditions are covered by an intact membrane (Mayer et al., 1992). Instead, in the presence of stressor agents the lysosomal stability membrane decreases as the membrane permeability increases (Mayer et al, 1992). Therefore, the neutral red dye assay utilizes the fact that lysosomes in healthy cells are able to retain the dye longer than perturbed lysosomes (Lowe et al., 1992; Wedderburn et al., 1998). The retention time gives a health index in terms of contaminant exposure.

### 1.2.1. Classification of Biomarkers

Biomarkers may be classified in three classes (Kammenga et al., 2000; Kendall et al., 2001; ATSDR, 2007).

*Biomarkers of exposure*: The simplest parameters of exposure are the presence of the contaminant or its metabolites detected within any biological tissue of the organism (blood, urine, feces, hair, etc.). Additionally, the resulting products of the interaction between a particular chemical and some target biomolecule are also included. In this way the formation of adducts with DNA or proteins by known chemical carcinogens are considered as biomarkers of exposure.

*Biomarkers of effect*: In this category are included any biochemical, physiological, or histological alterations that, depending on its magnitude, are associated with a potential or already established health impairment or disease. Some biomarkers of effect may be regarded as *selective* or *specific*, since they are elicited by the exposure to one or a few contaminants that are chemically related. In this way the substances may share or compete for the same target. The inhibition of the enzyme ALA-D by lead, the induction of metallothioneins by several toxic metals, and the inhibition of the AchE by organophosphate and carbamate pesticides are perhaps the most relevant examples. In addition, there are *general* or *non-specific* biomarkers, which are related to those responses elicited by the organism as a consequence of the chemical stress. In this case, these biomarkers are induced by a great variety of substances even those that are not chemically related (Dorigan and Harrisson, 1987; Viarengo and Canesi, 1991).

*Biomarkers of susceptibility*: These parameters are indicators of a genetic inherent or acquired decrease ability of the organism to respond to the challenge of the exposure to contaminants. Thus, some chemicals may adversely affect the resistance to other chemical, biological or physical stressors by inhibiting the activity of the immune system or increasing the susceptibility to infections and cancer.

### 1.2.2. The "Ideal" Biomarker

Before using in regular biomonitoring programs, biomarkers should meet the following criteria that have been resumed by several authors (Mayer et al., 1992; Kammenga et al., 2000; van der Oost et al., 2003):

- They should reflect, in either a qualitative or quantitative way, the interaction of the biological system with a given contaminant. Thus, a clear dose-effect relationship should be established between the environmental or the internal concentration and the biomarker response. In addition, deviations from this response should be distinguishable from other confounding stressors, either those of biological and physiological nature (sex, age, etc.) or different environmental stressors (changes in temperature, salinity, oxygen, etc.).
- They should be quali- and quantitatively reproducible with respect to time. Any transient response should be perfectly known.

- The biomarker assay should be reliable, easy to perform and relatively inexpensive.
- Ideally, they should have a known specificity and sensitivity to the interaction.

## 1.3. Advantages of Biomarkers

Biomarkers determined at the suborganism level have many advantages for assessing environmental quality. They are considered as "early warning" signals, since they are able to detect the deleterious effects of contaminants within cells/tissues before they may propagate to higher levels of biological organization (Moore et al., 2004). Biomarkers provide a measurement of the internal dose of chemicals that have been absorbed by all the possible routes of uptake (He, 1999). In addition, they allow evaluating the interactive effects caused by mixtures of chemicals (Walker, 1998; Wang and Fowler, 2008). On the other hand, they provide insight into the mechanisms of toxic action (Hyne and Maher, 2003).

Biomarkers may contribute to establish the health status of individuals and to identify already degraded sites from those with emerging problems (He, 1999; Walker et al, 2001; Hinck et al, 2006). Another important advantage is that for many contaminants, biomarker determinations are much less expensive than chemical analyses (Hagger et al., 2008).

Although it is still difficult to link the biomarker responses of individuals to population and community levels, several advances have been made in this direction (Hyne and Maher, 2003; Moore et al., 2004).

## 1.4. Limitations of Biomarkers

Since long, it has been recognized in the literature that biomarker responses may be affected not only by the effect of chemical pollutants but also by different natural and physiological factors (Peakall, 1994; Kammenga et al., 2000; van der Oost et al., 2003; Monserrat et al., 2006; Sanchez et al., 2008).

Aquatic systems and in particular estuarine environments are subjected to continuous fluctuations in the physico-chemical characteristics of the water according to the hydrology of the basin and the anthropogenic inputs. Changes in salinity, pH, redox potential, ionic strength, total suspended matter, dissolved organic carbon content, potential complexing agents, temperature are the most relevant. Many of these properties have an important impact on the speciation of metals, altering their bioavailability, bioaccumulation, and hence their toxicity to aquatic organisms (Miller et al., 2005).

Among the most important physiological factors that may alter biomarker responses are sex, age, reproductive stage and nutritional conditions (Peakall, 1994; Vasseur and Cossu-Leguille, 2003).

Seasonal variations in certain biomarkers have been attributed to both environmental and biological factors related mainly with the ambient temperature and the metabolic status of the animals (Falfushynska et al., *in press*).

Biomarkers like metallothioneins and stress proteins may be markedly influenced by changes in both environmental and biological conditions (Moffat and Denizeau, 1997; Vasseur and Cossu-Leguille, 2003; Amiard et al., 2006; Geffard et al., 2007).

EROD activity depends on variables such as sex, maturity and genetic factors of the organisms (Bonacci et al., 2003). In addition, resistance to EROD induction has been observed in certain fish populations living in highly contaminated areas (Meyer and Di Giulio, 2002). Similarly, glutathione-S-transferase may be involved on insect resistance to organophosphate insecticides (Hyne and Maher, 2003).

It has been shown that many different environmental contaminants, other than organophosphates and carbamates, are also able to inhibit the cholinesterase activity. Metal ions such as Hg(II), Cu(II), Zn(II), Cd(II) and Pb(II) depress the enzyme activity of fish and invertebrates either *in vivo* or *in vitro* conditions (Gill et al., 1990, 1991; Schmidt and Ibrahim, 1994; Devi and Fingerman, 1995; Labrot et al., 1996; Martinez-Tabche et al., 2001; Alves Costa et al., 2007). Organic compounds like surfactant agents and the bipyridylium herbicide paraquat were also found to be non-specific anti-cholinesterase agents (Nemcsók et al., 1984; Seto and Shinohara, 1987; Szabó et al., 1992; Payne et al., 1996; Láng et al., 1997; Guilhermino et al., 2000). Therefore, inhibition of cholinesterase may be a valuable tool for assessing the extent of pollution near agricultural areas, but it should not be used outside them (Peakall, 1994; van der Oost et al., 2003).

## 1.5. Test Organisms

We have discussed to some extent the principal advantages and limitations of the biomarkers most commonly used for field studies. However when considering the proper selection of the biomarkers, it should be taken in mind that the choice of the test organism is equally important.

In previous studies we have found that the freshwater gastropod *Biomphalaria glabrata* was much more sensitive to lead induced inhibition on ALA-D activity than the freshwater oligochaete *Lumbriculus variegatus* (Aisemberg et al., 2005). However, an opposite result was found in relation to the inhibition of cholinesterase enzymes by the organophosphate pesticide azinphos-methyl, the oligochaetes were much more sensitive than the gastropods (Kristoff et al., 2006). Differential responses in the antioxidant defense responses were also observed when these organisms were exposed to paraquat and to azinphos-methyl (Cochón et al., 2007; Kristoff et al., 2008). In all these cases, the animals used had been also reared in the laboratory, and the bioassays were performed under controlled conditions that are very unlikely to find in the nature.

For field studies the selection of the target species should satisfy the following criteria (Lehtonen et al., 2006):

- Availability throughout the year
- The species should largely be stationary
- The species should be indicative of bioaccumulation and/or biomagnification
- Enough tissue should be available to perform simultaneously analyses of several biomarkers.

Fish species have been regarded as suitable organisms for field monitoring studies (van der Oost et al., 2003). In general they meet most of the requirements excepting for their high mobility in the water course. However, they can be found virtually everywhere and they have an important ecological relevance (van der Oost et al., 2003).

### 1.5.1. Test Organism Selected

The organism selected for this work was the freshwater fish *Prochilodus lineatus* (Class Actinopterigios, Order: Characiformes). It is a neotropical fish, widely distributed in the La Plata basin, where it represents 50-60% of the total ichthyomass (Oldani, 1990; Parma de Croux, 1994). This species is a bottom feeder fish that is able to incorporate contaminants present not only in the water column but also in sediment particles. For this reason, this species has been considered as a suitable bioindicator organism to assess simultaneously the water and sediment quality (Almeida et al., 2005; Santos et al., 2005; Camargo et al., 2006). Nevertheless, the dynamic of its population may constitute a disadvantage. Under normal conditions, sexually mature fish migrate upstream (200 – 600 km) the La Plata River for reproductive purposes (Oldani, 1990). Apparently the temperature of the water would be the principal stimulus that elicits the migration. These migrations usually start in autumn-winter (south hemisphere). The reproductive cycle occurs during the spring-summer seasons (from October to March) and after the spawning the animals return to the La Plata River (Sverlij et al., 1993).

## 1.6. Δ-Aminolevulinic Acid Dehydratase (ALA-D)

δ-Aminolevulinic acid dehydratase (ALA-D), also called porphobilinogen sintetase (EC 4.2.1.24) is the second enzyme in the heme biosynthesis pathway (Figure 1). This cytosolic enzyme catalyses the condensation of two molecules of δ-aminolevulinic acid to yield porphobilinogen (PBG). The heme group is the precursor of vital macromolecules such as the hemoglobin pigment, cytocromes and chlorophyll (Jaffe, 2004), being present therefore in almost all phyla (Rocha et al., 2004).

It is considered that ALA-D activity was first detected in haemolysed erythrocytes from birds and algae (Dresel and Falk, 1953; Shemin and Russell, 1953; Granick, 1954). Since then, the enzyme has been purified from a large number of organisms, such as bacteria, plants, birds and mammals and it is probably the best known enzyme within the heme pathway.

According to its crystalline structure, the protein is formed by eight identical subunits. The sequence of amino acids that conforms the active site is phylogenetically invariable, excepting for those amino acids that are bound to the metal and act as cofactor for the enzyme (Jaffe, 2003; 2004). In fact, diverse metal ions may act as cofactors and they may be present at the active site or at the allosteric site depending on the phyla. Therefore, Jaffe (2003) has proposed the following classification for the enzyme:

Type 1: Zinc at the active site and magnesium at the allosteric site.
Type 2: Zinc at the active site and the absence of magnesium at the allosteric site.
Type 3: Magnesium at the active site and the absence of metals at the allosteric site.
Type 4: Absence of metals at the active site.

The enzyme from animals (metazoan), fungi, archaea and some bacterium species presents zinc at the active site (Jaffe, 2003). Other species may require magnesium or a monovalence metal ion, while some others do not require any metal at the active site (Jaffe, 2004). On the other hand, the enzyme from metazoan does not require magnesium at the allosteric site (Type 2) but it has another binding site to a second atom of zinc, besides the active site, that it is not essential for the enzyme activity (Conner and Fowler, 1994).

In vertebrates, the highest activity is observed in erythrocytes, liver and kidney (Abdulla et al., 1978). The enzyme is considerably stable at temperatures above 40°C, it has a molecular weight of 280,000 D, it contains 28 SH-residues, its affinity constant (Km) is in the range of $1 - 4 \times 10^{-4}$ M, and the optimal pH range is between 6.3 and 7.1 (Conner and Fowler, 1994; Farina et al., 2003). In general it is not a rate-limiting enzyme; only in a few species (some algae and mutant yeasts) it functions as a regulatory enzyme in the heme biosynthesis (Battle and Magnin, 1988).

In humans, the enzyme activity exhibits marked changes during the development. The highest activity levels are found in the fetal liver (Doyle and Schimke, 1969; Weissberg and Woytek, 1974). The values decrease several times immediately after the birth, but they increase again after a few weeks, remaining practically unchanged during the adult life (Doyle and Schimke, 1969). In humans and other animals, individual differences in the values of enzymatic activity (usually from 3 and 4 times) have been found (Sassa et al., 1973; Sassa et al., 1977). In relation to fish, there are still few works that have focused on the biochemical characterization of this enzyme and no register about the genome sequences (Rodrigues et al., 1989; Conner and Fowler, 1994; Nakagawa et al., 1995; Alves Costa et al., 2007). Nevertheless, the studies suggest that structural differences may be found between the ALA-D from mammalians and from fish. Conner and Fowler (1994) have found that the enzyme from fish liver has a lower affinity constant, higher resistance to high temperatures, and it does not require zinc as cofactor in comparison with human liver ALA-D.

### 1.6.1. The Use of ALA-D as Biomarker

According to Peakall (1994) the inhibition of ALA-D as biomarker was first investigated by Hernberg et al. (1970) to assess environmental exposure to lead in humans. Since then, analyses of enzyme activity are routinely used in Clinical Toxicology.

Its specificity as biomarker of lead exposure is considerable high for most of the species studied, including mammals, birds, aquatic organisms and bacteria (Stegeman et al., 1992; Walker et al., 2001). The enzyme may be maximally inhibited before other signs of toxicity become perceptible (Mayer et al., 1992).

Studies in vivo have shown that metals such as cadmium, copper, mercury and zinc have not effect on the enzyme activity in rainbow trout (Hodson et al., 1977). Similar results were reported for cadmium, copper, and both organic and inorganic mercury by in vitro studies in an avian species (Scheuhammer, 1987). According to this report lead was found to be 10-100-fold more potent inhibitor for ALA-D than the other metals. In agreement with these results, the metals/metalloids arsenic, cadmium and copper did not inhibit the enzyme of the whole body tissue of the freshwater oligochaete Lumbriculus variegatus either in vivo or in vitro assays (Verrengia Guerrero, unpublished results).

The high sensitivity of ALA-D as a specific biomarker of lead has opened the possibility to explore the use of a bacterial enzyme activity as a useful biosensor to monitor bioavailable lead in contaminated aquatic systems (Ogunseitan et al., 2000).

However, recently it was found that copper may also caused significant decreases of ALA-D activity in hepatic, muscle and renal tissues of the fish *Leporinus obtusidens* when exposing chronically to sublethal metal concentrations (Gioda et al., 2007). Similarly, zinc induced ALA-D inhibition in hepatic, renal and cerebral tissues (Gioda et al., 2007).

Apparently the sensitivity of ALA-D to toxic metals other than lead could depend on the identity of the metallic cofactors (Tanaka et al., 1995; Ogunseitan et al., 1999, 2000) and/or to the lead-binding affinity (Wetmur, 1994).

As we mentioned above, in most vertebrate tissues the enzyme requires zinc as cofactor. In this case, the affinity of lead for the enzyme is almost 25 times greater than for zinc (Rodrigues et al., 1989; Bergdahl et al., 1998). The high specificity of lead for the ALA-D enzyme, among many other zinc proteins, may be explained by the composition of amino acids present at the active site. The yeast and mammalian enzymes contain a unique zinc-binding site with three cysteine residues (Godwin, 2001). Consequently, lead can displace the zinc from the active site given an identical molecular structure (Godwin, 2001). The effect of lead on the enzyme is of a non-competitive nature (Erskine et al., 1997).

### *1.6.2. ALA-D Reactivation*

In several mammalian tissues, once ALA-D has been inhibited by lead, its activity can be reactivated by adding *in vitro* activator agents such as dithiothreitol (DDT) and zinc, either alone or in combination. Thus several authors have used only DDT (Granick et al., 1973; Sakai et al., 1980; Rodrigues et al., 1996), while others obtained better results with zinc (Polo et al., 1995), or a mixture of DDT plus zinc (Yagmines and Villeneuve, 1987) depending on the animal species and the particular tissue. In this way it is possible to fully restore the enzyme activity.

The reactivation index (RI) can be calculated using both values of enzyme activity, the original ALA-D activity and the reactivated ALA-D(r) activity, as:

$$RI = \frac{ALA-D(r) - ALA-D}{ALA-D(r)} x100$$

It is considered that the reactivation index is an even better sensitive parameter to evaluate ALA-D inhibition, especially for those samples that show a high variability or when low inhibition values are observed (Rodrigues et al., 1996).

The enzyme reactivation has been also investigated in both blood (Hodson et al., 1977; Rodrigues et al., 1989) and liver samples (Rodrigues et al., 1989; Conner and Fowler, 1994) of fish, but so far inconsistent results have been obtained.

## 1.7. The Selected Study Area

The basin of the La Plata River is one of the most important in South America. The La Plata River is formed by the confluence of the Paraná and Uruguay rivers and is the last lotic system of the basin (Figure 2). The river flows into the Atlantic Ocean as a big estuary, with a highly variable width ranging from about 40 km up to 200 km at its mouth, while its total length is approximately 300 km (Dagnino Pastore, 1973; Bazán and Arraga, 1993).

Figure 2. Map of the sampling area.

Buenos Aires, capital city of Argentina, is the biggest city located along the coast, approximately 30 km to the south from the origin of the River. This capital city and surroundings are the highest population density areas, with almost 11,500,000 inhabitants, representing nearly 32% of the country population (data from 2001). Therefore, many industries are located in this area to satisfy the demands of the continuously increasing population. In addition, one of the principal ports of the country is located there. Unfortunately, most of the industrial, domestic and urban discharges are released into the La Plata River without any previous treatment. As a consequence, it is believed that the aquatic system and its resources have been severely affected. On the other hand, the river provides the drinking water supply for the population.

Despite the extension of the basin, the high population density, the anthropogenic activities, together with the minor water effluents that are heavily contaminated, comparatively few studies have been conducted to assess the extent of contamination and the water quality of the La Plata River. Perhaps the most important monitoring program was conducted almost 20 years ago, sampling the principal navigation channel located about 4-5 km from the coast (CARP-SINH-SOHMA, 1990). According to this study, levels of many contaminants in water samples were in general within the range of concentrations allowed for superficial freshwaters. Many reasons can influence this behavior, such as the hydrological dynamic of the river and the content of suspended material (CARP-SINH-SOHMA, 1990). However, a trend to higher levels of contaminants was observed in coastal areas. In addition, only a few works have reported levels of metals in fish tissues from the La Plata River (Marcovecchio et al., 1991; Verrengia Guerrero and Kesten, 1993; Villar et al., 2001; Marcovecchio, 2004), but none of these studies sampled the areas of the sewage discharges.

At Buenos Aires city, total suspended matter may reach values of 70-80 mg L$^{-1}$. These values significantly increase towards the south, as a consequence of turbulences that result from the marine currents, but they decrease close to the river mouth, due to sedimentation processes (Bazán and Arraga, 1993). Therefore, most of the contaminants that are released to the aquatic system become associated with the suspended matter and bottom sediments, posing a risk for the benthic organisms. Sewage effluents are known to contain complex mixtures of chemicals, some of them release in large quantities (Matthiessen and Law, 2002). Many substances have been widely recognized as endocrine-disrupting chemicals for aquatic wildlife such as human hormone steroids, synthetic hormones (e.g. diethylstilbestrol, used as a contraceptive), other pharmaceuticals, several pesticides, detergent surfactants, aromatic hydrocarbons (derived from fossil fuels), polychlorinated byphenols (PCBs) and many plasticizers (Tolar et al., 2001; Ying et al, 2002; Matozzo et al., 2007). In addition to the endocrine disruption, aquatic animals exposed to sewage treatment works may suffer from oxidative stress processes (Carney Almroth et al., 2008).

Most of these substances originate from household use, but a certain proportion may also derive from industrial wastes that are released either accidentally or illegally. In fact, it is considered that urban sewerage systems rarely transport only domestic sewage, but also industrial and storm-water run offs (Singh and Agrawal, 2008). In addition, most of the compounds listed above are hydrophobic and persistent organic chemicals of low volatility. Therefore, it is expected that they concentrate in sediment particles (Tolar et al., 2001).

Even when sewage is treated, the insoluble residue remaining, the sewage sludge, still contains large amounts of non-easily degradable organic compounds and different toxic metals (García-Delgado et al., 2007; Singh and Agrawal, 2008). It has been estimated that almost 700,000 pounds of lead per year has been released from the sewage treatment plants to the atmosphere and aquatic systems in Chicago, USA (Johnson, 1998). When analyzing the lead content in sewage sludge from several municipal treatment plants, in general, levels correlated positively with the population density of the city (García-Delgado et al., 2007).

## 1.8. Objectives

The first hypothesis for this work is that lead may be present at levels of environmental concern in water and sediments of the La Plata River, especially in the zone of Berazategui where the sewage ducts are discharged without any treatment. The metal may pose a risk for the aquatic biota, including the iliophagous fish *P. lineatus*. In fact, detectable levels of lead were already quantified in samples of liver and gills (Lombardi et al., *in press*). Therefore, the main purpose of this work was to use a well known biomarker of lead exposure, the activity of the enzyme ALA-D, to investigate the extent of metal contamination. Firstly, some basic characteristics of the enzyme were determined. Then, analyses of ALA-D activity were performed both in blood and liver samples of adult organisms collected each 2 or 3 months during June 2002 and May 2004. The influence of sex and seasonal variations on the biomarker response was also investigated. In addition, the condition factor and the hepatosomatic index were calculated in order to estimate the physiological status of the organisms. The values of ALA-D activity were correlated with the levels of lead in the tissues.

However the information provides by a biomarker measured in organisms collected from the same site is difficult to interpret. While most of the field monitoring programs is based on comparisons between polluted and relatively unpolluted areas, this approach could not be applied because of the absence of non-polluted areas in the course of the La Plata River. In addition, baseline levels of ALA-D activity in tissues of *P. lineatus* were not available. In view of this problem, two strategies were used. Firstly, the data were compared with values of ALA-D activity in blood and liver samples of fish collected in another water course, the Paraná River, where much lower levels of lead contamination are expected. Secondly, a technique of enzyme reactivation was optimized to assess the extent of ALA-D inhibition.

## 2. MATERIALS AND METHODS

### 2.1. Organism and Tissues Selected

Samples of *P. lineatus* were collected in Berazategui, a sampling station along the coast of the La Plata River (34° 43' 450'' S; 58° 10' 250'' W). Samplings were performed each 1-3 months, from June 2002 to May 2004. The fish were obtained from local fishermen and transported to the laboratory in plastic aquaria (50 L) containing river water. Once the animals reached the laboratory their length and weight were determined, their average values are detailed in Table 2. The sex was determined by observation of the gonads. Only adult organisms were used.

**Table 2. Mean (± S.D.) values of body length and weight of *P. lineatus* collected in Berazategui**

|        |        | $n^a$ | *Length (cm)* | *Weight (g)* |
|--------|--------|-------|---------------|--------------|
| Jun-02 | $F^b$  | 5     | 47.6 ± 7.0    | 3396 ± 1679  |
|        | $M^c$  | 7     | 49.1 ± 1.0    | 3526 ± 111   |
| Jul-02 | F      | 4     | 46.8 ± 2.4    | 3089 ± 633   |
|        | M      | 3     | 47.0 ± 1.0    | 3545 ± 126   |
| Oct-02 | F      | 6     | 42.5 ± 7.8    | 2855 ± 1831  |
|        | M      | 3     | 45.0 ± 1.4    | 2718 ± 82    |
| Dec-02 | F      | 4     | 51.0 ± 2.2    | 3629 ± 596   |
|        | M      | 5     | 43.8 ± 3.6    | 2630 ± 773   |
| Feb-03 | F      | 5     | 42.5 ± 1.9    | 2257 ± 336   |
|        | M      | 3     | 41.4 ± 3.2    | 2339 ± 785   |
| Apr-03 | F      | 7     | 49.8 ± 3.1    | 3903 ± 852   |
|        | M      | 3     | 48.5 ± 1.8    | 3391 ± 1265  |
| Jun-03 | F      | 5     | 48.6 ± 3.8    | 3574 ± 743   |
|        | M      | 3     | 44.3 ± 4.0    | 2492 ± 630   |
| Sep-03 | F      | 4     | 47.6 ± 2.6    | 2989 ± 636   |
|        | M      | 3     | 45.1 ± 4.1    | 2923 ± 696   |
| Dec-03 | F      | 4     | 47.3 ± 6.8    | 3720 ± 1609  |
|        | M      | 3     | 42.8 ± 3.0    | 2617 ± 242   |
| Feb-04 | F      | 4     | 43.8 ± 6.0    | 2561 ± 1229  |
|        | M      | 3     | 42.8 ± 10.3   | 2469 ± 2010  |
| May-04 | F      | 4     | 49.4 ± 1.6    | 3490 ± 308   |
|        | M      | 4     | 43.6 ± 4.1    | 2628 ± 957   |

[a] number of fish samples, [b] Female and [c] Male organisms.

A few number of fish organisms (3 males and 2 females) were collected also in Corrientes city (27° 27' S 58° 49' W), Argentina, located along the coast of the Paraná River, one of the main contributors to the La Plata River, about 1000 km to the north from Buenos Aires city. They were treated in a similar way to those collected in Berazategui.

The fish condition factor (CF) was calculated according to Bagenal and Tesch (1978):

$$CF = \frac{Wt}{L^3} \times 100$$

where $Wt$ is the total weight of the fish (g) and $L$ is the total length (cm).

The hepatic somatic index (HSI) was calculated as (Sloff et al., 1983):

$$HSI = \frac{Wl}{Wt} \times 100$$

where $Wl$ is the weight of the fish liver (g).

Analyses of ALA-D activity were performed in blood and liver tissues. Analyses of lead bioaccumulation were performed in blood, liver and also gill tissues.

Blood samples were extracted from the caudal vein using 5 ml heparinized syringes (0.30 mg heparin per ml of blood). One aliquot was kept in liquid nitrogen until the enzyme activity was measured. The other was kept at -20 °C for lead analysis. Then, the animals were killed by excision of the spinal cord behind the operculum. Liver and gill tissues were dissected at 0-2 °C. The whole liver was washed in 50 mM Tris-HCl buffer at pH = 7.4, dried on filter paper and weighed. About 5 g was used for lead analysis. Another portion was homogenized in 50 mM Tris-HCl buffer at pH = 7.4 (ratio 1:3 w/v). Homogenates were centrifuged at 11700 $g$ during 20 minutes at 2 °C and the pellets discarded. Supernatants were kept in liquid nitrogen for the analysis of the enzyme activity.

## 2.2. ALA-D Activity

Blood samples (0.1 ml) were treated with distilled water (0.65 ml) and incubated at 38 °C during 15 min to break red cell membranes (Berlin et al., 1974). Then, 0.5 ml of buffered substrate, prepared just before using (10 mM δ-aminolevulinic acid dissolved in 50 mM phosphate buffer, pH = 6.2), were added. Similarly, the liver ALA-D activity was measured in an aliquot of 0.2 ml of the supernatants and adding 0.5 ml of buffered substrate.

The tubes were incubated for 1 h at 38 °C. Finally, 0.5 ml of TCA were added to stop the reaction (final concentration = 10%). The tubes were centrifuged at 4000 rpm x 10 min. An aliquot of 1 ml of the supernatant was transferred to another tube where 1.0 ml of Ehrlich's reagent was added. The Ehrlich's reagent was prepared by dissolving 0.24 g of *p*-dimethylaminobenzaldehyde in 1.92 ml of perchloric acid and 12 ml of glacial acetic acid. The reaction product (PBG) was measured at λ = 555 nm in a Metrolab UV/visible spectrophotometer. Each sample was analyzed by duplicate.

An enzymatic unit (U) of ALA-D was defined as the amount of enzyme catalysing the formation of 1 nmol of product (PBG) in 1 h at 38 °C. The blood enzymatic activity (EA) was calculated as:

$$EA = \frac{U}{Hematocrit} \times 100$$

The hematocrit was determined by a capillary method. The EA for liver samples was expressed as U per mg of protein. Protein concentrations were determined using the method of Lowry et al. (1951) and bovine serum albumin as standard.

The enzyme affinity constant (Km) was determined using six different concentrations of substrate. Its value was calculated from Lineweaver-Burk plots.

To investigate the enzyme reactivation, different concentrations of either $Zn^{2+}$, dithiothreitol (DTT), or a combination of both were added to the samples. Then, the tubes were incubated for 30 min prior to the addition of the buffered substrate.

## 2.3. Lead Analyses

Blood samples were treated with 0.05% Tritón X-100 and 0.5% phosphate solution. Lead was quantified by electrothermal atomic absorption spectrophotometry.

Approximately 1-2 g (wet weight) of either liver or gill tissue was place in acid prewashed borosilicate tubes. The digestion procedure was performed in the presence of 5-10 ml of ultrapure concentrate nitric acid (Merck Laboratories, Argentina) at 100-120 °C until the complete destruction of the organic material (Verrengia Guerrero, 1995). Then the samples were diluted to a final volume of 5 ml with 1% (v/v) ultrapure nitric acid, and centrifuged to remove any residue. Each tissue was analyzed at least by duplicate. For each 6-8 samples, a blank was performed and processed simultaneously.

Lead concentrations were measured in a 575 AA Varian atomic absorption spectrophotometer with background correction, applying the method of direct flame atomization in an air-acetylene flame. Values of metal accumulation were expressed as μg metal per g wet tissue.

All the glassware was prewashed with 5% nitric acid, thoroughly rinsed with double-distilled water, and dried. Blank values were negligible. To check the accuracy of the analyses, standard addition methods were used to overcome matrix effects.

## 2.4. Laboratory Bioassays

Organisms of P. lineatus collected from the study area (Berazategui) were acclimated to laboratory conditions for 10 days in aerated aquaria of 1000 L, at a temperature of $21 \pm 2$ °C, and at a photoperiod regime 12:12 (l/d). The tanks contained 800 L of tap water, pH = 8.4, hardness = $53 \pm 12$ mg CaCO$_3$ L$^{-1}$, alkalinity = $406 \pm 22$ mg CaCO$_3$ L$^{-1}$. Half of the water medium was renewed each 48 h.

Similar aquaria were used for the bioassays, where adult organisms (average length = 35.4 ± 2.2 cm; average weight = 1037 ± 195 g) were exposed to 0.1 mg Pb L$^{-1}$ (prepared from a stock solution containing 500 mg Pb L$^{-1}$ using lead acetate in distilled water) for 7 and 15 days by duplicate. Control organisms were exposed in the same conditions without lead. During the treatments, animals were not fed. There was not mortality in any of the experimental procedures.

## 2.5. Statistical Analyses

Pearson correlation coefficients were calculated using the Excel® software package (Microsoft, U.S.). Data of ALA-D activity had heterogeneous variances (Bartlett's test), therefore comparisons among sampling periods were performed using the non-parametric Tukey test (Sokal and Rohlf, 1997). Similarly, lead contents were compared using Kruskal-Wallis non-parametric tests (Sokal and Rohlf, 1997). The significance level α was set at p = 0.05.

# 3. RESULTS

## 3.1. Some Basic ALA-D Characteristics

Table 3 shows some basic characteristics of the ALA-D enzyme from blood and hepatic tissues of *P. lineatus* collected in Berazategui.

The range of optimal pH was slightly higher for the enzyme from liver than that found for the red blood cells. Nevertheless, the values are within the range usually reported for mammalian enzymes. In fact, a pH range of 6.0 – 6.4 is considered as optimal for measuring ALA-D activity in human red cells (Granick et al., 1973). An optimal pH = 6.3 was found for beef liver (Wilson et al., 1972); 6.6 for liver of the fish species *Ictalurus punctatus* (Conner and Fowler, 1994), and for red cells of the fish *Salmo gairdneri* (Hodson et al., 1977). Instead, in tissues from several invertebrate species slightly higher values have been reported. A range of pH = 6.8 – 7.0 was found in whole body tissues of the freshwater gastropod *Biomphalaria glabrata* (Verrengia Guerrero et al., 1997) and the oligochaete *Lumbriculus variegatus* (Aisemberg et al., 2005), while an optimal value of 8.0 was found for the gastropod *Gammarus pulex* (Kutlu and Sümer, 2008).

The affinity constant, Km, (apparent value) was 8 times higher for the enzyme from red cells than the value found for the hepatic ALA-D.

**Table 3. Some basic characteristics of ALA-D from tissues of *P. lineatus***

| | Red blood cells | Liver |
|---|---|---|
| Optimal pH | 5.4 – 6.2 | 5.8 – 6.6 |
| Km | 0.48 mM | 0.06 mM |
| Optimal temperature | 55 – 75 °C | 55 – 75 °C |

The data reflect that the hepatic enzyme has a higher affinity for the substrate than the ALA-D from red cells. Similar affinity constants values have been observed in samples of either red blood cells (Nakagawa K. et al., 1995; Alves Costa et al., 2007) or liver of different fish species (Conner and Fowler, 1994).

On the other hand, the optimal incubation temperature was considerable high. Similar high values were reported for the enzyme of fish hepatic tissue (65°C) (Conner and Fowler, 1994) and rat liver homogenates (55°C) (Sassa, 1982).

The data suggest that isoenzymes are present in red cells and hepatic tissue of *P. lineatus*. Finally, we also found that for both tissues the enzyme did not require of sulfhydryl protector groups (tested with DDT) to exhibit maximal activity and that the activity remained unchanged for at least 300 days if the samples were maintained in liquid nitrogen.

## 3.2. ALA-D Activity in Tissues of *P. Lineatus*

Values of ALA-D activity in blood samples of *P. lineatus* collected in Berazategui are shown in Table 4.

In general, the data showed a great variability within a given sampling period. It is not unusual to find a greater variability in biomarker responses measured in biological samples collected from natural environments in comparison with those derived from animals that were exposed under laboratory conditions. In field situations, inter-individual variations may be due to multiple factors, such as an increased genetic heterogeneity, diversity of age and size, nutritional conditions, different sex, and also to the influence of other natural and anthropogenic variables (Mayer et al., 1992).

The lowest values of enzyme activity were recorded in samples collected in June-2003, while the highest values were observed in February-2004 ($p < 0.05$). However, the data did not follow any clear seasonal pattern.

**Table 4. ALA-D activity (mean values ± SD) in red blood cells (RBC) of *P. lineatus* collected in Berazategui**

| | n | Female U (ml RBC)$^{-1}$ | n | Male U (ml RBC)$^{-1}$ | n | Both sex U (ml RBC)$^{-1}$ |
|---|---|---|---|---|---|---|
| Jun-02 | 4 | 683 ± 278 | 3 | 702 ± 202 | 7 | 691 ± 229 [a b] |
| Jul-02 | 4 | 705 ± 95 | 3 | 648 ± 133 | 7 | 680 ± 106 [a b] |
| Oct-02 | 6 | 624 ± 151 | 3 | 612 ± 392 | 9 | 621 ± 196 [a b] |
| Dec-02 | 4 | 805 ± 108 | 5 | 777 ± 214 | 9 | 789 ± 166 [b] |
| Feb-03 | 5 | 807 ± 365 | 3 | 572 ± 156 | 8 | 708 ± 294 [a b] |
| Apr-03 | 7 | 600 ± 127 | 3 | 633 ± 244 | 10 | 610 ± 156 [a b] |
| Jun-03 | 5 | 562 ± 91 | 3 | 480 ± 91 | 8 | 531 ± 94 [a] |
| Sep-03 | 4 | 613 ± 204 | 3 | 618 ± 165 | 7 | 615 ± 173 [a b] |
| Dec-03 | 4 | 705 ± 265 | 3 | 671 ± 194 | 7 | 691 ± 219 [a b] |
| Feb-04 | 4 | 937 ± 171 | 3 | 938 ± 286 | 7 | 938 ± 192 [b] |
| May-04 | 4 | 892 ± 347 | 3 | 636 ± 200 | 7 | 764 ± 294 [a b] |

n = number of fish samples. Different letters indicate significant differences ($p < 0.05$).

**Table 5. ALA-D activity (mean values ± SD) in
hepatic tissue of *P. lineatus* collected in Berazategui**

|  | n | Female U (mg protein)$^{-1}$ | n | Male U (mg protein)$^{-1}$ | n | Both sex U (mg protein)$^{-1}$ |
|---|---|---|---|---|---|---|
| Jun-02 | 4 | 3.7 ± 1.0 | 3 | 3.5 ± 0.9 | 7 | 3.6 ± 0.9 |
| Jul-02 | 4 | 3.3 ± 0.3 | 3 | 4.4 ± 1.3 | 7 | 3.8 ± 1.0 |
| Oct-02 | 6 | 3.6 ± 0.5 | 3 | 3.3 ± 0.8 | 9 | 3.5 ± 0.5 |
| Dec-02 | 4 | 3.1 ± 0.7 | 5 | 3.2 ± 0.6 | 9 | 3.1 ± 0.6 |
| Feb-03 | 5 | 3.3 ± 0.5 | 3 | 3.4 ± 0.6 | 8 | 3.3 ± 0.5 |
| Apr-03 | 7 | 3.4 ± 0.6 | 3 | 3.8 ± 1.0 | 10 | 3.3 ± 0.9 |
| Jun-03 | 5 | 2.9 ± 1.1 | 3 | 3.7 ± 0.7 | 8 | 3.3 ± 0.8 |
| Sep-03 | 4 | 4.3 ± 1.1 | 3 | 3.5 ± 0.3 | 7 | 3.9 ± 0.8 |
| Dec-03 | 4 | 3.0 ± 0.9 | 3 | 3.2 ± 0.6 | 7 | 3.1 ± 0.7 |
| Feb-04 | 4 | 3.6 ± 0.6 | 3 | 3.8 ± 0.6 | 7 | 3.7 ± 0.5 |
| May-04 | 4 | 3.3 ± 0.6 | 4 | 3.3 ± 0.5 | 8 | 3.3 ± 0.5 |
| Total | 51 | 3.4 ± 0.7 | 36 | 3.5 ± 0.7 | 87 | 3.4 ± 0.7 |

n = number of fish samples.

Values of ALA-D activity in liver samples are presented in Table 5.

In general, these data showed a lower variability among individuals but not significant differences were observed among the different sampling periods ($p > 0.05$).

In any case, no significant differences were found between male and female fish ($p > 0.05$). Curiously very few studies have investigated the influence of sex on ALA-D activity and, to our best knowledge, none of them in fish species. According to the available reports, no sex-associated differences were observed in plasma levels of this enzyme in several species of birds (Dieter et al., 1976; Finley et al., 1976; Grue et al., 1984).

## 3.3. Relationships between ALA-D Activity and Lead Content

In the previous section we have discussed some aspects of ALA-D activity in tissues of *P. lineatus*. While the enzyme activity from blood samples showed some differences during the studied period, the enzyme activity from liver samples remained unchanged. The next step was to investigate what happened with the lead levels.

The metal was detected in all the blood samples analyzed, proving that the organisms had actually been exposed and incorporated the metal. The values are presented in Figure 3.

As no significant differences were observed between male and female fish ($p > 0.05$), the data were pooled for each sampling period. In general the data presented a high variability and in some periods statistically significant differences were observed ($p < 0.05$). However, lead levels in samples collected in June 2003 and February 2004 (for which the lowest and highest ALA-D activity values were observed) did not show significant differences between them ($p > 0.05$). Values of lead concentrations in samples of liver and gills, shown in Figure 4, presented a greater variability than that observed for blood samples.

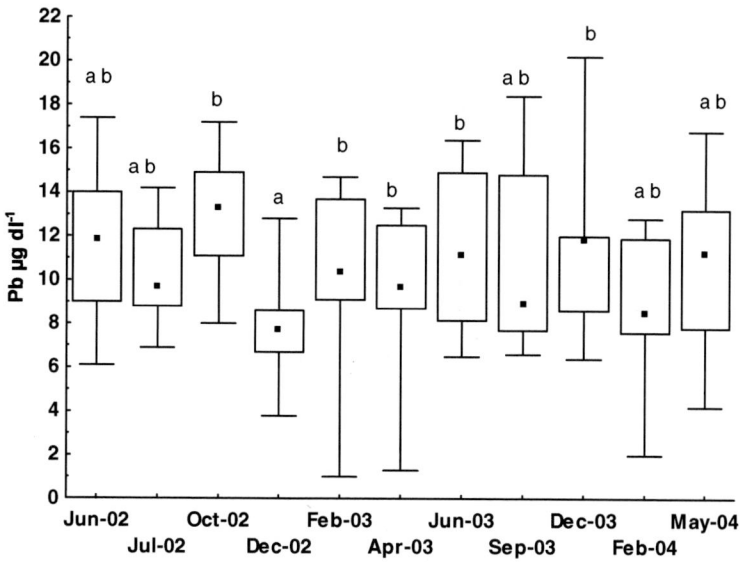

Figure 3. Lead levels in blood samples of *P. lineatus* collected in Berazategui. Different letters indicate significant differences (p < 0.05).

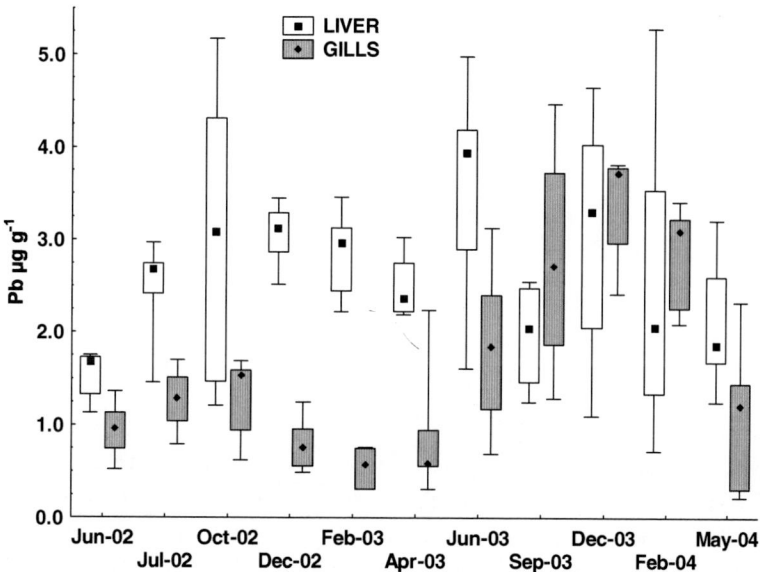

Figure 4. Lead levels in liver and gills samples of *P. lineatus* collected in Berazategui.

As for blood, the data for male and female organisms were pooled since no significant differences were found (p > 0.05). From June 2002 to June 2003 and during May 2004, the liver presented higher levels of lead than the gills. Instead, during September 2003 up to February 2004, lead levels in liver and gills overlapped each others. On the other hand, lead levels in either liver or gill samples collected in June 2003 and February 2004 did not show significant differences between them (p > 0.05).

**Table 6. Pearson correlation coefficients between ALA-activity and (log) lead levels**

| | Blood ALA-D | Liver ALA-D | Pb blood (Log) | Pb liver (Log) | Pb gills (Log) |
|---|---|---|---|---|---|
| Blood ALA-D | 1.00 | | | | |
| Liver ALA-D | -0.00 | 1.00 | | | |
| Pb blood (Log) | -0.30* | -0.02 | 1.00 | | |
| Pb liver (Log) | -0.26 | -0.01 | 0.02 | 1.00 | |
| Pb gills (Log) | -0.12 | 0.12 | 0.17 | 0.15 | 1.00 |

* significant correlations at p = 0.05.

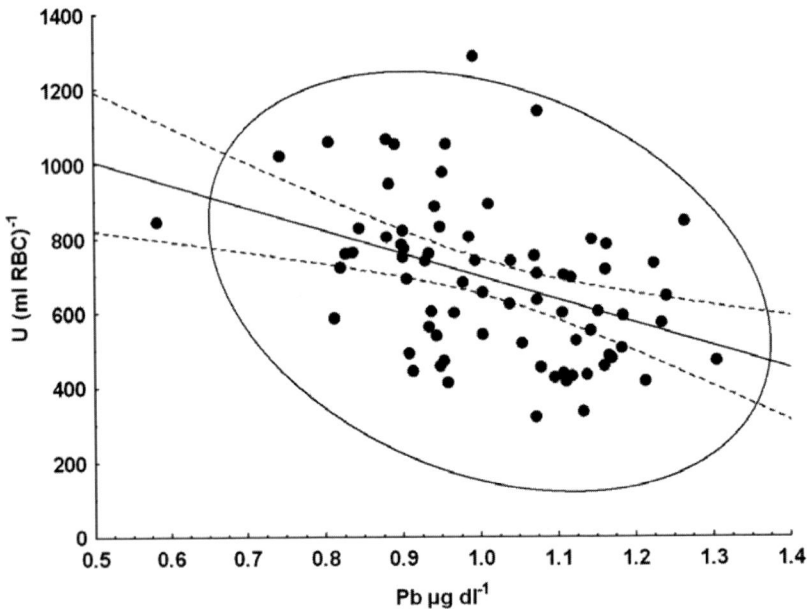

Figure 5. ALA-D activity versus (log) lead content in blood samples of *P. lineatus* collected in Berazategui (ALA-D = 1314.51 – 616.72 x $Log_{10}$ Pb; r = - 0.38; p = 0.0007).

Therefore, by looking at the results, the lowest value of blood ALA-D activity could not be attributed to a higher lead content in any of the tissues analyzed. In fact, when correlations between ALA-D activity and lead levels in samples of liver and gills were statistically investigated, the values were very low, reflecting the lack of any significant correlation (Table 6), excepting for blood samples as it is shown in Figure 5. But even in this case the regression coefficient was considerable low, although being statistically significant.

It should be expected that the biomarker responds in a proportional relationship to the contaminant. According to the literature very good negative correlations have been found between ALA-D activity and the blood lead concentrations (on a log basis) in humans exposed to the metal (Peakall, 1994; Gurer-Orhan et al., 2004). Similarly, good correlations have been reported in blood samples from several fish species (Hodson et al., 1977; Hodson et al., 1978; Nakagawa H. et al., 1995a, 1995b, 1997; Nakagawa K. et al., 1995). In addition, very good correlations have been found between ALA-D activity and lead concentrations in tissues of freshwater invertebrates (Verrengia Guerrero et al., 1997; Aisemberg et al., 2005).

It should be noted that most of the animal data have been obtained by laboratory bioassays exposing the organisms to the metal (the only contaminant) under normalized conditions and even in some cases using organisms that had been also reared under laboratory conditions. Such controlled conditions are by far much less complex than natural environments. Instead, in field studies significant correlations are not always so evident, at least for all the tissues (Schmitt et al., 1984, 1993, 2007; Vanparys et al., 2008).

## 3.4. Interpretation of the Data

Most of the field monitoring programs based on biomarker determinations considers upstream-downstream comparisons to assess the effect of a specific point source (Schmitt et al., 1984, 1993, 2007; Sanchez et al., 2008). Following this criterion several studies have been performed. Andersson (2003) analyzed the ALA-D activity in spleen and kidney samples of two fish species collected at the vicinity of an old lead mine and at a site located upstream that was selected as a reference. The author found that the values were 50% lower near the mine than those found in animals collected upstream (Andersson, 2003).

Alternatively, the samples can be collected along a well characterized pollution gradient (Pyle et al., 2005; Vanparys et al., 2008). However, those approaches are not easily applicable to the La Plata River, where both upstream and downstream areas are almost equally contaminated by lead and many other chemicals from different sources.

At this point we know for sure that the values of ALA-D activity in fish collected in Berazategui are difficult to interpret. Many and opposite questions arise such as:

Is ALA-D activity actually inhibited?
Is the enzyme activity so inhibited that we are looking at the minimum basal activity levels?
Are any biological or environmental confounding factors influencing the results?
Is the fish species selected sensitive enough to reflect environmental lead levels?
Have the organisms developed some kind of tolerance to the metal?

When suitable upstream values are not available, it might be of great interest to assess biomarker responses in contaminant streams in relation to background levels recorded in "clean" areas which may be accepted as reference values of fish biomarkers (Sanchez et al., 2008).

Accordingly to this alternative, ALA-D activity was measured in blood and liver samples of *P. lineatus* collected from Corrientes, an area located at the Paraná River and about 1000 km to the north from Buenos Aires city.

Although this area cannot be strictly considered as an actual "clean zone", at least the fish collected might help with the interpretation of the results. Only five animals (3 males and 2 females) were collected and the mean values of enzyme activity and the ranges of lead concentrations in blood and liver samples are shown in Table 7.

**Table 7. ALA-D activity and lead levels in tissues of *P. lineatus* collected in Corrientes**

|                  | n | ALA-D (mean values ± SD)      | [Pb] (range)                  |
|------------------|---|-------------------------------|-------------------------------|
| Red blood cells  | 5 | $1229 \pm 31$ U (ml RBC)$^{-1}$ | $<0.1 - 4.4$ µg dl$^{-1}$   |
| Hepatic tissue   | 5 | $3.6 \pm 1.1$ U (mg protein)$^{-1}$ | $0.47 - 0.68$ µg g$^{-1}$ |

n = number of fish samples.

While the mean value for blood ALA-D in samples from Corrientes was in general significantly higher (p < 0.05), the mean value for liver samples was not significantly different (p > 0.05) from the samples collected in Berazategui.

During all the sampling periods, the median values of lead concentrations in both blood and liver samples of fish from Berazategui (Figure 3 and 4, respectively) were much higher than the upper limit of concentrations found in the fish from Corrientes. Consequently, we can reasonably assumed that the fish from Corrientes were exposed to lower levels of lead as reflecting by the lower levels of bioaccumulation. Most important, these data suggested that the ALA-D activity in blood samples of *P. lineatus* collected from the La Plata River at the area of influence of the sewer ducts could be inhibited, but not the activity in liver samples.

## 3.5. Morphometric Indices

Different morphometric parameters were used to estimate the physiological status of the animals. Firstly, only adult organisms of similar length were selected. A strong correlation was found between the length and the whole body weight of the fish, with a correlation coefficient r = 0.91.

The condition factor and the hepatic somatic index (gross indices) were calculated and the values are shown in Table 8.

The condition factor is a good indicator of the fish shape and its energy reserves (Goede and Burton, 1990; Sanchez et al., 2008). This index may decrease as result of several environmental stressors, due to a depletion of the energy reserves such as stored liver glycogen or body fat. However, it should be taken in mind that the energy reserves may also fluctuate seasonally, as a consequence of changes in feeding activity and nutrient availability, or any other biological change affecting the susceptibility to contaminants (Dissanayake et al., 2008). For instances, this index may change with the physiological development and sexual maturation of the fish (Goede and Burton, 1990). According to Table 8, the condition factor for *P. lineatus* did not present significant differences either seasonally or by the sex of the organisms. Therefore, the data could be pooled and a total mean value of 3.0 ± 0.4 was calculated.

The hepatic somatic index is another common parameter that may be regarded as a general health indicator. It is sensitive to stress induced by environmental pollutants (van der Oost et al., 2003; Almeida et al., 2005). This index may decrease, as a consequence of depletion of the energy reserves stored in the liver to cope with stressors. However, its value may also increase due to several chemical contaminants. The increase is commonly seen for those chemicals that induce hyperplasia or hypertrophy of the liver (Goede and Burton, 1990; Almeida et al., 2005). According to the data, the hepatic somatic index varied between 0.9

and 1.4 (Table 8). However, no significant differences were found among the sampling periods and mean values were $1.2 \pm 0.4$ for both male and female fish.

**Table 8. Condition factors and hepatic somatic indices of *P. lineatus* collected in Berazategui (mean values ± SD)**

| | Sex[a] | n[b] | Condition Factor | Hepatic somatic index |
|---|---|---|---|---|
| Jun-02 | F | 5 | $2.9 \pm 0.5$ | $1.3 \pm 0.9$ |
| | M | 7 | $3.0 \pm 0.5$ | $1.1 \pm 0.4$ |
| Jul-02 | F | 4 | $3.0 \pm 0.2$ | $1.3 \pm 0.4$ |
| | M | 3 | $3.4 \pm 0.3$ | $0.9 \pm 0.4$ |
| Oct-02 | F | 6 | $3.2 \pm 0.7$ | $1.1 \pm 0.3$ |
| | M | 3 | $3.0 \pm 0.2$ | $1.3 \pm 0.2$ |
| Dec-02 | F | 4 | $2.7 \pm 0.3$ | $1.1 \pm 0.3$ |
| | M | 5 | $3.1 \pm 0.2$ | $1.1 \pm 0.2$ |
| Feb-03 | F | 5 | $2.9 \pm 0.1$ | $1.1 \pm 0.2$ |
| | M | 3 | $3.0 \pm 0.5$ | $0.9 \pm 0.3$ |
| Apr-03 | F | 7 | $3.1 \pm 0.3$ | $1.3 \pm 0.3$ |
| | M | 3 | $2.9 \pm 0.8$ | $1.2 \pm 0.2$ |
| Jun-03 | F | 5 | $3.1 \pm 0.5$ | $1.2 \pm 0.3$ |
| | M | 3 | $2.9 \pm 0.6$ | $1.3 \pm 0.6$ |
| Sep-03 | F | 4 | $2.8 \pm 0.3$ | $1.3 \pm 0.5$ |
| | M | 3 | $3.1 \pm 0.1$ | $1.2 \pm 0.5$ |
| Dec-03 | F | 4 | $3.3 \pm 0.5$ | $1.0 \pm 0.2$ |
| | M | 3 | $3.3 \pm 0.4$ | $1.4 \pm 0.2$ |
| Feb-04 | F | 4 | $2.9 \pm 0.3$ | $1.2 \pm 0.2$ |
| | M | 3 | $2.7 \pm 0.5$ | $1.1 \pm 0.8$ |
| May-04 | F | 4 | $2.9 \pm 0.3$ | $1.2 \pm 0.2$ |
| | M | 4 | $3.0 \pm 0.3$ | $1.1 \pm 0.2$ |
| Total | F | 51 | $2.9 \pm 0.4$ | $1.2 \pm 0.4$ |
| | M | 36 | $3.0 \pm 0.4$ | $1.2 \pm 0.4$ |
| | BS | 87 | $3.0 \pm 0.4$ | $1.2 \pm 0.4$ |

[a] Sex F: Female; M: Male; BS: Both sex.
n[b]: number of fish samples.

In summary, the condition factor and the hepatic somatic index did not follow any seasonal pattern. These results suggest that both the natural environmental factors and the contaminant loads remained rather constant during the studied period. Although these results may seem surprising, they may be explained by several reasons. Firstly, the study area is the recipient of high amounts of sewage discharges during the whole year. Therefore, it may be reasonably assumed that the availability of nutrients, specifically organic matter, remained temporally unchanged, so the organisms did not experience any deprivation condition. On the contrary, specimens collected in Berazategui were considerable fatter than those collected in other areas as it is shown in Figure 6.

At the same time, it could be assumed that due to the great amount and diversity of substances released, the toxicity of the effluents (in terms of "toxicity equivalents") it is high enough to overlap any temporal pattern. Indeed, fish collected from this area presented high levels of organic contaminants such as aliphatic hydrocarbons (Colombo et al., 2007a) and polychlorinated biphenyls (Colombo et al., 2007b) without observing any seasonal pattern.

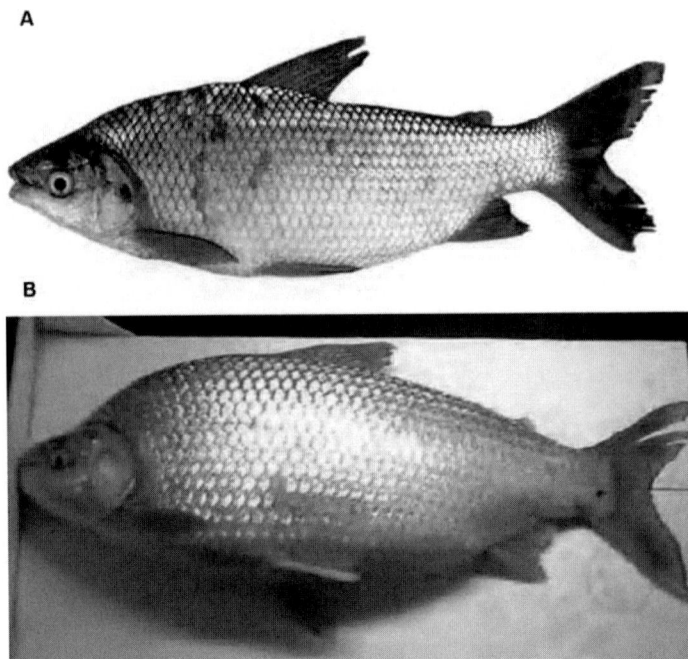

Figure 6. Physical aspect of *P. lineatus* collected in A) a relatively clean area and B) in Berazategui.

The morphometric indices did not present significant differences between male and female organisms ($p > 0.05$). However, it is worth noting that even when the specimens sampled for this study can be classified as male or female, they did not show evidence of sexual maturity. At least partly, it could be hypothesized that the total contaminant burden could induce some kind of endocrine disruption, preventing or delaying the sexual maturity of the animals.

## 3.6. ALA-D Activity and Lead Content in Fish Tissues Exposed Under Laboratory Conditions

In an attempt to clarify the problem of interpreting the data, laboratory bioassays were performed, using fish collected from the field, acclimated to the laboratory, and exposed to 0.1 mg Pb L$^{-1}$ (see Materials and Methods). Analyses of ALA-D activity and lead levels in blood samples are presented in Figure 7A and B respectively.

Values of ALA-D activity in samples of unexposed fish (control organisms) did not change within this depuration period in clean water ($p > 0.05$). These results suggest that the enzyme did not recover its normal activity level for at least 15 days. Similar results were reported in several fish species, such as *Oncorhynchus mykiss* (Hodson et al., 1977;

Johansson-Sjöbeck and Larsson, 1979); *Salvelinus fontalis* (Hodson et al., 1977); *Cyprinus carpio* (Nakagawa H. et al., 1995a), and *Carassius auratus* (Nakagawa H. et al., 1997). Instead, after 7 and 15 days of exposure, ALA-D activity in blood samples significantly decreased with respect to control organisms (Figure 7A) ($p < 0.05$).

On the other hand, lead levels in samples of control organisms remained unchanged ($p > 0.05$). Therefore, the lead absorbed by *P. lineatus* after the chronic exposure in the field was not rapidly eliminated from the blood tissue within the period studied. For this reason, the enzyme activity in blood samples of control fish remained practically constant.

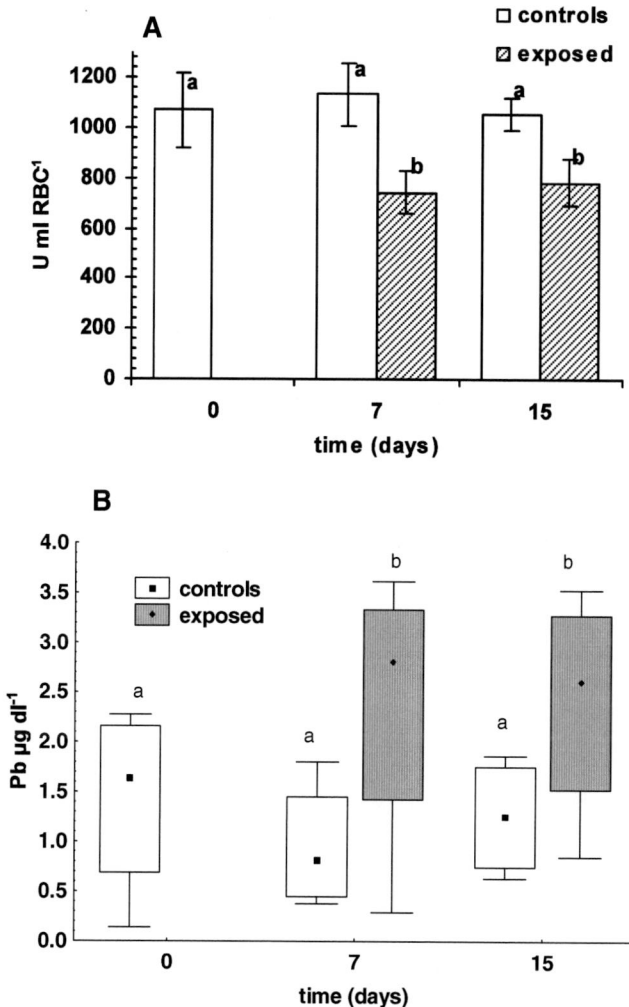

Figure 7. (A) ALA-D activity and (B) lead levels in blood samples of *P. lineatus* exposed to 0.1 mg Pb L$^{-1}$. Different letters indicate significant differences ($p < 0.05$).

The decreases in blood ALA-D activity were correlated with increases in lead levels in both blood and liver samples for the treated organisms, as it is shown in Figure 7B and 8B respectively ($p < 0.05$).

However, a rather different pattern was observed for the enzyme activity in liver samples (Figure 8A).

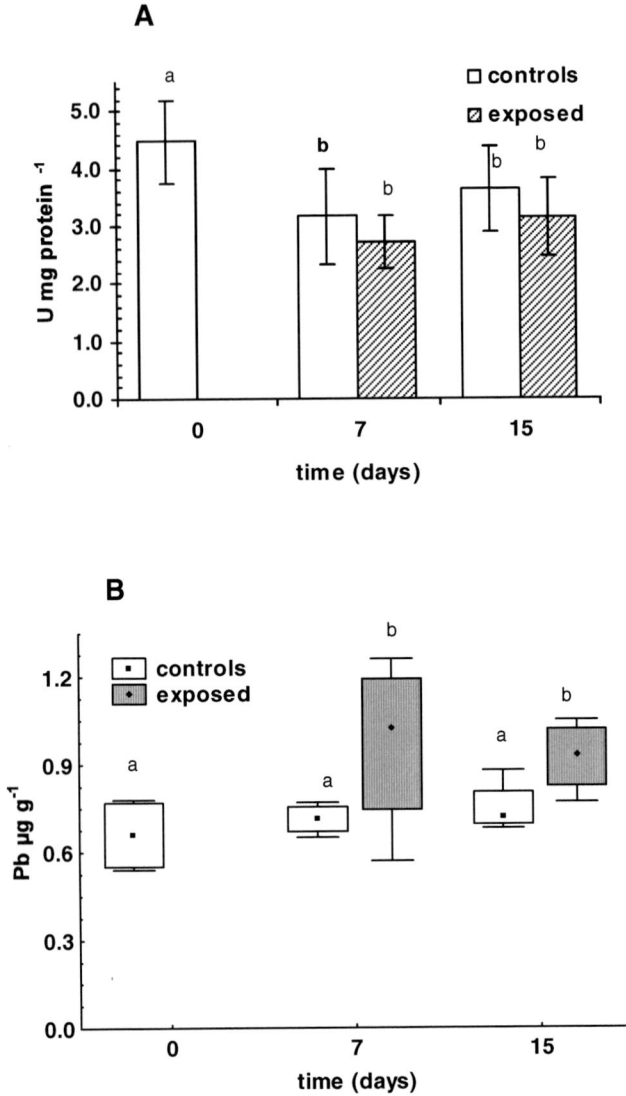

Figure 8. (A) ALA-D activity and (B) lead levels in liver samples of *P. lineatus* exposed to 0.1 mg Pb L$^{-1}$. Different letters indicate significant differences ($p < 0.05$).

In unexposed control organisms, the levels of enzyme activity after 7 and 15 days in clean water, significantly decreased with respect to those at $t = 0$, presenting similar values than the samples of fish that had been exposed to lead. However, the levels of the metal in unexposed organisms did not show any significant modifications ($p > 0.05$) (Figure 8B). In addition, the lead content in exposed fish effectively increased with respect to control fish. No reasonably explanation could be found for these anomalous results.

The results of the laboratory experiments proved that only the blood enzyme of fish was still able to respond to an additional lead exposure by increasing its extent of inhibition. Accordingly, the data of the blood enzyme activity derived from the field monitoring study did not represent merely minimum basal levels.

## 3.7. ALA-D Reactivation

As the blood enzyme could be further inhibited by increasing lead concentrations, reactivation of its activity was investigated. Firstly, the reactivation was studied by adding different zinc concentrations to the incubation media. The results are presented in Figure 9A.

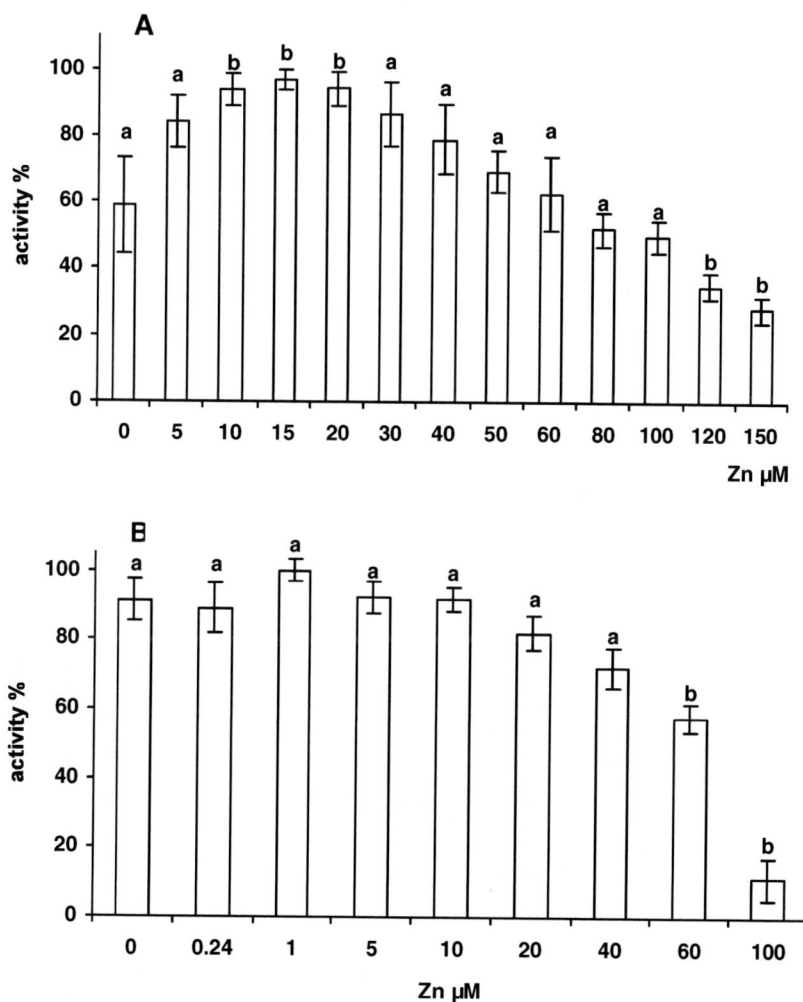

Figure 9. Reactivation of blood ALA-D activity by (A) different Zn concentrations and (B) by the mixture containing 0.024 mM DDT plus different Zn concentrations. Different letters indicate significant differences with respect to control ($p < 0.05$).

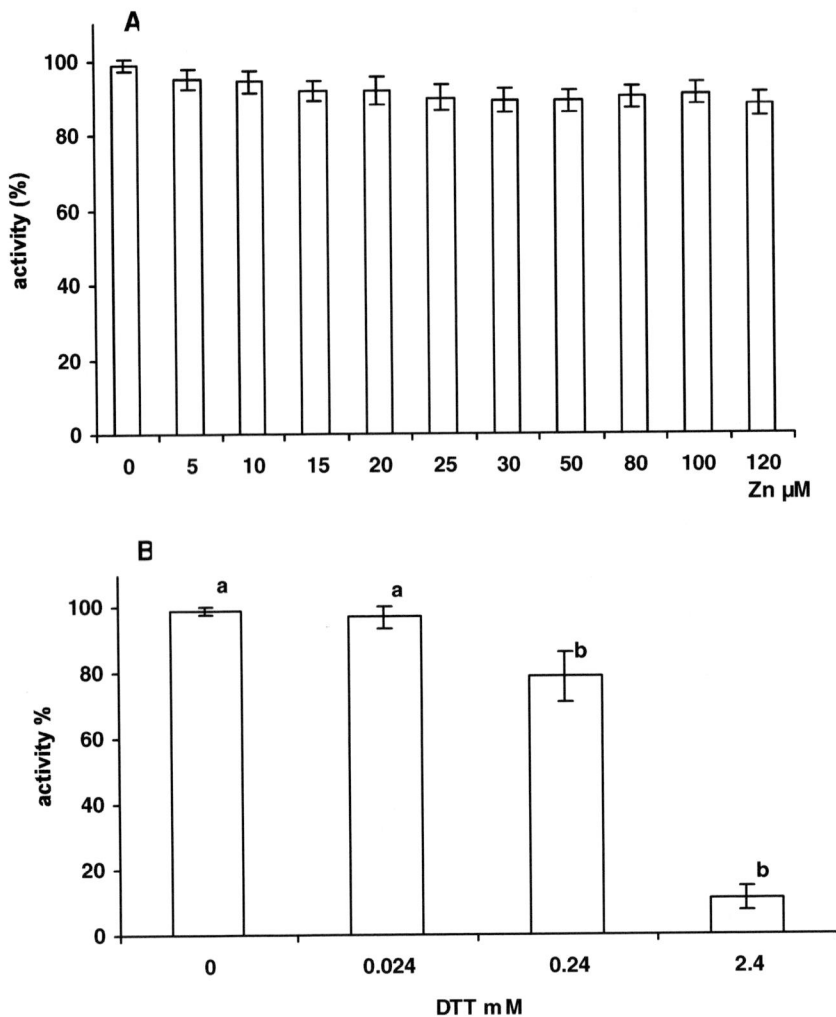

Figure 10. Reactivation of liver ALA-D activity by (A) different Zn concentrations and (B) different DDT concentrations. Different letters indicate significant differences with respect to control. (p <0 .05).

Maximum blood ALA-D activity values were observed after adding zinc in the range between 10 and 20 μM Zn. Lower and higher zinc concentrations induced decreases in the activity (p < 0.05).

The influence of mixtures of zinc (from 0.24 to 100 μM Zn) and 0.024 mM of DDT on blood ALA-D activity was also investigated. Higher concentrations of DDT were not tested since they had shown to induce enzyme inhibition. The results are presented in Figure 9B. ALA-D activity remained unchanged while the incubation medium contained the binary mixture of Zn-DDT from 0 up to 20 μM of zinc concentrations (p > 0.05). At higher zinc concentrations, ALA-D activity decreased with respect to controls (p < 0.05). Therefore, the simultaneous presence of DDT and zinc did no prove to be useful to reactivate the enzyme activity, on the contrary it seems that the DDT abolished the positive effect of zinc.

The optimal concentrations of zinc needed for reactivating blood ALA-D activity in *P. lineatus* are considerably lower than the values reported for mammals, where the range varies between 90 and 100 µM Zn in blood samples (Granick et al., 1973; Finelli et al., 1974; Yagminas and Villeneuve, 1987). On the contrary, at this range an inhibitory effect was observed for the blood samples of *P. lineatus*.

The possible effects of zinc and DDT were also investigated in liver samples and the results are shown in Figure 10A and B respectively.

At any zinc or DDT concentration tested, the liver ALA-D activity in *P. lineatus* could not be reactivated. This result agrees with previous reports for liver samples from different fish species (González et al., 1987; Rodrigues et al., 1989; Conner and Fowler, 1994).

## 3.8. Values of Reactivated Blood ALA-D Activity

We showed in the previous section that only the ALA-D activity in blood samples of *P. lineatus* could be reactivated and that zinc alone demonstrated to be the best activator agent. A concentration of 15 µM Zn was selected to calculate the reactivated values. As it was observed for the direct activity values (Table 4) no significant differences were observed by sex (p > 0.05). In consequence the data were pooled and presented in Table 9. Reactivation indices for each sampling period were also included.

According to our best knowledge, this is the first report that includes reactivated ALA-D values in blood samples of fish collected in the field.

Interestingly, for the reactivated ALA-D values no significant differences were observed among the different sampling periods (p > 0.05). For this reason the data could be pooled and a total mean value was calculated. In addition, it is worth noting that all these values were significant higher than those obtained for the original enzyme activity presented in Table 4.

**Table 9. Values of reactivated ALA-D activity (mean values ± SD) and reactivation indices in red blood cells (RBC) of *P. lineatus* collected in Berazategui**

| | *n* | *Reactivated ALA-D U (ml RBC)$^{-1}$* | *Reactivation Index* |
|---|---|---|---|
| Jun-02 | 7 | 1146 ± 305 | 40 ± 9 [a b] |
| Jul-02 | 7 | 1084 ± 127 | 37 ± 13 [a b] |
| Oct-02 | 9 | 1057 ± 282 | 41 ± 10 [a b] |
| Dec-02 | 9 | 976 ± 108 | 20 ± 12 [a] |
| Feb-03 | 8 | 1068 ± 301 | 35 ± 11 [a b] |
| Apr-03 | 10 | 895 ± 148 | 32 ± 11 [a] |
| Jun-03 | 8 | 1097 ± 140 | 51 ± 7 [b] |
| Sep-03 | 7 | 986 ± 190 | 36 ± 13 [a b] |
| Dec-03 | 7 | 940 ± 218 | 27 ± 11 [a] |
| Feb-04 | 7 | 1180 ± 160 | 21 ± 14 [a] |
| May-04 | 7 | 1103 ± 432 | 28 ± 12 [a] |
| Total | 87 | 1044 ± 241 | - - |

n = number of fish samples. Different letters indicate significant differences (p < 0 .05).

Another important fact is that the total mean value for the reactivated enzyme in samples from Berazategui did not show any significant difference with respect to the mean value of ALA-D activity found in fish collected in Corrientes (p > 0.05) (Table 7).

Initially, the use of specimens collected in Corrientes belonging to a different aquatic system could be considered as an unsuitable option for comparison purposes. In fact, although the animals are of the same species, their natural habitat, especially in terms of climate, hydrological regime, salinity, availability and composition of nutrients, and the inputs of contaminants are very different from those found in Berazategui. In addition only one sampling was performed in Corrientes, collecting a few organisms. Probably, the most favorable factor when selecting these animals was to verify the low levels of lead in their tissues. Therefore, the sampling in Corrientes might support the hypothesis that the ALA-D activity in *P. lineatus* from Berazategui may be actually inhibited by lead, but it was not enough as a conclusive proof. Instead, by using the reactivated ALA-D values the previous hypothesis could be confirmed. Furthermore, the reactivation was performed also in the blood samples of *P. lineatus* collected in Corrientes. The values of activity for reactivated enzyme did not differ from the values for the non-reactivated enzyme [1273 ± 328 and 1229 ± 318 U (ml RC)$^{-1}$, respectively], confirming that the enzyme was not inhibited (p > 0.05). Therefore, we can now reasonably assume that these values represent the maximal activity level of blood ALA-D for adult specimens of *P. lineatus*.

As it is defined (see 1.5.2) the reactivation index represents the percentage of enzyme inhibition. So we could demonstrated that in all the blood samples of fish collected in Berazategui the enzyme presented lower levels than normal, with percentages of inhibition varying between 20 and 51%. In addition, the reactivation index could be positively correlated with the lead content, as it is shown in Figure 11.

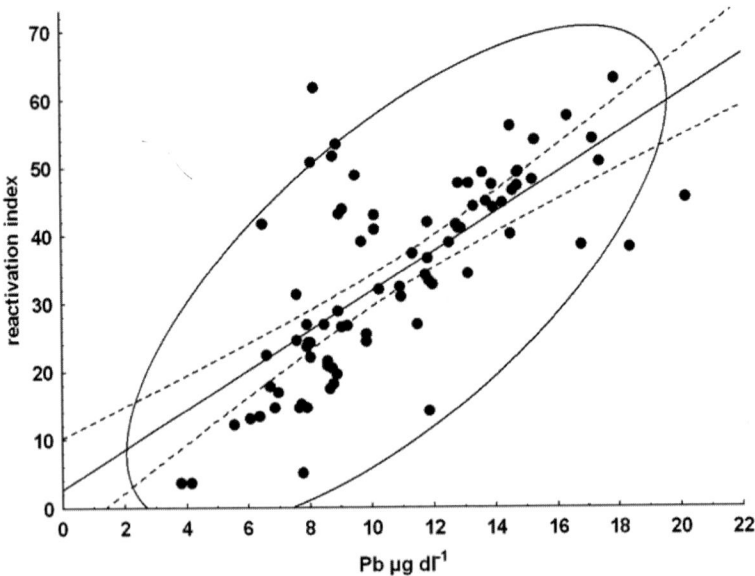

Figure 11. Reactivation index (RI) versus lead content in blood samples of *P. lineatus* collected in Berazategui (RI = 2.66 + 2.91 x Pb; r$^2$ = 0.49; r = 0.70, p = 0.00001).

This correlation was statistically much more significant than that observed for the original ALA-D values. Other authors have also proposed that the plot of the reactivation indices and the human blood lead levels provides the best alternative, even to estimate from this plot the blood lead levels with a reasonable accuracy (Polo et al., 1995).

## CONCLUSION

According to the results, the ALA-D activity in blood samples collected from Berazategui showed a great variability and significant differences among the sampling periods. However, the data did not show any seasonal pattern. On the contrary, the enzyme activity in liver samples showed in general a lower variability and no significant differences among the sampling periods. Only for blood samples, the enzyme activity could be correlated with the lead levels in that tissue.

Usually high variability in biomarker responses are observed when they are measured in animals collected from natural environments. Among all the factors that may affect biomarker responses, variables like sex, size, nutritional status and season did not seem to be the principal responsible for such variations in *P. lineatus* collected in Berazategui. More likely, the genetic heterogeneity together with other natural and anthropogenic factors could explain the inter-individual variations.

In comparison with the levels obtained for ALA-D activity in blood samples of fish collected in Berazategui, higher levels were found in fish from Corrientes. Instead, similar levels were found in liver samples. On the other hand, lower lead levels were recorded in blood and liver samples of the organisms from Corrientes. Therefore, only the blood enzyme would be actually inhibited in the samples from Berazategui.

The laboratory bioassays greatly contributed to the interpretation of the results. Firstly, because they showed that the blood enzyme could be further inhibited by an additional exposure of the fish to the metal, in parallel with an increase in the lead level. Secondly, as a consequence of this inhibition, they provided the biological material to investigate the reactivation of the enzyme activity. Only zinc appeared to be useful for reactivating the enzyme, suggesting that the blood ALA-D would be a Type 2 isoenzyme, according to the classification of Jaffe (2003), containing this metal at the active site.

Interestingly, the values of the reactivated enzyme in blood samples of Berazategui were similar to the values measured directly in samples of Corrientes, which contained lower levels of lead. Both values together with the reactivated enzyme of blood samples from Corrientes, allow the estimation of the maximal basal ALA-D activity for this fish species. In addition, the reactivation index could be calculated and positively correlated with the levels of lead in the samples.

Instead, for the liver enzyme the results were not conclusive. After the laboratory bioassays for 7 and 15 days, both control and exposed organisms presented similar activity values between them, which for unknown reasons were lower than the control value at $t = 0$. However, the lead content in liver samples of exposed *P. lineatus* increased as result of the laboratory exposure. On the other hand, the liver ALA-D activity enzyme in exposed fish could not be reactivated by using either zinc or DDT.

In summary, two relevant conclusions may be withdrawn from this work. From a biochemical point of view, both the field and laboratory results confirm that the liver and blood present different isoenzymes.

The liver ALA-D was not so easily inhibited by lead, at least by the levels of metal found in that tissue. In fact, similar ALA-D activity values were observed in liver samples of fish collected from Berazategui and Corrientes, even when the concentrations of lead were significant different. Different ALA-D enzymes may also account for the anomalous results where ALA-D may be inhibited for metals other than lead as reported by some authors (Gioda et al., 2007; Kalman et al, 2008). It would be advisable to establish some basic enzyme characteristics, especially in relation to optimal pH and requirements of sulfhydryl protectors groups before attempting measuring its activity.

For environmental and risk assessment purposes, the reactivation index is proposed as a better tool than the classical ALA-D activity values as biomarkers of lead contamination. Once the methodology for reactivating the enzyme has been optimized in a given biological tissue (in every cases that it is possible), a doubling of information is achieved. In this way, it is unnecessary to monitor a reference unpolluted area, which in some cases may be difficult or even impossible to find, since the data could be obtained simultaneously from the same biological sample. It implies only an additional enzyme determination.

## ACKNOWLEDGEMENTS

This manuscript was supported by the University of Buenos Aires, Argentina (grants X-147 and X-233) and the Agencia Nacional de Promoción Científica y Tecnológica, Argentina (PICTR 2002-00203). Thanks to Prof. Dr. Ana M.P. de D'Angelo and to Lic. M.N. Piol for their valuable encouragement, advice and friendship. Our gratitude to the local fishermen.

## REFERENCES

Abdulla, M.; Svensson, S.; Haeger-Aronsen, B.; Mathur, A. and Wallwnius, K. (1978). Effect of age and diet on delta-aminolevulinic acid dehydratase in red blood cells. *Enzyme, 23*, 170-175.

Aisemberg, J.; Nahabedian, D.E.; Wider, E.A. and Verrengia Guerrero, N.R. (2005). Comparative study on two freshwater invertebrates for monitoring environmental lead exposure. *Toxicology, 210*, 45-53.

Almeida, J.S.; Meletti, P. and Martinez, C.B.R. (2005). Acute effects of sediments taken from an urban stream on physiological and biochemical parameters of the neotropical fish *Prochilodus lineatus. Comparative Biochemistry and Physiology C, 140*, 356-363.

Alves Costa, J.R.M.; Mela, M.; da Silva de Assis, H.C.; Pelletier, E.; Ferreira Randi, M. A. and de Oliveira Ribeiro, C.A. (2007). Enzymatic inhibition and morphological changes in *Hoplias malabaricus* from dietary exposure to lead (II) or methylmercury. *Ecotoxicology and Environmental Safety, 67*, 82–88.

Amiard, J.C.; Amiard-Triquet, C.; Barka, S.; Pellerin, J. and Rainbow, P.S. (2006). Metallothioneins in aquatic invertebrates: Their role in metal detoxification and their use as biomarkers. *Aquatic Toxicology, 76,* 160-202.

Andersson, L. (2003). Detecting Effects of Exposure to Pesticides and Lead in Atlantic Rain Forest Rivers in Brazil, Using Biomarkers in Fish. PhD thesis. Göteborg University, Gothenburg, Sweden.

ATSDR: Agency for Toxic Substances and Disease Registry (2007). *Toxicological Profile for Lead.* Atlanta, Georgia.

Bagenal, T.B. and Tesch, F.W. (1978). Methods for assessment of fish production in fresh water. In: T.B. Bagenal (Ed.), *Age and Growth* (pp. 101-136). Oxford, UK: Blackwell Scientific Publications.

Batlle, A.M. del C. and Magnin, P.H. (1988). Porfirias y porfirinas. Aspectos clínicos y bioquímicos. *Acta Bioquímica Clínica Latinoamericana Superior 2,* 109 pp.

Bazán, J.M. and Arraga, E. (1993). In A. Boltovskoy, and H.L.López (Eds.), *Conferencias de Limnología* (pp. 71-82). La Plata, Argentina: Instituto de Limnología Dr. R. A. Ringuelet.

Bergdahl, I.A.; Sheveleva, M.; Schütz, A.; Artamonova, V.G. and Skerfving, S. (1998). Plasma and blood lead in humans: Capacity-limited binding to δ-aminolevulinic acid dehydratase and other lead-binding components. *Toxicological Sciences, 46,* 247–253.

Berlín, A. and Schaller, K.H. (1974). European standardized method for the determination of delta-aminolevulinic acid dehydrase activity in blood. *Zeitschrift Klinische Biochemie, 12,* 389-390.

Bonacci, S.; Corsi, I.; Chiea, R.; Regoli, F. and Focardi, S. (2003). Induction of EROD activity in European eel (*Anguilla anguilla*) experimentally exposed to benzo[*a*]pyrene and β-naphthoflavone. *Environment International, 29,* 467-473.

Camargo, M.M.P. and Martinez, C.B.R. (2006). Biochemical and physiological biomarkers in *Prochilodus lineatus* submitted to *in situ* test in an urban stream in southern Brazil. *Environmental Toxicology and Pharmacology, 21,* 61-69.

Carney Almroth, B.; Sturve, J. and Förlin, L. (2008). Oxidative damage in rainbow trout caged in a polluted river. *Marine Environmental Research, 66,* 90-91.

CARP-SINH-SOHMA (1990). Comisión Administradora del Río de la Plata - Servicio de Hidrografía Naval Argentina – Servicio de Oceanografía, Hidrografía y Meteorología de la Armada, Uruguay. *Estudio para la evaluación de la contaminación en el río de la Plata.* Buenos Aires, Argentina.

Cochón, A.C.; Della Penna, A.B.; Kristoff, G.; Piol, M.N.; San Martín de Viale, L.C. and Verrengia Guerrero, N.R. (2007). Differential effects of paraquat on oxidative stress parameters and polyamine levels in two freshwater invertebrates. *Ecotoxicology and Environmental Safety, 68,* 286-292.

Colombo, J.C.; Cappelletti, N.; Migoya, M.C. and Speranza, E. (2007a). Bioaccumulation of anthropogenic contaminants by detritivorous fish in the Rio de la Plata estuary: 1- Aliphatic hydrocarbons. *Chemosphere, 68,* 2128-2135.

Colombo, J.C.; Cappelletti, N.; Migoya, M.C. and Speranza, E. (2007b). Bioaccumulation of anthropogenic contaminants by detritivorous fish in the Rio de la Plata estuary: 2- Polychlorinated biphenyls. *Chemosphere, 69,* 1253-1260.

Conner, E.A. and Fowler, B.A. (1994). Biochemical and inmunological properties of hepatic delta-aminolevulinic acid dehidratase in channnel catfish (*Ictalurus-punctatus*). *Aquatic Toxicology, 28,* 37-52.

Dagnino Pastore, L. (1973). *Mi Galaxia* (volúmen 2). Buenos Aires, Argentina: Editorial Nobis.

Dalle-Donne, I.; Rossi, R.; Giustarini, D.; Milzani, A. and Colombo, R. (2003). Protein carbonyl groups as biomarkers of oxidative stress. *Clinica Chimica Acta, 329,* 23-38.

Depledge, M.H.; Aagaard, A. and Györkös, P. (1995). Assessment of trace metal toxicity using molecular, physiological and behavioral biomarkers. *Marine Pollution Bulletin, 31,* 19-27.

Devi, M. and Fingerman, M. (1995). Inhibition of acethylcholinesterase activity in the central nervous system of the red swamp crayfish, *Procambarus clarkii,* by mercury, cadmium, and lead. *Bulletin of Environmental Contamination and Toxicology, 55,* 746-750.

Dieter, M.P.; Perry, M.C. and Mulhern, B.M. (1976). Lead and PCB's in Canvasback Ducks: Relationship between enzyme levels and residues in blood. *Archives of Environmental Contamination and Toxicology, 5,* 1-13.

Dissanayake, A.; Galloway, T.S. and Jones, M.B. (2008). Nutritional status of *Carcinus maenas* (Crustaces: Decapoda) influences susceptibility to contaminant exposure. *Aquatic Toxicology, 89,* 40-46.

Dorigan, J.V. and Harrison, F.L. (1987). *Physiological Responses of Marine Organisms to Environmental Stress.* Department of Energy, United States of America, DOE/ER-0317, 501 pp.

Doyle, D. and Schimke, R.T. (1969). The genetic and developmental regulation and hepatic δ-aminolevulinate dehydratase in mice. *Journal Biology Chemical, 244,* 5449-5459.

Dresel, E.I.B. and Falk, J.E. (1953). Conversion of delta-aminolevulinate to porphobilinogen in a tissue system. *Nature, 172,* 1185-1187.

Eason, C. and O'Halloran, K. (2002). Biomarkers in toxicology versus ecological risk assessment. *Toxicology, 181-182,* 517-521.

Engel, D.W. and Vaughan, D.S. (1996). Biomarkers, natural variability and risk assessment: Can they co-exist? *Human and Ecological Risk Assessment, 2,* 257-262.

Erskine, P.T.; Senior, N.; Awan, S.; Lambert, R.; Lewis, G.; Tickle, I.J.; Sarwar, M.; Spencer, P.; Thomas, P.; Warren, M.J.; Shoolingin-Jordan, P.M.; Wood, S.P. and Cooper, J.B. (1997). X-ray structure of 5-aminolaevulinate dehydratase, a hybrid aldolase. *Nature Structural Biology, 4,* 1025-1031.

Falfushynska, H.I. and Stolyar, O.B. (*in press*). Responses of biochemical markers in carp *Cyprinus carpio* from two field sites in Western Ukraine. *Ecotoxicology and Environmental Safety.*

Farina, M.; Brandao, R.; Lara, F.S.; Soares, F.A.A.; Souza, D.O. and Rocha, J.B.T. (2003). Mechanisms of the inhibitory effects of selenium and mercury on the activity of δ-aminolevulinate dehydratase from mouse liver, kidney and brain. *Toxicology Letters, 139,* 55-66.

Finelli, V.N., Murthy, L., Peirano, W.B., and Petering, H.G. (1974). δ-aminolevulinate dehydratase, a Zinc dependent enzyme. *Biochemical and Biophysical Research communications, 60,* 1418-1424.

Finley, M.T.; Dieter, M.P. and Locke, L.N. (1976). Delta-aminolevulinic acid dehydratase: inhibition in ducks dosed with lead shot. *Environmental Research, 12,* 243-249.

Ford J. (1989). The effects of chemical stress on aquatic species composition and community structure. In S.A.Levin, M.A. Harwell, J.R. Kelly and K.D. Kimball, (Eds.),

*Ecotoxicology: Problems and Approaches* (pp. 99-144). New York, USA: Springer-Verlag.

Förlin, L.; Goksøyr, A. and Mette-Husøy, A. (1994). Cytochrome P450 monooxygenase as indicator of PCB/dioxin like compounds in fish. In K.J.M. Kramer (Ed.), *Biomonitoring of Coastal Waters and Estuaries* (pp. 135-150). Boca Raton, FL: CRC Press, Inc.

García-Delgado; M., Rodríguez-Cruz, M.S.; Lorenzo, L.F.; Arienzo, M. and Sánchez-Martín, M.J. (2007). Seasonal and time variability of heavy metal content and its chemical forms in sewage sludges from different wastewater treatment plants. *Science of the Total Environment, 382,* 82-92.

Geffard, A.; Quéau, H.; Dedourge, O.; Biagianti-Risboug, S. and Geffard, O. (2007). Influence of biotic and abiotic factors on metallothionein level in *Gammarus pulex. Comparative Biochemical and Physiology C, 145,* 632-640.

GESAMP (1991). Joint Group of Experts on the Scientific Aspects of Marine Pollution Statement of the Intergovernmental Meeting of Experts on Land Based Sources of Marine Pollution. World Health Organization. Geneva.

Gill, T.S.; Tewari, H. and Pande, J. (1990). Use of the fish enzyme system in monitoring water quality: Effects of mercury on tissue enzymes. *Comparative Biochemical Physiology C, 97,* 287-292.

Gill, T.S.; Tewari, H. and Pande, J. (1991). *In vivo* and *in vitro* effects of cadmium on selected enzymes in different organs of the fish *Barbus conchonius* ham (Rosy Barb). *Comparative Biochemical and Physiology C, 100,* 501-505.

Gioda, C.R.; Lissner, L.A.; Pretto, A.; da Rocha, J.T.B.; Schetinger, M.R.C.; Neto, J.R.; Morsch, V.M. and Loro, V.L. (2007). Exposure to sublethal concentrations of Zn(II) and Cu(II) changes biochemical parameters in *Leporinus obtusidens. Chemosphere, 69,* 170-175.

Godwin, H.A. (2001). The biological chemistry of lead. *Current Opinion in Chemical Biology, 5,* 223-227.

Goede, R.W. and Burton, B.A. (1990). Organismic Indices and Autopsy-Based Assessment as Indicators of Health and Condition of Fish. In S. Marshal Adams (Ed.), *Biological Indicator of Stress in Fish* (First edition, pp. 93-108). Maryland, USA: American Fisheries Society.

González, O.; Fernández, J. and Martín, M. (1987). Inhibition of trout (*Salmo gairdneri* R.) PBG-synthase by some metals ions ($Mg^{2+}$, $Pb^{2+}$, $Zn^{2+}$). *Comparative Biochemical and Physiology C, 86,* 163-167.

Granick, S. (1954). Enzymatic conversion of delta-aminolevulinic acid to porphobilinogen. *Science, 120,* 1105-1107.

Granick, S.; Sassa, S.; Granick, J.L.; Levere, R.D. and Kappas, A. (1973). Studies in lead poisoning, II Correlation between the ratio of activated to inactivated δ-aminolevulinic acid dehydratase of whole blood and the blood lead level. *Biochemical Medicine, 8,* 149-159.

Grue, C.E.; O'Shea, T.J. and Hoffamn, D.J. (1984). Lead concentration and reproduction in highway-nesting Barn Swallows. *Condor, 86,* 383-389.

Guilhermino, L.; Lacerda, M N.; Nogueira, A.J.A. and Soares, A.M.V.M. (2000). *In vitro* and *in vivo* inhibition of *Daphnia magna* acetylcholinesterase by surfactant agents: possible implications for contamination biomonitoring. *Science of the Total Environment, 247,* 137-141.

Gurer-Orhan, H.; Sabir, H.U. and Özgünes, H. (2004). Correlation between clinical indicators of lead poisoning and oxidative stress parameters in control and lead-exposed workers. *Toxicology, 195,* 147-154.

Hagger, J.A.; Jones, M.B.; Lowe, D.; Leonard, P.D.R.; Owen, R. and Galloway, T.S. (2008). Application of biomarkers for improving risk assessments of chemicals under the Water Framework Directive: A case study. *Marine Pollution Bulletin, 56,* 1111-1118.

Hahn, M.E. (1998). The aryl hydrocarbon receptor: A comparative perspective. *Comparative Biochemistry and Pyisiology C, 121,* 23-53.

Handy, R.D. and Depledge, M.H. (1999). Physiological responses: Their measurement and use as environmental biomarkers in ecotoxicology. *Ecotoxicology, 8,* 329-349.

He, F. (1999). Biological monitoring of exposure to pesticides: current issues. *Toxicology Letters, 108,* 277-283.

Hemelraad, J.; Holwerda, D.A.; Herwing, H.J. and Zandee, D.I. (1990). Effects of cadmium in freshwater clams. III. Interaction with energy metabolism in *Anodonta cygnea. Archives of Environmental Contamination and Toxicology, 19,* 699-703.

Hernberg, S.; Nikkanen, J.; Millen, G. and Lilius H. (1970). Δ-aminolevulinic acid dehydrase as a measure of lead exposure. *Archives of Environmental Health, 21,* 140-145.

Hinck, J.E.; Schmitt, C.J.; Blazer, V.S.; Denslow, N.D.; Bartish, T.M.; Anderson, P.J.; Coyle, J.J.; Dethloff, G.M. and Tillitt, D.E. (2006). Environmental contaminants and biomarker responses in fish from the Columbia River and its tributaries: Spatial and temporal trends. *Science of the Total Environment, 366,* 549-578.

Hodson, P.V.; Blunt, B.R.; Spry, D.J. and Austen, K. (1977). Evaluation of erythrocyte δ-aminolevulinic acid dehydratase activity as a short-term indicator in fish of a harmful exposure to lead. *Journal of the Fisheries Research Board of Canada, 34,* 501-508.

Hodson, P.V.; Blunt, B.R. and Spry, D.J. (1978). Chronic toxicity of water-borne and dietary lead to rainbow trout (*Salmo gardneri*) in Lake Ontario water. *Water Research, 12,* 869-878.

Hwang, E. and Kim, G. (2007). Biomarkers for oxidative stress status of DNA, lipids, and proteins *in vitro* and *in vivo* cancer research. *Toxicology, 229,* 1-10.

Hyne, R.V. and Maher, W.A. (2003). Invertebrate biomarkers: Links to toxicosis that predict population decline. *Ecotoxicology and Environmental Safety, 54,* 366-374.

Jaffe, E.K. (2003). An unusual phylogenetic variation in the metal ion binding sites of porphobilinogen synthase. *Chemistry and Biology, 10,* 25–34.

Jaffe, E.K. (2004). The porphobilinogen synthase catalyzed reaction mechanism. *Bioorganic Chemistry, 32,* 316–325.

Johanson-Sjöbeck, M. and Larsson, A. (1979). Effects of inorganic lead on delta-aminolevulinic acid dehydratase activity and hematological variables in the rainbow trout, *Salmo gairdnerii. Archives of Environmental Contamination and Toxicology, 8,* 419-431.

Johnson, F.M. (1998). The genetic effects of environmental lead. *Mutation Research, 410,* 123-140.

Kalman, J.; Riba, I.; Blasco, J. and DelValls, T. (2008). Is δ-aminolevulinic acid dehydratase activity in bivalves from south-wet Iberian Peninsula a good biomarker of lead exposure? *Marine Environmental Research, 66,* 38-40.

Kammenga, J.E.; Dallinger, R.; Donker, M.H.; Köhler, H.; Simonsen, V.; Triebskorn, R. and Weeks, J.M. (2000). Biomarkers in terrestrial invertebrates for ecotoxicological soil risk assessment. *Reviews of Environmental Contamination and Toxicology, 164,* 93-147.

Kendall, R.J.; Anderson, T.A.; Baker, R. J.; Bens, C.M.; Carr, J.A.; Chiodo, L.A.; Cobb III, G.P.; Dickerson, R.L.; Dixon, K.R.; Frame, L.T.; Hooper, M.J.; Martin, C.F.; McMurry, S.T.; Patino, R.; Smith, E.E. and Theodorakis, C.W. (2001). Ecotoxicology. In C.D. Klaassen (Ed.), *Casarett and Doull's Toxicology: The Basic Science of Poisons* (6th edition, pp. 1013-1045). Washington, USA: Mc Graw-Hill.

Ketchum, B.H. (1983). Ecosystem of the World 26: Estuaries and Enclosed Seas. Elsevier Scientific Publishing Co., 481 p.

Kinder, A.; Sierts-Herrmann, A.; Biselli, S.; Henzel, N.; Hühnerfuss, H.; Kammann, U.; Reineke, N.; Theobald, N. and Steinhart, H. (2007). Expression of heat shock protein 70 in a cell line (EPC) exposed to sediment extracts from the North Sea and the Baltic Sea. *Marine Environmental Research, 63,* 506-515.

Koeman, J.H. (1991). From comparative physiology to toxicological risk assessment. *Comparative Biochemistry and Physiology C, 100,* 7-10.

Kristoff, G.; Verrengia Guerrero, N.R. and Cochón, A.C. (2008). Oxidative stress parameters in *Biomphalaria glabrata* and *Lumbriculus variegatus* exposed to azinphos-methyl. *Chemosphere, 72,* 1333-1339.

Kristoff, G.; Verrengia Guerrero, N.R.; Pechen de D'Angelo, A.M. and Cochón, A.C. (2006). Inhibition of cholinesterase activity by azinphos-methyl in two freshwater invertebrates: *Biomphalaria glabrata* and *Lumbriculus variegatus. Toxicology, 222,* 185-194.

Kutlu, M. and Sümer, S. (2008). Some biochemical properties of δ-aminolevulinic acid dehydratase in *Gammarus pulex. Food and Chemical Toxicology, 46,* 115–118.

Labrot, F.; Ribera, D., Saint Denis, M. and Narbonne, J.F. (1996). *In vitro* and *in vivo* studies of potential biomarkers of lead and uranium contamination: lipid peroxidation, acetylcholinesterase, catalase and glutathione peroxidase activities in three non-mammalian species. *Biomarkers, 1,* 21-28.

Lam, P.K.S. and Gray, J.S. (2003). The use of biomarkers in environmental monitoring programmes. *Marine Pollution Bulletin, 46,* 182-186.

Láng, G.; Kufcsák, O.; Szegletes, T. and Nemcsók, J. (1997). Quantitative distributions of different cholinesterases and inhibition of acetylcholinesterase by metidathion and paraquat in alimentary canal of common carp. *General Pharmacology, 29,* 55-59.

Lehtonen, K.K.; Schiedek, D.; Köhler, A.; Lang, T.; Vuorinen, P.J.; Förlin, L.; Baršienė, J.; Pempkowiak, J. and Gercken, J. (2006). The BEEP project in the Baltic Sea: Overview of results and outline for a regional biological effects monitoring strategy. *Marine Pollution Bulletin, 53,* 523-537.

Lombardi, P.E.; Peri, S. and Verrengia Guerrero, N.R. (*in press*). Trace metal levels in *Prochilodus lineatus* collected from the La Plata River, Argentina. *Environmental Monitoring and Assessment.*

Lowe, D.M., Moore, M., and Evans, B. (1992). Contaminant impact on interactions of molecular probes with lysosomes in living hepatocytes from dab *Limanda limanda. Marine Ecological Progress Series, 91,* 135-140.

Lowry, O.H.; Rosebrough, N.L.; Farr, A.L. and Randall, R.J. (1951). Protein measurement with the folin-phenol reagent. *Journal of Biological Chemistry, 193,* 265-275.

Malins, D.C. and Ostrander G.K. (1991). Perspectives in aquatic toxicology. *Annual Review of Pharmacology and Toxicology, 31*, 371-399.

Mansuy, D. (1998). The great diversity of reactions catalized by cytochromes P450. *Comparative Biochemistry and Physiology C, 121*, 5-14.

Marcovecchio, J.E.; Moreno, V.J. and Perez, A. (1991). Metal accumulation in tissues of sharks from the Bahía Blanca Estuary, Argentina. *Marine Environmental Research, 31*, 263-274.

Marcovecchio, J.E. (2004). The use of *Micropogonias furnieri* and *Mugil liza* as bioindicators of heavy metals pollution in La Plata river estuary, Argentina. *Science of the Total Environment, 323*, 219-226.

Martinez-Tabche, L.; Grajeday Ortega, M.A.; Ramirez Mora, B.; German Faz, C.; Lopez Lopez, E. and Galar Martinez, M. (2001). Hemoglobin concentration and acetylcholinesterase activity of oligochaetes in relation to lead concentration in spiked sediments from Ignacio Ramirez Reservoir. *Ecotoxicology and Environmental Safety, 49*, 76-83.

Matozzo, V.; Gagne, F.; Marin, M.G.; Ricciardi, F. and Blaise, C. (2007). Vitellogenin as a biomarker of exposure to estrogenic compounds in aquatic invertebrates: A review. *Environment International. 34*, 531-545.

Matthiessen, P. and Law, R.J. (2002). Contaminants and their effects on estuarine and coastal organism in the United Kingdom in the late twentieth century. *Environmental Pollution, 120*, 739-757.

Mayer, F.L.; Versteeg, D.J.; McKee, M.J.; Folmar, L.C.; Graney, R.L.; McCume, D.C. and Rattner, B.A. (1992). Physiological and Nonspecific Biomarkers. In R.J. Huggett, R.A. Kimerle, P.M. Mehrie Jr., H.L. Bergman (Eds.), *Biomarkers: Biochemical, Physiological, and Histological Markers of Anthropogenic Stress* (pp. 5-85). London, UK: Lewis Publishers.

Meyer, J. and Di Giulio, R. (2002). Patterns of heriatability of decresead EROD activity and resistance to PCB 126-induced teratogenesis in laboratory-reared offspring of killfish (*Fundulus heteroclitus*) from a creosote-contaminated site in the Elizabeth River, VA, USA. *Marine Environmental Research, 54*, 621-626.

Miller, J.R.; Anderson, J.B; Lechler, P.J.; Kondrad, S.L; Galbreath, P.F. and Salter, E.B. (2005). Influence of temporal variations in water chemistry on the Pb isotopic composition of rainbow trout (*Oncorhynchus mykiss*). *Science of the Total Environment, 350*, 204-224.

Moffatt, P. and Denizeau, F. (1997). Metallothionein in physiological and physiopathological processes. *Drug Metabolism Reviews, 29*, 261-307.

Monserrat, J.M.; Martínez, P.E.; Geracitano, L.A.; Lund Amado, L.; Martínez Gaspar Martins, C.; Lopes Leães Pinho, G.; Soares Chaves, I.; Ferreira-Cravo, M.; Ventura-Lima, J. and Bianchini, A. (2006). Pollution biomarkers in estuarine animals: Critical review and new perspectives. *Comparative Biochemical and Physiology C, 146*, 221-234.

Moore, M.N.; Depledge, M.H.; Readman, J. W. and Paul, L.D.R. (2004). An integrated biomarker-based strategy for ecotoxicological evaluation of risk in environmental management. *Mutation Research, 552*, 247-268.

Nakagawa, H.; Nakagawa, K. and Sato, T. (1995a). Evaluation of erythrocyte 5-aminolevulinic acid dehydratase activity in the blood of carp *Cyprinus carpio*. *Fisheries Science, 61*, 91-95.

Nakagawa, H.; Sato, T. and Kubo, H. (1995 b). Evaluation of chronic toxicity of water lead for carp *Cyprinus carpio* using its blood 5-aminolevulinic acid dehydratase. *Fisheries Science, 61,* 956-959.

Nakagawa, H.; Toshihiro, T.; Sato, T. and Watanabe, M. (1997). Evaluation of erythrocyte 5-aminolevulinic acid dehydratase activity in the blood of crucian carp *Carassius auratus langsdorfii*, as an indicator in fish of water lead pollution. *Journal of the Faculty of Agronomy, Kyushu University, 41,* 205-213.

Nakagawa, K.; Nakagawa, H. and Aso, Y. (1995). The type of inhibition of erythrocyte 5-aminolevulinic acid dehydrase activity in the blood of carp *Cyprinus carpio* caused by lead and cadmium. *Science Bulletin of the Faculty of Agronomy, Kyushu University, 50,* 51-57.

Nemcsók, J.; Németh, A.; Buzás, Z. S. and Boross, L. (1984). Effect of copper, zinc and paraquat on acetylcholinesterase activity in carp (*Cyprinus carpio* L.). *Aquatic Toxicology, 5,* 23-31.

NRC: Committee on Biological Markers of the National Research Council. (1987). Biological markers in environmental health research. *Environmental Health Perspectives, 74,* 3-9.

Nriagu, J.O.; Lawson, G.; Wong, H.K.T. and Azcue, J.M. (1993). A Protocol for minimizing contamination in the analysis of trace metals in Great Lakes Waters. *Journal of Great Lakes Research, 19,* 175-182.

Ogunseitan, O.A.;Yang, S. and Scheinbach, E. (1999).The delta- aminolevulinate dehydratase activity of Vibrio alginolyticus is resistant to lead (Pb). *Biological Bulletin, 197,* 283-284.

Ogunseitan, O.A.; Yang, S. and Ericson, J. (2000). Microbial δ-aminolevulinate dehydratase as a biosensor of lead bioavailability in contaminated environments. *Soil Biology and Biochemistry, 32,* 1899-1906.

Oldani, N. (1990). Variaciones de la abundancia de peces del valle del río Paraná. *Revista de Hydrobiología Tropical, 23,* 67-76.

Pampanin, D.M.; Andersen, O.K.; Viarengo, A. and Garrigues, P. (2006). Background for the BEEP Stavanger workshops: Biological effects on marine organism in two common, large, laboratory experiments and in a field study. Comparison of the value (sensitivity, specificity, etc.) of core and new biomarkers. *Aquatic Toxicology, 78S,* S1-S4.

Parma de Croux, M.J. (1994). Some haematological parameters in *Prochilodus lineatus* (pisces, Curimatidae). *Revista de Hydrobiología Tropical, 27,* 113-119.

Payne, J.F.; Mathieu, A.; Melvin, W. and Francey, L.L. (1996). Acetylcholinesterase, an old biomarker with a new future? Field trials in association with two urban rivers and a paper mill in Newfoundland. *Marine Pollution Bulletin, 32,* 225-231.

Peakall, D. (1994). *Animal Biomarkers as Pollution Indicators*. M. Depledge and B. Sanders (Eds.). London, UK: Chapman and Hall, Ecotoxicological Series 1.

Polo, C.F.; Afonso, S.G.; Navone, N.M.; Rossetti, M.V. and Batlle, A.M. del C. (1995). Zinc aminolevulinic acid dehydratase reactivation index as a tool or diagnosis of lead exposure. *Ecotoxicology and Environmental Safety, 32,* 267-272.

Pyle, G.C.; Rajotte, J.W. and Couture, P. (2005). Effects of industrial metals on wild fish populations along a metal contamination gradient. *Ecotoxicology and Environmental Safety, 61,* 287-312.

Rand, G.M. (1995) (Ed.). *Fundamentals of Aquatic Toxicology*. (2nd edition), Washington, USA: Taylor and Francis.

Regoli, F. and Winston, G.W. (1998). Applications of a new method for measuring the total oxyradical scavenging capacity in marine invertebrates. *Marine Environmental Research, 46,* 493-442.

Regoli, F. and Winston, G.W. (1999). Quantification of total oxidant scavenging capacity of antioxidants for peroxynitrite, peroxyl radicals, and hydroxyl radicals. *Toxicology and Applied Pharmacology, 156,* 96-105.

Rocha, J.B.T.; Tuerlinckx, S. M.; Schetinger, M. R.C. and Folmer, V. (2004). Effect of Group 13 metals on porphobilinogen synthase in vitro. *Toxicology and Applied Pharmacology, 200,* 169– 176.

Rodrigues, A.L.; Bellinaso, M.L. and Dick, T. (1989). Effect of some metal ions on blood and liver delta-aminolevulinate dehydratase of *Pimelodus maculatus* (Pisces, Pimelodidae). *Comparative Biochemical and Physiology B, 94,* 65-69.

Rodrigues, A.L.S.; Rocha, J.B.T.; Pereira, M.E. and Souza, D.O. (1996). δ-aminolevulinic acid dehydratase activity in weanling and adult rats exposed to lead acetate. *Bulletin of Environmental Contamination and Toxicology, 57,* 47-53.

Rodriguez-Castellanos, L. and Sanchez-Hernandez, J.C. (2007). Earthworm biomarkers of pesticide contamination: Current status and perspectives. *Journal of Pesticide Science, 32,* 360-371.

Roos, P.H.; Tschirbs, S.; Pfeifer, F.; Welge, P.; Hack, A.; Wilhelm, M. and Bolt, H.M. (2004). Risk potentials for humans of original and remediated PAH-contaminated soils: application of biomarkers of effect. *Toxicology, 205,* 181-184.

Sakai, T.; Yanagihara, S. and Ushio, K. (1980). Restoration of lead-inhibited 5-aminolevulinate dehydratase activity in whole blood by heat, zinc ion, and (or) dithiothreitol. *Clinical Chemistry, 26,* 652-628.

Sanchez, W.; Piccini, B.; Ditche, J.M. and Porcher, J.M. (2008). Assessment of seasonal variability of biomarkers in three-spined stickleback (*Gasterosteus aculeatus* L.) from a low contaminated stream: Implication for environmental biomonitoring. *Environment International,34,* 791-798.

Santos, R.S.; Martins, M.L.; Marengoni, N.G.; Francisco, C.J.; Piazza, R.S.; Takahashi, H.K. and Onaka, E.M. (2005). *Neoechinorhynchus curemai* (Acanthocephala: Neoechinorhynchidae) in *Prochilodus lineatus* (Osteichthyes: Prochilodontidae) from the Paraná River; Brazil. *Veterinaria Parasitològica, 134,* 111-115.

Sassa, S. (1982). Delta-aminolevulinic acid dehydratase assay. *Enzyme, 28,* 133–145.

Sassa, S. and Bernstein, S.E. (1977). Levels of δ-aminolevulinate dehydratase uroporphyrinogen I synthase and protoporphyrin IX in erythrocytes from anemic mutant mice. *Proceedings of the National Academy of Sciences of the United States of America, 74,* 1181-1184.

Sassa, S.; Granick, S.; Bickers, D.R.; Levere, R.D. and Kappas, A. (1973). Studies on the inheritance of human erythrocyte δ-aminolevulinate dehydratase and uroporphyrinogen synthase. *Enzyme, 16,* 326-333.

Schäfer, S.G., Dawes, R.L.F., Elsenhans, B., Forth, W. and Schümann, K. (1999). Metals. In H. Marquardt, S.G. Schäfer, R. McClellan and F. Welsch (Eds.), *Toxicology* (755-804). Elsevier Inc.

Scheuhammer, A.M. (1987). Erythrocyte δ-aminolevulinic acid dehydratase in birds. I. The effects of lead and other metals *in vitro. Toxicology, 45,* 155-163.

Schmidt, G.H. and Ibrahim, N.H.M. (1994). Heavy metal content (Hg $^{2+}$, Cd $^{2+}$, Pb $^{2+}$) in various body parts: its impact on cholinesterase activity and binding glycoproteins in the grasshopper *Aiolopus thalassimus* adults. *Ecotoxicology and Environmental Safety, 29,* 148-164.

Schmitt, C.J.; Dwyer, F.G. and Finger, S.E. (1984). Biovailability of Pb and Zn from mine tailings as indicated by erythrocyte δ-aminolevulinic acid dehydratase (ALA-D) activity in suckers (Pisces: Catostomidae). *Canadian Journal of Fisheries and Aquatic Science, 41,* 1030-1040.

Schmitt, C.J.; Wildhaber, M.L.; Hunn, J.B.; Nash, T.; Tieger, M.N. and Steadman, B.L. (1993). Biomonitoring of lead-contaminated Missouri streams with an assay for erythrocyte δ-aminolevulinic acid dehydratase activity in fish blood. *Archives of Environmental Contamination and Toxicology, 25,* 464-475.

Schmitt, C.J.; Whyte, J.J.; Roberts, A.P.; Annis, M.L.; May, T.W. and Tillitt D.E. (2007). Biomarkers of metals exposure in fish from lead-zinc mining areas of Southeastern Missouri, USA. *Ecotoxicology and Environmental Safety, 67,* 31-47.

Seto, Y. and Shinohara, T. (1987). Inhibitory effects of paraquat and its related compounds on the acetylcholinesterase activities of human erythrocytes and electric eel *Electrophorus electricus. Agriculture, Biology and Chemistry, 51,* 2131-2138.

Seto, Y. and Shinohara, T. (1988). Structure-activity relationship of reversible cholinesterase inhibition including paraquat. *Archives of Toxicology, 62,* 37-40.

Shemin, D. and Russel, C.S. (1953). δ-aminolevulinic acid, its role on the biosynthesis of porphyrins and purines. *Journal American Chemical Society, 75,* 4873-4874.

Shinohara, T. and Seto, Y. (1986). *In vitro* inhibition of acetylcholinesterase by paraquat. *Agriculture, Biology and Chemistry, 50,* 255-256.

Singh, R.P. and Agrawal, M. (2008). Potential benefits and risks of land application of sewage sludge. *Waste Management, 28,* 347-358.

Sloff, W.; van Kreijl, C.F. and Baars, A.J. (1983). Relative liver weights and xenobiotic-metabolizing enzymes of fish from polluted surface waters in the Netherlands. *Aquatic Toxicology, 4,* 1-14.

Sokal, R.R. and Rohlf, F.J. (1997). *Biometry: The Principles and Practice of Statistics in Biological Research.* San Francisco, CA: Wh. Freeman and Co.

Stegeman, J.J; Brouwer, M.; Di Giulio, R.T.; Förlin, L.; Fowler, B.A.; Sanders, B.M. and Van Veld, P.A. (1992). Molecular responses to environmental contamination: Enzyme and protein systems as indicators of chemical exposure and effect. In R.J. Huggett, R.A. Kimerle, P.M. Mehrie Jr. and H.L. Bergman (Eds.), *Biomarkers: Biochemical, Physiological, and Histological Markers of Anthropogenic Stress* (pp. 235-335). London, UK: Lewis Publishers.

Stegeman, J.J. and Livingstone, D.R. (1998). Forms and functions of cytochrome P450. *Comparative Biochemistry and Physiology C, 121,* 1-3.

Sumpter, J.P. (1995). Feminized responses in fish to environmental estrogens. *Toxicology Letters, 82-83,* 737-742.

Sumpter, J.P. (1998). Xenoendocrine disrupters-environmental impacts. *Toxicology Letters, 102-103,* 337-342.

Sverlig, S.A.; Espinach Ros, A. and Orti, G. (1993). Sinopsis de los datos biológicos y pesqueros del sábalo *Prochilodus lineatus* (Valenciennes, 1847). FAO, Sinopsis sobre la

Pesca (FAO), no. 154. Organización de las Naciones Unidas para la Agricultura y la Alimentación. Rome, Italy, 64 pp.

Szabó, A.; Nemcsók, J.; Asztalos, B.; Rakonczay, Z.; Kása, P. and Huu Hieu, L. (1992). The effect of pesticides on carp (*Cyprinus carpio* L.). Acetylcholinesterase and its biochemical characterization. *Ecotoxicology and Environmental Safety, 23*, 39-45.

Tanaka, T.; Kakizono, T.; Nishikawa, S.; Watanabe, K.; Sasaki, K.; Nishino, N. and Nagai, S. (1995). Screening of 5-aminolevulinic acid dehydratase inhibitors. *Seibutsu-Kogaku Kaishi, 73*, 13-19.

Timbrell, J.A. (2000). *Principles of Biochemical Toxicology*. (3rd edition). London, UK: Taylor and Francis.

Tolar, J.F.; Mehollin, A.R.; Watson, R.D. and Angus, R.A. (2001). Mosquito fish (*Gambusia affinis*) vitellogenin: identification, purification, and immunoassay. *Comparative Biochemical and Physiology C, 128*, 237-245.

Valavidinis, A.; Vlahogianni, T.; Dassenakis, M. and Scoullos, M. (2006). Molecular biomarkers of oxidative stress in aquatic organisms in relation to toxic environmental pollutants. *Ecotoxicology and Environmental Safety, 64*, 178-189.

van der Oost, R.; Beyer, J. and Vermeulen, N.P.E. (2003). Fish bioaccumulation and biomarkers in environmental risk assessment: a review. *Environmental Toxicology and Pharmacology, 13*, 57-149.

Vanparys, C.; Dauwe, T.; Campenhout, K.V.; Bervoets, L.; De Coen, W.; Blust, R. and Eens, M. (2008). Metallothioneins (MTs) and δ-aminolevulinic acid dehydratase (ALAd) as biomarkers of metal pollution in great tits (*Parus major)* along a pollution gradient. *Science of the Total Environment, 401*, 184-193.

Vasseur, P. and Cossu-Leguille, C. (2003). Biomarkers and community indices as complementary tools for environmental safety. *Environment International, 28*, 711-717.

Verrengia Guerrero, N.R. (1995). Contaminantes Metálicos en el Río de La Plata: Monitoreo del Sistema Acuático y Estudio de Algunos Efectos Tóxicos en Moluscos Bivalvos por medio de Bioensayos. PhD thesis. Facultad de Ciencias Exactas y Naturales, Universidad de Buenos Aires, Buenos Aires, Argentina.

Verrengia Guerrero, N.R. and Kesten, E.M. (1993). Levels of heavy metals in biota from the La Plata river. *Environmental Toxicology and Water Quality, 8*, 335-344.

Verrengia Guerrero, N.R.; Mozzarelli, M.N.; Giancarlo, H.; Nahabedian, D. and Wider, E.A. (1997). *Biomphalaria glabrata*: relevance of albino organisms as a useful tool for environmental lead monitoring. *Bulletin of Environmental Contamination and Toxicology, 59*, 822-827.

Viarengo, A. and Canesi, L. (1991). Mussels as biological indicators of pollution. *Aquaculture, 94*, 225-243.

Villar, C.; Stripeikis, J.; Colautti, D.; D'Huicque, L.; Tudino, M. and Bonetto, C. (2001). Metals contents in two fishes of different feeding behavior in the Lower Paraná River and Río de la Plata Estuary. *Hydrobiología, 457*, 225-233.

Walker, C.H. (1998). The use of biomarkers to measure the interactive effects of chemicals. *Ecotoxicology and Environmental Safety, 40*, 65-70.

Walker, C.H.; Hopkin, S.P.; Sibly, R.M. and Peakall, D.B. (2001). *Principles of Ecotoxicology* (2nd edition). London, UK: Taylor and Francis.

Wang, G. and Fowler, B.A. (2008). Roles of biomarkers in evaluating interactions among mixtures of lead, cadmium and arsenic. *Toxicology and Applied Pharmacology, 233*, 92-99.

Wedderburn, J.; Cheung, V.; Bambre, S.; Bloxhan, M. and Depledge, M.H. (1998). Biomarkers of biochemical and cellular stress in *Carcinus maenas*: an *in situ* field study. *Marine Environmental Research, 46,* 321-324.

Weissberg, J.B. and Woytek, P.E. (1974). Liver and red cell porphobilinogen synthasa in the adult and fetal guinea pig. *Biochemical Biophysical Acta, 364,* 304-319.

Wetmur, J.G. (1994). Influence of the common human δ-aminolevulinate dehydratase polymorphism on lead body burden. *Environmental Health Perspective, 102* (Suppl. 3), 215-219.

Wilson, E.L.; Burger, P. E. and Dowdle, E.B. (1972). Beef-liver 5-aminolevulinic acid dehydratase: purification and properties. *European Journal of Biochemistry, 29,* 563–571.

Wu, R.S.S.; Siu, W.H.L. and Shin, P.K.S. (2005). Induction, adaptation and recovery of biological responses: Implications for environmental monitoring. *Marine Pollution Bulletin, 51,* 623-634.

Yagminas, A.P. and Villeneuve, D.C. (1987). Kinetic parameters of the inhibition of red blood cell aminolevulinic acid dehydratase by triethyl lead and its reversal by dithiothreitol and zinc. *Journal of Biochemical and Toxicology, 2,* 115-124.

Ying, G.; Kookana, R.S. and Ru, Y. (2002). Occurrence and fate of hormone steroids in the environment. *Environment International, 28,* 545-551.

Younes, M. (1999). Free radicals and reactive oxygen species. In H. Marquardt, S.G. Schäfer, R. McClellan and F. Welsch (Eds.), *Toxicology* (pp. 111-125). Elsevier Inc.

In: Aquatic Ecosystem Research Trends
Editor: George H. Nairne

ISBN 978-1-60692-772-4
© 2009 Nova Science Publishers, Inc.

*Chapter 8*

# ECOLOGY OF TEMPERATE PORT COMMUNITIES

## *Chariton Charles Chintiroglou* and *Chryssanthi Antoniadou*

Aristotle University, School of Biology,
Department of Zoology, Thessaloniki, Greece

## ABSTRACT

Temperate port communities constitute a modified coastal environment, whose ecology has been poorly studied. Lately, the scientific interest on this subject has been increased for two main reasons: (1) the low ecological quality of their waters that severely influences nearby areas through the diffusion of organic matter, given that ports constitute a hot spot area of organic pollution and (2) the increased presence of allochthonous species, several of which proved to be invasive, that are probably transferred via coastal shipping. Accordingly, the biomonitoring of port communities has become a subject of priority. This study was aimed to assess the structure and function of the benthic communities that develop on artificial hard substratum at a temperate port, namely Thessaloniki Port, with high levels of commercial shipment. Sampling was carried out by diving at three sites and at three depth levels. Two separate assemblages were detected: the blocks of various Serpulids in the midlittoral and the upper sublittoral zone, and the beds of the common Mediterranean mussel *Mytilus galloprovincialis* in the sublittoral zone. The low ecological quality of port communities was evident in their structure, as most of the recorded species were opportunistic and tolerant to organic pollution. At a functional level, the fauna was classified at few trophic groups, among which suspension feeders dominated. An extensive comparison with similar, previously collected, data was also performed, and revealed an apparent degradation of the ecosystem. Algal-dominated communities that have been reported from the same area, are currently replaced by animal-dominated ones, possibly as a result of the extensive deployment of mussel-cultures at the nearby coastal zone that have caused a rapid expand of mussel populations, which now dominate on the harbor piers.

---

* Corresponding Author: Chariton Charles Chintiroglou, Aristotle University, School of Biology, Department of Zoology, Thessaloniki, Greece, Gr -54124. Tel. +302310998405. Fax. +302310998269. Email. chintigl@bio.auth.gr

**Keywords:** Port, Hard substratum, Ecology, Zoobenthos, Organic pollution, Aegean

# INTRODUCTION

Benthic communities that settle on the piers of ports constitute a special entity according to the typology of benthos in temperate seas (Peres and Picard 1964). This is mostly because they develop on very sheltered coastal environments, where water renewal is minimal in contrast to sedimentation, which is very intense often creating the need for regular dredging operations (Karalis et al. 2003; Antoniadou et al. 2005; Simonini et al. 2005). Furthermore, additional technical works and constructions, such as, quay extensions, new breakwaters and pontoons, are often carried out. These constructions, recently termed as physical pollutants (Elliott 2003), can severely affect both the structure and function of benthic communities, principally by altering the local water currents (Baxevanis and Chintiroglou 2002; Antoniadou et al. 2002).

Temperate port communities constitute a modified coastal environment, whose ecology has been poorly studied (Bellan-Santini 1968; Leung Tack Kit 1972; Desrosiers et al. 1986; Kocak et al. 1999; Urkiaga-Alberdi et al. 1999; Damianidis and Chintiroglou 2000; Karalis et al. 2003; Chintiroglou et al. 2004a, 2004b; Cinar et al. 2008). Lately, the scientific interest on this subject has been increased for two main reasons: (1) the low ecological quality of their waters that severely influences nearby areas through the diffusion of heavy metals, hydrocarbons and organic matter, given that ports constitute a hot spot area of pollution (Fichet et al. 1998; Gupta et al. 2005) and (2) the increased presence of allochthonous species, several of which proved to be invasive, that are probably transferred via coastal shipping (Zibrowius 1991; Boudouresque and Verlaque 2002). Accordingly, the need to establish environmental management plans for each port has been recognized and, consequently, the biomonitoring of port communities has become a subject of priority (Gupta et al. 2005).

Considering all the above, this study was aimed: (1) to analyze the structure of the benthic communities that develop on artificial hard substratum at a temperate port with high levels of commercial shipment, namely Thessaloniki Port, (2) to analyze the fauna at a functional level and (3) to compare the above data with previous reports in order to assess the temporal evolution of the ecosystem.

# MATERIALS AND METHODS

## Study Area

The study area was Thessaloniki Port (40°38' N, 22°56' E), located in the NW Thermaikos Gulf also known as Thessaloniki Bay (Figure 1). Thermaikos is a shallow-water embayment in the NW Aegean Sea (Eastern Mediterranean), which is considered among the most disturbed marine areas in Greece. It receives discharges from large river systems (Axios, Loudias, Aliakmonas) and also sewage and industrial effluents from the city of Thessaloniki (Nikolaidis and Moustaka-Gouni 1992; Antoniadou et al. 2004). Water circulation follows a cyclonic pattern and is driven mainly by the prominent northward winds of the area (Hyder et al. 2002; Krestenitis et al. 2007).

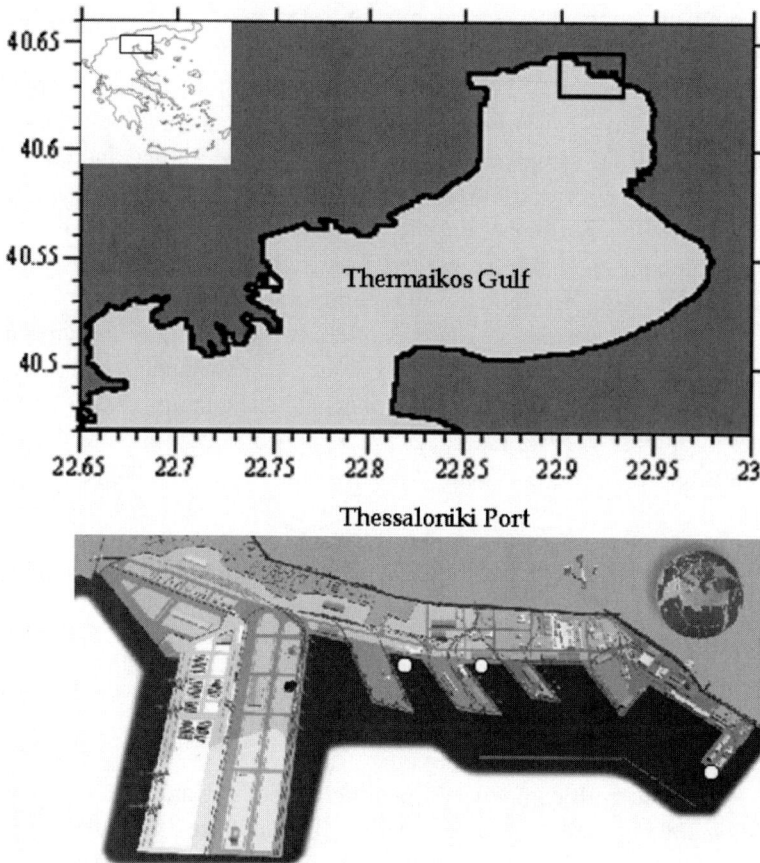

Figure 1. Map of the study area showing sampling stations in Thessaloniki Port.

The abiotic parameters follow a seasonal pattern. In general, the water column is homogenous from fall to spring, whereas a thermocline with variable depth range appears during the intermediate period. Salinity generally decreases in spring, where the inflow of the adjacent rivers is maximized (Hyder et al. 2002; Krestenitis et al. 2007). These hydrological features result in large concentrations of organic matter and nutrients, which often cause algal blooms, especially to the more sheltered inner part, i.e. Thessaloniki Bay (Pagou 2005).

Thessaloniki Port is the second major port in Greece, from which a large number of container vessels, passenger ships, flying dolphins and cruisers transports. The Port handles an estimated annual average of over 16,000,000 tons of cargo, 370,000 TEUs containers, 3,000 ships and 220,000 passengers. It has 26 quays with a total length of 6,200 m and a depth down to 12 m. It is a very sheltered port, exposed mostly on southward winds, whereas water renewal follows low rates. For the purposes of the study three quays (Figure 1) were selected as sampling stations that differ in terms of exposure ranked as follows: Q1 > Q3 > Q2.

## Field Sampling

Samplings were carried out in August 2004 at three depth levels: (1) 0.5 m, (2) 3 m and (3) 7m. Three replicate samples were randomly collected from each site using SCUBA diving by totally scrapping of the artificial hard substratum with a quadrate sampler covering a surface of 400 cm$^2$ (Karalis et al. 2003; Antoniadou and Chintiroglou 2005a). The previous data have been gathered in 1994 and 1995, also in summer and applying the same techniques (Karalis et al. 2003). All the samples obtained were sieved (mesh opening 0.5 mm) and fixed in 9% formaldehyde. After sorted all living specimens were identified at species level, using a binocular stereoscope or microscope and the relevant identification-keys for each taxa, and counted. At a functional level, the fauna was classified at trophic groups according to the nature and origin of food, as follows: (1) herbivores (H), feeding on macroalgae, (2) carnivores (C), feeding on various sessile or motile invertebrates, (3) suspension feeders (S), feeding on suspended organic particles in the water column and (4) deposit feeders (D), feeding on deposit particles at the sea bottom (Fauchald and Jumars 1979; Karalis et al. 2003; Chintiroglou et al. in press). Also, the fauna was assigned in ecological categories according to its tolerance to pollution (Borja et al. 2000; Antoniadou et al. 2005b; Chintiroglou et al. 2006).

At the three sampling sites, the main abiotic factors, i.e. temperature, salinity, dissolved oxygen and pH were also measured in the water column, with a CTD (SeaBird SBE-19) on a monthly basis and water clarity was estimated with a Secchi disc.

## Data Analyses

Data were analyzed with common biocoenotic methods (Karalis et al. 2003; Antoniadou and Chintiroglou 2005a), including the estimation of abundance as population density (N/m$^2$) and the calculation of diversity indices (i.e. Margalef's richness, Shannon-Wiener and Pielou's evenness, based on log$_2$).

Analysis of variance (one-way ANOVA) was used to test for temporal effects (year of sampling, three-level random factor) on the average abundance of the fauna and of the dominant taxonomic groups separately, through a general linear model (Underwood, 1997). Prior to the analyses, the homogeneity of variances was tested with Cohran's test and, when necessary, data were transformed to log(x+1). The Fisher LSD test was used for post hoc comparisons when appropriate. The faunistic diversity, expressed as the number of species S, and through diversity indices, i.e. Margalef's richness, Shannon-Wiener and Pielou's evenness, was also tested with the same model of ANOVA.

Hierarchical cluster analysis and non-metric multidimensional scaling (nMDS) via Bray-Curtis distances on square-root transformed numerical abundances data were used to visualize temporal changes in the composition of the fauna. Analysis of similarity (ANOSIM) was used to assess the significance of the multivariate results. Similarity of percentage (SIMPER) was used to identify the species which were responsible for any temporal pattern found in the composition of the fauna. All multivariate analyses were performed with the PRIMER software package (Clarke and Warwick 2001).

# RESULTS AND DISCUSSION

## Abiotic Factors

The values of the measured abiotic factors were similar among sampling sites and sampling years. As regards the seasonal pattern observed, temperature ranged from 10.9 to 28.7 °C, salinity from 35.8 to 36.5 psu, dissolved oxygen from 4.8 to 7.5 mg/l and pH varied around 8.7. Therefore, the values of the abiotic parameters in the water column were within the expected seasonal limits (Hyder et al. 2002; Krestenitis et al. 2005). Water clarity reached 5 m in wintertime, and only 3 m in summertime.

## Community Structure

A total of 44,594 individuals were collected, identified to 131 zoobenthic species, while four higher taxa, namely Nematoda, Nemertea, Oligochaeta and Phoronida were not identified at species level (Table 1). The most dominant in terms of abundance was the taxon of Polychaeta, in particular the family Serpulidae, followed by Peracarida, Nematoda and Ascidiacea (Figure 2). The percent contribution of the higher taxonomic groups showed a similar pattern between 1994 and 1995, whereas strongly deviates in 2004 with polychaetes covering almost the three quarters of total abundance. In all samplings the majority of the recorded fauna was constituted by opportunistic species related to organic pollution (Table 1), a fact indicative of the low ecological quality of port communities. However, it should be noted that many hard substratum species has not yet be assigned according to their tolerance to organic pollution, since the relevant data refers almost exclusively on soft substratum (Antoniadou and Chintiroglou 2005b; Chintiroglou et al. 2006).

The abundance of the total fauna, as well as of the taxonomic groups Polychaeta and Crustacea, didn't vary significantly between the years of sampling (F = 0.75 p = 0.48, F = 0.11 p = 0.89, F = 1.46 p = 0.24, respectively). In contrast, the abundance of Nematoda, Mollusca and Ascidiacea showed significant temporal variations (F = 11.56 $p<0.05$, F = 15.09 $p<0.05$, F = 84.11 $p<0.05$, respectively).

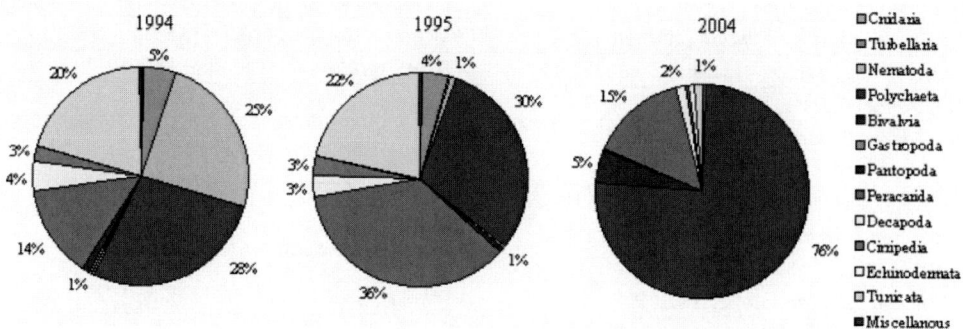

Figure 2. Percent contribution of the taxonomic groups at the benthic communities deployed on the quays of Thessaloniki Port, in 1994, 1995 and 2004.

**Table 1. Taxonomic list of the recorded benthic species from Thessaloniki Port**

| Taxa - Species | |
|---|---|
| Porifera | Gastropoda |
| *Cliona* sp. | *Nucula nucleus* (Linnaeus, 1758) |
| Cnidaria | *Plagiocardium papillosum* (Poli, 1795) |
| *Aiptasiogeton pellucidus* (Hollard, 1848) | *Scrobicularia cottardi* (Payraudeau, 1826) |
| *\*Anemonia viridis* (Forskal, 1775) | *Striarca lactea* (Linnaeus, 1758) |
| *Cereus pedunculatus* (Pennant, 1777) | Gastropoda |
| *Obelia dichotoma* (Linnaeus, 1758) | *Bittum reticulatum* (daCosta, 1778) |
| *Obelia geniculata* (Linnaeus, 1758) | *Chrysallida brusinai* (Cossmann, 1921) |
| Nematoda | *Chrysallida juliae* (deFolin, 1872) |
| Nemertea | *Diodora giberulla* (Lamarck, 1822) |
| Turbelaria | *Gibbula philberti* (Récluz, 1843) |
| *Leptoplana* sp. | *Nassarius corniculum* (Olivi, 1792) |
| *Stylochus* sp. | *Nassarius incrassatus* (Strom, 1768) |
| Oligochaeta | *Puncturella noachina* (Linnaeus, 1771) |
| Polychaeta | *Pusillina radiata* (Philippi, 1836) |
| *\*Ancistrosyllis* sp. | *Rissoa lia* (Monterosato, 1884) |
| *\*Aphelochaeta marionii* (de Saint Joseph, 1894) | Copepoda |
| *\*Capitella capitata* (Fabricius, 1780) | *Cyclops* sp. |
| *\*Caulleriella alata* (Southern, 1914) | *\*Nebalia bipes* Fabricius, 1780 |
| *\*Caulleriella bioculata* (Keferstein, 1862) | Cirripedia |
| *\*Ceratonereis costae* (Grube, 1840) | *Balanus eburneus* Gould, 1841 |
| *\*Chaetozone setosa* Malmgren, 1867 | *Balanus perforatus* Bruguiere, 1789 |
| *\*Cirriformia tentaculata* (Montagu, 1808) | *\*Balanus trigonus* Darwin, 1854 |
| *\*Dodecaceria concharum* Oersted, 1843 | Amphipoda |
| *\*Euclymene lumbricoides* (Quatrefages, 1865) | *\*Caprella acanthifera* Leach, 1814 |
| *Eunice oerstedii* Stimpson, 1853 | *\*Corophium acutum* Chevreux, 1908 |
| *\*Eunice vittata* (Delle Chiaje, 1828) | *\*Elasmopus rapax* Costa, 1853 |
| *Harmothoe areolata* (Grube, 1860) | *\*Erichthonius brasiliensis* (Dana, 1855) |
| *Harmothoe reticulata* (Claparède, 1870) | *\*Jassa marmorata* (Holmes, 1903) |
| *Harmothoe spinifera* (Ehlers, 1864) | *Lyssianassa* sp. |
| *\*Heteromastus filiformis* (Claparède, 1864) | Anisopoda |
| *\*Hydroides elegans* (Haswell, 1883) | *\*Leptochelia savignyi* (Kroyer, 1842) |
| *Hydroides pseudouncinata* Zibrowius, 1968 | *Pseudoparatanais batei* (G.O. Sars, 1882) |
| *Kefersteinia cirrata* (Keferstein, 1862) | *\*Tanais dulongii* (Audouin, 1826) |
| *Lepidonotus clava* (Montagu, 1808) | Isopoda |
| *\*Lumbrineris latreilli* Audouin and MilneEdwards, 1834 | *\*Cyathura carinata* (Kroyer, 1847) |
| *\*Malacoceros fuliginosus* (Claparède, 1868) | *Cymodoce truncata* Leach, 1814 |
| *\*Magalia perarmata* Marion and Bobretzky, 1875 | *Dynamene bidentatus* (Adams, 1800) |
| *\*Melinna palmata* Grube, 1870 | *Sphaeroma serratum* (Fabricius, 1787) |
| *\*Marphysa sanguinea* (Montagu, 1815) | Decapoda |
| *Nainereis laevigata* (Okuda, 1946) | Natantia |
| *\*Neanthes caudata* (Delle Chiaje, 1827) | *Athanas nitescens* (Leach, 1814) |
| *\*Neanthes succinea* (Frey and Leuckart, 1847) | *Palaemon elegans* Rathke, 1837 |

## Table 1. (Continued)

| Taxa - Species | |
|---|---|
| *Ophiodromus pallidus* (Claparède, 1864) | *Processa edulis* (Risso, 1816) |
| *Phyllodoce mucosa* Oersted, 1843 | *Thoralus cranchii* (Leach, 1817) |
| *Pirakia punctifera* (Grube, 1860) | Anomura |
| *Platynereis dumerilii* (Audouin and Milne-Edwards, 1833) | *Pisidia longimana* (Risso, 1816) |
| *Polydora caeca* (Oersted, 1843) | Brachyura |
| *Polyophthalmus pictus* (Dujardin, 1839) | *Pachygrapsus marmoratus* (Fabricius, 1787) |
| *Prionospio malmgreni* Claparède, 1869 | *Pilumnus hirtellus* (Linnaeus, 1761) |
| *Prionospio steenstrupi* Malmgren, 1867 | Mysidacea |
| *Pseudopotamilla reniformis* (Bruguiere, 1789) | *Siriella clause* G.O. Sars, 1877 |
| *Sabellaria spinulosa* Leuckart, 1849 | Pantopoda |
| *Schistomeringos rudolphi* (Delle Chiaje, 1828) | *Callipalene* sp. |
| *Scoletoma funchalensis* (Kinberg, 1865) | *Nymphon* sp. |
| *Serpula concharum* Langerhans, 1880 | Bryozoa |
| *Serpula lobiancoi* Rioja, 1917 | *Bowerbankia imbricata* (Adams, 1798) |
| *Serpula vermicularis* Linnaeus, 1767 | *Bowerbankia* sp. |
| *Serpula* sp. | *Bugula fulva* Ryland, 1960 |
| *Syllidia armata* Quatrefages, 1865 | *Bugula neritina* (Linnaeus, 1758) |
| *Syllis cornuta* Rathke, 1843 | *Bugula stolonifera* Ryland, 1960 |
| *Syllis prolifera* Krohn, 1852 | *Conopeum seurati* (Canu, 1928) |
| *Syllis vittata* Grube, 1840 | *Cryptosula pallasiana* (Moll, 1803) |
| *Terebella lapidaria* Linnaeus, 1767 | *Electra* sp. |
| *Vermiliopsis infundibulum* (Philippi, 1844) | Nolellidae sp.1 |
| Phoronida | *Schizoporella errata* (Waters, 1878) |
| Bivalvia | *Schizoporella unicornis* (Johnston in Wood, 1844) |
| *Anomia ephippium* Linnaeus, 1758 | *Scrupocellaria reptans* (Linnaeus, 1767) |
| *Arca noae* Linnaeus, 1758 | Echinodermata |
| *Chlamys varia* (Linnaeus, 1758) | *Ophiothrix fragilis* (Abildgaard, in O.F. Muller, 1789) |
| *Gastrana fragilis* (Linnaeus, 1758) | Tunicata |
| *Gregariella petagnae* (Scacchi, 1832) | *Ciona intestinalis* (Linnaeus, 1758) |
| *Hiatella arctica* (Linnaeus, 1767) | *Clavelina lepadiformis* Muller, 1776 |
| *Hiatella rugosa* (Linnaeus, 1767) | *Phallusia mamillata* (Cuvier, 1815) |
| *Kellia suborbicularis* (Montagu, 1803) | *Pyura microcosmus* (Savigny, 1816) |
| *Modiolula phaseolina* (Philippi, 1844) | *Microcosmus savignyi* Monniot, 1962 |
| *Modiolus barbatus* (Linnaeus, 1758) | *Styela canopus* (Savignyi, 1816) |
| *Mytilus galloprovincialis* Lamarck, 1819 | *Styela plicata* (Lesueur, 1823) |

* indicates species that are typical in organically polluted coastal areas.

Fisher LSD post-hoc tests showed that the abundance of Nematoda showed increased values only in 1994; the abundance of Mollusca showed the opposite pattern with very high values in 2004, while the abundance of ascidians decreased at the same time (Figure 3).

The diversity of the fauna, expressed as the total number of species and through diversity indices showed significant temporal variations in all cases (F = 13.99 p<0.05, F = 10.73

p<0.05, F = 25.93 p<0.05, F = 18.08 p<0.05, for S, d, H and J, respectively). These variations can be synopsized to the decreased values recorded in 2004 (Figure 3).

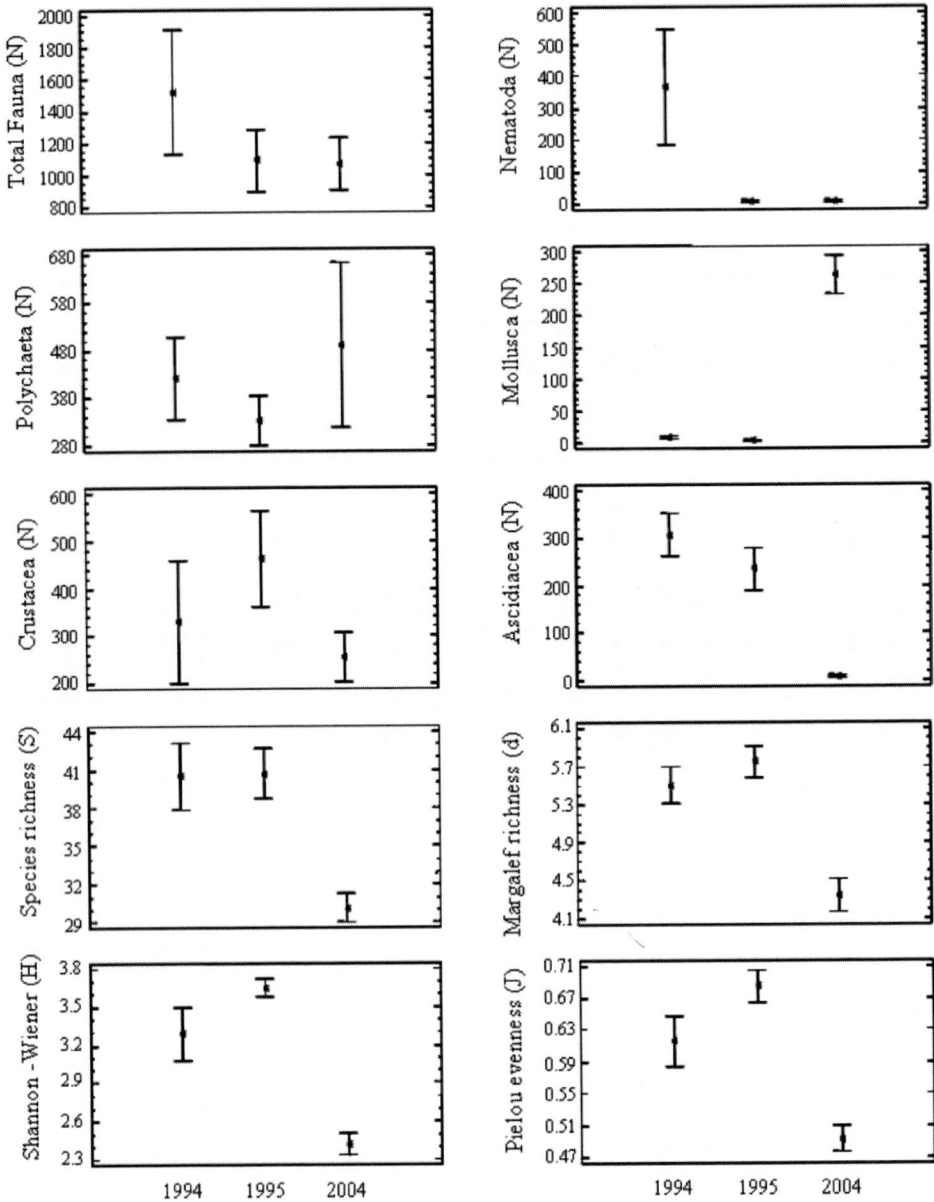

Figure 3. Mean value and variance of the abundance and the diversity of the fauna and of the dominant taxonomic groups (S = number of species, d = Margalef's richness, H = Shannon-Wiener, J = Pielou's evenness) at a temporal scale.

Multidimensional analyses of the assemblage structure discriminated all samples from 2004, while the relevant samples from 1994 and 1995 were placed in the same group (Figure 4). The above discrimination was significant (ANOSIM results, R = 0,907, p<0,1). SIMPER analysis showed that 13 species contribute to 60% of the average similarity of each group, while 20 species contribute to 60% of the average dissimilarity between the two groups (Table 2).

**Table 2. Species contributing on about 60% of the average in-group similarity or among groups dissimilarity resulted from SIMPER analysis**

| Taxa | Within Group Similarity | | Between Group Dissimilarity |
|---|---|---|---|
| | Group I | Group II | Group I / II |
| | 1994 and 1995 | 2004 | 1994-95 / 2004 |
| | 58.49% | 57.91 % | 72.02 % |
| *Balanus trigonus* | 3.39 | | 1.66 |
| *Callipalene* sp. | | | 1.58 |
| *Clavelina lepadiformis* | | | 3.58 |
| *Corophium acutum* | 3.84 | | 2.15 |
| *Cymodoce truncata* | | | 1.81 |
| *Elasmopus rapax* | 6.15 | 5.79 | 3.50 |
| *Erichthonius brasiliensis* | | | 2.84 |
| *Harmothoe reticulata* | | 2.73 | |
| *Harmothoe spinifera* | | | 1.72 |
| *Hydroides elegans* | 2.93 | | 7.11 |
| *Hydroides pseudouncinata* | 3.17 | | 1.86 |
| *Mytilus galloprovincialis* | | 17.95 | 7.69 |
| Nematoda | 3.64 | 1.96 | 3.36 |
| *Ophiodromus pallidus* | 7.48 | | 4.99 |
| *Ophiothrix fragilis* | | 6.16 | 2.43 |
| *Pachygrapsus marmoratus* | 3.89 | | 1.67 |
| *Pisidia longimana* | 3.29 | 5.82 | |
| *Platynereis dumerilii* | | 4.35 | |
| *Polydora caeca* | | 3.34 | |
| *Prionospio steenstrupi* | | | 1.71 |
| *Pseudoparatanais batei* | | 1.83 | |
| *Serpula concharum* | | | 2.06 |
| *Staurocephalus rudolphii* | 8.18 | 2.84 | 3.11 |
| *Styela partita* | 5.77 | | 4.21 |
| *Stylochus* sp. | 5.40 | 2.81 | 2.27 |
| *Syllidia armata* | | 2.18 | |
| *Terebella lapidaria* | 3.83 | 3.74 | |

According to the presented results, it is evident that the previously reported (i.e. in 1994 and 1995) homogenous algal-dominated community that expanded to the entire bathymetric range of the Port quays (Karalis et al. 2003) has been currently substituted by an animal-dominated community, in which two separate assemblages can be detected: the blocks of various Serpulids in the midlittoral and the upper sublittoral zone, and the beds of the common Mediterranean mussel *Mytilus galloprovincialis* in the sublittoral zone. At a functional level, deposit and suspension feeding organisms dominated benthic communities in

1994 and 1995, followed by herbivores and carnivores. In 2004, the percentage of deposit feeders has been strongly reduced and suspension feeders clearly dominated, in both Seprulid blocks and mussel beds, followed by herbivores and deposit feeders, respectively (Figure 5).

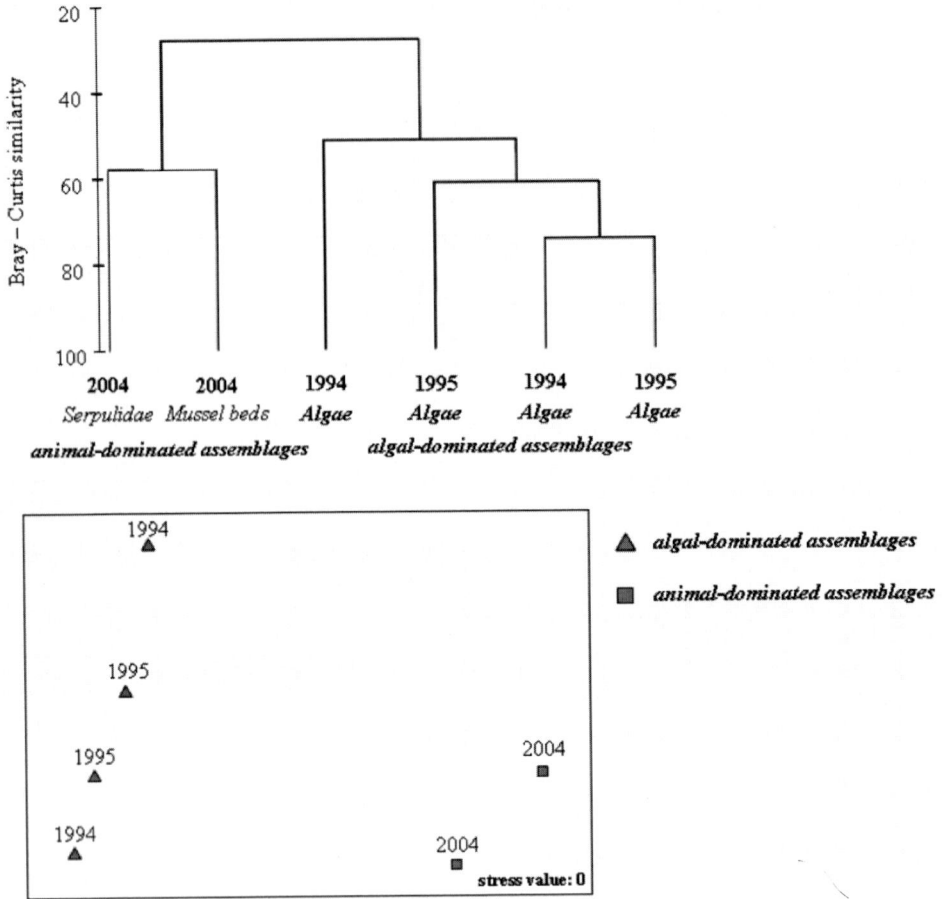

Figure 4. Hierarchical cluster analysis and non-metric multidimensional scaling showing temporal changes in the assemblage structure, based on Bray-Curtis similarity index calculated from root transformed numerical abundances data.

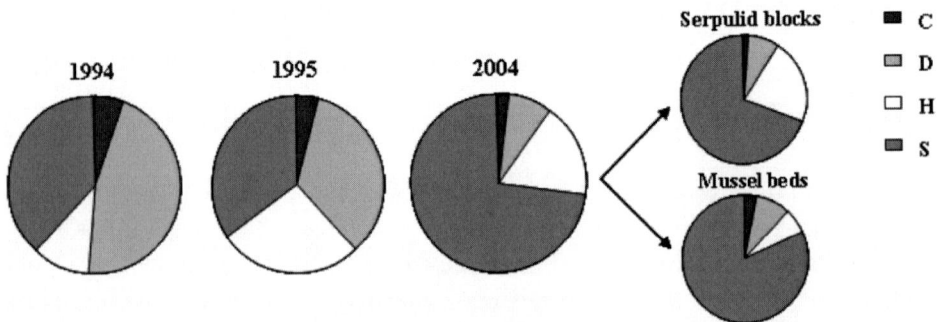

Figure 5. Percent contribution of the trophic groups at the benthic communities deployed on the quays of Thessaloniki Port, in 1994, 1995 and 2004.

*M. galloprovincialis* can colonize various artificial substrata in clean as well as organically polluted waters, forming dense beds with a diverse associated fauna (Bellan-Santini 1968; Kocatas 1978; Damianidis and Chintiroglou 2000; Chintiroglou et al. 2004a, 2004b). Consequently, it has been widely reported that the biomonitoring of mussel beds constitute a valuable tool for the assessment of the ecological quality in coastal environments (Chintiroglou et al. 2004b; Cinar et al. 2008). The mussel *M. galloprovincialis* is intensively cultured over the western Thessaloniki Bay, and the total annual production has been rapidly increased reaching 35,000 tons after 2000 (Arsenoudi et al. 2003). As a consequence, and taking into account the large reproductive abilities of the species in the area (Topoglidi 2003), *M. galloprovincialis* has colonized any available substrata in the entire Thessaloniki Bay, including the Port quays. The expansion of Serpulid blocks is probably also linked with the development of mussel-cultures in Thermaikos during the mid nineties. Various Serpulid species successfully settle on the mussel shells, especially on cultured ones, as well as on the pillars and lines of the culture, taking advantage of the availability of biological and artificial substrata (Rayan 2004). Thus, the reproductive output of both Serpulids and mussels has been considerably increased, while the spread of larvae in the entire bay was a matter of time, according to the circulation pattern of water masses (Krestenitis et al. 2007). It seems therefore, that besides the biomonitoring and the establishment of specific management plans for each port (Gupta et al. 2005), it is necessary to develop integrated coastal zone management plans, with emphasis on landplanning policies and inclusion of all potential uses.

## CONCLUSION

The present study has detected important changes in the structure of hard substratum benthic communities in Thessaloniki Port. In the decade of 1990 the photophilic algae community occurred in the midlittoral and sublittoral zone with a significant temporal stability, since only Nematodes showed a significant change in their abundance between 1994 and 1995. In contrast, in 2004 this community has been substituted by two separate assemblages: (1) the blocks of Serpulids in the midlittoral and the upper sublittoral zone, and (2) the mussel beds in the sublittoral zone. Serpulids and mussels are typical suspension feeding organisms that thrives in eutrophic environments. Various Serpulids, such as the species *Hydroides elegans*, *Serpula vermicularis* and *Vermiliopsis infundibulum*, are typical fouling species (OCDE 1967; Bianchi 1981), while mussel beds are frequently found on temperate ports, persisting under severe pollution events (Bellan 1980; Urkiaga-Alberdi et al. 1999; Chintiroglou et al. 2004b; Cinar et al. 2008). All the performed analyses showed the degradation of port communities in 2004, assessed by the impoverishment of the fauna and the increase presence of species tolerant to organic pollution. The intense culture of *M. galloprovincialis* at the west Thessaloniki Bay, is possibly involved, since its population boomed and colonized any available substrata over the entire Bay.

## ACKNOWLEDGEMENTS

We wish to thank Dr. John Halley and Dr. DimitrisVafidis for their critical advice and contribution to this article.

## REFERENCES

Antoniadou, C.; Chintiroglou, C. Biodiversity of zoobenthic hard-substrate sublittoral communities in the Eastern Mediterranean (North Aegean Sea). *Estuar. Coast. Shelf S.* 2005a, *62*, 637-653.

Antoniadou, C.; Chintiroglou, C. Response of polychaete populations to disturbance: an evaluation of methods in hard substratum. *Fresenius Env. Bull.* 2005b, *14*, 1066-1073.

Antoniadou, C.; Karalis, P.; Chintiroglou, C. A new dock in a fishery port (N. Michaniona, N. Aegean Sea) and its influence on the biota. In: Dassenakis, M. (ed), *Oceanographical Aspects for a Sustainable Mediterranean,* Athens, Greece.

Antoniadou, C.; Krestenitis, Y.; Chintiroglou, C. Structure of the "amphioxus sand" community in Thermaikos Bay (Eastern Mediterranean). *Fresenius Env. Bull.* 2004, *13*, 1122-1128.

Antoniadou, C.; Sarantidis, S.; Krestenitis, Y.; Chintiroglou, C. Ecological quality of hard substratum benthic communities in the main port of Thessaloniki. *Proc. 2$^{nd}$ Environm. Symposium,* Thessaloniki, Greece.

Arsenoudi, P.; Scouras, Z.; Chintiroglou, C. First evaluation of *Mytilus galloprovincialis* LMK, natural populations in Thermaikos Gulf: structure and distribution. *Fresenius Env. Bull.* 2003, *12*, 1384-1393.

Baxevanis, A.; Chintiroglou, C. Peracarida crustaceans of the artificial hard substratum in N. Mihaniona (N. Aegean). *Belg. J. Zool.* 2000, *130*, 11-16.

Bellan, G. Annelides polychetes des substrats solides de trois milieux pollues sur les cotes de Provence (France): Cortiou, Golfe de Fos, Vieux port de Marseille. *Téthys* 1980, *9*, 267-278.

Bellan-Santini, D. Contribution a l'étude des milieux portuaires (Le Vieux Port de Marseille). *Rapp. Comm. Int. Mer Médit.* 1968, *19*, 93-95.

Bianchi, C.N. *Guide per il riconoscimento delle specie animali delle acque lagunari e costiere italiane. Policheti Serpuloidei.* Consiglio Nazionale delle Ricerche, Pavia, Italy 1981, pp 187.

Borja, A.; Franko, J.; Perez, V. A marine biotic index to establish the ecological quality of soft-bottom benthos within European estuarine and coastal environment. *Mar. Pollut. Bull.* 2000, *40*, 1100-1114.

Boudouresque, C.F.; Verlaque, M. Biological pollution in the Mediterranean Sea : invasive versus introduced macrophytes. *Mar. Pollut. Bull.* 2002, *44*, 32-38.

Chintiroglou, C.; Antoniadou, C.; Baxevanis, A.; Damianidis, P.; Karalis, P.; Vafidis, D. Peracarida populations of hard substrate assemblages in ports of the NW Aegean Sea (Eastern Mediterranean). *Helgoland Mar. Res.* 2004a, *58*, 54-61.

Chintiroglou, C.; Damianidis, P.; Antoniadou, C.; Lantzouni, M.; Vafidis, D. Macrofauna biodiversity of mussel bed assemblages in Thermaikos Gulf (northern Aegean Sea). *Helgoland Mar. Res.* 2004b, *58*, 62-70.

Chintiroglou, C.; Antoniadou, C.; Krestenitis, Y. Can polychaetes be used as a surrogate group in assessing ecological quality in soft bottom communities (NE Thermaikos Gulf)? *Fresenius Env. Bull.* 2006, *15*, 1199-1207.

Chintiroglou, C.; Antoniadou, C.; Damianidis, P. Spatio-temporal variability of zoobenthic communities in a tectonic lagoon (Lake Vouliagmeni, Attika, Greece). *J. Mar. Biol. Ass. UK.* 2008, *in press*.

Cinar, M.E.; Katagan, T.; Kocak, F.; Ozturk, B.; Ergen, Z.; Kocatas, A.; Onen, M.; Kirkim, F.; Bakir, K.; Kurt, G.; Dagli, E.; Acik, S.; Dogan, A.; Ozcan, T. Faunal assemblages of the mussel *Mytilus galloprovincialis* in and around Alsancak Harbour (Izmir Bay, eastern Mediterranean) with special emphasis on alien species. *J. Mar. Sys.* 2008, *71*, 1-17.

Clarke, K.R.; Warwick, M.R. *Change in marine communities: an approach to statistical analysis and interpretation.* Natural Environment Research Council, Plymouth, UK, 2001; 2nd edition, pp 175.

Damianidis, P.; Chintiroglou, C. Structure and function of Polychaetofauna living in *Mytilus galloprovincialis* assemblages in Thermaikos Gulf (north Aegean Sea). *Oceanol Acta*, 2000, *23*, 323-339.

Desrosiers, G.; Bellan-Santini, D.; Brethes, J.C. Organisation trophique de quatre peuplements se substrats rocheux selon un gradient de pollution industrielle (Golfe de Fos, France). *Mar. Biol.* 1986, *9*, 107-120.

Elliott, M. Biological pollutants and biological pollution - an increasing cause for concern. *Mar. Pollut. Bull.* 2003, *46*, 275-280.

Fauchald, K.; Jumars, P. The diet of worms: a study of polychaete feeding guilds. *Oceanogr. Mar. Biol. Annu. Rev.* 1979, *17*, 193-284.

Fichet, D. ; Radenac, G. ; Miramand, P. Experimental studies of impacts of harbour sediments resuspension to marine invertebrate larvae: bioavailability of Cd, Cu, Pb and Zn toxicity. *Mar. Pollut. Bull.* 1998, *36*, 509-518.

Gupta, A.K.; Gupta, S.K. ; Patil, R.S. Environmental management plan for port and harbour projects. *Clean Techn Envir Policy.* 2005, *7*, 133-141.

Hyder, P.; Simpson, J.H.; Christopoulos, S.; Krestenitis, Y. The seasonal cycles of stratification and circulation in the Thermaikos Gulf Region of Freshwater Influence (ROFI), northwest Aegean. *Cont. Shelf Res.* 2002, *22*, 2573-2597.

Karalis, P.; Antoniadou, C.; Chintiroglou, C. Structure of the artificial hard substrate assemblages in ports, in Thermaikos gulf (North Aegean Sea). *Oceanol. Acta* 2003, *26*, 215-224.

Kocak, F.; Ergen, Z.; Cinar, M.E. Fouling organisms and their developments in a polluted and an unpolluted marina in the Aegean Sea (Turkey). *Ophelia* 1999, *50*, 1-20.

Kocatas, A. Contribution a l'étude des peuplements des horizons superieurs de substrat rocheux de Golfe d'Izmir (Turquie). *Egée Univ. Fen. Fak. Monogr. Ser.* 1978, *12*, 1-93.

Krestenitis, Y. ; Kombiadou, K. ; Savvidis, Y. Modelling the cohesive sediment transport in the marine environment : the case of Thermaikos Gulf. *Ocean Sci.* 2007, *3*, 91-104.

Leung Tack Kit, D. Etude d'un milieu pollue: le Vieux Port de Marseille. Influence des conditions physiques et chimiques sur la physionomie du peuplement des quais. *Téthys* 1972, *3*, 767-826.

Nikolaidis, G. ; Moustaka-Gouni, M. Nutrient distribution and eutrophication effects in a polluted coastal area of Thermaikos Gulf, Macedonia, Greece. *Fresenius Env. Bull.* 1992, *1*, 250-255.

O.C.D.E. *Principales salissures marines. Serpules tubicoles.* Publications de l' O.C.D.E. Paris, France, 1967, pp 80.

Pagou, K. *Eutrophication in Hellenic coastal areas.* i: Papathanassiou and Zenetos (eds) *State of the Hellenic Marine Environment.* Hellenic Centre of Marine Research, Athens, 311-317.

Rayann, A.N. Biotic interactions of the cultured edible bivalve mussel Mytilus galloprovincialis Lam. in Thermaikos Gulf. Doctorate Thesis, Aristotle University, Thessaloniki, Greece, 2003, pp 187.

Simonini, R. ; Ansaloni, I. ; Cavallini, F. ; Graziosi, F.; Iotti, M.; MassambaN'Siala, G.; Mauri, M.; Preti, M.; Prevedelli, D. Effects of long term dumping of harbor-dredged material on macrozoobenthos at four disposal sites along the Emilia-Romagna coast (Northern Adriatic Sea, Italy). *Mar. Pollut. Bul.* 2005, *50*, 1595-1605.

Topoglidi, A. *Spatio-temporal analysis of the structure and the physiological condition of Mytilus galloprovincialis LMK natural stocks in Thermaikos Gulf (NW Aegean).* Master Thesis, Aristotle University, Thessaloniki, Greece, 2003, pp 75.

Underwood, A.J. *Experiments in Ecology. Their logical design and interpretation using analysis of variance.* Cambridge University Press, UK, 1997, pp 504.

Urkiaga-Alberdi, J.; Pagola-Carte, S.; Saiz-Salinas, J.I. Reducing effort in the use of benthic bioindicators. *Acta Oecologica* 1999, *20*, 489-497.

Zibrowius, H. Ongoing modification of the Mediterranean marine fauna and flora by the establishment of exotic species. *Mésogée* 1991, *51*, 83-107.

In: Aquatic Ecosystem Research Trends
Editor: George H. Nairne

ISBN 978-1-60692-772-4
© 2009 Nova Science Publishers, Inc.

*Chapter 9*

# CHEMICAL ECOLOGY AND MEDICINAL CHEMISTRY OF MARINE NF-κB INHIBITORS

## *F. Folmer, M. Schumacher, M. Jaspars, M. Dicato and M. Diederich*

Laboratoire de Biologie Moléculaire et Cellulaire du Cancer, Luxembourg

## ABSTRACT

NF-κB is an inducible transcription factor found in virtually all types of vertebrate cells, as well as in some invertebrate cells. While normal activation of NF-κB is required for cell survival and immunity, its deregulated expression is characteristic of cancer, inflammation, and numerous other diseases. Hence, NF-κB has recently become one of the major targets in drug discovery.

Several marine organisms use NF-κB (or analogues thereof), NF-κB inducers, or NF-κB inhibitors as chemical defence mechanisms, for parasitic invasion, for symbiosis, or for larval development. In particular, a wide range of marine natural products have been reported to possess NF-κB inhibitory properties, and some of these marine metabolites are currently in clinical trials as anticancer or anti-inflammatory drugs.

In the present review, we discuss the role of NF-κB inhibitors in marine chemical ecology, as well as in biomedicine. We also describe synthetic modifications that have been made to a range of highly promising marine NF-κB inhibitors, including the macrolide bryostatin 1 isolated from the bryozoan *Bugula neritina*, the lactone-γ-lactam salinosporamide A isolated from the actinomycete *Salinispora tropica*, the alkaloid hymenialdisine isolated from various sponges, the sesquiterpenoid hydroquinone avarol isolated from the sponge *Dysidea avara*, and the sesterterpene lactone cacospongonolide B isolated from the sponge *Fasciospongia cavernosa*, to increase their bioactivity and bioavailability, to decrease their level of toxicity or to lower the risk of other detrimental side-effects, and to increase the sustainability of their pharmaceutical production by facilitating their chemical synthesis.

## ABBREVIATIONS

| | |
|---|---|
| CDK | cyclin dependent kinase; |
| CK | casein kinase; |
| DAG | 1,2-diacyl-*sn*-glycerol; |
| GFP | green fluorescent protein; |
| GSK-3b | glycogen synthase kinase; |
| HIF-1 | hypoxia-inducible transcription factor-1; |
| HIV-1 | human immuno-deficiency virus-1; |
| IkB | inhibitor of kappa B |
| IKK | kinase of IκB; |
| IL-1β | interleukin-1β; |
| $K_i$ | kinase inhibitory activity; |
| NF-κB | nuclear factor-κB; |
| PKC | protein kinase C; |
| PMA | phorbol 12-myristate 13-acetate; |
| RBL | rat basophilic leukemia; |
| ROS | reactive oxygen species; |
| SAR | structure-activity relationship; |
| TNF-α | tumour necrosis factor-α; |
| V-ATP-ase | vacuolar-type $H^+$-ATPase. |

## 1. INTRODUCTION

The sea covers seventy percent of the earth's surface, and it is an even more important ecological entity in terms of the total volume of habitable space than terrestrial habitats [1]. Therefore, it is not surprising that the oceans have drawn the attention of numerous scientists from a wide range of disciplines, and that new discoveries about ecology and biodiversity of marine organisms are made every day. Even though a vast majority of the oceans remains unexplored and our knowledge about marine ecology is mostly restricted to the littoral, which is a relatively narrow section of the oceans at the land-sea interface, marine scientists are well aware that the oceans form a unique source of very high biodiversity [2-8]. This high level of biodiversity can be attributed in part to the sheer complexity of the marine ecosystem [3, 4, 8]. Marine habitats cover a wide range of temperatures, air exposure levels, UV exposure levels, pH, dissolved gas concentrations, sun light availability, nutrient availability, pressure levels, current prevalence, turbulence levels, and sediment compositions [5]. Along with the latter abiotic factors, a wide range of biotic factors also govern the marine ecosystem. These include predation, parasitism, fouling, competition for resources, and competition for space [1, 3, 4, 6] (figure 1).

To cope with the biotic and abiotic factors reigning in the marine realm, marine organisms have developed various physical and chemical defence mechanisms. Physical defences include locomotion, camouflage, and the development of a hard external shell [1], while chemical defences rely on the production of toxic or deterrent natural products [3, 4, 8, 9].

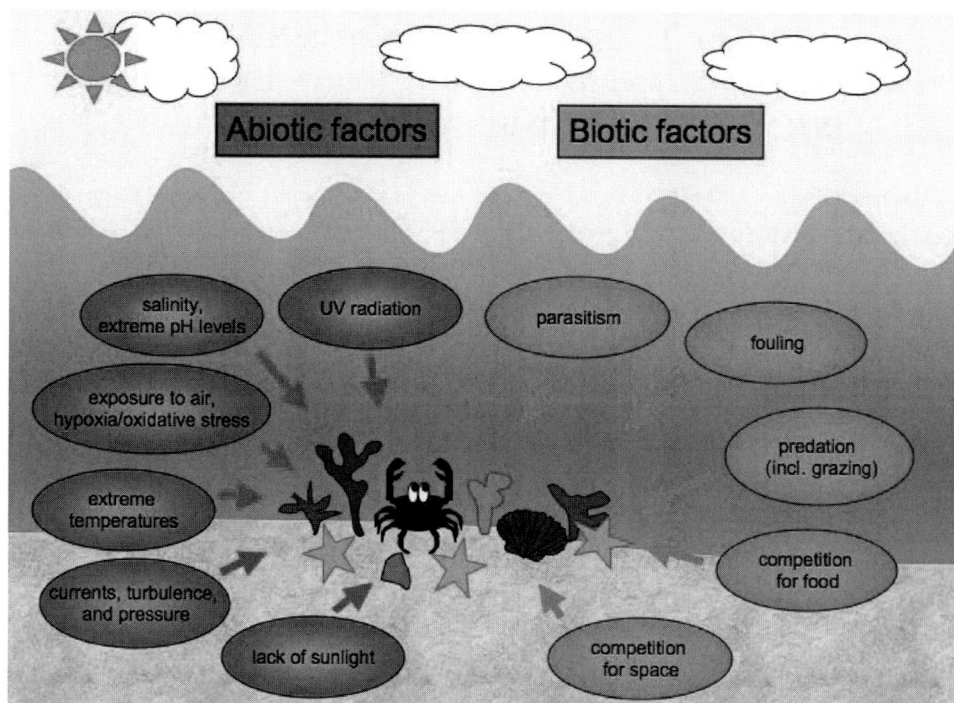

Figure 1. The biotic and abiotic factors governing the marine ecosystem.

In the 1940's, a new branch of ecology termed *marine chemical ecology* was established based on the observation that sessile, soft-bodied marine plants and animals rely heavily on natural products as a defence mechanism, as well as for communication in the ocean [3, 4, 8-11]. Heterotrophs can either produce the secondary metabolites used as chemical defences themselves or acquire them through dietary intake or symbiosis. Algae, which are sessile autotrophs, do not have any dietary intake of secondary metabolites, and the opportunities of evolving symbiotic relationships that might provide defence are quite limited. Furthermore, their sessile lifestyle prevents them from moving away from grazers or fouling organisms or spatial competitors. Hence, the autonomous production of natural products as chemical defences is a particularly important mechanism in algae [6].

The probing of the biomedical properties of marine natural products started in the 1960's [5]. This research avenue, which is often referred to as the search for "*Drugs from the Seas*", has led to the recognition of the extraordinary biomedical potential of marine natural products [7, 12-14].

In the present review, we discuss the ecological and biomedical roles of marine natural products that inhibit the nuclear transcription factor-κB (NF-κB), which has recently been recognized as a key target in cancer- and inflammation-related drug discovery [15-17]. Unfortunately, the natural production of bioactive secondary metabolites by wild or cultured marine organisms is generally unsustainable for pharmaceutical applications, and chemical synthesis is, in most cases, a necessity in order to complete clinical trials [14, 18, 19]. In the final section of the present review, we describe how the synthesis of a range of highly promising marine NF-κB inhibitors and their analogues has provided invaluable information about the structure-activity relationship of marine natural products targeting NF-κB.

## 2. THE TRANSCRIPTION FACTOR NF-κB:
## ITS PATHOLOGICAL IMPLICATIONS IN CANCER DEVELOPMENT AND INFLAMMATION, AND ITS ACTIVATION PATHWAY

The transcription of DNA into RNA is the first stage in the expression of genes. Transcription is orchestrated by a group of proteins called transcription factors, which bind to DNA on specific binding sites within the genes' promoter regions and initiate the expression of the genes [20, 21]. The transcription factor nuclear factor-κB (NF-κB), which was discovered in 1986 by Sen and Baltimore, has been shown to be implicated in the regulation of over 150 different human genes [22-24]. Although NF-κB is required for both innate and adaptive immunity [25-27], most of the genes regulated by NF-κB encode for proteins that play critical roles in cancer development, inflammation, and other pathologies [23, 24, 28]. Up-regulated activity of NF-κB has been observed primarily in patients with various types of lymphomas [29-35], with several types of solid cancers [36, 37], with asthma [38, 39], with rheumatoid arthritis [40-42], with various inflammatory bowel diseases [43], with multiple sclerosis [44, 45], and with osteoporosis [46]. Furthermore, aberrant NF-κB activation has been recognized as one of the major resistance factors to chemotherapy and radiotherapy, as it impairs the ability of damaged, malfunctioning, or cancerous cells to undergo apoptosis [47-54]. Because of its numerous pathological implications in cancer development and inflammation, NF-κB has become a major target in drug discovery, and the activation pathway of the transcription factor has been studied in depth in several laboratories around the world [15, 29, 55-58].

NF-κB is a dimer of proteins belonging to the Rel family, which includes RelA (p65), RelB, c-Rel, p50 (NF-κB1), and p52 [59, 60]. All five Rel family proteins contain a well-conserved Rel homology domain (RHD) responsible for the dimerization of NF-κB, for the interactions of NF-κB with its cytoplasmic inhibitory protein IκB, and for the binding of NF-κB to DNA [60, 61].

The nuclear localization signal (NLS) at the C-terminus of the RHD plays an important role in the nuclear translocation of activated NF-κB [59, 60]. p65, RelB, and c-Rel also contain a terminal transactivation domain (TAD) at their C-terminal end which is required for the activation of transcription [60]. p50 and p52 lack a terminal transactivation domain and are therefore transcriptionally inactive [60] (figure 2). The present paper focuses on the inhibition of the heterodimer p50/p65, which is the most common form of NF-κB [62-64].

NF-κB is normally found in the cytoplasm, in an inactive form as it is bound to its cytoplasmic inhibitor IκB [60, 61, 65]. IκB is a family of a handful of closely related molecules, the most common ones being IκBα and IκBβ. All known IκB proteins contain a 30-33 amino acid long sequence called ankyrin repeat, which binds, via non-covalent interactions, and with extremely high affinity, to the RHD of NF-κB [65], and which masks the NLS of NF-κB [61], thereby preventing the latter's nuclear translocation [59] (figure 3). The N-terminal region of IκB proteins is known to play a major role in the signal-dependent degradation of IκB by the proteasome, whereas the C-terminus domain is implicated in the regulation of NF-κB-DNA binding, and in the nuclear export of NF-κB [61].

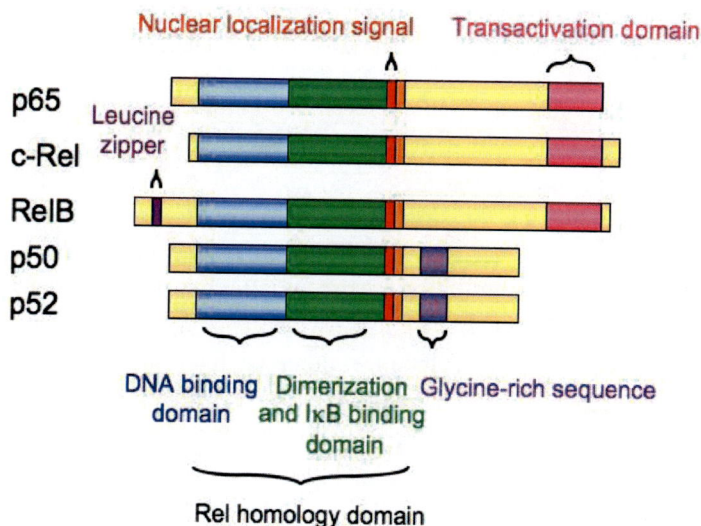

Figure 2. Schematic representation of the five Rel family proteins RelA (p65), c-Rel, RelB, p50, and p52, showing the highly conserved Rel homology domain (RHD) including a DNA binding domain and a dimerization and IκB binding domain on the N-terminus of each protein. The transactivation domain (TAD) of RelA, c-Rel, and RelB, as well as the Leucine zipper (LZ) in the case of RelB are required for the transcriptional activation. p50 and p52, which lack a transactivation domain, are transcriptionally inactive. The glycine-rich region towards the C-terminus in p50 and p52 is involved in the generation of the latter from their precursors (p105 and p100, respectively) [60]. (figure modified from Keutgens *et al.* [60]).

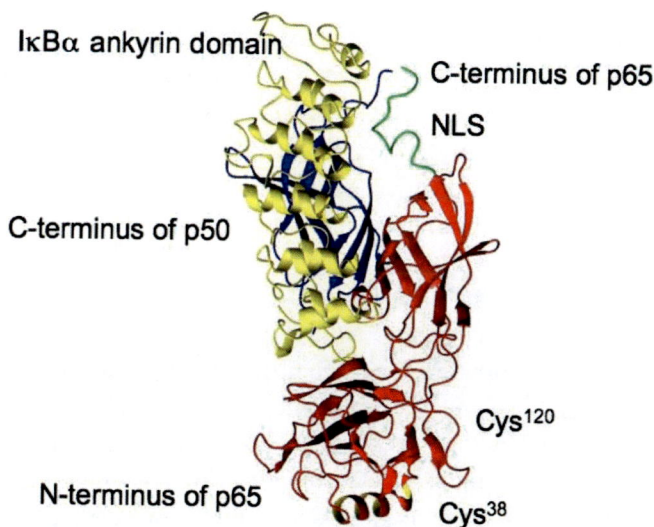

Figure 3. X-ray diffraction structure of NF-κB (p50/p65 heterodimer) bound to IκB (*Mus musculus*, synthetic construct, PDB 1IKN, Huxford *et al.*, 1998 [66]). Cartoon shows sheet strands (arrows pointing towards the C-terminus end of the chains) and α-helices. The red chain represents the N-terminus of p65. The C-terminus of p65 is shown in green. The blue chain represents the C-terminus of p50. The N-terminus of p50 is not shown in this figure. The ankyrin domain of IκBα (all α-helices) is shown in yellow. The N-terminus and C-terminus regions of IκBα are not shown in this figure. The image was generated using the programme MOLMOL [67].

In most cases, the activation of NF-κB is triggered by extra-cellular stimulations including ionizing radiation, oxidative stress, various toxins, and signalling through pro-inflammatory cytokines such as interleukin-1β (IL-1β), tumour necrosis factor-α (TNF-α), or receptor activator for nuclear factor-κB ligand (RANKL) [68-71]. RANKL, which binds to RANK (receptor activator for nuclear factor-κB) to induce the NF-κB activation pathway, a membrane protein expressed on the surface of osteoclasts, is, itself, activated by the ATP-dependent proton pump vacuolar-type $H^+$-ATPase (V-ATPase) [72]. The extra-cellular stimuli culminate, through a complex upstream signalling cascade, in the activation of the IκB kinase complex IKK [59, 60, 73-75]. IKK, which is composed of two catalytic subunits (IKKα and IKKβ) and of one regulatory subunit (IKKγ, which is also called NEMO), phosphorylates IκBα [59, 74, 75]. The phosphorylation of the N-terminal serines $Ser^{32}$ and $Ser^{36}$ of IκBα leads to conformational changes within the protein and to a consequent exposure of the Lys-containing region of IκBα for ubiquitination [59, 74, 75]. The $Lys^{48}$-linked polyubiquitin chains that attach to IκBα at $Lys^{21}$ and/or $Lys^{22}$ target IκBα for rapid degradation by the 26S proteasome [59, 76-78]. Once freed from IκB, NF-κB translocates into the nucleus, where it activates its target genes [79]. The NF-κB activation pathway described above, which is generally referred to as the "classical" or "canonical" NF-κB activation pathway, is illustrated in figure 4. It is the most common NF-κB activation pathway [62, 73, 80], and the only one discussed in the present paper.

The activation of NF-κB can be inhibited at different levels along the activation cascade. The major targets of NF-κB inhibitors are the binding of NF-κB to its DNA binding sites, the degradation of IκB by the 26S proteasome, and the phosphorylation of IκB by IKK (figure 4).

The binding of NF-κB to DNA can be inhibited by molecules that mask either the NF-κB binding sites on DNA or the DNA binding domain on NF-κB [81]. However, because of the large interaction surface mediating the binding of NF-κB to DNA, it is unusual to find natural products that are large enough to block the binding of NF-κB to DNA [81]. The majority of natural products reported to date to interfere directly with NF-κB-DNA binding are terrestrial sesquiterpene lactones [81, 82]. Sesquiterpene lactones possessing α,β-unsaturated carbonyl groups are thought to interfere with the binding of NF-κB to its binding sites on DNA by undergoing Michael-type conjugate additions to the nucleophilic cysteine sulfhydryl groups $Cys^{38}$ and $Cys^{120}$ in the p65 monomer of NF-κB [82-84]. Helenalin (1) isolated from the terrestrial plant *Arnica* sp. is one of the first natural sesquiterpene lactones reported to inhibit NF-κB [85]. Another approach to inhibit the binding of NF-κB to DNA involves the creation of steric hindrance in the DNA binding region of p50 through the formation of hydrogen bonds with amino acids in that region. One example of a natural product interfering with the binding of NF-κB to DNA in that way is gallic acid (2) isolated from garlic. Gallic acid 2 has been shown to form strong hydrogen bonds with $Ser^{66}$ in the DNA-binding region of p50.

Further upstream, NF-κB activation can be inhibited by compounds targeting the proteasomal degradation of IκB [86-88]. Proteasome inhibitors include the peptide boronate Velcade® (3) (also known as bortezomib or PS-341) , which is already on the market as an anti-cancer drug [86, 88], and the marine γ-lactam-β-lactone salinosporamide A (4)[89], which is currently in clinical trials [86].

A further step upstream, NF-κB activation can be inhibited by compounds interfering with the kinase activity of IKK [81, 90].

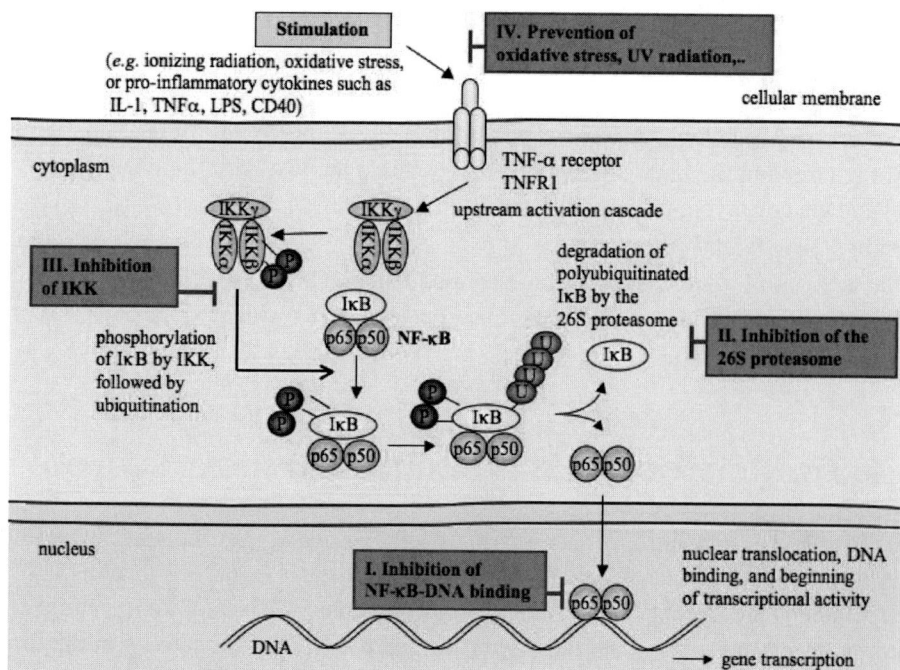

Figure 4. The NF-κB activation pathway ("classical pathway"). Upon stimulation by pro-inflammatory cytokines, ionizing radiation, or oxidative stress, a complex upstream activation cascade leads to the phosphorylation and activation of the IκB kinase complex IKK. IKK phosphorylates IκB. Once phosphorylated, IκB is polyubiquitinated, and the polyubiquitin chains target IκB for degradation by the 26S proteasome. After the degradation of IκB, NF-κB is free to translocate into the nucleus, and to induce the transcription of its target genes (modified after Keutgens *et al.* (2006))[60]. The major targets for NF-κB inhibitors are highlighted in green: NF-κB inhibitors can target the binding of NF-κB to DNA (I), the degradation of polyubiquitinated IκB by the 26S proteasome (II), or the phosphorylation of IκB by IKK (III). NF-κB activation can also be slowed down by compounds that protect against UV radiation or that prevent oxidative stress which would otherwise induce the NF-κB activation cascade.

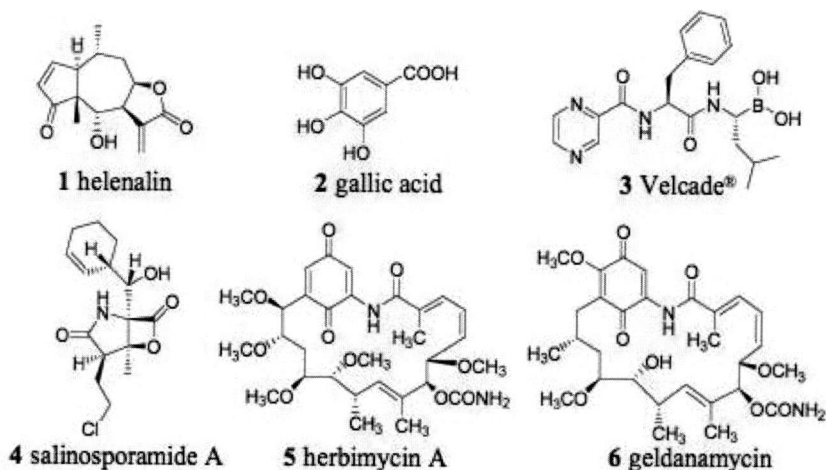

**1 helenalin**          **2 gallic acid**          **3 Velcade®**

**4 salinosporamide A**     **5 herbimycin A**          **6 geldanamycin**

Scheme 1.

The inhibition of IKK is viewed as the most specific way of interfering with the NF-κB activation pathway, as IKK does not phosphorylate any protein outside the NF-κB signalling cascade, while the proteasome degradation of proteins and the binding of transcription factors to DNA are implicated in a wide range of biological processes [90]. Natural products specifically targeting the kinase activity of IKK include the two benzoquinones herbimycin A (5) [91] and geldanamycin (6) [92] isolated from bacteria.

Finally, NF-κB activation can be slowed down by antioxidants such as vitamins C (ascorbic acid) and E (α-tocopherol), co-enzyme Q10, and a variety of polyphenolics [81, 90, 93-95]. Antioxidants reduce reactive oxygen species (ROS) which could otherwise activate NF-κB [75, 81, 90, 93, 96].

# 3. The Role of NF-κB In Marine Chemical Ecology

As described in the introduction, marine chemical ecology is the field of science that investigates the ecological role of natural products in the marine ecosystem [3, 4, 8-11]. Although there is a coherent link between the anti-predatory effects of toxic natural products released by sessile soft-bodied marine organisms and the anti-cancer activity of the former, in many cases, there is no easy explanation for the production of secondary metabolites by a particular organism. If the ecological reason behind the production of a secondary metabolite of interest can be understood, this can be of tremendous value for the large-scale production of the latter. For instance, once the ecological role of the compound of interest is understood, bioprospecting for the compound can be enhanced by culturing the source organism under conditions that trigger the production of the compound. The chemical ecology behind the production of a natural product can also provide valuable cues about various environments or organisms which can potentially serve as an alternative source of the compound of interest, or of analogues thereof. In this way, the sustainability of the large-scale production of the compound can be increased, and the natural biodiversity of the source organisms can be protected. It is generally accepted that natural products have evolved under the pressure of natural selection to bind to specific receptors in ecological targets, and that evolutionary pressures may have compelled marine organisms to produce substances that cause growth inhibition or mortality in their competitors [6]. In the case of NF-κB inhibition, there are several potential evolutionary and ecological explanations for the finding of NF-κB inhibiting natural products in marine organisms (figure 5).

## 3.1. Evolution of the Presence of NF-κB and NF-κB Analogues in Marine Organisms

From an evolutionary point of view, one interesting potential explanation for the finding of NF-κB inhibitors in marine organisms is the fact that marine invertebrates and fish, no matter how distantly related to us they appear to be, possess, in many cases, NF-κB or closely related analogues. As a matter of fact, the NF-κB/IκB cascade has a very ancient evolutionary origin reaching all the way back to the "living fossil" and most ancient arthropod

*Carcinoscorpius rotundicauda* (horseshoe crab) [97, 98]. Members of the Rel/NF-κB family have been reported in the sea hare *Aplysia californica* [99], in the sea urchin *Strongylocentrotus purpuratus* [100], in the mussel *Mytillus galloprovincialis* [101], in the crab *Chasmagnathus granulatus* [102], in ascidians [103], in the oyster *Crassostera gigas* [104], in the hydrothermal vent mussel *Bathymodiolus azoricus* [105], in marine cnidarians [106], and in sponges [106, 107]. In marine bivalves, including the mussels *Mytillus galloprovincialis* and *Bathymodiolus azoricus* and the oyster *Crassostera gigas,* NF-κB appears to function primarily as an inducer of the innate immune system, as it is involved in the transcription of genes encoding for antibacterial molecules such as mytilin [105].

Oysters (*Crassostera gigas* and *Pinctada fucata*) have been shown to possess an IKK-like protein oIKK that share structural and functional properties with their mammalian homologues. When transfected into human cells, oIKK has been shown to activate NF-κB-controlled reporter genes [108, 109].

Another mollusc, the Hawaiian squid *Euprymna scolopes,* has also been shown to possess analogues of several elements of the NF-κB activation pathway, including Rel, IκB, IKKγ, and TRAF6 [110]. Goodson *et al.* (2005) [110] showed that the NF-κB activation pathway of is modulated by the squid's beneficial symbiont *Vibrio fischeri.*

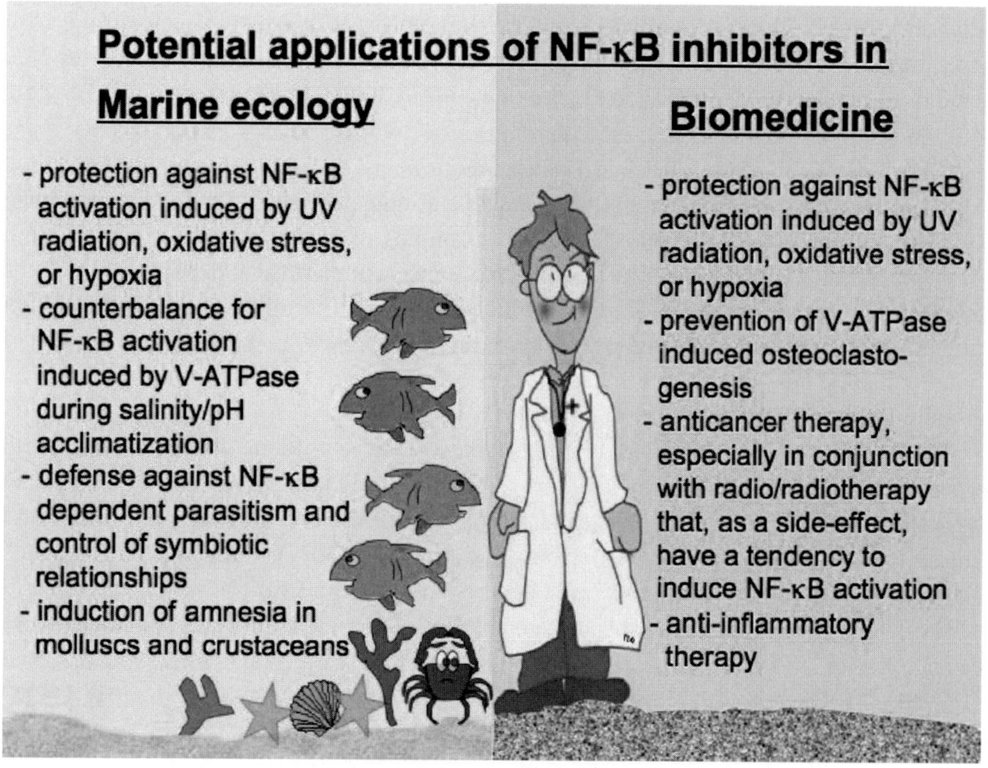

**Potential applications of NF-κB inhibitors in Marine ecology**

- protection against NF-κB activation induced by UV radiation, oxidative stress, or hypoxia
- counterbalance for NF-κB activation induced by V-ATPase during salinity/pH acclimatization
- defense against NF-κB dependent parasitism and control of symbiotic relationships
- induction of amnesia in molluscs and crustaceans

**Biomedicine**

- protection against NF-κB activation induced by UV radiation, oxidative stress, or hypoxia
- prevention of V-ATPase induced osteoclasto-genesis
- anticancer therapy, especially in conjunction with radio/radiotherapy that, as a side-effect, have a tendency to induce NF-κB activation
- anti-inflammatory therapy

Figure 5. Potential applications of NF-kB inhibitors in marine ecology and in biomedicine.

In the interesting case of ascidians, the NF-κB proteins As-rel1 and As-rel2 have been shown to be involved in the regulation of the formation and degradation of the notochord [103]. Ascidians are unique in the sense that they start their larval life as vertebrates, bearing, like all other chordates, a notochord at the centre of the larval tail, and then gradually losing their notochord to become invertebrate adults [103].

With the presence of NF-κB or NF-κB analogues in marine organisms [97, 99, 100, 102, 111-115], it can be hypothesized that the same organisms also produce antagonists of the activation of NF-κB. Such antagonists might play critical roles in the switching-off of positive feed-back loops within the NF-κB activation pathway, or in the halting of aberrant NF-κB production.

## 3.2. Potential Ecological Reasons for the Production of NF-κB Inhibitors by Marine Organisms

### 3.2.1. Defence Against UV Radiation, Oxidative Stress, and Hypoxia
Ultraviolet (UV) radiation mainly affects intertidal plants and animals, but planktonic and subtidal benthic organisms, and sessile coral reefs organisms in particular, are also exposed to harmful levels of UV radiation in surface waters and at shallow depths [3]. Absorption of UV photons causes organic molecules to undergo conformational changes that can interfere with their vital metabolic functions, and it is now widely accepted that short wavelength UV radiation induces NF-κB activation [116-118]. In addition to it's own effects on the NF-κB activation cascade, UV radiation mediates the formation of reactive oxygen species (ROS) including the superoxide anion radical peroxides, ($O_2^-$), the hydroxyl radical ($^\cdot$OH), hydrogen peroxide ($H_2O_2$), and other free radicals that cause NF-κB-inducing oxidative stress [117, 118]. As a chemical defence against UV radiation, many marine organisms produce potent UV-absorbing sunscreens and antioxidants capable of quenching photo-oxidative reactions [3, 119]. Marine sunscreens, the most common examples of which include mycosporine-like amino acids (MAAs), scytonemin (7), phlorotannins, coumarins, and polyphenolics, have been documented in depth by Karentz (2001) [119]. Marine antioxidants include several carotenoids, tocopherols, phycocyanins, and anthocyanins [95, 119, 120]. Amongst the strongest marine antioxidants are the algal derivatives cymopol (8) and avrainvilleol (9) and the spongean metabolite puupehenone (10) [95].

Antioxidants play a particularly important role in sessile or slow-moving intertidal organisms. As a matter of fact, the latter are not just exposed to ROS formed by UV radiation. They are also continuously experiencing aerobiosis-anaerobiosis transitions, and the abrupt reintroduction of oxygen into their metabolism during the return from an anoxic state back to an oxygen-rich state often leads to a rapid generation of ROS [121].

Oxidative stress is not the only oxygen-related environmental condition that has been associated with NF-κB induction. At the other extreme, hypoxia, which impairs energy availability and hence cell viability [122], is also linked to NF-κB activation [123]. In humans, it has recently been discovered that NF-κB activation is necessary for the transcription of *Hif1α* mRNA under hypoxic conditions [123, 124]. Hypoxia-inducible transcription factor-1 (HIF-1) upregulates pro-angiogenic factors to restore nutrient, oxygen, and energy supply in order to maintain tissue integrity and homeostasis [123]. HIF-1 activation has also been associated with the innate immune response against bacterial infection [123]. However, aberrant expression and activation of HIF-1α can be detrimental [125, 126], and it can be hypothesized that marine organisms have developed antagonists to the expression of HIF-1, including compounds targeting the NF-κB-induced transcription of the protein. Several marine natural products, including the diterpenoid laurenditerpenol (11) isolated from the tropical alga *Laurencia intricata* [127], the triterpenoids sodwanone A (12) and yardenone A (13) isolated from the South African sponge *Axinella* sp. [128], norsesterterpene peroxides from the marine sponge *Diacarnus levii* [125], various benzochromenones isolated from the marine crinoid *Comantheria rotula* [129], the phenolic pyrrole 7-hydroxyneolamellarin A (14) isolated from the sponge *Dendrilla nigra* [130], various strongylophorines isolated from the sponge *Petrosia (Strongylophora) strongylata* [131], and the macrolide latrunculin A (15) isolated from the Red Sea sponge *Negombata magnifica* [132] have been reported as HIF-1 inhibitors.

### 3.2.2. Osmoregulation and Adaptation to Changes in pH

Marine organisms are frequently exposed to salinity or pH changes, and in order to acclimatize to those changes, they rely on enzymes such as $Na^+$, $K^+$-ATPase and vacuolar-type $H^+$-ATPase (V-ATPase) [133]. V-ATPase is found in vacuoles, lysosomes, and plasma membranes of many different types of cells. Plasma membrane V-ATPase is generally associated with pH homeostasis. In the intercalated cells of kidneys, V-ATPase is implicated in the pumping of protons into the urine, to allow the reabsorption of bicarbonate into the blood [134, 135].

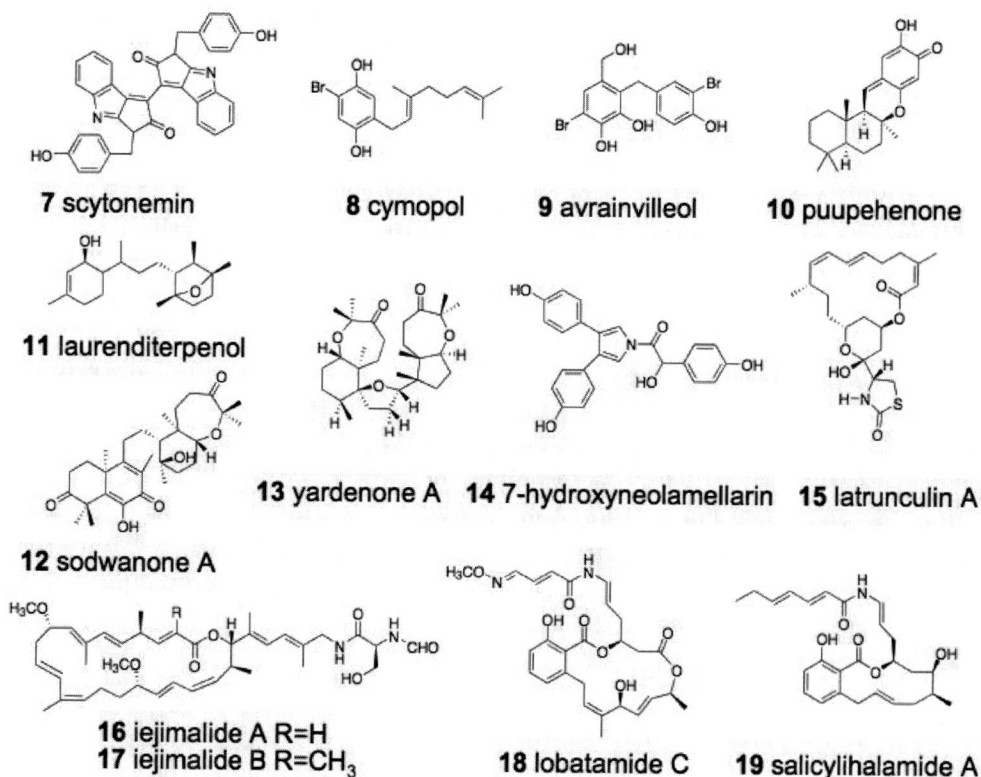

**7 scytonemin**   **8 cymopol**   **9 avrainvilleol**   **10 puupehenone**

**11 laurenditerpenol**   **13 yardenone A**   **14 7-hydroxyneolamellarin**   **15 latrunculin A**

**12 sodwanone A**

**16 iejimalide A R=H**
**17 iejimalide B R=CH$_3$**   **18 lobatamide C**   **19 salicylihalamide A**

Scheme 2.

As mentioned in the introduction, V-ATPase-mediated acidosis activates RANKL in osteoclasts [72]. The binding of RANKL to RANK on the membrane surface of osteoclasts induces the NF-κB activation pathway and promotes NF-κB-dependent osteoclastogenesis [72]. Again, it can be hypothesized that marine organisms are likely to produce V-ATPase inhibitors to counteract aberrant activation of V-ATPase. Alternatively, they may produce compounds that target NF-κB activity resulting from V-ATPase activation induced during the process of osmoregulation or of adaptation to changes in pH. Marine natural products inhibiting V-ATPase include the macrolides iejimalide A (16) and B (17) isolated from the tunicate *Eudistoma rigida*, lobatamide C (18) isolated from the tunicate *Aplidium lobatum* [134], and salicylihalamide A (19) isolated from the sponge *Haliclona* sp. [134, 135].

Examples of marine sunscreens, antioxidants, HIF-1 inhibitors, and V-ATPase inhibitors are presented in scheme 2.

### 3.2.3. Defence against parasites and interference with learning processes and memory in other marine organisms

The natural response of cells infected by parasites is often to commit apoptosis and deprive the parasite of an opportunity to proliferate within the infected organism. Although this particular aspect of ecology has, so far, been mainly investigated amongst terrestrial organisms and there is hardly any literature covering marine examples, there are many reports of viral, bacterial, and protozoan parasites taking advantage of the anti-apoptotic pathways to escape host defence mechanisms [136-140]. Parasites reported to induce the activation of NF-κB include human immunodeficiency virus type 1 (HIV-1), adenovirus 5, human T cell lymphotropic virus type 1 (HTLV-1), Epstein-Barr virus (EBV), hepatitis B virus (HBV), hepatitis C virus (HCV), the influenza virus, the tick *Theileria* sp., and the protista *Leishmania major*, *Cryptosporidium parvum* and *Toxoplasma gondii* [136, 137, 139, 141]. The tick *Theileria* sp. has been shown to depend on the constitutive activation of NF-κB for survival in its host. The constitutive activation of NF-κB is achieved through the recruitment of IKK into large IKK signalosome assemblies at the surface of the parasite [142]. IKK has also been shown to be constitutively activated in HIV-1, adenovirus 5, or HTLV-1 infected cells [137, 139, 143]. EBV, on the other hand, produces an oncoprotein, LMP1, that interacts with the TNF-α receptor associated factor 2 (TRAF2) [139], while the polypeptide HBx produced by HBV is directly implicated in the degradation of IκBα [139]. Noteworthy, it is now well established that constitutive activation of NF-κB as a consequence of viral infection often leads to malignancies in humans [137] (figure 5).

Some parasites, including the bacteria *Salmonella* sp., *Yersinia* sp., and *Pseudomonas* sp., are also producing NF-κB inhibitors, in order to attenuate inflammatory responses that would otherwise lead to their clearance from the host cells [138, 144] (figure 5).

One of the earliest reports of the presence of NF-κB analogues in marine invertebrates describes the role of ApNF-κB, an axoplasmic protein found in the mollusc *Aplysia californica*. ApNF-κB is involved in learning processes of the sea hare and is rapidly inactivated after nerve injury [99]. Similarly, the crab *Chasmagnathus granulatus* has been reported to possess an NF-κB analogue that plays a crucial role in long-term memory [102]. The inhibition of NF-κB in *Chasmagnathus granulatus* by inhibitors of IKK induces amnesia in the crab [102, 145] (figure 5). To date, there is no reported evidence for ecological roles of interferences with learning processes or long-term memory amongst marine organisms.

# 4. MARINE INHIBITORS OF NF-κB

Several marine natural products with NF-κB inhibitory properties have been described during the last decade. However, the exact targets and mechanisms of action of most marine NF-κB inhibitors remain poorly understood [81, 146, 147].

Marine natural products inhibiting the 26S proteasome include the γ-lactam-β-lactone salinosporamide A (4) isolated from the actinomycete *Salinispora tropica* [148] and the spongean metabolites agosterol C (20) and secomucalolide A (21) [149] (scheme3). The macrolide bryostatin 1 (22) isolated from the marine bryozoan *Bugula neritina* and from its γ–proteobacterial symbiont *Candidatus endobugula sertula* [150] has never been reported as a direct inhibitor of NF-κB, but it has potent synergetic anti-cancer effects with several chemotherapeutical drugs, and with the prototypical proteasome inhibitor lactacystin in particular [151]. The major target of bryostatin 1 (22), which is currently in phase I and phase II clinical trials as an anti-cancer drug, is the serine/threonine protein kinase C (PKC) [152, 153]. The PKC signalling pathway, which occurs upstream of the NF-κB activation cascade, has been shown to be essential for the activation of NF-κB in various cell-lines [62, 154-156].

Inhibitors of IKK include the two naphtopyrones 6-methoxy-comaparvin (23) and 6-methoxy-comaparvin-5-methyl ether (24) isolated from the crinoid *Comanthus parvicirrus* [157] (scheme 3). The carotenoid astaxanthin (25) isolated from various bacteria and algae by [158]. and the indole-alkaloid sunscreen scytonemin (7) isolated from cyanobacteria [159], which have antioxidative and UV-absorbing properies, respectively, are known to inhibit NF-κB upstream of the activation pathway (scheme 3).

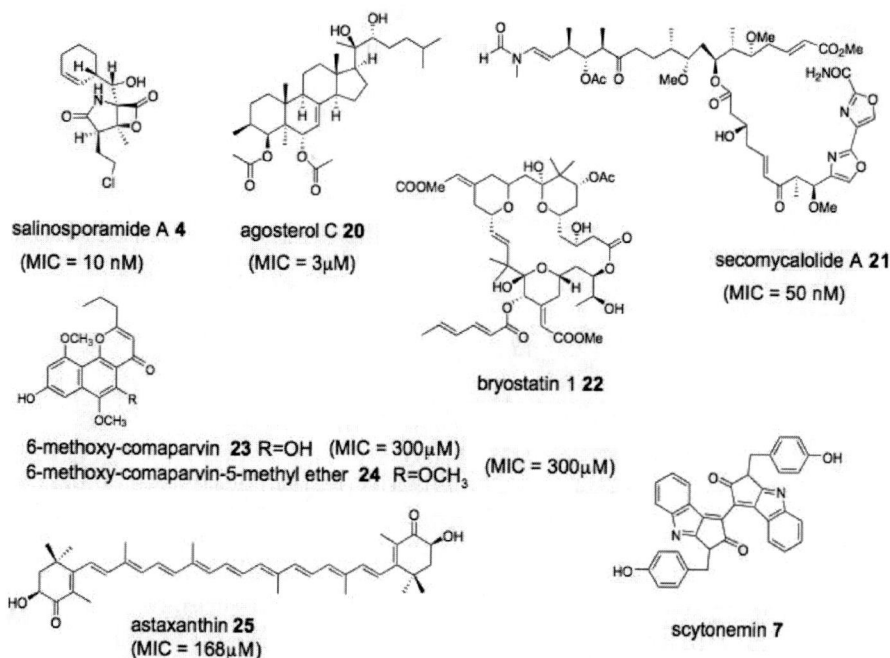

salinosporamide A 4
(MIC = 10 nM)

agosterol C 20
(MIC = 3μM)

secomycalolide A 21
(MIC = 50 nM)

bryostatin 1 22

6-methoxy-comaparvin 23 R=OH   (MIC = 300μM)
6-methoxy-comaparvin-5-methyl ether 24  R=OCH₃   (MIC = 300μM)

astaxanthin 25
(MIC = 168μM)

scytonemin 7

Scheme 3. Marine NF-kB inhibitors with identified targets. Where available, the MIC values are provided.

Marine natural products with unidentified or poorly established molecular targets include the bacterial secondary metabolites cycloprodigiosin hydrochloride (26) [160, 161] and streptochlorin (27) [162], as well as the fungal secondary metabolite verracurin (28) [163] and several natural products isolated from sponges. NF-κB inhibitors isolated from marine sponges include the sesterterpene lactones cacospongiolide B (29) [164], petrosaspongiolide M (30) [165], and cyclolinteinone (31) [166], the sesquiterpene hydroquinone avarol (32) [167], the bromopyrrol alkaloid hymenialdisin (33) [168-170], the sesquiterpene benzoquinone ilimaquinone (34) [171], and the diterpene cycloamphilectene (35) [172] (scheme 4).

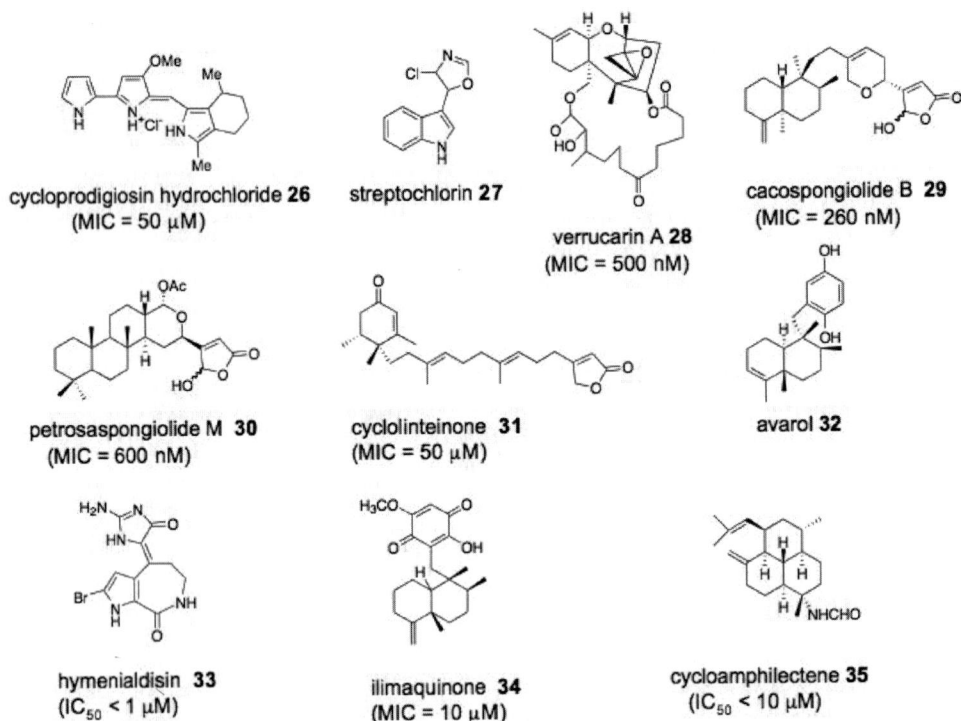

cycloprodigiosin hydrochloride 26 (MIC = 50 μM)

streptochlorin 27

verrucarin A 28 (MIC = 500 nM)

cacospongiolide B 29 (MIC = 260 nM)

petrosaspongiolide M 30 (MIC = 600 nM)

cyclolinteinone 31 (MIC = 50 μM)

avarol 32

hymenialdisin 33 (IC$_{50}$ < 1 μM)

ilimaquinone 34 (MIC = 10 μM)

cycloamphilectene 35 (IC$_{50}$ < 10 μM)

Scheme 4. Marine NF-kB inhibitors without identified targets. Where available, the MIC or the IC$_{50}$ values are provided.

## 5. EXAMPLES OF SYNTHETIC MODIFICATIONS THAT HAVE BEEN ATTEMPTED ON MARINE NATURAL PRODUCTS IN THE CONTEXT OF NF-κB INHIBITORY ACTIVITY

Marine natural products are now widely recognized as highly promising drug candidates and, as mentioned in the introduction, several marine natural products are currently in clinical trials [2, 14, 173-177]. With the number of marine natural products identified as NF-κB inhibitors steadily increasing, marine metabolites will doubtlessly gain further importance in the field of anti-cancer and anti-inflammatory drug discovery. However, marine natural products are, in general, an unsustainable source of druggable compounds [13, 19, 146, 178,

179]. The natural abundance of the source organisms is, in most cases, too low to yield sufficient material for clinical trials without destroying the marine ecosystem, and the access to the collection sites is often very limited. Aquaculture of marine invertebrates and algae for pharmaceutical purposes has been revealed to be extremely challenging, and the large-scale fermentation of marine microbes, although more successful than other strategies used in the production of marine drugs, can be troublesome as well [7, 11, 13, 58, 179-184]. For these reasons, chemical synthesis is, in most cases, an unavoidable step in the large-scale production of marine drugs. In addition to serving as an approach to get around the hurdles associated with the harvest of marine drugs from their natural sources, the chemical synthesis of bioactive marine natural products offers the possibility to produce analogues of the natural products, and to use the former for structure-activity relationship (SAR) studies. SAR studies on analogues of lead molecules play a crucial role in the drug development, as they can lead to the discovery of more potent, less toxic, and more bioavailable drug candidates [185].

Five marine natural products involved in the inhibition of NF-κB activation have been synthesized and have become the subject of detailed SAR studies. These are the proteasome inhibitor salinosporamide A (4) isolated from the marine actinomycete *Salinospora tropica*, the protein kinase C (PKC) inhibitor bryostatin 1 (22) isolated from the bryozoan *Bugula neritina*, and the three sponge-derived natural products hymenialdisine (33), avarol (32), and cacospongionide B (29). A brief survey of the strategies used for the chemical synthesis of these five compounds and their analogues, and the results obtained from the NF-κB-related SAR studies are presented below.

## 5.1. Chemical Synthesis and NF-κB-Related SAR of Salinosporamide A and Its Analogues

The γ-lactam-β-lactone salinosporamide A (4) isolated from the marine actinomycete *Salinospora tropica* is by far the most widely documented marine lactone with NF-κB inhibitory properties [148, 186-188]. The compound, which was discovered at the beginning of this decade by William Fenical *et al.* at the Scripps Institution of Oceanography (University of California, San Diego (U.S.A.)) [189] and by Barbara Potts *et al.* at Nereus Pharmaceuticals in San Diego (U.S.A.) [148], targets the degradation of IκB by the 26S proteasome [148, 186-188] and it has been reported as a suppressor of RANKL-induced osteoclastogenesis [188]. Salinosporamide A (4) entered phase I clinical trials in May of 2006, initially against solid tumors and leukemia, and in April of 2007 another phase I trial against multiple myeloma was initiated [190]. Salinosporamide A (4) is still in anti-cancer clinical trials at the moment[190].

The first total syntheses of salinosporamide A (4) and of some of its analogues were achieved by Corey *et al.* at Harvard University (U.S.A.) [191, 192] and by Danishefsky *et al.* at Columbia University (U.S.A.) [193] in 2004-2005. One of the major challenges for both research groups was the stereo-controlled assembly of the γ-lactam group. Another challenge was the water-sensitive β-lactone moiety [191, 193]. Corey *et al.* started the synthesis from L-threonine methyl ester [191], while Danishefsky *et al.* started from a bicyclo derivative of L-glutamic acid [193]. In Corey's synthetic route, the L-threonine methyl ester underwent condensation and silyation before cyclizing into a *cis*-fused ring under Stork conditions.

Finally, the addition of cyclohexenyl zinc led to the diastereo-controlled formation of salinosporamide A (4) [191]. The first step of Danishefsky's synthetic route involved the formation of the γ-lactam frame through conjugate addition and alkylation of the starting block. After two cyclization steps, the synthesis of salinosporamide A (4) was completed through the addition of cyclohexenyl zinc as in Corey's synthetic route [193]. Following their initial synthesis of salinosporamide A (4) in 2004, Corey et al. have developed a new, slightly simpler synthetic route for the compound [194]. Recently, an asymmetric synthetic pathway has been described for (-)-salinosporamide A [195], and several derivatives of salinosporamide A (4) have been synthesized over the last three years [186, 192, 194]. Additionally, the large-scale fermentation of the source organism *Salinispora tropica* under various culture conditions has led to the isolation of several novel salinosporamide derivatives [186, 187].

The SAR of salinosporamide A (4) analogues has been studied in depth by Barbara Potts *et al.* at Nereus Pharmaceuticals (San Diego, U.S.A.). The results, which have been published by Macherla *et al.* (2005) [186] and by Reed *et al.* (2007) [187], are summarized in scheme 5.

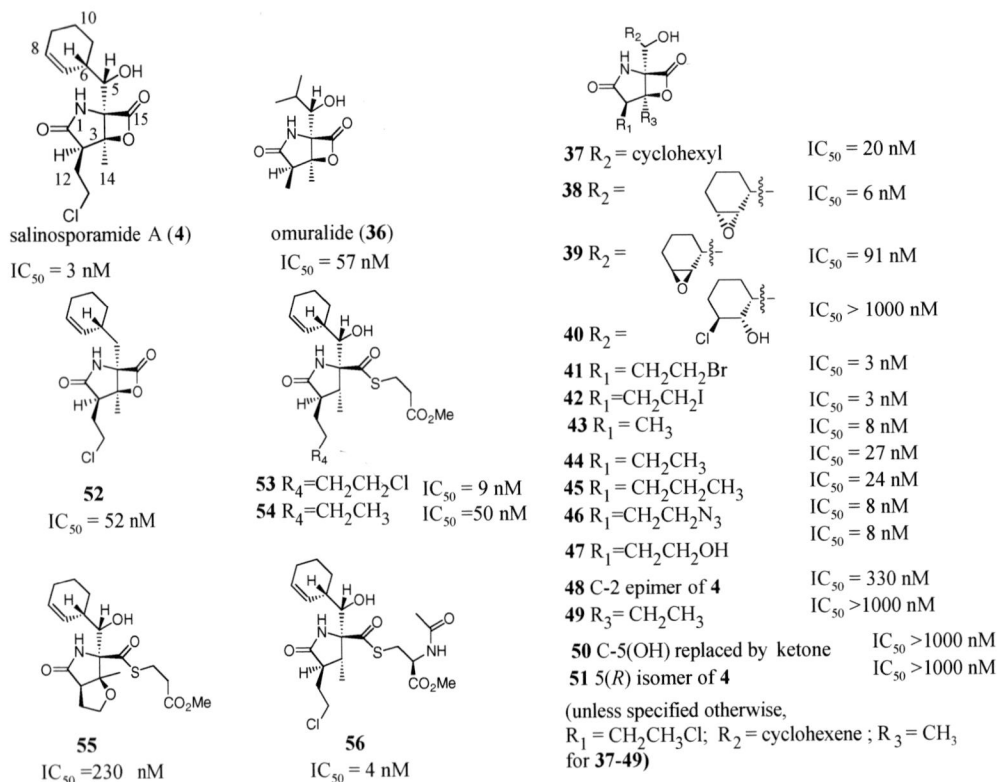

Scheme 5. Analogues of salinosporamide A (4) investigated by Macherla *et al.* (2005) [186]for their effects on the proteolytic activity of the 26S proteasome. IC$_{50}$ values (in nM) refer to the inhibiton of the enzymatic activity of the chymotrypsin-like (CT-L) subunit of the 26S proteasome in rabbit (*in vitro*).

Amongst the salinosporamide A (4) analogues included in the SAR study performed by Macherla et al. (2005) [186], the strongest inhibitors of the proteolytic activity of the chymotrypsin-like (CT-L) catalytic unit of the 26S proteasome where salinosporamide A (4) itself and the analogues 38, 41, and 42 (scheme 5). These compounds were also the most potent inhibitors of NF-κB activation in luciferase reporter gene assays performed by Macherla *et al.* (2005) (with IC$_{50}$ values $\leq 34$ nM) [186].

The results of the SAR study performed by Potts *et al.* [186] show that modification to the cyclohexene ring is only moderately tolerated in terms of NF-κB inhibition potential and 20S proteasome inhibition activity. In the SAR studies reported by Macherla *et al.* (2005) [186], the hydrogenation of the cyclohexene ring in (37) resulted in up to 10-fold loss of activity. Furthermore, the (7*S*,8*S*)-epoxide (39) of salinosporamide A (4) was less active than its (7*R*,8*R*) diastereomer (38). This observation and the fact that chlorohydrin (40) was shown to be 3-log units less active than salinosporamide A (4) proved to Macherla *et al.* that steric hindrance at either side of the cylcohexene moiety has a negative effect on bioactivity. However, halogen exchange such as in 41 and 42 resulted in equivalent bioactivity, underlining the crucial role of the chlorine leaving group in salinosporamide A (4). The inversion of the C-2 or C-5 stereocenters (48; 51) resulted in a dramatic loss of the observed activity, providing further support for the hypothesis that the chlorine leaving group of salinosporamide A (4) plays a critical role in the bioactivity of the natural product. The role of the chlorine leaving group and of other halogenated leaving groups in the proteasome inhibitory activity of salinosporamide (4) analogues has been investigated by Brad Moore *et al.* at the Scripps Institution of Oceanography (University of California, San Diego, U.S.A.) [196, 197], and the results have revealed the opening up of the γ-lactam β-lactone ring and the covalent tethering of salinosporamide analogues to the proteasome is a mechanism that is reversible if fluoride is present in the leaving group, but irreversible if chloride is present in the leaving group [148, 196, 197]. The oxidation of the C-5 alcohol, which has been identified as a pharmacophore involved in the binding of salinosporamide A (4) to the threonine residue Thr[21] of the chymotrypsin-like catalytic subunit of the 26S proteasome [186], to a ketone in (50), was shown by Macherla et al. (2005) [186] to result in a significant loss of activity. Finally, the substitution of the C-3 methyl group by an ethyl in (49) was accompanied by a significant loss of activity. This loss of activity was attributed by Macherla *et al.* to an increase in steric hindrance [186]. As described in Reed *et al.* [187], the presence of the thioester group is very likely to be responsible for the weaker 26S proteasomal inhibition potential of the salinosporamide A (4) analogues 53-56 [187].

The conclusions drawn from the SAR studies performed by Barbara Potts *et al.* at Nereus Pharmaceuticals (San Diego. U.S.A) and by Brad Moore *et al.* at the Scripps Institution of Oceanography (University of California, San Diego, U.S.A.) are presented in scheme 6. The results of the SAR studies, together with crystallography studies published by Groll *et al.* (2006) [148], have resulted in the identification of the pharmacophores of salinosporamide A (4) [186, 187], which are highlighted in blue and in magenta in scheme 6. The SAR and crystallography studies revealed that the mode of action of salinosporamide A (4) involves covalent ester linkages between the 26S proteasome and salinosporamide A (4) resulting from the nucleophilic addition of the threonine residue Thr[1] within the catalytic subunits of the proteasome to the carbonyl group of the β-lactone and from the opening of the lactone ring. The opening of the lactone ring is followed by the intramolecular nucleophilic addition of C-3O to the chloroethyl group of salinosporamide A (4), which leads to the formation of a cyclic ether [148, 186].

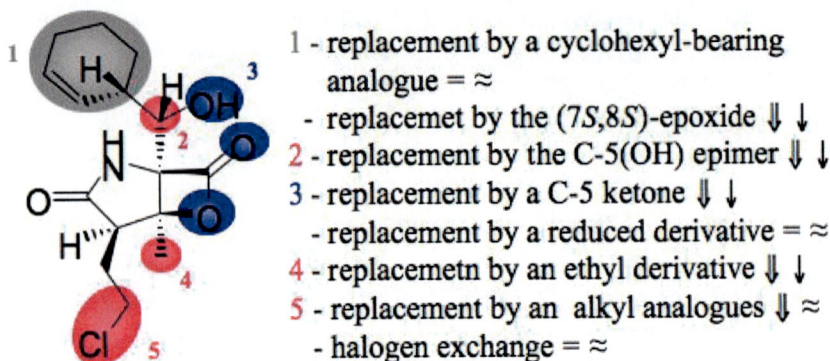

1 - replacement by a cyclohexyl-bearing
   analogue = ≈
 - replacemet by the (7S,8S)-epoxide ⇓ ↓
2 - replacement by the C-5(OH) epimer ⇓ ↓
3 - replacement by a C-5 ketone ⇓ ↓
 - replacement by a reduced derivative = ≈
4 - replacemetn by an ethyl derivative ⇓ ↓
5 - replacement by an alkyl analogues ⇓ ≈
 - halogen exchange = ≈

⇓: decrease in NF-κB inhibition activity

↓: decrease in the potential to inhibit the proteolytic activity of the chymotrypsin-like (CT-L) catalytic subunit of the 26S proteasome

=: no major effect on the NF-κB activation pathway

≈: no major effect on the proteolytic activity of the chymotrypsin-like (CT-L) catalytic subunit of the 26S proteasome

Scheme 6. Summary of the results obtained by Macherla *et al.* (2005) [186] in their SAR study of effects of salinosporamide A (**4**) analogues on the proteolytic activity of 26S proteasome and on the activation of NF-κB. The pharmacophores identified during the SAR study and through crystallography studies by Groll *et al.* (2006) [148] are highlighted in colour.

The protonation of the amine group of Thr[1] by C3-OH severely hinders the deacylation activity of the enzyme [148, 198]. The γ-lactam ring fixes the position of C-3O in the salinosporamide A (**4**) to the proteasome complex and precludes the regeneration of the β-lactone ring and the elimination of the compound from the proteasome [148, 186]. The alcohol function of C-5OH is involved in the formation of hydrogen bonds between salinosporamide A (**4**) and the amine of the proteasomal threonine residue Thr[21]. The amide of the γ-lactam ring binds to the glycine residue Gly[47]. The binding of salinosporamide to the proteasomal residues Thr[1] and Gly[47] hinders the enzymatic activity of the 26S proteasome by disturbing the proton shuttling through water molecules present in the vicinity of the latter residues in absence of salinosporamide A (**4**). The cyclohexenyl ring of salinosporamide A (**4**) offers additional hydrophobic interactions to the proteasome [148].

## 5.2. Chemical Synthesis and NF-κB-related SAR of Bryostatin 1 and its Analogues

Bryostatin 1 (**22**) is a macrolide isolated from the bryozoan *Bugula neritina*, and produced by the latter's γ-proteobacterial symbiont *Candidatus endobugula sertula* [150]. Bryostatin 1 (**22**) is a potent anti-neoplastic agent with activity mainly in leukaemic tumours

[199, 200]. The macrolide is currently in phase I and II clinical trials as an anti-cancer drug, but as part of a combination with other drug rather than as a single entity. Although bryostatin 1 (22) does not directly inhibit NF-κB, it has been shown to have synergetic anti-cancer effects together with the prototypical proteasome inhibitor lactacystin [151], as well as with other chemotherapeutic agents, including paclitaxel, and vincristine [184]. The mode of action of bryostatin 1 (22) is related to the macrolide's strong affinity for the regulatory domain of various kinases, and of the serine/threonine protein kinase C (PKC) in particular [152, 153]. In the context of the present paper, it is noteworthy that the PKC signalling pathway, which occurs upstream of the NF-κB activation cascade, has been shown to be essential for the activation of NF-κB in various cell-lines [62, 154-156].

Bryostatin 1 (22) and its analogues of it have been synthesized by several research groups around the world, in order to meet the clinical need for multi-gram quantities of compound [201-204]. However, due to the large number of synthetic steps and the low overall yield, total synthesis has been considered to be inappropriate for the supply of bryostatin 1 (22) for clinical trials. To this date, clinical material is mainly provided through aquaculture carried out by Mendola *et al.* at CalBioMarine (U.S.A.) [205]. Culturing the symbiont of the bryozoan, γ-proteobacteria *Candidatus Endobugula sertula* could dramatically increase the supply of bryostatin1 (22) [206], but *Candidatus Endobugula sertula* remains, to date, unculturable [207].

SAR studies of the inhibition of the kinase activity of PKC by bryostatin 1 (22) and its analogues have revealed that that the epoxidation of the C13-C30 double bond does not alter the binding activity, in contrast to the epoxidation of the two C16-C17 and C21-C34 alkenes which results in a lower activity [208]. The *R*-configuration is required for binding activity to PKC as the chemically modified 26-*epi*-analogue of bryostatin 1 (22) has been shown to exhibit a 25-fold lower activity, and the acetylating of the C26 hydroxyl has lead to a significant reduction in the binding capability of bryostatin 1 (22) to the C1 unit of PKC [208]. The absence of the C19 hydroxyl function in the naturally occurring bryostatin analogues has resulted in a 2 orders of magnitude reduction in activity [184].

Based on these observations, Wender *et al.* developed a computer model for the comparison of the calculated low-energy molecule of 1,2-diacyl-*sn*-glycerol (DAG) (57), the x-ray structures of phorbol 12-myristate 13-acetate (PMA) (58) and bryostatin 1 (22) [209, 210] (scheme 7). The computer model proposed an important role of the C19 and C26 hydroxyl functionalities and the C1 carboxyl group of bryostatin 1 (22), which corroborate spatially with the C4, C9, C20 oxygen atoms of PMA. Based on the study by Wender *et al.*, the C19, C26 and C1 oxygen groups are required for binding to the C1 domain of PKC, whereas the lipophilic regions influences the binding orientation of the molecule and the partition into the membrane [211]. Moreover, the region of the two rings A and B, referred to as "spacer domain", is needed for the control of the conformation of the recognition domain (C4-C16) [210, 212] (scheme 8).

Wender *et al.* pursued their studies by designing two derivatives with simplified A- and B-rings, an intact C-ring, except for a deleted methyl group in analogue 2 (59) [210]. The hydropyranyl (B-ring) was substituted by a dioxane moiety, allowing a simplified synthesis. The coupling of the recognition domain and spacer domain, though macroacetalization, prior to lactonization, led to a convergent synthesis resulting in a reduced number of steps, an improved overall yield and the possibility to easily introduce synthetic modifications.

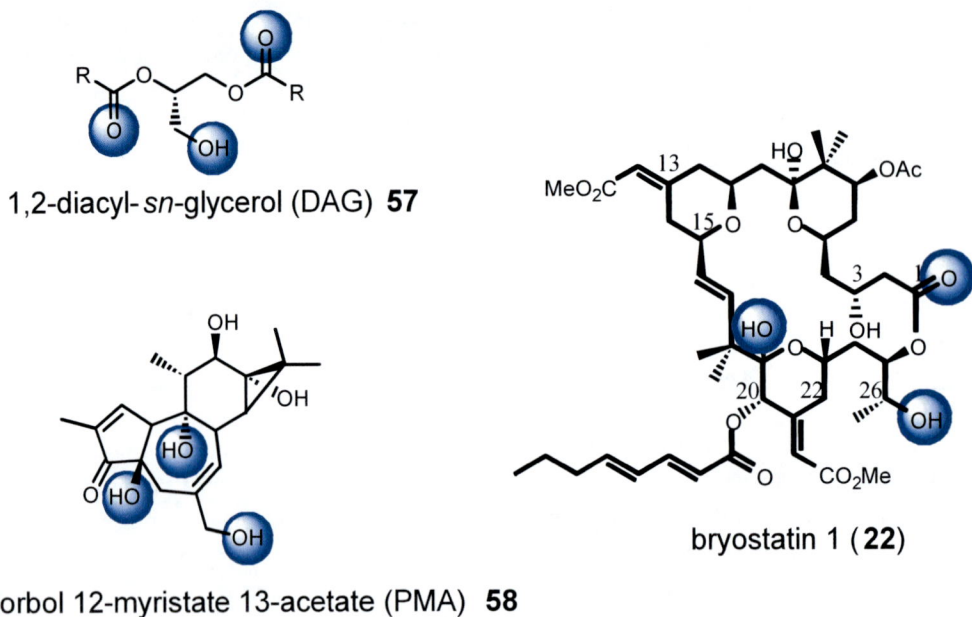

1,2-diacyl-*sn*-glycerol (DAG) **57**

bryostatin 1 (**22**)

phorbol 12-myristate 13-acetate (PMA) **58**

Scheme 7. Chemical structures of bryostatin 1 (22), DAG (57), and PMA (58) used in computer model-based SAR studies of the inhibition of PKC by Wender *et al.* (1998) [210]. The pharmacophores are highlighted in blue.

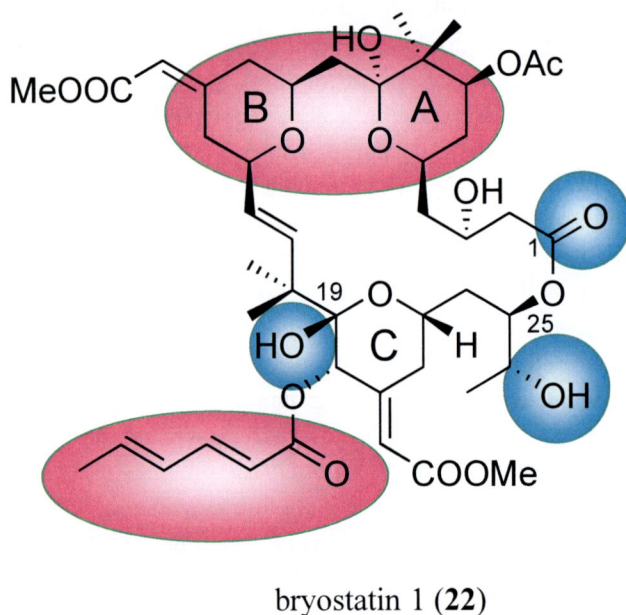

bryostatin 1 (**22**)

Scheme 8. Pharmacophores of bryostatin 1 (22) as identified by Wender *et al.* (1998) [210].

Step economy is one of the hallmarks of the research group of Wender *et al.* permitting the economic supply of sufficient bryostatin analogues (also referred to as "bryologs") for future clinical trials [210, 212, 213] (schemes 9 and 10).

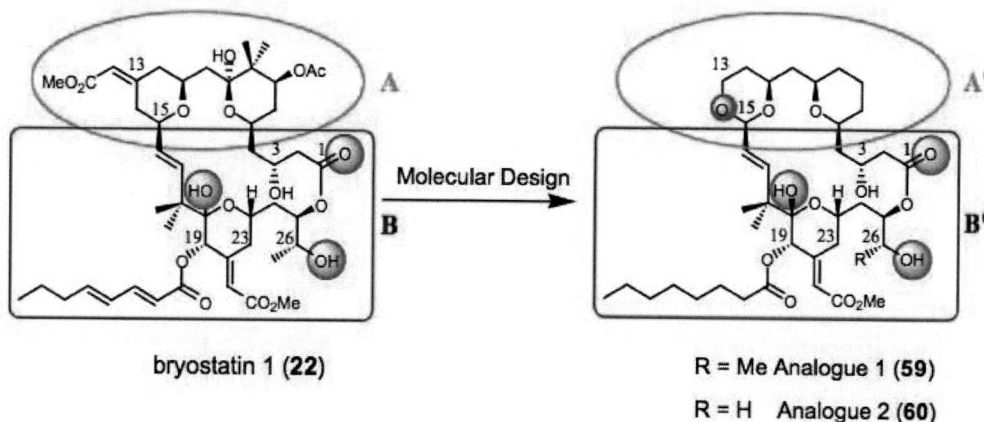

bryostatin 1 (22)

R = Me Analogue 1 (59)

R = H    Analogue 2 (60)

Scheme 9. The first bryostatin 1 analogues 59 and 60 synthesized by Wender *et al.* in 1998 [210]. In comparison with the natural product bryostatin 1 (22), the functionalities along C7-C13 have been eliminated, and the carbon atom in position C14 has been replaced by an oxygen atom.

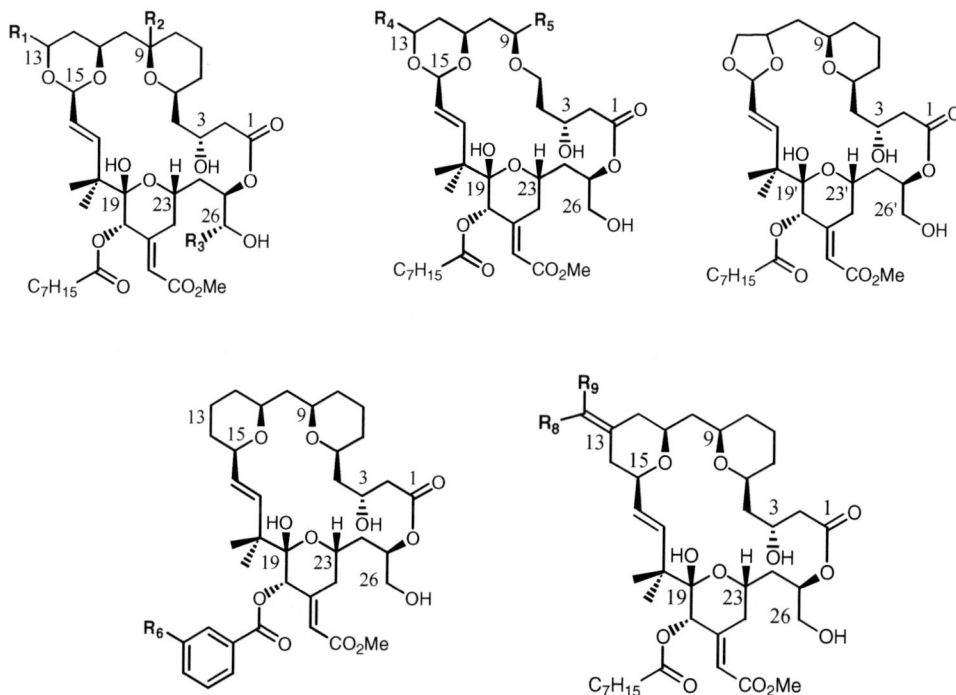

Scheme 10. Chemical structures of the "bryolog" analogues of bryostatin 1 (22) synthesized by Wender *et al.* (1998) [210-217, 220-226]. $R_1$-$R_9$ are listed in tables 1 to 5.

The two bryostatin 1 (22) analogues 59 and 60 were assessed by Wender *et al.* (1998-2008) [210, 214-216] and by Baryza *et al.* (2004) [217] for their binding affinity to a mixture of PKC isozymes isolated from rat brain and the translocation of PKCδ coupled to green-fluorescent protein (PKCδ-GFP) in rat basophilic leukaemia (RBL) cells. The binding affinity to PKC, a thermodynamic parameter, indicates biological interactions, but it is not associated with a certain biological function, a kinetic parameter. For this reason, PKCδ-GFP translocation is considered to be a more appropriate method, than the PKC binding affinity, to assess the biological function of the examined compounds. Furthermore, activated PKCδ is linked to crucial cellular functions such as cell growth control, differentiation, and apoptosis [218, 219]. In the first set of experiments, analogues 59 and 60 exhibited an inhibitory concentration ($K_i$) of 3.0 nM and 0.25 nM respectively to a mixture of PKC isolated from rat brain in comparison to 1.35 nM for bryostatin 1 (22) [214]. It was reported in a later publication that variation in $K_i$ values was observed depending on the specific batch of PKC [214]. Moreover, the synthesised analogues 59 and 60 possess higher translocation potencies and rates of PKCδ-GFP translocation to the nucleus. Their minimum dose was lower (1 nM and 100 pM respectively) than bryostatin 1 (22) (5 nM). Wender *et al.* (2002) [220] showed that, in the cellular membrane, analogues 59 and 60 form a tertiary complex with PKC. The distinct lipophilicities of the derivatives might alter thermodynamic or kinetic parameters of this interaction and result in more potent PKC-translocation activity. These results correlated with the more potent growth inhibition activity of the "bryologs" in 24 out of 35 cancer cell lines assays compared to bryostatin 1 (22) [220].

**Table 1. PKC inhibitory activity of "bryologs" modified at position C9, C13 or C26. The PKC kinase inhibitory concentrations ($K_i$), obtained from the listed references, are given in nM**

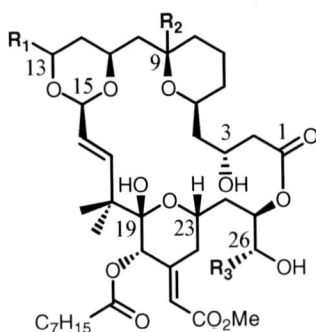

| structure | $R_1$ = | $R_2$ = | $R_3$ = | $K_i$ (nM) | Reference |
|-----------|---------|---------|---------|-----------|-----------|
| 61 | H | H | Me | 3.4 | [220] |
| 62 | H | H | H | 0.25 | [214, 220] |
| 63 |  | H | H | 1.2 | [275] |
| 64 |  | H | H | 0.67 | [275] |
| 65 | H | OH | H | 2 | [276] |
| 66 | H | OH | Me | 4 | [276] |

**Table 2. PKC inhibitory activity "bryologs", lacking an A-ring, and modifications at C13. The PKC kinase inhibitory concentrations (K$_i$), obtained from the listed references, are given in nM**

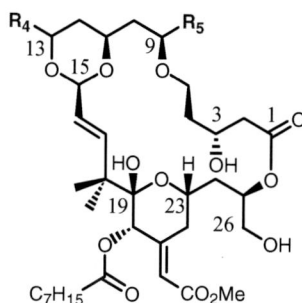

| structure | R$_4$ = | R$_5$ = | K$_i$ (nM) | Reference |
|---|---|---|---|---|
| 67 | H |  | 2.3 | [226] |
| 68 | H |  | 1.6 | [226] |
| 69 | H |  | 3.1 | [226] |
| 70 | H |  | 3.8 | [226] |
| 71 | H |  | 4.6 | [226] |
| 72 | H |  | 30 | [226] |
| 73 | H |  | 6.5 | [221] |
| 74 | H |  | 1.9 | [221] |
| 75 | H | H | 47 | [210] |
| 76 |  | H | 3.0 | [275] |
| 77 |  | H | 2.6 | [275] |

**Table 3. PKC inhibitory activity of "bryologs" with a five-membered B-ring. The PKC kinase inhibitory concentration (K$_i$), obtained from [226], is given in nM**

| structure | | K$_i$ (nM) | Reference |
|---|---|---|---|
| 78 |  | 5.4 nM | [215] |

**Table 4. PKC inhibitory activity of "bryologs" with different side chains at C20. The PKC kinase inhibitory concentrations (K$_i$), obtained from the listed references, are given in nM**

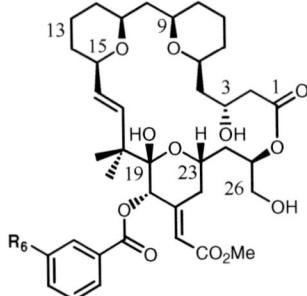

| structure | R$_6$ = | K$_i$ (nM) | Reference |
|---|---|---|---|
| 79 | NO$_2$ | 10 | [216] |
| 80 | NH$_2$ | 60 | [216] |
| 81 |  | 91 | [216] |
| 82 |  | 18 | [216] |
| 83 |  | 12 | [216] |
| 84 |  | 77 | [216] |
| 85 |  | 21 | [216] |

**Table 5. PKC inhibitory activity of "bryologs" with a hydropyranyl A-ring. The PKC kinase inhibitory concentrations (K$_i$), obtained from the listed references, are given in nM**

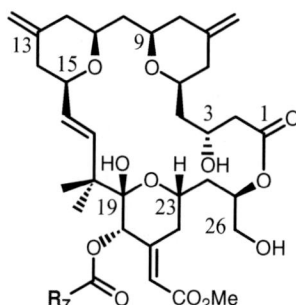

| structure | R$_7$ = | K$_i$ (nM) | Reference |
|-----------|---------|------------|-----------|
| 86 | | 0.70 | [227] |
| 87 | | 1.05 | [227] |
| 88 | | 0.70 | [227] |

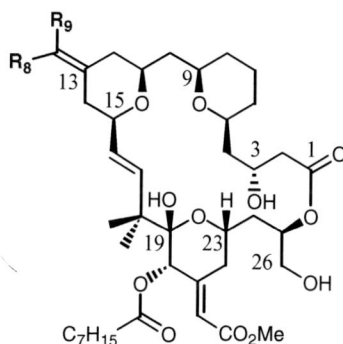

| structure | R$_8$ = | R$_9$ = | K$_i$ (nM) | Reference |
|-----------|---------|---------|------------|-----------|
| 89 | H | H | 1.6 | [214] |
| 90 | H | CO$_2$Me | 2.5 | [214] |
| 91 | CO$_2$Me | H | 0.9 | [214] |
| 92 | / | / | 3.1 | [214, 220] |

Interestingly, the missing of the A-ring does not alter the binding affinity of bryostatin analogues to PKC, even the stericallly less demanding *p*-bromo-phenylpropyl derivative **74** expressed a higher binding affinity than the bulkier *tertio*-butyl substituent **73** [221]. However, a single hydrogen at this position in **75** resulted in a significant reduction of binding affinity [210]. A five-membered B-ring analogue (**78**) exhibited a binding affinity of

5.4 nM to a mixture of PKC isozymes isolated from rat brain and the translocation from PKCδ was rapid and complete, similar to the lead the bryostatin 1 analogue 60. However, the selectivity for PKCβI was significantly decreased in comparison to bryostatin 1 (22) [222].

Finally, lipophilic side-chains do not significantly affect the binding affinity to an isozyme mixture of PKC, except for the highly lipophilic derivative 74, for which the lower binding affinity is explained by reduced ability to partition to the phospholipid vesicles, as already observed in the case of phorbol esters [215, 223].

New SAR studies on the PKC inhibitory activity of brystatin 1 (22) are currently in progress. But the bryostatin 1 (22) analogues synthesized so far by Wender *et al.* (1998) [224, 225] have already been shown to have highly promising *in vitro* biological activities, warranting a very exciting future for this field of research.

## 5.3. Chemical Synthesis and NF-κB-related SAR of Hymenialdisine and its Analogues

The bromopyrrole alkaloid hymenialdisine (33) isolated from various marine sponges was first reported as an NF-κB inhibitor in 1997 by Breton *et al.* and by Roshak *et al.*, but no information about the molecular targets of the compound was provided at that time [228, 229]. The first synthesis of hymenialdisine (33), involving 8 steps starting from commercially available pyrrole-2-carboxylic acid, had already been reported two years earlier by Annoura and Tatsuoka [230].

The mode of action of hymenialdisine (33) was investigated in depth in 2000 by Meijer *et al.* [231]. Meijer *et al.* (2000) [231] showed that hymenialdisine targets the kinases glycogen synthase kinase-3b (GSK-3b), cyclin dependent kinase 1 (CDK1/cyclin1) and 2 (CDK5/p25), and casein kinase 1 (CK1). These kinases are not directly involved in the canonical NF-kB activation pathway, but casein kinases have been reported to induce IκBα degradation through a non-canonical pathway in mammalian cells exposed to UV radiation [74, 117]. GSK-3b, CDK1, CDK5, and CK1 are also known to be involved in hyperphosphorylation of tau proteins, resulting in neurofibrillary tangles, one of the pathological symptom of Alzheimer's disease [232-234], and in cell cycle regulation [235]. Hymenialdisine (33) was reported to inhibit the latter kinases at nanomolar concentrations [231]. To date, a wide range of natural and synthetic analogues of hymenialdisine (33) have been reported [236-239], and have been the subject of several SAR studies [240-242] (scheme 11).

Meijer *et al.* (2000) [231] reported that the diacetyl- and diacetyldebromo- derivatives of hymenialdisine 93 and 94, and the structurally related metabolites odiline 95 axinohydantoin 96 were less active than the natural compound. The crystal structure of the CDK2-hymenialdisine complex revealed that hymenialdisine (33) and the kinase interacted through direct hydrogen bonds and through van der Waals interactions at the His[84], Phe[82] and Leu[134] residues of the enzyme. Because of the nature of these interactions, even small modifications to the core structure of hymenialdisne (33) can lead to dramatic losses of activity, as observed by Meijer *et al.* [243] for the analogues 94-96. The results were consistent with the findings of Tasdemir *et al.* (2002) [242] evaluated the MEK1 inhibition potency and the cytotoxicity of the hymenialdisine derivatives 97-100 isolated from the marine sponge *Stylissa massa*.

33 hymenialdisine   93 R₁=Br diacetylhymenialdisine   95 odiline   96 axinohydantoine   97 R = H aldisine
94 R₂=H diacetyldebromohymenialdisine                                                      98 R = Br 2-bromoaldisine

99 debromohymenialdisine   100 hymenin

| 101 R₁ = Br; R₂-R₄ = H |
| 102 R₁ -R₂ = Br; R₃-R₄ = H |
| 103 R₁-R₄ = H |
| 104 R₁ , R₃, R₄ = H; R₂ = Br |
| 105 R₁ - R₂ = Cl; R₂ - R₄ = H |
| 106 R₁ - R₂ = Br; R₂ = Me R₄ = H |
| 107 R₁ - R₂ = Br; R₂ = H R₄ = Ac |
| 108 R₁ - R₂ = Br; R₃ = H; R₄ = Et |

| 109 R₁-R₄ = H |
| 110 R₁ , R₃, R₄ = H; R₂ = F |
| 111 R₁ , R₃, R₄ = H; R₂ = Cl |
| 112 R₁ , R₃, R₄ = H; R₂ = Br |
| 113 R₁ , R₃, R₄ = H; R₂ = SO₂Me |
| 114 R₁ , R₃, R₄ = H; R₂ = NO₂ |
| 115 R₁ , R₃, R₄ = H; R₂ = NH₂ |
| 116 R₁ = NO₂ ; R₂ - R₄ = H |
| 117 R₁ = NH₂ ; R₂ - R₄ = H |
| 118 R₁ - R₂ = Br; R₂ = Me R₄ = H |
| 119 R₁ - R₂ = Br; R₂ = H R₄ = Ac |

| 121 R₁ = F; R₂ = A |
| 122 R₁ = Cl; R₂ = A |
| 123 R₁ = Br; R₂ = A |
| 124 R₁ = Cl; R₂ = B |
| 125 R₁ = Cl; R₂ = C |
| 126 R₁ = Cl; R₂ =D |
| 127 R₁ = F; R₂ = B |
| 128 R₁ = F; R₂ = C |
| 129 R₁ = F; R₂ = D |

120       130

A =       B =    Cl    C =    Cl    D =

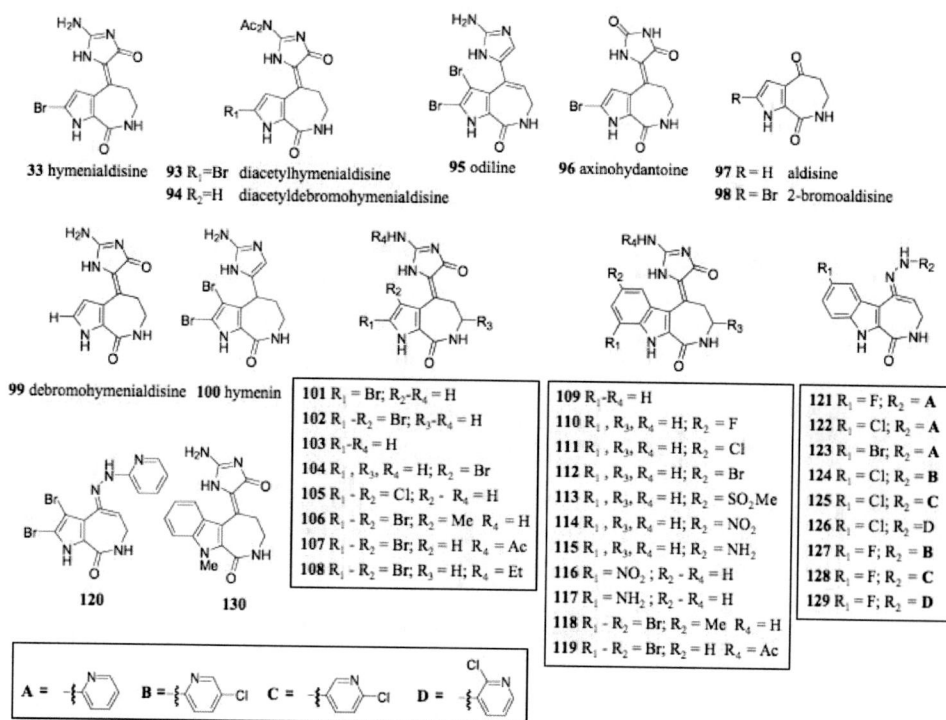

Scheme 11. Natural and synthetic analogues of hymenialdisine (33). Because of the high number of existing hydrazone analogues of hymenialdisine (33), only a selection of them are presented in the scheme. (Ac = acetyl group; Et = ethyl group; Me = methyl group).

Wan *et al.* (2004) [241] carried out an extensive SAR study using synthetic hymenialdisine analogues 101-129. As a first step, they evaluated structural modifications at the pyrroloazepine ring. Halogenation at the C2 and C3 positions in the pyrrole ring did not significantly affect inhibition potency against CDK5/p25 or GSK3β, but led to a 3-4 fold lowered activity against CDK1/cyclin B. CDK5/CDK1 inhibition selectivity was increased in case of a bromine in the C3 position (101). For all three kinases, methylation of the azepine ring (106) resulted in clearly reduced inhibition potential. Acylation or ethylation of the free guanidine amine moiety, which is involved in the hydrogen bonds between hymenialdisine and the kinases was shown to lower the inhibition potential significantly (107 and 108) [241]. The two novel hydrazone-indole analogues 128 and 129 were shown to have a particularly high CDK1/cyclinβ potential. Further, they were shown to selectively inhibit CDK5/p25 over GSK3β. The three analogues 121-124 were shown to significantly inhibit cell proliferation [244], and the three compounds 125-127 (all containing a pyridin-3-yl-hydrazine moiety) induced G2/M cell cycle arrest at a ten-fold lower concentration than hymenialdisine (33).

Sharma *et al.* (2004) [245] synthesized the two indoloazepine analogues 109 and 130 and investigated their effects on IL-2 and TNF-α production and on cell growth. To their surprise, the two compounds showed weaker bioactivity than hymenialdisine (13) [170, 245]. The weakness of the bioactivity of 130 may be explained by the methylation of the free indole-amine, which has been recognized as a major pharmacophore in hymenialdisine (33) since the first SAR studies [170, 245].

The results of the SAR studies performed on hymenialdisine (33) and its analogues are summarized in scheme 12. In brief, with the exception of a few hydrazone-indole analogues, even small modifications of the core unit of hymenialdisine (33) lead to dramatic decreases in bioactivity. None of the analogues of hymenialdisine (33) has entered clinical trials to date, but their potential use as medicinal drugs has been patented (US Patent 7098204).

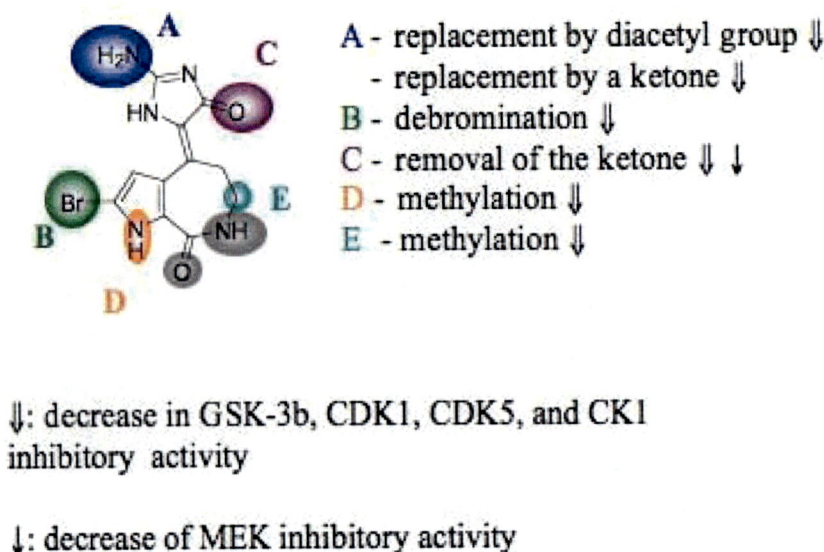

A - replacement by diacetyl group ⇓
  - replacement by a ketone ⇓
B - debromination ⇓
C - removal of the ketone ⇓ ↓
D - methylation ⇓
E - methylation ⇓

⇓: decrease in GSK-3b, CDK1, CDK5, and CK1 inhibitory activity

↓: decrease of MEK inhibitory activity

Scheme 12. Summary of the SAR of hymenialdisine (13) and its analogues in the context of the inhibition of the kinases GSK-3b, CDK1, CDK5, CK1, and MEK. The major pharmacophores are highlighted in colour. Some essential elements of the core structure are highlighted in grey.

## 5.4. Chemical Synthesis and NF-κB-related SAR of Avarol and Its Analogues

The sesquiterpene hydroquinone avarol (32) has been isolated from marine sponges since the mid 1970's, and it has, since then, been thoroughly investigated for anti-viral and anti-cancer activity [246-252]. The anti-viral and anti-cancer properties of avarol (32) have been patented under US Patent 4946869 and US patent 5082865, respectively, and the compound has advanced into clinical trials in HIV-infected patients in Germany [181]. The NF-κB inhibitory activity of avarol (32), which might play a critical role in the observed anti-viral and anti-cancer bioactivity of the compound, has only been described very recently [246]. The precise target of avarol (32) along the NF-κB activation pathway remains to be determined [246].

Avarol (32) is a rather unique marine natural product in the sense that it can be produced in large amounts through mariculture or through the ex situ culture of primary sponge cells [180] [181, 253]. Nevertheless, chemical synthesis, which was achieved for the first time in 1982 by Sarma et al. [254], still plays a critical role in the production of avarol and of its analogues [255]. The stereocontrolled synthesis of (±)-avarol by Sarma et al. (1982) was realised in an eight-step linear synthesis starting from a ene ketol [254, 256]. The synthetic analogues of avarol (32) are shown in scheme 13, together with some natural analogues of the compound.

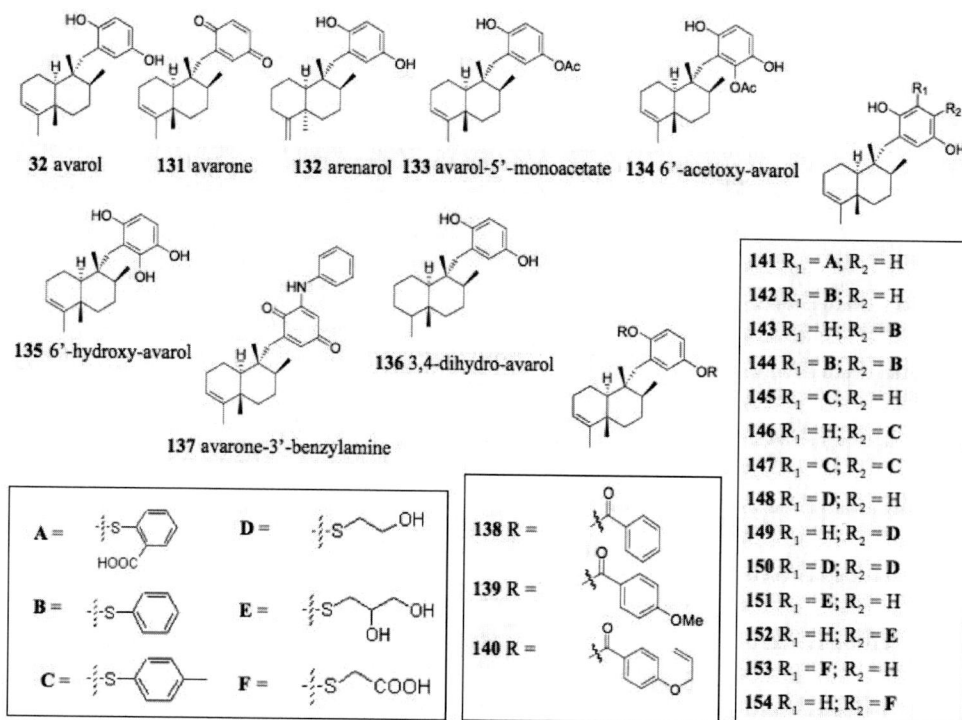

Scheme 13. Synthetic and natural analogues of avarol (32). Ac represents an acetyl group and Me represents a methyl group.

SAR studies on the anti-cancer activity of avarol (32) and its analogues have shown that some analogues, including 133 [257], are much more potent than the lead compound, while others, including the spongean natural product 132, which possess a *cis*-decaline moiety are significantly weaker anti-cancer compounds [258] . The cytotoxiciy of avarol (32) and its analogues is thought to be caused mainly by DNA damage. The damage of DNA is induced by the hydroxyl radicals produced by avarol (32) in the presence of oxygen and of low levels of superoxide dismutase [251]. This hypothesis has been corroborated by the finding that tryptophan, which is known to be a potent radical scavenger, inhibits avarol-induced DNA damage (32) [249]. Thiosalicylate analogues of avarol (32) have been shown to stabilize the produced radical and decreased its cytotoxicity [248, 249]. Belisario *et al.* (1992) reported that the radical scavenger property of avarol (32) is due to the presence of an easily donatable proton [259]. The departure of the proton is hampered by the esterification of the hydroxyl functionalities within the molecule [259].

SAR studies related to the anti-inflammatory activity of avarol (32) and its analogues revealed that the monophenyl-thio and thiosalicylate analogues (141-143 in particular) and the avarone-3'-benzylamine derivative 137 were the strongest inhibitors of UVB-induced NF-κB activation and TNF-α production [260, 261]. These results suggest that a benzylamine or thiosalicylate moiety at the C3' position plays a crucial role in the anti-inflammatory activity of avarol analogues.

In terms of HIV-1 inhibition, the presence of a 6'-hydroxyl-group (135) has been shown to dramatically increase the bioactivity of the lead compound.

The results obtained from the various SAR studies performed on avarol (32) and its analogues are summarized in scheme 14.

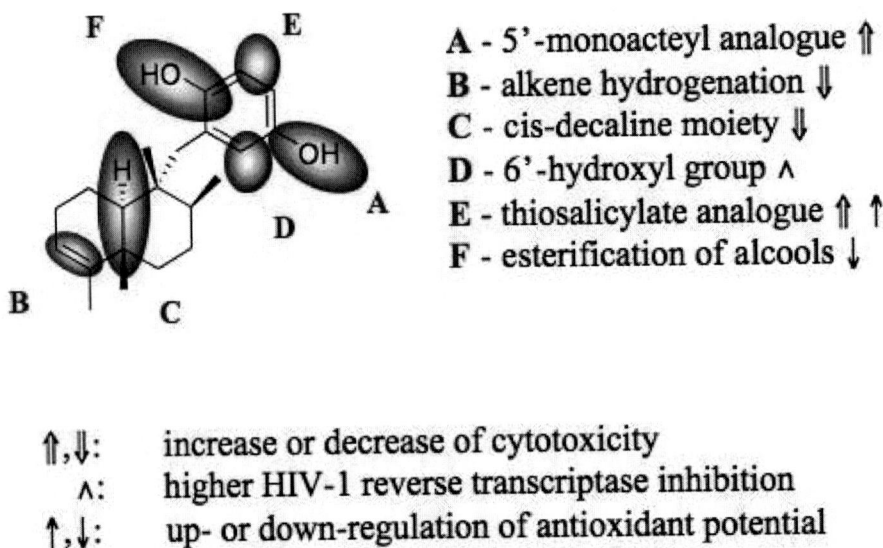

**A** - 5'-monoacteyl analogue ⇑
**B** - alkene hydrogenation ⇓
**C** - cis-decaline moiety ⇓
**D** - 6'-hydroxyl group ∧
**E** - thiosalicylate analogue ⇑ ↑
**F** - esterification of alcools ↓

⇑,⇓:   increase or decrease of cytotoxicity
∧:     higher HIV-1 reverse transcriptase inhibition
↑,↓:   up- or down-regulation of antioxidant potential

Scheme 14. Summary of the results of the various SAR studies performed on avarol (**32**) and its analogues. The pharmacophores identified through the SAR studies are highlighted in colour.

## 5.5. Chemical Synthesis and NF-κB-related SAR of Cacospongiolide B and its Analogues

Cacospongionolide B (29), a potent anti-inflammatory natural product isolated from various marine sponges, is closely related to avarol (32). The anti-inflammatory activity of 29 has been associated with its inhibition of the pro-inflammatory secretory phospholipase $A_2$ (PLA$_2$) [262-264]. PLA$_2$ induces the production of arachidonic acid, which is a precursor of leukotrienes and other cytokines implicated in the activation of NF-κB [265], PLA$_2$ inhibitors can hence be considered as indirect NF-kB inhibitors [266]. The NF-κB inhibitory potential of cacospongiolide B (29) was first reported by D'Acquisto et al. [166] and Palanki et al. [267] in 2000. Further investigation of the NF-κB inhibitory potential of cacospongiolide B (29) was performed by Posadas et al. in 2003 [164]. The synthesis of cacospongiolide B (29) analogues of it has been available since 1998 [268]. The first total synthesis of cacospongionolide B was achieved by Cheung and Snapper in 2002, through Michael addition of an enolate to an enone [269]. This synthetic approach by Cheung et al. enabled the access to several cacospongionolide analogues [269, 270]. The synthetic analogues of cacospongiolide B (29) are presented in scheme 15, together with some natural analogues.

The PLA$_2$ inhibitory activity of cacospongiolide B (29) has been attributed to the masking of an aldehyde group in the enzyme by the γ-hydroxy-butenolide moiety of 29 [271, 272].

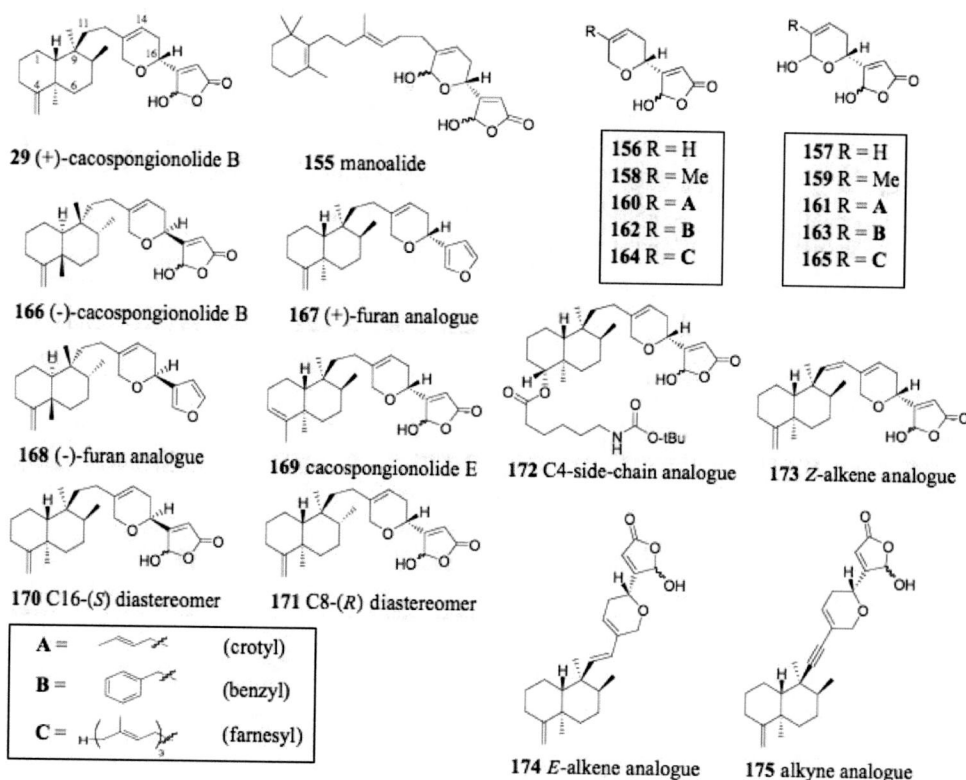

29 (+)-cacospongionolide B

155 manoalide

| 156 R = H | 157 R = H |
| 158 R = Me | 159 R = Me |
| 160 R = A | 161 R = A |
| 162 R = B | 163 R = B |
| 164 R = C | 165 R = C |

166 (-)-cacospongionolide B

167 (+)-furan analogue

168 (-)-furan analogue

169 cacospongionolide E

172 C4-side-chain analogue

173 Z-alkene analogue

170 C16-(S) diastereomer

171 C8-(R) diastereomer

| A = | (crotyl) |
| B = | (benzyl) |
| C = | (farnesyl) |

174 E-alkene analogue

175 alkyne analogue

Scheme 15. Synthetic and natural analogues of cacospongiolide B (29). Me represents a methyl group.

Like its closely related analogue manoalide (155) isolated from marine sponges, cacospongiolide B (29) has been shown to interfere with the enzymatic activity of PLA$_2$ through the formation of a Schiff base-like covalent imine bond between the aldehyde generated upon the opening of the γ-hydroxybutenolide of the natural product and a lysine residue at the enzyme-lipid interface of PLA$_2$ [271, 272]. The hydrophobic region of manoalide (155) and cacospongionolide (29) allows the formation of non-covalent bonds between the enzyme and the natural products, thereby facilitating the formation of the Schiff base-like covalent bond [273].

SAR studies on the PLA$_2$ inhibitory activity of cacospongiolide B (29) and its analogues have revealed that the presence of a certain level of lipophilicity is a pre-requisite for the inhibition of the enzymatic activity of PLA$_2$. Cacospongiolide B analogues with short alkyl side-chains, such as methyl-, crotyl- and benzyl- functionalities (160-163), possess a significantly lower inhibition potential than analogues with a farnesyl side-chain (164-165), for which the IC$_{50}$ values are up to five times lower than the IC$_{50}$ of cacospongiolide B (29) [268].

The PLA$_2$ inhibition-related SAR studies on cacospongiolide B (29) and its analogues have also highlighted a crucial role of the stereochemistry of the compounds in the latter's bioactivity [269, 270]. The PLA$_2$ inhibitory activity of the synthetic cacospongiolide B (29) enantiomer 166 has been shown to be half the bioactivity of cacospongiolide B (29) [269, 270]. The substitution of the γ-hydroxybutenolide by a furan ring (167-168) led to a slightly lower bioactivity, but did not result in a complete loss of bioactivity, suggesting that the pyranofuranone ring is one, but not the unique key element for PLA$_2$ inhibition activity [269, 270]. The stereochemical inversion of the C16 and C8 stereocentres has been shown to boost the PLA$_2$ inhibitory activity in 170 and 171, respectively [269, 270]. An SAR study on the conformationally restricted cacospongionolide B analogues 173-175 recently synthesised by Murelli *et al.* [264] has investigated the impact of the three-dimensional orientation of cacospongiolide B analogues on the latter's PLA$_2$ inhibitory potential. The study revealed that rigidified three-dimensional structure such as in 173 are the most favourable in terms of PLA$_2$ inhibitory activity [264, 269, 270].

Finally, the noteworthy structural relatedness of cacospongiolide B (29) to avarol (32) within the *trans*-decaline core could potentially provide some explanations for the NF-κB inhibitory activity shared by the two marine products.

The results of the PLA$_2$ inhibitory activity-related SAR studies performed on cacospongiolide B (29) and its analogues are summarized in scheme 16. Briefly, the results of the SAR studies indicate that the PLA$_2$ inhibitory activity of cacospongionolide B (29), which is involves the formation of a Schiff-base-like covalent bond formation between the masked aldehyde generated upon the opening of the and a lysine of the enzyme, is both enantio- and diastereo- selective [264, 269, 270]. The most potent cacospongiolide B analogue (164) has been shown to possess an anti-inflammatory potential equal to the one of indomethacin (Indocin®), which is an approved nonsteroidal anti-inflammatory drug [274]. This finding suggests a highly promising future for cacospongiolide B analogues in the field of anti-inflammatory drug discovery.

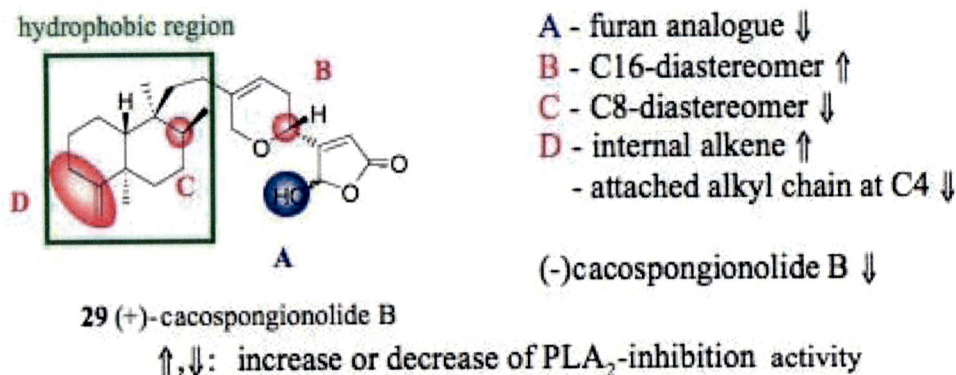

A - furan analogue ⇓
B - C16-diastereomer ⇑
C - C8-diastereomer ⇓
D - internal alkene ⇑
    - attached alkyl chain at C4 ⇓

(-)cacospongionolide B ⇓

29 (+)-cacospongionolide B

⇑,⇓: increase or decrease of PLA$_2$-inhibition activity

Scheme 16. Summary of the results obtained in the PLA$_2$ inhibition-related SAR studies performed on cacospongiolide B (29) and its analogues. The pharmacophores identified through the SAR studies are highlighted in colour.

# CONCLUSION

There is constantly growing evidence for the important role of NF-κB inhibitors both in marine ecology and in biomedicine. Marine organisms have been shown to be a rich source of highly diverse natural products with very promising NF-κB inhibitory properties. While the detailed mode of action of marine NF-κB inhibitors remains, in many cases, poorly understood, structure activity relationship (SAR) studies on natural and synthetic analogues of lead marine NF-kB inhibitors have provided a tremendous amount of information on the molecular targets of marine NF-κB inhibitors. As described in the present review, the molecular target of marine natural products shown to inhibit NF-κB activation is, in many cases, upstream of the NF-κB activation cascade. NF-κB inhibition-related SAR studies have led to the discovery of several synthetic analogues with significantly higher bioactivities than the corresponding natural products. Some of these recently discovered compounds have reached clinical trials. The chemical synthesis of marine NF-κB inhibitors has played a very important role in marine drug discovery, not just in terms of identifying potent analogues of the bioactive natural products, but also in terms of providing a sustainable source of the bioactive molecules for clinical trials and for the future marketing of the compounds as medicinal drugs.

## ACKNOWLEDGMENTS

MJ is the recipient of a BBSRC Research Development Fellowship, and MS is a recipient of a postdoctoral research fellowship (BFR06/016) from the Luxembourg government. Research at the Laboratoire de Biologie Moléculaire et Cellulaire du Cancer (LBMCC) is financially supported by "Recherche Cancer et Sang" foundation, by the "Recherches Scientifiques Luxembourg" association, by "Een Häerz fir kriibskrank Kanner" a.s.b.l. (Luxembourg), and by Télévie Luxembourg.

## REFERENCES

[1]    Barnes RSK, Hughes R. *An introduction to marine ecology*. Oxford, U.K.: Blackwell Science, 1988.

[2]    Simmons TL, Andrianasolo E, McPhail K, Flatt P, Gerwick WH. Marine natural products as anticancer drugs. *Mol. Cancer Ther.* 2005;4:333-42.

[3]    Pawlik JR. Marine Invertebrate Chemical Defenses. *Chem. Rev.* 1993;93:1911-22.

[4]    Paul VJ, Puglisi MP, Ritson-Williams R. Marine chemical ecology. *Nat. Prod. Rep.* 2006;23:153-80.

[5]    Scheuer PJ. Some Marine Ecological Phenomena - Chemical Basis and Biomedical Potential. *Science* 1990;248:173-7.

[6]    Garson M. The biosynthesis of marine natural products. *Chem. Rev.* 1993;93:1699-733.

[7]    Faulkner DJ. Biomedical Uses for Natural Marine Chemicals. *Oceanus* 1992;35:29-35.

[8]    Paul VJ, Ritson-Williams R. Marine chemical ecology. *Nat. Prod. Rep.* 2008;*in press* (DOI: 10.1039/b702742g).

[9]    Williams DH, Stone MJ, Hauck PR, Rahman SK. Why are secondary metabolites (natural products) biosynthesized? *J. Nat. Prod.* 1989;52:1189-208.

[10]   Hay MF. Marine chemical ecology: What's known and what's next? *Journal of experimenta marine biology and ecology* 1996;200:103-34.

[11]   Fenical W. Marine pharmaceuticals - Past, present, future. *Oceanus* 2006;19:111-9.

[12]   Jensen PR, Fenical W. Strategies for the discovery of secondary metabolites from marine bacteria: ecological perspectives. *Annu. Rev. Microbiol.* 1994;48:559-84.

[13] Battershill CN, Jaspars M, Long P. Marine biodiscovery: new drugs from the ocean depths. *Biologist* 2005;52:107-14.

[14] Butler M. Natural products to drugs: natural product-derived compounds in clinical trials. *Nat. Prod. Rep.* 2008;25:475-516.

[15] Aggarwal BB, Takada Y, Shishodia S, Gutierrez AM, Oommen OV, Ichikawa H, et al. Nuclear transcription factor NF-kappa B: role in biology and medicine. *Indian J. Exp. Biol.* 2004;42:341-53.

[16] Ichikawa H, Nakamura Y, Kashiwada Y, Aggarwal BB. Anticancer drugs designed by mother nature: ancient drugs but modern targets. *Curr. Pharm. Des.* 2007;13:3400-16.

[17] Aggarwal BB, Ichikawa A, Garodia P, Weerasinghe P, Sethi G, Bhatt I, et al. From traditional Ayurvedic medicine to modern medicine: identification of therapeutic targets for suppression of inflammation and cancer. *Expert Opin. Ther Targets* 2006;10:87-118.

[18] Kingston DG, Newman DJ. Natural products as drug leads: an old process or the new hope for drug discovery? *IDrugs* 2005;8:990-2.

[19] Faulkner DJ. Marine pharmacology. *Antonie Van Leeuwenhoek* 2000;77:135-45.

[20] Latchman DS. Transcription factors: an overview. *Int. J. Biochem. Cell Biol.* 1997;29:1305-12.

[21] Karin M. Too many transcription factors: positive and negative interactions. *New Biol.* 1990;2:126-31.

[22] Sen R, Baltimore D. Multiple nuclear factors interact with the immunoglobulin enhancer sequences. *Cell* 1986;46:705.

[23] Ghosh S, May MJ, Kopp EB. NF-kappa B and Rel proteins: evolutionarily conserved mediators of immune responses. *Annual Review on Immunology* 1998;16:225-60.

[24] Pahl HL. Activators and target genes of Rel/NF-kappaB transcription factors. *Oncogene* 1999;18:6853-66.

[25] Li Q, Verma IM. NF-kappaB regulation in the immune system. *Nature Rev. Immunol.* 2002;2:725-34.

[26] Gerondakis S, Grossmann M, Nakamura Y, Pohl T, Grumont R. Genetic approaches in mice to understand Rel/NF-kappaB and IkappaB function: transgenics and knockouts. *Oncogene* 1999;18:6888-95.

[27] Gerondakis S, Grumont R, Rourke I, Grossmann M. The regulation and roles of Rel/NF-kappa B transcription factors during lymphocyte activation. *Curr. Opin. Immunol.* 1998;10:353-9.

[28] Sen R, Baltimore D. Inducibility of kappa immunoglobulin enhancer-binding protein Nf-kappa B by a posttranslational mechanism. *Cell* 1986;47:921-8.

[29] Jost PJ, Ruland J. Aberrant NF-kappaB signaling in lymphoma: mechanisms, consequences, and therapeutic implications. *Blood* 2007;109:2700-7.

[30] Bargou RC, Leng C, Krappmann D, Emmerich F, Mapara MY, Bommert K, et al. High-level nuclear NF-kappa B and Oct-2 is a common feature of cultured Hodgkin/Reed-Sternberg cells. *Blood* 1996;87:4340-7.

[31] Zhou H, Du MQ, Dixit VM. Constitutive NF-kappaB activation by the t(11;18)(q21;q21) product in MALT lymphoma is linked to deregulated ubiquitin ligase activity. *Cancer Cell* 2005;7:425-31.

[32] Davis RE, Brown KD, Siebenlist U, Staudt LM. Constitutive nuclear factor kappaB activity is required for survival of activated B cell-like diffuse large B cell lymphoma cells. *J. Exp. Med.* 2001;194:1861-74.

[33] Keller SA, Schattner EJ, Cesarman E. Inhibition of NF-kappaB induces apoptosis of KSHV-infected primary effusion lymphoma cells. *Blood* 2000;96:2537-42.

[34] Sun SC, Ballard DW. Persistent activation of NF-kappaB by the tax transforming protein of HTLV-1: hijacking cellular IkappaB kinases. *Oncogene* 1999;18:6948-58.

[35] Xiao G, Cvijic ME, Fong A, Harhaj EW, Uhlik MT, Waterfield M, et al. Retroviral oncoprotein Tax induces processing of NF-kappaB2/p100 in T cells: evidence for the involvement of IKKalpha. *Embo. J.* 2001;20:6805-15.

[36] Rayet B, Gelinas C. Aberrant rel/nfkb genes and activity in human cancer. *Oncogene* 1999;18:6938-47.

[37] Spano JP, Bay JO, Blay JY, Rixe O. Proteasome inhibition: a new approach for the treatment of malignancies. *Bull. Cancer* 2005;92:E61-6, 945-52.

[38] Kumar A, Lnu S, Malya R, Barron D, Moore J, Corry DB, et al. Mechanical stretch activates nuclear factor-kappaB, activator protein-1, and mitogen-activated protein kinases in lung parenchyma: implications in asthma. *Faseb. J.* 2003;17:1800-11.

[39] Hart LA, Krishnan VL, Adcock IM, Barnes PJ, Chung KF. Activation and localization of transcription factor, nuclear factor-kappaB, in asthma. *Am. J. Respir. Crit. Care Med.* 1998;158:1585-92.

[40] Marok R, Winyard PG, Coumbe A, Kus ML, Gaffney K, Blades S, et al. Activation of the transcription factor nuclear factor-kappaB in human inflamed synovial tissue. *Arthritis Rheum.* 1996;39:583-91.

[41] Gilston V, Jones HW, Soo CC, Coumbe A, Blades S, Kaltschmidt C, et al. NF-kappa B activation in human knee-joint synovial tissue during the early stage of joint inflammation. *Biochem. Soc. Trans* 1997;25:518S.

[42] Miyazawa K, Mori A, Yamamoto K, Okudaira H. Constitutive transcription of the human interleukin-6 gene by rheumatoid synoviocytes: spontaneous activation of NF-kappaB and CBF1. *Am. J. Pathol.* 1998;152:793-803.

[43] Ellis RD, Goodlad JR, Limb GA, Powell JJ, Thompson RP, Punchard NA. Activation of nuclear factor kappa B in Crohn's disease. *Inflamm Res.* 1998;47:440-5.

[44] Bonetti B, Stegagno C, Cannella B, Rizzuto N, Moretto G, Raine CS. Activation of NF-kappaB and c-jun transcription factors in multiple sclerosis lesions. Implications for oligodendrocyte pathology. *Am. J. Pathol.* 1999;155:1433-8.

[45] Gveric D, Kaltschmidt C, Cuzner ML, Newcombe J. Transcription factor NF-kappaB and inhibitor I kappaBalpha are localized in macrophages in active multiple sclerosis lesions. *J. Neuropathol. Exp. Neurol.* 1998;57:168-78.

[46] Strait K, Li Y, Dillehay DL, Weitzmann MN. Suppression of NF-kappaB activation blocks osteoclastic bone resorption during estrogen deficiency. *Int. J. Mol. Med.* 2008;21:521-5.

[47] Aggarwal BB, Takada Y, Oommen OV. From chemoprevention to chemotherapy: common targets and common goals. *Expert Opin. Invest Drugs* 2004;13:1327-38.

[48] Feinman R, Koury J, Thames M, Barlogie B, Epstein J, Siegel DS. Role of NF-kappaB in the rescue of multiple myeloma cells from glucocorticoid-induced apoptosis by bcl-2. *Blood* 1999;93:3044-52.

[49] Ahn KS, Sethi G, Aggarwal BB. Reversal of chemoresistance and enhancement of apoptosis by statins through down-regulation of the NF-kappaB pathway. *Biochem. Pharmacol.* 2008;75:907-13.

[50] Perona R, Sanchez-Perez I. Signalling pathways involved in clinical responses to chemotherapy. *Clin. Transl. Oncol.* 2007;9:625-33.

[51] Wu M, Lee H, Bellas RE, Schauer SL, Arsura M, Katz D, et al. Inhibition of NF-kappaB/Rel induces apoptosis of murine B cells. *Embo J.* 1996;15:4682-90.

[52] Wang CY, Mayo MW, Baldwin AS, Jr. TNF- and cancer therapy-induced apoptosis: potentiation by inhibition of NF-kappaB. *Science* 1996;274:784-7.

[53] Winkler JD, Eris T, Sung CM, Chabot-Fletcher M, Mayer RJ, Surette ME, et al. Inhibitors of coenzyme A-independent transacylase induce apoptosis in human HL-60 cells. *J. Pharmacol. Exp. Ther.* 1996;279:956-66.

[54] Coussens LM, Werb Z. Inflammation and cancer. *Nature* 2002;420:860-7.

[55] Courtois G, Gilmore TD. Mutations in the NF-kappaB signaling pathway: implications for human disease. *Oncogene* 2006;25:6831-43.

[56] Kumar A, Takada Y, Boriek AM, Aggarwal BB. Nuclear factor-kappa B: its role in health and disease. *J. Mol. Med.* 2004;82:434-48.

[57] Karin M, Greten FR. NF-kappaB: linking inflammation and immunity to cancer development and progression. *Nature Rev. Immunol.* 2005;5:749-59.

[58] Folmer F, Jaspars M, Diederich M. Marine natural products as targeted modulators of the transcription factor NF-κB. *Biochem. Pharmacol.* 2008;75:603-17.

[59] May MJ, Ghosh S. Signal transduction through NF-kappa B. *Immunol. Today* 1998;19:80-8.

[60] Keutgens A, Robert I, Viatour P, Chariot A. Deregulated NF-κB activity in haemological malignancies. *Biochem. Pharmacol.* 2006;72:1069-80.

[61] Jacobs MD, Harrison SC. Structure of an IkappaBalpha/NF-kappaB complex. *Cell* 1998;95:749-58.

[62] Ghosh S. Handbook of Transcription Factor NF-kappaB. Boca Raton, FL (U.S.A.): CRC Press, 2006.

[63] Baldwin AS, Jr. The NF-kappa B and I kappa B proteins: new discoveries and insights. *Annual Review on Immunology* 1996;14:649-83.

[64] Chen FE, Huang DB, Chen YQ, Ghosh G. Crystal structure of p50/p65 heterodimer of transcription factor NF-kappaB bound to DNA. *Nature* 1998;391:410-3.

[65] Bergqvist S, Croy C, Kjaergaard M, Huxford T, Ghosh G, Komives E. Thermodynamics reveal that helix four in the NLS of NF-κB p65 anchors IκBα, forming a very stable complex. *J. Mol. Biol.* 2006;360:421-34.

[66] Huxford T, Huang DB, Malek S, Ghosh G. The crystal structure of the IkappaBalpha/NF-kappaB complex reveals mechanisms of NF-kappaB inactivation. *Cell* 1998;95:759-70.

[67] Koradi R, Billeter M, Wüthrich K. MOLMOL: a program for display and analysis of macromolecular structures. *Journal of Molecular Graphics* 1996;14:51-5.

[68] Bours V, Bonizzi G, Bentires-Alj M, Bureau F, Piette J, Lekeux P, et al. NF-kappaB activation in response to toxical and therapeutical agents: role in inflammation and cancer treatment. *Toxicology* 2000;153:27-38.

[69] Garg AK, Aggarwal BB. Reactive oxygen intermediates in TNF signaling. *Mol. Immunol.* 2002;39:509-17.

[70]  Karin M, Takahashi T, Kapahi P, Delhase M, Chen Y, Makris C, et al. Oxidative stress and gene expression: the AP-1 and NF-kappaB connections. *Biofactors* 2001;15:87-9.

[71]  Ichikawa H, Aggarwal BB. Guggulsterone inhibits osteoclastogenesis induced by receptor activator of nuclear factor-kappaB ligand and by tumor cells by suppressing nuclear factor-kappaB activation. *Clin. Cancer Res.* 2006;12:662-8.

[72]  Kim JM, Min SK, Kim H, Kang HK, Jung SY, Lee SH, et al. Vacuolar-type H+-ATPase-mediated acidosis promotes in vitro osteoclastogenesis via modulation of cell migration. *Int. J. Mol. Med.* 2007;19:393-400.

[73]  Karin M. How NF-kappaB is activated: the role of the IkappaB kinase (IKK) complex. *Oncogene* 1999;18:6867-74.

[74]  Karin M. Nuclear factor-kB in cancer development and progression. *Nature* 2006;441:431-6.

[75]  Bowie A, O'Neill LA. Oxidative stress and nuclear factor-kappaB activation: a reassessment of the evidence in the light of recent discoveries. *Biochem. Pharmacol.* 2000;59:13-23.

[76]  Hochstrasser M. Lingering mysteries of ubiquitin-chain assembly. Cell 2006;124:27-34.

[77]  Mukhopadhyay D, Riezman H. Proteasome-independent functions of ubiquitin in endocytosis and signaling. *Science* 2007;315:201-5.

[78]  Evans P. Regulation of pro-inflammatory signalling networks by ubiquitin: identification of novel targets for anti-inflammatory drugs. *Expert Reviews in Molecular Medicine* 2005;7:1-19.

[79]  Haefner B. NF-kappa B: arresting a major culprit in cancer. *Drug Disc. Today* 2002;7:653-63.

[80]  Greten FR, Karin M. The IKK/NF-kappaB activation pathway-a target for prevention and treatment of cancer. *Cancer Lett.* 2004;206:193-9.

[81]  Bremner P, Heinrich M. Natural products as targeted modulators of the nuclear factor-kappa B pathway. *J. Pharm. Pharmacol.* 2002;54:453-72.

[82]  Rungeler P, Castro V, Mora G, Goren N, Vichnewski W, Pahl HL, et al. Inhibition of transcription factor NF-kappa B by sesquiterpene lactones: a proposed molecular mechanism of action. *Bioorg. Med. Chem.* 1999;7:2343-52.

[83]  Drahl C, Cravatt BF, Sorensen EJ. Protein-reactive natural products. *Angew. Chem.* 2005;44:5788-809.

[84]  Garcia-Pineres AJ, Castro V, Mora G, Schmidt TJ, Strunck E, Pahl HL, et al. Cysteine 38 in p65/NF-kappa B plays a crucial role in DNA binding inhibition by sesquiterpene lactones. *J. Biol. Chem.* 2001;276:39713-20.

[85]  Lyss G, Schmidt TJ, Merfort I, Pahl HL. Helenalin, an anti-inflammatory sesquiterpene lactone from *Arnica*, selectively inhibits transcription factor NF-κB. *Biol. Chem.* 1997;378:951-61.

[86]  Adams J. The proteasome as a novel target for the treatment of breast cancer. *Breast Disease* 2002;15:61-70.

[87]  Kisselev AF, Goldberg AL. Proteasome inhibitors: from research tools to drug candidates. *Chem. Biol.* 2001;8:739-58.

[88]  Delcros JG, Baudy Floc'h M, Prigent C, Arlot-Bonnemains Y. Proteasome inhibitors as therapeutic agents: Current and future strategies. *Curr. Med. Chem.* 2003;10:479-503.

[89]  Ahn KS, Sethi G, Chao TH, Neuteboom ST, Chaturvedi MM, Palladino MA, et al. Salinosporamide A (NPI-0052) potentiates apoptosis, suppresses osteoclastogenesis,

and inhibits invasion through down-modulation of NF-kappaB regulated gene products. *Blood* 2007;110:2286-95.

[90] Karin M, Yamamoto Y, Wang QM. The IKK NF-kappa B system: A treasure trove for drug development. *Nature Rev. Drug Disc.* 2004;3:17-26.

[91] Ogino S, Tsuruma K, Uehara T, Nomura Y. Herbimycin A abrogates nuclear factor-kappaB activation by interacting preferentially with the IkappaB kinase beta subunit. *Mol. Pharmacol.* 2004;65:1344-51.

[92] Sasaki K, Rinehart KL, Jr., Slomp G, Grostic MF, Olson EC. Geldanamycin. I. Structure assignment. *J. Am. Chem. Soc.* 1970;92:7591-3.

[93] Shrivastava A, Aggarwal BB. Antioxidants differentially regulate activation of Nuclear Factor-κB, Activator Protein-1, c-Jun Amino-Terminal Kinases, and apopoptosis induced by Tumor Necrosis Factor: Evidence that JNK and NF-κB activation are not linked to apoptosis. *Antiox Redox Signalling* 1999;1:181-91.

[94] Palladino MA, Bahjat FR, Theodorakis EA, Moldawer LL. Anti-TNF-alpha therapies: The next generation. *Nature Rev. Drug Disc.* 2003;2:736-46.

[95] Takamatsu S, Hodges TW, Rajbhandari I, Gerwick WH, Hamann MT, Nagle DG. Marine natural products as novel antioxidant prototypes. *J. Nat. Prod.* 2003;66:605-8.

[96] Strivastava A, Aggarwal BB. Antioxidants differentially regulate activation of NFKB, APA, CJNK and apoptosis induced by TNF. *Antiox Redox Signalling* 1999;1:181-91.

[97] Wang XW, Tan NS, Ho B, Ding JL. Evidence for the ancient origin of the NF-{kappa}B/I{kappa}B cascade: Its archaic role in pathogen infection and immunity. *Proc. Natl. Acad. Sci. USA* 2006;103:4204-9.

[98] Friedman R, Hughes AL. Molecular evolution of the NF-kappaB signaling system. *Immunogenetics* 2002;53:964-74.

[99] Povelones M, Tran K, Thanos D, Ambron RT. An NF-kappaB-like transcription factor in axoplasm is rapidly inactivated after nerve injury in *Aplysia*. *J. Neurosci.* 1997;17:4915-20.

[100] Pancer Z, Rast JP, Davidson EH. Origins of immunity: transcription factors and homologues of effector genes of the vertebrate immune system expressed in sea urchin coelomocytes. *Immunogenetics* 1999;49:773-86.

[101] Mitta G, Vandenbulcke F, Roch P. Original involvement of antimicrobial peptides in mussel innate immunity. *FEBS Lett* 2000;486:185-90.

[102] Merlo E, Freudenthal R, Romano A. The I kappa B kinase inhibitor sulfasalazine impairs long-term memory in the crab *Chasmagnathus*. *Neuroscience* 2002;112:161-72.

[103] Shimada M, Satoh N, Yokosawa H. Involvement of Rel/NF-kappaB in regulation of ascidian notochord formation. Dev, Growth, *Differentiation* 2001;43:145-54.

[104] Gueguen Y, Cadoret JP, Flament D, Barreau-Roumiguiere C, Girardot AL, Garnier J, et al. Immune gene discovery by expressed sequence tags generated from hemocytes of the bacteria-challenged oyster, *Crassostrea gigas*. *Gene* 2003;303:139-45.

[105] Bettencourt R, Roch P, Stefanni S, Rosa D, Colaço A, Santos R. Deep sea immunity: Unveiling immune constituents from the hydrothermal vent mussel *Bathymodiolus azoricus*. *Marine Environmental Research* 2007;64:108-27.

[106] Waterhouse RM, Kriventseva EV, Meister S, Xi Z, Alvarez KS, Bartholomay LC, et al. Evolutionary dynamics of immune-related genes and pathways in disease-vector mosquitoes. *Science* 2007;316:1738-43.

[107] Gauthier MJ, Degnan BM. The transcription factor NF-kappaB in the demosponge *Amphimedon queenslandica*: insights on the evolutionary origin of the Rel homology domain. *Development Genes and Evolution* 2008;doi 10.1007/s00427-007-0197-5.

[108] Escoubas JM, Briant L, Montagnani C, Hez S, Devaux C, Roch P. Oyster IKK-like protein shares structural and functional properties with its mammalian homologues. *FEBS Lett.* 1999;453:293-8.

[109] Xiong X, Feng Q, Chen L, Xie L, Zhang R. Cloning and characterization of an IKK homologue from pearl oyster, *Pinctada fucata*. *Developmental and Comparative Immunology* 2008;32:15-25.

[110] Goodson MS, Kojadinovic M, Troll JV, Scheetz TE, Casavant TL, Soares MB, et al. Identifying components of the NF-κB pathway in the beneficial *Euprymna scolopes-Vibrio fischeri* light organ symbiosis. *Applied and Environmental Microbiology* 2005;71:6934-46.

[111] Schlezinger JJ, Blickarz CE, Mann KK, Doerre S, Stegeman JJ. Identification of NF-kappaB in the marine fish *Stenotomus chrysops* and examination of its activation by aryl hydrocarbon receptor agonists. *Chem. Biol. Interactions* 2000;126:137-57.

[112] Muller WE, Gamulin V, Rinkevich B, Spreitzer I, Weinblum D, Schroder HC. Ubiquitin and ubiquitination in cells from the marine sponge *Geodia cydonium. Bio Chem. Hoppe-Seyler* 1994;375:53-60.

[113] Muller WE, Koziol C, Muller IM, Wiens M. Towards an understanding of the molecular basis of immune responses in sponges: the marine demosponge *Geodia cydonium* as a model. *Microscopy Res. Tech.* 1999;44:219-36.

[114] Muller WE, Wiens M, Adell T, Gamulin V, Schroder HC, Muller IM. Bauplan of urmetazoa: basis for genetic complexity of metazoa. *Int. Rev. Cytol.* 2004;235:53-92.

[115] Muller WE, Wiens M, Muller IM, Schroder HC. The chemokine networks in sponges: potential roles in morphogenesis, immunity and stem cell formation. *Prog. Mol. Subcell Biol.* 2004;34:103-43.

[116] Li N, Karin M. Ionizing radiation and short wavelength UV activate NF-κB through two distinct mechanisms. *Proc. Natl. Acad. Sci. USA* 1998;95:13012-7.

[117] Kato T, Delhase M, Hoffmann A, Karin M. CK2 is a C-terminal IκB kinase responsible for NF-κB activation during the UV response. *Mol. Cell* 2003;12:829-39.

[118] Bender K, Gottlicher S, Whiteside H, Rahmsdorf H, Herrlich P. Sequential DNA damage-independent and -dependent activation of NF-κB by UV. *EMBO Journal* 1998;17:5170-81.

[119] Karentz D. Chemical defenses of marine organisms against solar radiation exposure: UV-absorbing mycosporine-like amino acids and scytonemin. *in*: Marine Chemical Ecology. Boca Raton: CRC Press, 2001.

[120] Patel A, Mishra S, Ghosh PK. Antioxidant potential of C-phycocyanin isolated from cyanobacterial species Lyngbya, Phormidium and Spirulina spp. *Indian Journal of Biochemistry and Biophysics* 2006;43:25-31.

[121] Pannunzio T, Storey K. Antioxidant defenses and lipid peroxidation during anoxia stress and aerobic recovery in the marine gastropod *Littorina littorea. J. Exp. Mar. Biol. Ecol.* 1997;221:277-92.

[122] Michiels C, Minet E, Mottet D, Raes M. Regulation of gene expression by oxygen: NF-kappaB and HIF-1, two extremes. *Free Radicals in Biology and Medicine* 2002;33:1231-42.

[123] Rius J, Guma M, Schachtrup C, Akassoglou K, Zinkernagel A, Nizet V, et al. NF-κB links innate immunity to the hypoxic response through transcriptional regulation of HIF-1α. *Nature* 2008;453:807-11.

[124] van Uden P, Kenneth NS, Rocha S. Regulation of hypoxia inducible factor-1 alpha by NF-kappaB. *Biochem J.* 2008;412:477-84.

[125] Dai J, Liu Y, Zhou YD, Nagle DG. Hypoxia-selective anti-tumor agents: norsesterterpene peroxides from the marine sponge *Diacarnus levii* preferentially suppresses the growth of tumor cells under hypoxic conditions. *J. Nat. Prod.* 2007;70:130-3.

[126] Seo SB, Jeong D, Chung HS, Lee JD, You YO, Kajiuchi T, et al. Inhibitory effect of high molecular weight water-soluble chitosan an hypoxia-induced inflammatory cytokine production. *Biol. Pharm. Bull* 2003;26:717-21.

[127] Mohammed KA, Hossain CF, Zhang L, Bruick RK, Zhou YD, Nagle DG. Laurenditerpenol, a new diterpene from the tropical marine alga *Laurencia intricata* that potently inhibits HIF-1 mediated hypoxic signaling in breast tumor cells. *J. Nat. Prod.* 2004;67:2002-7.

[128] Dai J, Fishback JA, Zhou YD, Nagle DG. Sodwanone and yardenone triterpenes from a South African species of the marine sponge *Axinella* inhibit hypoxia-inducible factor-1 (HIF-1) activation in both breast and prostate tumor cells. *J. Nat. Prod.* 2006;69:1715-20.

[129] Dai J, Liu Y, Zhou YD, Nagle DG. Benzochromenones from the marine crinoid *Comantheria rotula* inhibit hypoxia-inducible factor-1 (HIF-1) in cell-based reporter assays and differentially suppress the growth of certian tumor cell lines. *J. Nat. Prod.* 2007;70:1462-6.

[130] Liu R, Liu Y, Zhou YD, Nagle DG. Molecular-targeted antitumor agents. 15. Neolamellarins from the marine sponge Dendrilla nigra inhibit hypoxia-inducible factor-1 activation and secreted vascular endothelial growth factor production in breast tumor cells. *J. Nat. Prod.* 2007;70:1741-5.

[131] Mohammed KA, Jaduico RC, Bugni TS, Harper MK, Sturdy M, Ireland C. Strongylophorines: natural product inhibitors of hypoxia-inducible factor-1 transcriptional pathway. *J. Med. Chem.* 2008;51:1402-5.

[132] El Sayed KA, Khanfar MA, Shallal HM, Muralidharan A, Awate B, Youssef DT, et al. Latrunculin A and Its C-17-O-carbamates Inhibit prostate tumor cell invasion and HIF-1 activation in breast tumor cells. *J. Nat. Prod.* 2008;71:396-402.

[133] Tsai JR, Lin HC. V-type H$^+$-ATPase and Na$^+$,K$^+$-ATPase in the gills of 13 euryhaline crabs during salinity acclimation. *J. Exp. Biol.* 2007;210:620-7.

[134] Beutler JA, McKee TC. Novel marine and microbial natural product inhibitors of vacuolar ATPase. *Curr. Med. Chem.* 2003;10:787-96.

[135] Xie XS, Padron D, Liao X, Wang J, Roth MG, De Brabander JK. Salicylihalamide A Inhibits the V$_0$ Sector of the V-ATPase through a Mechanism Distinct from Bafilomycin A$_1$. *J. Biol. Chem.* 2004;279:19755-63.

[136] James ER, Green DR. Manipulation of apoptosis in the host-parasite interaction. *Trends Parasitol.* 2004;20:280-7.

[137] Lisowska K, Witkowski JM. Viral strategies in modulation of NF-kappaB activity. *Arch. Immunol. Ther. Exp.* 2003;51:367-75.

[138] Neish AS. Bacterial inhibition of eukaryotic pro-inflammatory pathways. *Immunol. Res.* 2004;29:175-86.

[139] Hiscott J, Kwon H, Genin P. Hostile takeovers: viral appropriation of the NF-kappa B pathway. *J. Clin. Invest* 2001;107:143-51.

[140] Lane N. Marine microbiology: Origins of Death. *Nature* 2008;453:583-5.

[141] Jayakumar A, Donovan MJ, Tripathi V, Ramalho-Ortigao M, McDowell MA. *Leishmania major* infection activates NF-κB and interferon regulatory factors 1 and 8 in human dendritic cells. *Infection and immunity* 2008;76:2138-48.

[142] Heussler VT, Rottenberg S, Schwab R, Kuenzi P, Fernandez PC, McKellar S, et al. Hijacking of host cell IKK signalosomes by the transforming parasite *Theileria. Science* 2002;298:1033-6.

[143] Umezawa K, Chaicharoenpong C. Molecular design and biological activities of NF-kappaB inhibitors. *Molecules and Cells* 2002;14:163-7.

[144] Le Negrate G, Faustin B, Welsh K, Loeffler M, Krajewska M, Hasegawa P, et al. Salmonella secreted factor L deubiquitinase of *Salmonella typhimurium* inhibits NF-κB, supresses IκBα ubiquitination and modulates innate immune responses. *Journal of Inmmunology* 2008;180:5045-56.

[145] Merlo E, Freudenthal R, Maldonado H, Romano A. Activation of the transcription factor NF-kappaB by retrieval is required for long-term memory reconsolidation. *Learning and Memory* 2005;12:23-9.

[146] Folmer F, Jaspars M, Dicato M, Diederich M. Marine natural products as targeted modulators of the transcription factor NF-kappaB. *Biochem. Pharmacol.* 2008;75:603-17.

[147] Terracciano S, Aquino M, Rodriquez M, Monti MC, Casapullo A, Riccio R, et al. Chemistry and biology of anti-inflammatory marine natural products: molecules interfering with cyclooxygenase, NF-kappaB and other unidentified targets. *Curr. Med. Chem.* 2006;13:1947-69.

[148] Groll M, Huber R, Potts BC. Crystal structures of Salinosporamide A (NPI-0052) and B (NPI-0047) in complex with the 20S proteasome reveal important consequences of beta-lactone ring opening and a mechanism for irreversible binding. *J. Am. Chem. Soc.* 2006;128:5136-41.

[149] Tsukamoto S. The search for inhibitors of the ubiquitin–proteasome system from natural resources for drug development. *Journal of Natural Medicines* 2006;60:273-8.

[150] Sharp KH, Davidson SK, Haygood MG. Localization of 'Candidatus *Endobugula sertula*' and the bryostatins throughout the life cycle of the bryozoan *Bugula neritina*. *The ISME Journal* 2007;1:693-702.

[151] Banerjee S, Wang Z, Mohammad M, Sarkar F, Mohammad R. Efficacy of selected natural products as therapeutic agents against cancer. *J. Nat. Prod.* 2008;71:492-6.

[152] Wender P, Cribbs C, Koehler M, Sharkey N, Herald CL, Kamano Y, et al. Modeling of the bryostatins to the phorbol ester pharmacophore on protein kinase C. *Proc. Natl. Acad. Sci. USA* 1988;85:7197-201.

[153] Stone RM, Sariban E, Pettit GR, Kufe DW. Bryostatin 1 activates protein kinase C and induces monocytic differentiation of HL-60 cells. *Blood* 1988;72:208-13.

[154] La Porta CA, Comolli R. PKC-dependent modulation of IκBα-NFκB pathway in low metastatic B16F1 murine melanoma cells and in highly metastatic *BL6 cells. Anticancer Res.* 1998;18:2591-7.

[155] Mortenson MM, Galante JG, Gilad O, Schlieman MG, Virudachalam S, Kung HJ, et al. BCL-2 functions as an activator of the AKT signaling pathway in pancreatic cancer. *J. Cell Biochem.* 2007;102:1171-9.

[156] Silberman DM, Zorrilla-Zubilete M, Cremaschi GA, Genaro AM. Protein kinase C-dependent NF-κB activation is altered in T cells by chronic stress. *Cell Mol. Life Sci.* 2005;62:1744-54.

[157] Folmer F, Harrison WT, Tabudravu JN, Jaspars M, Aalbersberg W, Feussner K, et al. NF-kappaB-inhibiting naphthopyrones from the Fijian echinoderm Comanthus parvicirrus. *J. Nat. Prod.* 2008;71:106-11.

[158] Lee SJ, Bai SK, Lee KS, Namkoong S, Na HJ, Ha KS, et al. Astaxanthin inhibits nitric oxide production and inflammatory gene expression by suppressing I kappa B kinase-dependent NF-kappa B activation. *Molecules and Cells* 2003;16:97-105.

[159] Stevenson CS, Capper EA, Roshak AK, Marquez B, Eichman C, Jackson JR, et al. The identification and characterization of the marine natural product scytonemin as a novel antiproliferative pharmacophore. *J. Pharmacol. Exp. Ther.* 2002;303:858-66.

[160] Terracciano S, Rodriquez M, Aquino M, Monti MC, Casapullo A, Riccio R, et al. Chemistry and biology of anti-inflammatory marine natural products: molecules interfering with cyclo-oxygenase, NF-κB and other unidentified targets. *Curr. Med. Chem.* 2006;13:1947-69.

[161] Kamata K, Okamoto S, Oka S, Kamata H, Yagisawa H, Hirata H. Cycloprodigiosin hydrocloride suppresses tumor necrosis factor (TNF) alpha-induced transcriptional activation by NF-kappa B. *FEBS Lett* 2000;507:74-80.

[162] Choi IK, Shin HJ, Lee HS, Kwon HJ. Streptochlorin, a marine natural product, inhibits NF-kappaB activation and supresses angiogenesis in vitro. *Journal of Microbiology and Biotechnology* 2007;17:1338-43.

[163] Oda T, Namikoshi K, Akamo K, Honma Y, Kasahara T. Verrucarin A inhibition of MAP kinase activation in a PMA-stimulated promyelocytic leukemia cell line. *Mar. Drugs* 2005;3:64-73.

[164] Posadas I, De Rosa S, Terencio MC, Paya M, Alcaraz MJ. Cacospongionolide B suppresses the expression of inflammatory enzymes and tumour necrosis factor-alpha by inhibiting nuclear factor-kappa B activation. *Br. J. Pharmacol.* 2003;138:1571-9.

[165] Posadas I, Terencio MC, Randazzo A, Gomez-Paloma L, Paya M, Alcaraz MJ. Inhibition of the NF-kappa B signaling pathway mediates the anti-inflammatory effects of petrosaspongiolide M. *Biochem. Pharmacol.* 2003;65:887-95.

[166] D'Acquisto F, Lanzotti V, Carnuccio R. Cyclolinteinone, a sesterterpene from sponge *Cacospongia linteiformis*, prevents inducible nitric oxide synthase and inducible cyclo-oxygenase protein expression by blocking nuclear factor-kappa B activation in J774 macrophages. *Biochem. J.* 2000;346:793-8.

[167] Amigo M, Paya M, Braza-Boïls A, De Rosa M, Terencio MC. Avarol inhibits TNF-alpha generation of NF-kappaB activation in human cells and in animal models. *Life Sci.* 2008;82:256-64.

[168] Breton JJ, ChabotFletcher MC. The natural product hymenialdisine inhibits interleukin-8 production in U937 cells by inhibition of nuclear factor-kappa B. *J. Pharmacol. Exp. Ther.* 1997;282:459-66.

[169] Roshak A, Jackson JR, Chabot Fletcher M, Marshall LA. Inhibition of NF kappa B-mediated interleukin-1 beta-stimulated prostaglandin E-2 formation by the marine natural product hymenialdisine. *J. Pharmacol. Exp. Ther.* 1997;283:955-61.

[170] Sharma V, Lansdell TA, Jin G, Tepe JJ. Inhibition of cytokine production by hymenialdisine derivatives. *J. Med. Chem.* 2004;47:3700-3.

[171] Lu PH, Chuch SC, Kung FL, Pan SL, Shen YC, Guh JH. Ilimaquinone, a marine metabolite, displays anticancer activity via GADD153-mediated pathway. *Eur. J. Pharmacol.* 2007;556:45-54.

[172] Lucas R, Casapullo A, Ciasullo L, Gomez-Paloma L, Paya M. Cycloamphilectenes, a new type of potent marine diterpenes: inhibition of nitric oxide production in murine macrophages. *Life Sci.* 2003;72:2543-52.

[173] Rawat DS, Joshi MC, Atheaya H. Marine peptides and related compounds in clinical trial. *Anticancer Agents and Medicinal Chemistry* 2006;6:33-40.

[174] D'Incalci M, Simone M, Tavecchio M, Damia G, Garbi A, Erba E. New drugs from the sea. *J. Chemotherapy* 2004;16 Suppl 4:86-9.

[175] Newman DJ, Cragg GM. Advanced preclinical and clinical trials of natural products and related compounds from marine sources. *Curr. Med. Chem.* 2004;11:1693-713.

[176] Schwartsmann G, Da Rocha AB, Mattei J, Lopes R. Marine-derived anticancer agents in clinical trials. *Expert Opin. Invest Drugs* 2003;12:1367-83.

[177] Jimeno J, Faircloth G, Fernandez Sousa-Faro JM, Scheuer PJ, Rinehart KL. New marine derived anticancer therapeutics - a journey from the sea to clinical trials. *Mar. Drugs* 2004;2:14-29.

[178] Haefner B. Drugs from the deep: marine natural products as drug candidates. *Drug Disc. Today* 2003;8:536-44.

[179] Jaspars M. Pharmacy of the Deep - Marine organisms as sources of anticancer agents. In: Harvey A, editor. *Advances in Drug Discovery Techniques.* London, UK: John Wiley and Sons Ltd., 1998. p. 65-84.

[180] Sipkema D, Osinga R, Schatton W, ., Mendola D, Tramper J, Wijffels RH. Large-scale production of pharmaceuticals by marine sponges: sea, cell, or synthesis? *Biotechnol. Bioeng.* 2005;90:201-22.

[181] De Rosa S, De Caro S, Iodice C, Tommonaro C, Stefanov K, Popov S. Development in primary cell culture of demosponges. *J. Biotechnol.* 2003;100:119-25.

[182] Jensen PR, Mincer TJ, Williams PG, Fenical W. Marine actinomycete diversity and natural product discovery. *Antonie Van Leeuwenhoek* 2005;87:43-8.

[183] Newman D, Hill R. New drugs from marine microbes: The tide is turning. *J. Ind. Microbiol. Biotechnol.* 2006;33:539-44.

[184] Mutter R, Wills M. Chemistry and clinical biology of the bryostatins. *Bioorg. Med. Chem.* 2000;8:1841-60.

[185] Neidle S, Thurston DE. Chemical approaches to the discovery and development of cancer therapies. *Nat. Rev. Cancer* 2005;5:285-96.

[186] Macherla VR, Mitchell SS, Manam RR, Reed KA, Chao TH, Nicholson B, et al. Structure-activity relationship studies of salinosporamide A (NPI-0052), a novel marine derived proteasome inhibitor. *J. Med. Chem.* 2005;48:3684-7.

[187] Reed KA, Manam RR, Mitchell S, Xu J, Teisan S, Chao T, et al. Salinosporamides D-J from the marine actinomycete *Salinispora tropica*, bromosalinosporamide, and

thioester derivatives are potent Inhibitors of the 20S proteasome. *J. Nat. Prod.* 2007;70:269-76.

[188] Ahn KS, Sethi G, Chao TH, Neuteboom ST, Chaturvedi MM, Palladino MA, et al. Salinosporamide A (NPI-0052) potentiates apoptosis, suppresses osteoclastogenesis, and inhibits invasion through down-modulation of NF-κB-regulated gene products. *Blood* 2007;110:2286-95.

[189] Feling RH, Buchanan GO, Mincer TJ, Kauffman CA, Jensen PR, Fenical W. Salinosporamide A: a highly cytotoxic proteasome inhibitor from a novel microbial source, a marine bacterium of the new genus *Salinospora. Angew Chem.* 2003;42:355-7.

[190] Newman D, Cragg G. Recent progress of natural marine products as anticancer drugs. *International Oncology Updates: Marine anticancer compounds in the Era of targetet therapies* 2008.

[191] Reddy RL, Saravan P, Corey EJ. A simple stereocontrolled synthesis of salinosporamide A. *J. Am. Chem. Soc.* 2004;126:6230-1.

[192] Hogan P, Corey EJ. Proteasome Inhibition by a Totally Synthetic β-Lactam Related to Salinosporamide A and Omuralide. *J. Am. Chem. Soc.* 2005;127:15386-7.

[193] Endo A, Danishefsky SJ. Total synthesis of salinosporamide A. *J. Am. Chem. Soc.* 2005;127:8298-9.

[194] Reddy LR, Fournier JF, Reddy BV, Corey EJ. New synthetic route for the enantioselective total synthesis of salinosporamide A and biologically active analogues. *Org. Lett 2005*;7:2699-701.

[195] Takahashi K, Midori M, Kawano K, Ishihara J, Hatakeyama S. Entry to heterocycles based on indium-catalyzed conia-ene reactions: asymmetric synthesis of (-)-salinosporamide A. *Angew Chem.* 2008;*in press* (PMID 18600821).

[196] Udwary DW, Zeigler L, Asolkar RN, Singan V, Lapidus A, Fenical W, et al. Genome sequencing reveals complex secondary metabolome in the marine actinomycete *Salinospora tropica. Proc. Natl. Acad. Sci. USA* 2007;104:10376-81.

[197] Eustaquio AS, Pojer F, Noel JP, Moore BS. Discovery and characterization of a marine bacterial SAM-dependent chlorinase. *Nature Chem. Biol.* 2008;4:69-74.

[198] Aikawa S, Matsuzawa F, Satoh Y, Kadota Y, Doi H, Itoh K. Prediction of the mechanism of action of omuralide (clasto-lactacystin beta-lactone) on human cathepsin A based on a structural model of the yeast proteasome beta5/PRE2-subunit/omuralide complex. *Biochim. Biophys Acta* 2006;1764:1372-80.

[199] Wender P, Hinkle K, Koehler M, Lippa M. The rational design of potential chemotherapeutic agents: Synthesis of bryostatin analogues. *Med. Res. Rev.* 1999;19:388-407.

[200] Wall NR, Mohammad RM, Reddy KB, Al-Katib AM. Bryostatin 1 induces ubiquitination and proteasome degradation of Bcl-2 in the human acute lymphoblastic leukemia cell line, Reh. *Int. J. Mol. Med.* 2000;5:165-71.

[201] Evans DAC, P. H.; Carreira, E. M.; Charette, A. B.;, Prunet JAL, M. Total Synthesis of Bryostatin 2. *J. Am. Chem. Soc.* 1999;121:7540-52.

[202] Kageyama MT, T.; Nantz, M. H.; Roberts, J. C.; Somfai, P.; Whritenour DCM, S. Synthesis of bryostatin 7. *J. Am. Chem. Soc.* 1990;112:7407-8.

[203] Manaviazar S, Frigerio M, Bhatia GS, Hummersone MG, Aliev AE, Hale KJ. Enantioselective formal total synthesis of the antitumor macrolide bryostatin 7. *Org. Lett* 2006;8:4477-80.

[204] Ohmori KO, Y.; Obitsu, T.; Ishikawa, Y.; Nishiyama, S.;, Yamamura S. Total Synthesis of Bryostatin 3. *Angew Chem,* Int Ed 2000;39:2290-4.

[205] Mendola D. Aquaculture of three phyla of marine invertebrates to yield bioactive metabolites: process developments and economics. *Biomol. Eng.* 2003;20:441-58.

[206] Piel J. Bacterial symbionts: prospects for the sustainable production of invertebrate-derived pharmaceuticals. *Curr. Med. Chem.* 2006;13:39-50.

[207] Hildebrand M, Waggoner LE, Liu H, Sudek S, Allen S, Anderson C, et al. bryA: an unusual modular polyketide synthase gene from the uncultivated bacterial symbiont of the marine bryozoan Bugula neritina. *Chem. Biol.* 2004;11:1543-52.

[208] Bogi K, Lorenzo PS, Szallasi Z, Acs P, Wagner GS, Blumberg PM. Differential selectivity of ligands for the C1a and C1b phorbol ester binding domains of protein kinase Cdelta: possible correlation with tumor-promoting activity. *Cancer Res.* 1998;58:1423-8.

[209] Statsuk AV, Bai R, Baryza JL, Verma VA, Hamel E, Wender PA, et al. Actin is the primary cellular receptor of bistramide A. *Nature Chem. Biol.* 2005;1:383-8.

[210] Wender P, DeBrabander J, Harran P, Jimenez J, Koehler M, Lippa B, et al. The design, computer modeling, solution structure, and biological evaluation of synthetic analogs of bryostatin 1. *PNAS* 1998;95:6624-9.

[211] Wender PA, Koehler KF, Sharkey NA, Dell'Aquila ML, Blumberg PM. Analysis of the phorbol ester pharmacophore on protein kinase C as a guide to the rational design of new classes of analogs. *Proc. Natl. Acad. Sci. USA* 1986;83:4214-8.

[212] Wender PA, Cribbs CM, Koehler KF, Sharkey NA, Herald CL, Kamano Y, et al. Modeling of the bryostatins to the phorbol ester pharmacophore on protein kinase C. *Proc. Natl. Acad. Sci. USA* 1988;85:7197-201.

[213] Wender PA, Mayweg AV, VanDeusen CL. A concise, selective synthesis of the polyketide spacer domain of a potent bryostatin analogue. *Org. Lett.* 2003;5:277-9.

[214] Wender PA, Dechristopher BA, Schrier AJ. Efficient synthetic access to a new family of highly potent bryostatin analogues via a Prins-driven macrocyclization strategy. *J. Am. Chem. Soc.* 2008;130:6658-9.

[215] Wender PA, Verma VA. Design, synthesis, and biological evaluation of a potent, PKC selective, B-ring analog of bryostatin. *Org. Lett.* 2006;8:1893-6.

[216] Wender PA, Baryza JL. Identification of a tunable site in bryostatin analogs: C20 Bryologs through late stage diversification. *Org. Lett.* 2005;7:1177-80.

[217] Baryza JL, Brenner SE, Craske ML, Meyer T, Wender P. Simplified analogs of bryostatin with anticancer activity display greater potency for translocation of PKCδ-GFP. *Chemistry and Biology* 2004;11:1261-7.

[218] Brodie C, Blumberg PM. Regulation of cell apoptosis by protein kinase c delta. *Apoptosis* 2003;8:19-27.

[219] Kikkawa U, Matsuzaki H, Yamamoto T. Protein kinase C delta (PKC delta): activation mechanisms and functions. *J. Biochem.* 2002;132:831-9.

[220] Wender PA, Baryza JL, Bennett CE, Bi FC, Brenner SE, Clarke MO, et al. The practical synthesis of a novel and highly potent analogue of bryostatin. *J. Am. Chem. Soc.* 2002;124:13648-9.

[221] Wender PA, Clarke MO, Horan JC. Role of the A-ring of bryostatin analogues in PKC binding: synthesis and initial biological evaluation of new A-ring-modified bryologs. *Org. Lett.* 2005;7:1995-8.

[222] Lallemend F, Hadjab S, Hans G, Moonen G, Lefebvre PP, Malgrange B. Activation of protein kinase Cbetal constitutes a new neurotrophic pathway for deafferented spiral ganglion neurons. *J. Cell Sci.* 2005;118:4511-25.

[223] Wang QJ, Fang TW, Fenick D, Garfield S, Bienfait B, Marquez VE, et al. The lipophilicity of phorbol esters as a critical factor in determining the pattern of translocation of protein kinase C delta fused to green fluorescent protein. *J. Biol. Chem.* 2000;275:12136-46.

[224] Wender PA, DeBrabander J, Harran PG, Jimenez JM, Koehler MF, Lippa B, et al. The design, computer modeling, solution structure, and biological evaluation of synthetic analogs of bryostatin 1. *Proc. Natl. Acad. Sci. USA* 1998;95:6624-9.

[225] Wender PA, DeBrabander J, Harran PG, I-Iinke KW, Lippa B, Pettit GR. Synthesis and Biological Evaluation of Fully Synthetic Bryostatin Analogues. *Tetrahedron Lett* 1998;39:8625-8.

[226] Wender PA, Horan JC. Synthesis and PKC binding of a new class of a-ring diversifiable bryostatin analogues utilizing a double asymmetric hydrogenation and cross-coupling strategy. *Org. Lett.* 2006;8:4581-4.

[227] Keck GE, Kraft MB, Truong AP, Li W, Sanchez CC, Kedei N, et al. Convergent assembly of highly potent analogues of bryostatin 1 via pyran annulation: bryostatin look-alikes that mimic phorbol ester function. *J. Am. Chem. Soc.* 2008;130:6660-1.

[228] Breton JJ, Chabot-Fletcher MC. The natural product hymenialdisine inhibits interleukin-8 production in U937 cells by inhibition of nuclear factor-kappaB. *J. Pharmacol. Exp. Ther.* 1997;282:459-66.

[229] Roshak A, Jackson JR, ChabotFletcher M, Marshall LA. Inhibition of NF kappa B-mediated interleukin-1 beta-stimulated prostaglandin E-2 formation by the marine natural product hymenialdisine. J. *Pharmacol. Exp. Ther.* 1997;283:955-61.

[230] Annoura H, Tatsuoka T. Total Syntheses of Hymenialdisine and Debromohymenialdisine: Stereospecific Construction of the 2-Amino-4-Oxo-2-Imidazolin-5(Z)-Disubstituted Ylidene Ring System. *Tetrahedron Letters* 1995;36:413-6.

[231] Meijer L, Thunnissen AM, White AW, Garnier M, Nikolic M, Tsai LH, et al. Inhibition of cyclin-dependent kinases, GSK-3beta and CK1 by hymenialdisine, a marine sponge constituent. *Chem. Biol.* 2000;7:51-63.

[232] Imahori K, Uchida T. Physiology and pathology of tau protein kinases in relation to Alzheimer's disease. J. *Biochem.* 1997;121:179-88.

[233] Mandelkow EM, Mandelkow E. Tau in Alzheimer's disease. *Trends Cell Biol.* 1998;8:425-7.

[234] Singh TJ, Grundke-Iqbal I, Iqbal K. Phosphorylation of tau protein by casein kinase-1 converts it to an abnormal Alzheimer-like state. *J. Neurochem.* 1995;64:1420-3.

[235] Sielecki TM, Boylan JF, Benfield PA, Trainor GL. Cyclin-dependent kinase inhibitors: useful targets in cell cycle regulation. *J. Med. Chem.* 2000;43:1-18.

[236] Chacun-Lefèvre L, Joseph B, Mérour J-Y. Synthesis and reactivity of azepino[3,4-b]indol-5-yl trifluoromethanesulfonate. *Tetrahedron* 2000;56:4491-9.

[237] Kaiser HM, Zenz I, Lo WF, Spannenberg A, Schroder K, Jiao H, et al. Preparation of novel unsymmetrical bisindoles under solvent-free conditions: synthesis, crystal structures, and mechanistic aspects. *J. Org. Chem.* 2007;72:8847-58.

[238] Papeo G, Posteri H, Borghi D, Varasi M. A new glycociamidine ring precursor: syntheses of (Z)-hymenialdisine, (Z)-2-debromohymenialdisine, and (+/-)-endo-2-debromohymenialdisine. *Org. Lett.* 2005;7:5641-4.

[239] Xu Yz Y, Yakushijin K, Horne DA. Synthesis of C(11)N(5) Marine Sponge Alkaloids: (+/-)-Hymenin, Stevensine, Hymenialdisine, and Debromohymenialdisine. *J. Org. Chem.* 1997;62:456-64.

[240] Meijer L, Raymond E. Roscovitine and other purines as kinase inhibitors. From starfish oocytes to clinical trials. *Acc. Chem. Res.* 2003;36:417-25.

[241] Wan Y, Hur W, Cho CY, Liu Y, Adrian FJ, Lozach O, et al. Synthesis and target identification of hymenialdisine analogs. *Chem. Biol.* 2004;11:247-59.

[242] Tasdemir D, Mallon R, Greenstein M, Feldberg LR, Kim SC, Collins K, et al. Aldisine alkaloids from the Philippine sponge Stylissa massa are potent inhibitors of mitogen-activated protein kinase kinase-1 (MEK-1). *J. Med. Chem.* 2002;45:529-32.

[243] Meijer L, Thunnissen AM, White AW, Garnier M, Nikolic M, Tsai LH, et al. Inhibition of cyclin-dependent kinases, GSK-3beta and CK1 by hymenialdisine, a marine sponge constituent. *Chem. Biol.* 2000;7:51-63.

[244] Wan Y, Hur W, Cho CY, Liu Y, Adrian FJ, Lozach O, et al. Synthesis and target identification of hymenialdisine analogs. *Chem. Biol.* 2004;11:247-59.

[245] Sharma V, Tepe JJ. Potent inhibition of checkpoint kinase activity by a hymenialdisine-derived indoloazepine. *Bioorg. Med. Chem. Lett.* 2004;14:4319-21.

[246] Amigo M, Paya M, Braza-Boils A, De Rosa S, Terencio MC. Avarol inhibits TNF-alpha generation and NF-kappaB activation in human cells and in animal models. *Life Sci.* 2008;82:256-64.

[247] Amigo M, Paya M, De Rosa S, Terencio MC. Antipsoriatic effects of avarol-3'-thiosalicylate are mediated by inhibition of TNF-alpha generation and NF-kappaB activation in mouse skin. *Br. J. Pharmacol.* 2007;152:353-65.

[248] Amigo M, Terencio MC, Mitova M, Iodice C, Paya M, De Rosa S. Potential antipsoriatic avarol derivatives as antioxidants and inhibitors of PGE(2) generation and proliferation in the HaCaT cell line. *J. Nat. Prod.* 2004;67:1459-63.

[249] Schroder HC, Wenger R, Gerner H, Reuter P, Kuchino Y, Sladic D, et al. Suppression of the modulatory effects of the antileukemic and anti-human immunodeficiency virus compound avarol on gene expression by tryptophan. *Cancer Res.* 1989;49:2069-76.

[250] Loya S, Hizi A. The inhibition of human immunodeficiency virus type 1 reverse transcriptase by avarol and avarone derivatives. *FEBS Lett.* 1990;269:131-4.

[251] Muller WE, Sladic D, Zahn RK, Bassler KH, Dogovic N, Gerner H, et al. Avarol-induced DNA strand breakage in vitro and in Friend erythroleukemia cells. *Cancer Res.* 1987;47:6565-71.

[252] Muller WE, Maidhof A, Zahn RK, Schroder HC, Gasic MJ, Heidemann D, et al. Potent antileukemic activity of the novel cytostatic agent avarone and its analogues in vitro and in vivo. *Cancer Res.* 1985;45:4822-6.

[253] Müller WEG, Böhm M, Batel R, De Rosa S, Tommonaro G, Müller IM, et al. Application of cell culture for the production of bioactive compounds from sponges: synthesis of avarol by primmorphs from *Dysidea avara. J. Nat. Prod.* 2000;63:1077-81.

[254] Sarma AS, Chattopadhyay P. Synthetic studies of trans-clerodane diterponoids and congeners: stereocontrolled total synthesis of (±)-avarol. *J. Org. Chem.* 1982;47:1727-31.

[255] Sipkema D, Osinga R, Schatton W, Mendola D, Tramper J, Wijffels RH. Large-scale production of pharmaceuticals by marine sponges: sea, cell, or synthesis? *Biotechnol. Bioeng.* 2005;90:201-22.

[256] Dutcher JS, Macmillan JG, Heathcock CH. Pentacyclic Triterpene Synthesis. 5. Synthesis of Optically Pure Ring AB Precursors. *J. Org. Chem.* 1976;41:2663-9.

[257] Crispino A, de Giulio A, de Rosa S, Strazzullo G. A new bioactive derivative of avarol from the marine sponge Dysidea avara. *J. Nat. Prod.* 1989;52:646-8.

[258] Schmitz FJ, Lakshmi V, Powell DR, Van der Helm D. Arenarol and arenarone: sesquiterpenoids with rearranged drimane skeletons from the marine sponge Dysidea arenaria. *J. Org. Chem.* 1984;49:241-4.

[259] Belisario MA, Maturo M, Pecce R, De Rosa S, Villani GR. Effect of avarol and avarone on in vitro-induced microsomal lipid peroxidation. *Toxicology* 1992;72:221-33.

[260] Amigo M, Terencio MC, Paya M, Iodice C, De Rosa S. Synthesis and evaluation of diverse thio avarol derivatives as potential UVB photoprotective candidates. *Bioorg Med. Chem. Lett.* 2007;17:2561-5.

[261] Amigo M, Schalkwijk J, Olthuis D, De Rosa S, Paya M, Terencio MC, et al. Identification of avarol derivatives as potential antipsoriatic drugs using an in vitro model for keratinocyte growth and differentiation. *Life Sci.* 2006;79:2395-404.

[262] De Rosa S, Crispino A, De Giulio A, Iodice C, Benrezzouk R, Terencio MC, et al. A new cacospongionolide inhibitor of human secretory phospholipase A2 from the Tyrrhenian sponge *Fasciospongia cavernosa* and absolute configuration of cacospongionolides. *J. Nat. Prod.* 1998;61:931-5.

[263] Garcia Pastor P, De Rosa S, De Giulio A, Paya M, Alcaraz MJ. Modulation of acute and chronic inflammatory processes by cacospongionolide B, a novel inhibitor of human synovial phospholipase A2. *Br. J. Pharmacol.* 1999;126:301-11.

[264] Murelli RP, Cheung AK, Snapper ML. Conformationally restricted (+)-cacospongionolide B analogues. Influence on secretory phospholipase A2 inhibition. *J. Org. Chem.* 2007;72:1545-52.

[265] Anthonsen MW, Solhaug A, Johansen B. Functional coupling between secretory and cytosolic phospholipase A2 modulates tumor necrosis factor-alpha- and interleukin-1beta-induced NF-kappa B activation. *J. Biol. Chem.* 2001;276:30527-36.

[266] Potts BC, Faulkner DJ, Jacobs RS. Phospholipase A2 inhibitors from marine organisms. *J. Nat. Prod.* 1992;55:1701-17.

[267] Palanki MS, Erdman PE, Gayo-Fung LM, Shevlin GI, Sullivan RW, Goldman ME, et al. Inhibitors of NF-kappaB and AP-1 gene expression: SAR studies on the pyrimidine portion of 2-chloro-4-trifluoromethylpyrimidine-5-[N-(3', 5'-bis(trifluoromethyl) phenyl)carboxamide]. *J. Med. Chem.* 2000;43:3995-4004.

[268] De Rosa M, Giordano S, Scettri A, Sodano G, Soriente A, Pastor PG, et al. Synthesis and comparison of the antiinflammatory activity of manoalide and cacospongionolide B analogues. *J. Med. Chem.* 1998;41:3232-8.

[269] Cheung AK, Snapper ML. Total syntheses of (+)- and (-)-cacospongionolide B: new insight into structural requirements for phospholipase A(2) inhibition. *J. Am. Chem. Soc.* 2002;124:11584-5.

[270] Cheung AK, Murelli R, Snapper ML. Total syntheses of (+)- and (-)-cacospongionolide B, cacospongionolide e, and related analogues. Preliminary study of structural features required for phospholipase a2 inhibition. *J. Org. Chem.* 2004;69:5712-9.

[271] Potts BC, Faulkner DJ, Jacobs RS. Phospholipase A2 inhibitors from marine organisms. *J. Nat. Prod.* 1992;55:1701-17.

[272] Potts BC, Faulkner DJ, De Carvalho MS, Jacobs RS. Chemical mechanism of inactivation of bee venom phospholipase A2 by the marine natural products manoalide, luffariellolide, and scalaradial. *J. Am. Chem. Soc.* 1992;114:5093-100.

[273] Glaser KB, de Carvalho MS, Jacobs RS, Kernan MR, Faulkner DJ. Manoalide: structure-activity studies and definition of the pharmacophore for phospholipase A2 inactivation. *Mol. Pharmacol.* 1989;36:782-8.

[274] De Rosa M, Giordano S, Scettri A, Sodano G, Soriente A, Pastor PG, et al. Synthesis and comparison of the antiinflammatory activity of manoalide and cacospongionolide B analogues. *J. Med. Chem.* 1998;41:3232-8.

[275] Wender PA, Horan JC, Verma VA. Total Synthesis and Initial Biological Evaluation of New B-Ring-Modified Bryostatin Analogs. *Org. Lett.* 2006; 8: 5299-302.

[276] Wender PA, Hilinski MK, Mayweg AV. Late-Stage Intermolecular CH Activation for Lead Diversification: A Highly Chemoselective Oxyfunctionalization of the C-9 Position of Potent Bryostatin Analogues. *Org. Lett.* 2005; 7:79-82.

In: Aquatic Ecosystem Research Trends
Editor: George H. Nairne

ISBN 978-1-60692-772-4
© 2009 Nova Science Publishers, Inc.

*Chapter 10*

# IMPACT OF CLIMATE AND ENVIRONMENTAL FACTORS ON THE EPIDEMIOLOGY OF *VIBRIO CHOLERAE* IN AQUATIC ECOSYSTEMS

## *Violeta Trinidad Pardío Sedas[2]*

Laboratorio de Toxicología, Universidad Veracruzana, Apartado Postal 1380,
Veracruz, Veracruz, México CP. 91700

## ABSTRACT

The global and anthropogenic climate change and variability, mainly global warming, is having measurable effects on ecosystems, communities, and populations. The combination of climate change and environmental degradation has created ideal conditions for the emergence, resurgence and spread of infectious diseases, and has led to growing concerns due to the effects of climate on health. Diverse environmental factors affect the distribution, diversity, incidence, severity, or persistence of diseases and other health effects - something that has been recognized for millennia. An important risk of climate change is its potential impact on the evolution and emergence of infectious disease agents. Evidence is indicating that the atmospheric and oceanic processes that occur in response to increased greenhouse gases in the broad-scale climate system may already be changing the ecology of infectious diseases. Ecosystem instabilities brought about by climate change and concurrent stresses such as land use changes, species dislocation, and increasing global travel could potentially influence the genetics of pathogenic microbes through mutation and horizontal gene transfer, giving rise to new interactions among hosts and disease agents. Recent studies have shown that climate also influences the abundance and ecology of pathogens, and the links between pathogens and changing ocean conditions, including human diseases such as cholera. *Vibrio cholerae* is well recognized as being responsible for significant mortality and economic loss in underdeveloped countries, most often centered in tropical areas of the world. Generally, *V. cholerae* is transmitted through contaminated food and water in communities that do not have access to proper sewage and water treatment systems, and is thus called, "the disease of poverty". During the last three decades, extensive research has been carried out

---
[2] Correspondence address. Email: vpardio@yahoo.com.mx.

to elucidate the virulence properties and the epidemiology of this pathogen. Within the marine environment, *V. cholerae* is found attached to surfaces provided by plants, filamentous green algae, copepods, crustaceans, and insects. The specific environmental changes that amplified plankton and associated bacterial proliferation and govern the location and timing of plankton blooms have been elucidated. Several studies have demonstrated that environmental non-O1 and non-O139 *V. cholerae* strains and *V. cholerae* O1 El Tor and O139 are able to form a three-dimensional biofilm on surfaces which provides a microenvironment, facilitating environmental persistence within natural aquatic habitats during interepidemic periods. Revealing the influence of climatic/environmental factors in seasonal patterns is critical to understanding temporal variability of cholera at longer time scales to improve disease forecasting. Recently, researchers have also been elucidating the environmental lifestyle of *V. cholerae*, and ecologically based models have been developed to define the role of environment, weather, and climate-related variables in outbreaks of this disease. This chapter provides current evidence for the influence of environmental factors on *Vibrio cholerae* dynamics and virulence traits of this organism, and the urgent need for action to prevent the consequences of climate change contributing to cholera.

**Keywords:** Vibrio cholerae, environmental factors, climate change, temporal variability

# INTRODUCTION

Anthropogenic climate change is having measurable effects on ecosystems, communities, and populations (Walther et al., 2002). Diverse environmental factors affect the distribution, diversity, incidence, severity, or persistence of diseases and other health effects - something that has been recognized for millennia (Anthes et al., 2007). Health and climate have been linked since antiquity. In the fifth century B.C., Hippocrates observed that many specific human illnesses were linked to changes of season, local weather patterns, and other environmental factors such as temperature (Rees, 1996).

Climate change encompasses temperature changes on global, regional, and local scales, and also changes in the mean and variability of rainfall, winds, and possibly ocean currents. The reality of climate change and the associated role of anthropogenic activity are being debated in many forums. Human activities may have more explicit impacts on ecological balances, potentially leading to new diseases associated with environmental changes: temperature extremes and violent weather events, such as fewer frosts and more storms, floods, and droughts; abnormal seasonal conditions associated with increases in temperature or moisture.

The climate change and variability have lead to growing concerns due to the effects of climate on health. The combination of climate change and environmental degradation has created ideal conditions for the emergence, resurgence and spread of infectious diseases, killing millions of people annually (Anthes et al., 2007). In a quantitative climate-health assessment study, the World Health Organization (WHO) examined the global burden of disease attributable to anthropogenic climate change up to the year 2000 (WHO, 2002). This study indicated that the climatic changes that have occurred since the mid-1970s could already cause over 150,000 deaths and approximately five million 'disability-adjusted life years' (DALYs) per year through diseases such as diarrhoea (temperature effects only), malaria and malnutrition mainly in developing countries (McMichael, 2004).

Many factors, including environmental change, have contributed to the persistence and increase in the occurrence of infectious diseases, such as human behavior, public health infrastructure, food production, and microbial adaptation. An important risk of climate change is its potential impact on the evolution and emergence of infectious disease agents. Ecosystem instabilities brought about by climate change and concurrent stresses such as land use changes, species dislocation, and increasing global travel could potentially influence the genetics of pathogenic microbes through mutation and horizontal gene transfer, giving rise to new interactions among hosts and disease agents. The disease agents and their vectors each have particular environments that are optimal for growth, survival, transport, and dissemination (Huq and Colwell, 1996).

*Vibrio cholerae* is well recognized as the causative agent of the human intestinal disease cholera, being responsible for significant mortality and economic loss in underdeveloped countries, most often centered in tropical areas of the world. Generally, *V. cholerae* is transmitted through contaminated food and water in communities that do not have access to proper sewage and water treatment systems (WHO, 2005a), and is thus called, "the disease of poverty" (Bonilla et al., 2000).

Cholera provides one of the best examples of how an emerging infectious disease has evolved from an oral– fecal transmission linear model of a waterborne bacterium and a human host, to a more complex ecological model of an infectious disease. During the last three decades, extensive research has been carried out to elucidate the virulence properties and the epidemiology of this pathogen. Recently, researchers have also been elucidating the environmental lifestyle of *V. cholerae*, and ecologically based models have been developed to define the role of environment, weather, and climate-related variables in outbreaks of this disease. These models also include global weather patterns, aquatic reservoirs, zooplankton, the collective behavior of surface attached cells, an adaptable genome, and the deep sea, together with the bacterium and its host (Colwell and Patz, 1998; Lobitz et al., 2000; Pascual et al., 2000).

This update describes the context and current evidence for the influence of environmental factors on *Vibrio cholerae* dynamics and virulence traits of this organism. It brings an appreciation of the complexity of the linkages among temporal and spatial variations of climate, anthropogenic environmental degradation, and disease risk. From this perspective, it remarks the urgent action needed to prevent the consequences of climate change and to improve the socioeconomic conditions of life required for sustained reduction of inequalities contributing to cholera.

## CLIMATE AND INFECTIOUS DISEASES LINKS

For the last decades, human environmental impacts have been urban-industrial air pollution, chemically polluted waterways and the manifestations of a rapid growth in rich and poor countries and concurrent stresses such as land use changes. These health hazards are now being exacerbated with the strong recent increase in world average temperature and changes in some of the planet's great biophysical and ecological systems and hence generating additional and larger-scale environmental health problems. Climate is a key determinant of health. Climate constrains the range of infectious diseases, while weather affects the timing and intensity of outbreaks. Accelerated climate change and variability are already influencing the functioning of many ecosystems and the seasonal cycles and geographic range of communities and populations, destabilizing biological control of infectious diseases, and affecting food production and water availability. This warming and unstable climate is playing an ever-increasing role in driving the global emergence, resurgence and redistribution of infectious diseases, and driving health impacts in human populations (Epstein, 2001). Tropical climate, ecological changes, poor water and food security, low socio-economic status and political instability define the regions that would be most vulnerable to the health effects of climate change (Moreno, 2006). It is also important to recognize how parallel processes of environmental degradation can exacerbate these risks. There is a close relationship among climate change, social factors, and ecological degradation. For example, the destruction of coastal wetlands or mangrove swamps will lessen the protective capacity of these ecosystems against typhoons and storm surges (Patz et al., 2006).

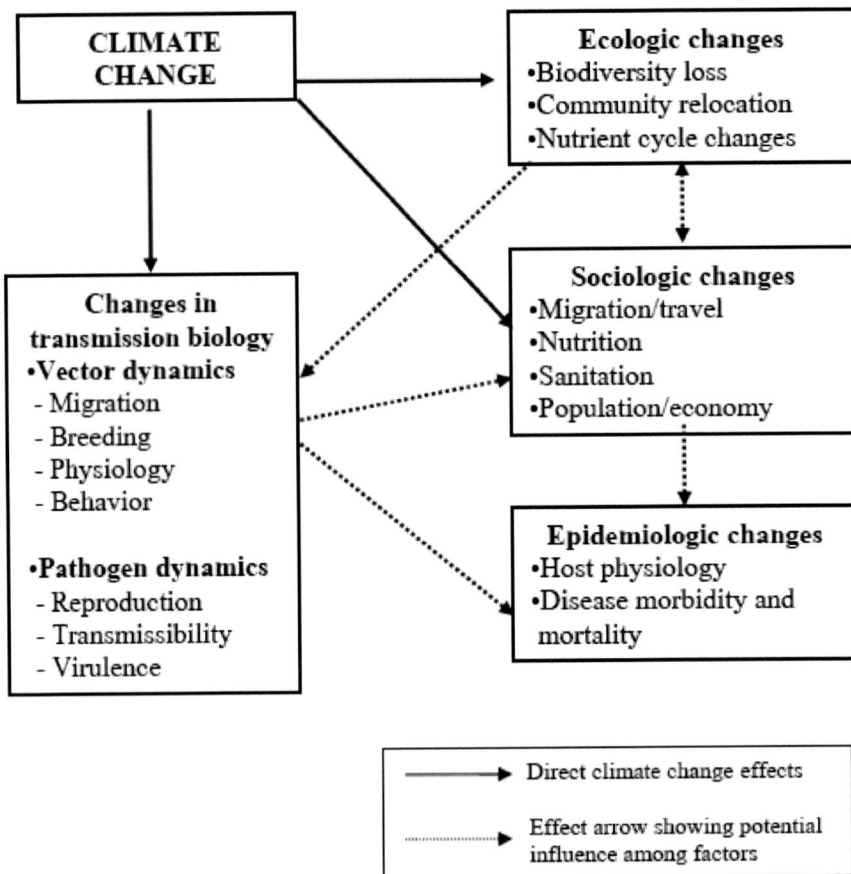

Figure 1. Integrated assessment framework for evaluating the association between climate change and infectious diseases (reproduced from Chan et al., 1999).

Global climate change is conceived as manifesting itself in each of three interrelated modules: changes in transmission biology, ecologic changes, and sociologic changes. Chan et al. (1999) developed an integrated assessment framework (IAF) for climate change and infectious diseases (Figure 1). These modules impact epidemiologic outcomes, including mortality and morbidity rates. Any effects of climate change will probably operate on the groups of factors in different ways and will likely be nonlinear, region specific, and time dependent. Most models developed for climate-sensitive health determinants and outcomes provided global or large regional estimates of changes in risk associated with climate change.

Three broad categories of health impacts are associated with climatic conditions depending to the direct impact of extremes in local weather conditions: impacts directly related to weather and climate variability, impacts resulting from environmental changes that occur in response to climate variability and change, and impacts resulting from consequences of climate-induced economic dislocation and environmental decline. The first two categories of climate-sensitive health determinants and outcomes include (1) changes in the frequency and intensity of thermal extremes and extreme weather events (floods and droughts) that directly affect population health, including deaths, injuries, psychological disorders, and exposure to thermal extremes alters rates of illness and death related to heat and cold; and (2) indirect impacts that occur through changes in the geographic range and intensity of transmission of infectious diseases and food- and waterborne diseases, and changes in the prevalence of adverse health outcomes associated with air pollutants and aeroallergens. Indirect effects include changes in local ecology of waterborne agents that modify the incidence of diarrhea and other infectious disease and food-borne infective agents, and alterations on range and activity of vectors that change geographical ranges and infective parasites and incidence of vector-borne disease. Estimating the consequences of indirect effects poses a challenge because those impacts typically result from changes in complex processes. They include alterations in the transmission of vector-borne infectious disease, alterations in water quality and quantity, and changes in the productivity of agroecosystems, with the potential for displacement of vulnerable populations as a result of local declines in food supply or sea level rise. There is a range of estimates of the risk of hunger reflecting different assumptions about future population's growth, international trade and adaptive agricultural technology. Such estimates, however, do not include the likely additional influence of extreme weather events or of increases in agricultural pests and pathogens (Haines et al., 2000; Kiska, 2000; Rivera, 2003; Ebi, 2007).

Studies of climatic influences on infectious diseases have mainly focused on the influence of El Niño-Southern Oscillation (ENSO), that has been found to be related to incidences of malaria in South America, rift valley fever in east Africa, dengue fever in Thailand, hantavirus pulmonary syndrome in the southwestern USA, childhood diarrhoeal disease in Peru. It is unclear at this stage whether global climate change will significantly increase the amplitude of ENSO variability, but if so, the regions surrounding the Pacific and Indian oceans are expected to be most vulnerable to the associated changes in health risks. Cholera provides an instructive example, as cholera outbreaks in Central and Southern Africa have recently been associated with the climatic changes of El Niño-Southern Oscillation events (Rodó et al., 2002). The ancestral home of cholera was apparently the Ganges delta, in India, where epidemics of a cholera-like disease have been described over the past four centuries. This seventh pandemic began in 1961 and is by far the longest lasting pandemic to date and that has reached further than ever before, affecting Asia, Europe, Africa, North

America and Latin America that began in the coastal regions of Peru, spreading to 21 countries including Mexico, and the new O139 outbreak first emerged as a pandemic threat in southeastern India and the Bay of Bengal in 1992, that may have arisen by genetic exchange with non-O1 *V. cholerae* strains as well as clinical strains of O1. Case reports of cholera-like disease involving the suspected O139 serotype from Thailand and Malaysia suggested the region could soon be overwhelmed by an eighth cholera pandemic. Moreover, other strains have since appeared in Latin America. At least one of these, a strain resistant to multiple antimicrobial drugs, was first identified in Mexico and elsewhere in the world in mid-1991 and has since spread widely throughout Central America (Evins et al., 1995). The extraordinary scale and persistence of these pandemics is thought largely to reflect the escalation in nutrient enrichment of coastal and estuarine waters by phosphates and nitrates in run-off wastewater, and the proliferation of urban slums without access to safe drinking water (Simanjuntak, 2001; Lipp et al., 2002; McMichael, 2004).

There was a sharp increase in the number of cholera cases reported to WHO in 2005. A total of 131 943 cases, including 2272 deaths, have been notified from 52 countries (Figure 2). Overall, this represents a 30% increase compared with the number of cases reported in 2004. The total number of countries reporting cases declined slightly (from 56 to 52), but there were a number of countries where cholera re-emerged after having been absent for several years [WHO, 2007].

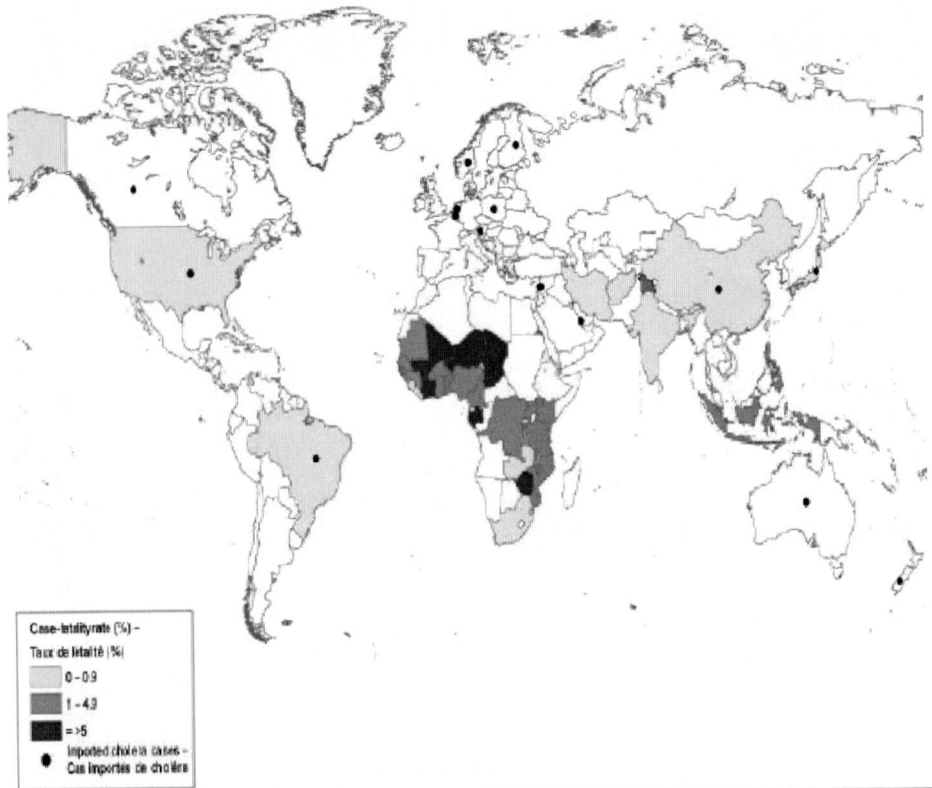

Figure 2. Countries/areas reporting cholera cases in 2005 (http://www.who.int/wer).

Several disease categories have been listed along with specific examples of diseases that had known or suspected relationships with weather-related variables (Rose et al., 2001):

- Airborne: influenza, meningococcal meningitis, coccidiodomycosis, respiratory syncytial virus (colds).
- Vector-borne: malaria, dengue, encephalitis, ehrlichiosis, leishmaniasis
- Water and food-borne: cholera, *Vibrio parahaemolitycus*, *V. vulnificus*, leptospirosis, human enteric viruses (enteroviruses, Norwalk and Norwal-like viruses), salmonellosis, shigellosis, campylobacteriosis, and over 30 species of parasites that infect the human intestines. Seven of these are distributed globally or cause serious illness: ameobiasis, giardiasis, *Taenia solium* taeniasis, ascariasis, hookworm, trichuriasis, and strongyloidiasis.

Understanding links between infectious disease and climate is difficult, given the multivariate nature of climate change, and nonlinear thresholds in both disease and climate processes. It is important to remember that the etiology of infectious diseases is a complex function of a number of factors, making it more difficult to generalize about the potential impacts of climate change. In order to relate environmental factors directly to disease or probability of infection, it is necessary to define the environmental factors related directly to the pathogen, and then to evaluate the influence of environmental factors in the concentration, distribution, prevalence, viability and virulence of the pathogen and the disease outcome. As an example, it has been reported that climate warming can affect host-pathogen interactions by increasing pathogen development rates, transmission, and number of generations per year, relaxing overwintering restrictions on pathogen life cycles, modifying host susceptibility to infection (Harvell et al., 2002). Not all infectious diseases are influenced by climate or weather variability. Vector-borne diseases were the first and most often to be associated with climate related variables (Rose et al., 2001).

## IMPACTS OF CLIMATE ON WATERBORNE DISEASES

Globally waterborne diseases are one of the major contributors to disease burden and mortality (Prüss and Havelaar, 2001). Indirectly transmitted water-borne anthroponoses are susceptible to climatic factors because the pathogens exist in the external environment during part of their life cycles (Wilson, 2001).

There are essentially four main water-related transmission routes for infections (Moren-Abat et al., 2000):

1) Water-borne infections; these occur when humans drink water containing infectious pathogens and consequently develop an infection, for example Cholera and Typhoid.
2) Water-washed infections (also known as water-scarce), are influenced by the quantity of water available. Within this category, we have Scabies and Trachoma.
3) Water based infections is where the pathogen spends its life-cycle in water, such as Schistosomiasis.

4) Water-related insect vector are those pathogens spread via insects which breed in water or bite near water. These include Malaria, Yellow Fever and Dengue.

The first three are most clearly associated with lack of improved domestic water supply. Waterborne diseases are spread through contaminated drinking water contaminated by human or animal faeces or urine containing pathogenic bacteria or viruses, exposure to contaminated water while swimming or other activities, or secondarily through food contaminated with bad water; include cholera, typhoid, amoebic and bacillary dysentery and other diarrheal diseases (Rose et al., 2001; Gleick, 2002).

All of these transmission patterns may be affected by climate variability and thus potentially by climate change. Occurrence of disease is dependent on three factors, all which may be critically mediated by climate. First, exposure is the extent to which a person is exposed to a climate related hazard such as floods. Second, sensitivity is the extent to which health outcomes are sensitive to climate change. Third, adaptive capacity is the ability of the individual to resist the health effects of climate change. Climate variability's effect on infectious diseases is determined largely by the unique transmission cycle of each pathogen. Transmission cycles that require a vector or non-human host are more susceptible to external environmental influences than those diseases which include only the pathogen and human (Moren-Abat et al., 2000).

Weather is often a factor in triggering waterborne disease outbreaks. If weather is a determinant of waterborne disease outbreaks, it is likely also a contributing factor to endemic cases of disease. Waterborne diseases are particularly sensitive to changes in the hydrological cycle. Thus, the impact of heavy rainfall on waterborne illness may be widespread (Charron et al., 2004). A major U.S. study by Curriero et al. (2001) reported a statistically significant association between excess rainfall and waterborne disease outbreaks over a long period of time and on a national scale. The study, based on 548 reported outbreaks in the United States from 1948 through 1994, was able to quantify the relationship between excessive rainfall and disease outbreaks. The study analyzed the relationship between rainfall and waterborne disease on a watershed level, stratified the outbreaks by groundwater and surface water contamination and controlled for effects due to season and hydrologic region. The results indicated that 51% of waterborne disease outbreaks were preceded by rainfall events above the 90th percentile and that 68% were preceded by events above the 80th percentile. Outbreaks due to surface-water contamination showed the strongest association with excessive rainfall during the month of the outbreak. Groundwater contamination events were preceded by a 2-month lag in rainfall accumulations.

For many waterborne diseases, the management and disposal of sewage, biosolids, and other animal wastes and the protection of watersheds and fresh water flows are critical variables that impact water quality and the risk of waterborne disease. Water-related diseases are a particular problem in poor countries and communities, where water supplies and sanitation often are inadequate. Many communities in developing countries continue to use combined sewer and storm water drainage systems; these may pose a health risk should the frequency or intensity of storms increase, because raw sewage bypasses treatment and is discharged into receiving surface waters during storms (Harvell et al., 2002; Hunter, 2003).

Major disturbances involving microorganisms began to occur more frequently in the mid-1970s, particularly between 1972 and 1976, which coincides with a relatively abrupt shift in the global climate regime. In order to explain disease emergence in the oceans various processes have been invoked, including warming of the earth's atmosphere, increased ultraviolet (UV) radiation resulting from ozone depletion and intensified pressures relating to anthropogenic impacts, such as overexploitation of fish and other higher trophic-level organisms, marine pollution, coastal eutrophication, oxygen depletion and sedimentation (Hayes et al., 2001). Research studies of the 1970s found strong evidence in deep-ocean sediments of variations in the Earth's global temperature during the past several hundred thousand years of the Earth's history. The analysis of components of Earth's heat balance quantitatively demonstrates that during the latter half of the 20th century, changes in ocean heat content dominate the changes in Earth's heat balance. Modeling results and observational estimates of ocean heat content supports the hypothesis that increases in radiative forcing are the source of the warming observed between 1955 and 1996. Because most of the increase in radiative forcing in the latter half of the 20th century is anthropogenic, this suggests a possible human influence on observed changes in climate system heat content (Levitus, 2001). Other subsequent studies have discovered that these temperature variations were closely correlated to the concentration of carbon dioxide in the atmosphere and variations in solar radiation received by the planet as controlled by the Milankovitch cycles. The amount of carbon dioxide that can be held in oceans is a function of temperature. Carbon dioxide is released from the oceans when global temperatures become warmer and diffuses into the ocean when temperatures are cooler. Initial changes in global temperature were triggered by changes in received solar radiation by the Earth through the Milankovitch cycles. The increase in carbon dioxide then amplified the global warming by enhancing the greenhouse effect (Pidwirny, 2007; Levitus, 2001; Barnett et al., 2001; Greene and Pershing, 2007). Nevertheless, functional linkages and mechanistic relationships have not been established between the parameters of change (temperature, UV, macronutrients, water chemistry) and the nature of outbreak responses (frequency, intensity, species diversity). Bacterial pathogens often respond to low concentrations of Fe in environments by enhancing the expression of genes responsible for the biosynthesis of exotoxins, specific outer membrane receptors, high-affinity Fe chelators (siderophores) and other virulence factors. When the supply of the limiting micronutrient Fe is increased in a marine environment, higher Fe levels favor outbreak species particularly when all other growth requirements (macronutrients, C substrates) are replete. The consequences are an increase in outbreak frequency, the evolution of highly virulent pathogens, shifts in the species composition and diversity of outbreak organisms, and a heightened prevalence of opportunistic species. This association provides the basis for understanding a functional linkage between climate variability and marine outbreaks (Hayes et al., 2001).

Oceanic and coastal waters are known to harbor and transport microorganisms that cause disease in humans and animals. As modulators of climate, oceans also indirectly influence disease patterns and distribution of many pathogens. While certain pathogenic or toxigenic microorganisms, including toxic phytoplankton and *Vibrio* spp., occur naturally in marine and estuarine waters, anthropogenic contaminants including enteric bacteria, protozoa and viruses may be introduced to coastal waters as sewage pollution. Despite the relatively unfavorable environment, these introduced organisms may survive for prolonged periods in the marine environment, often associated with sediments and other protective environments (Rose et al., 2001). The main mechanisms for regulation of bacterial populations in the aquatic environment include availability of nutrients, temperature, bacterivory, and lysis by viruses, with nutrient availability being the most important factor in limiting the size and abundance of bacterial populations. In eutrophic habitats of coastal ocean, estuary, and freshwater ecosystems, bacterial abundance, production, and growth rates are regulated by physical and chemical factors in addition to nutrient availability (Heidelberg, 2002).

The current evidence of the impact of climate on the epidemiology of waterborne disease is considered under the following headings: the impact of heavy rainfall events, the decrease in salinity, the sea level increase and subsequent changes in ocean circulation, the impact of flooding, and the impact of increased temperature (Figure 3).

The environmental factors that influence directly, or indirectly, the survival of bacterial populations and ability to produce disease are (Huq and Colwell, 1996): (i) temperature, where greater inactivation/death rates occur at higher temperatures; (ii) sunlight, which can affect the persistence and spread of a pathogen if it is associated with phytoplankton and/or algae; (iii) sunlight (UV), in which nucleic acids absorb the UV energy and are damaged; (iv) humidity, resulting from evaporation due to elevation of temperature; (v) moisture content, where low moisture inactivates /kills some microbes; (vi) pH, in which extreme pH inactivates microbes with important exceptions as enteric pathogens which survive at pH 3.0; (vii) weather, where warmer weather increases some microbes and wet weather carries microbes also resuspended in water resources; (viii) chemicals and nutrients as its levels influence microbe survival, in which lack of nutrients (e.g. carbon, nitrogen) will limit proliferation; (ix) biological factors such as attachment of bacteria to a host-specific or site-specific location before they can multiply; the factors that influence attachment include temperature, pH, and nutrient concentration; (x) rainfall, where heavy rainfall may lead to changes in the direction of flow of water systems and for surface water sources, heavy rainfall can lead to overflow of storm drains that may be combined with the sewage system,

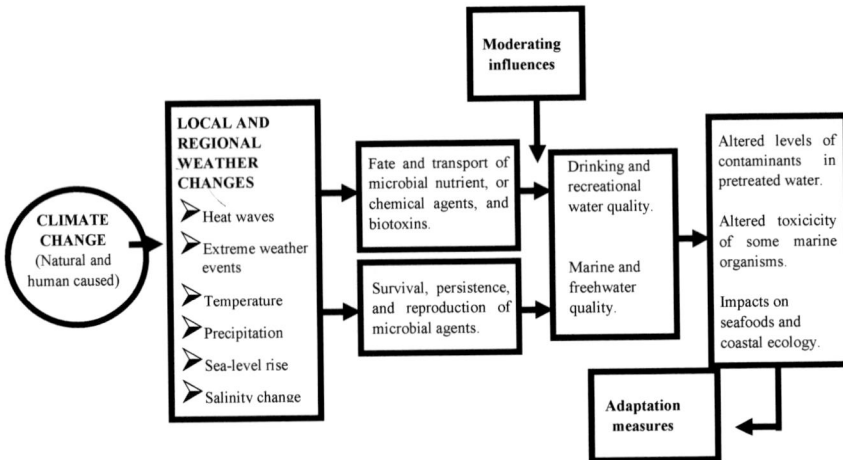

Figure 3. Potential waterborne and foodborne diseases. Foodborne diseases primarily related to marine or freshwater contamination. Moderating influences include non-climate factors that effect climate-related health outcomes, such as state of coastal wetlands, land use, and water treatment facilities. Finished drinking water must meet higher regulatory standards than recreational water or water used for irrigation. Adaptation measures include actions to reduce risks of water contaminations, such as watershed management, improved water treatment engineering, and enhanced surveillance of waterborne disease outbreaks (reproduced from Rose et al., 2001).

allowing faecally polluted water into rivers (Hunter, 2003); (xi) flooding, in developing nations where there is evidence of outbreaks following floods caused by heavy rains, and may eliminate the habitat for both vectors and vertebrate hosts. For example, developing countries located in tropical and subtropical regions with crowding and poverty, heavy rainfall and flooding may trigger outbreaks of diarrhea (McMichael et al., 2006). Each of these climatic factors can have markedly different impacts on the epidemiology of various waterborne diseases (Burke et al., 2001).

## ENVIRONMENTAL FACTORS INFLUENCING *VIBRIO CHOLERAE*

Cholera is an indirectly transmitted water-borne anthroponose that is transmitted by a water vehicle, the bacteria (*Vibrio cholerae*), appears to reside in marine ecosystems by attaching to zooplankton (Colwell, 1996). The search for explanations of climatic and environmental impacts on cholera began over 100 years ago in the Indian subcontinent. In spite of the development of new technologies to provide safer water and sewage treatment in industrialized nations, cholera remains a public health problem in large regions of the globe. The mechanistic basis for a climate–cholera connection, which involve multiple pathways, remains poorly understood. The marked seasonality of cholera and the simultaneous appearance of cases at different locations have been main reasons for the long-standing search for climatic and environmental drivers (Pascual et al., 2002).

Recent studies have shown that climate also influences the abundance and ecology of pathogens, as well as the links between pathogens and changing ocean conditions, including human diseases such as cholera (Harvell et al., 2002; Lipp et al. 2002). *Vibrio cholerae* is a motile, Gram-negative curved rod that belongs to the family Vibrionaceae, and is a well-known human pathogen that has caused cholera epidemics worldwide and continues to be prevalent in many developing countries (Villalpando et al., 2000; WHO, 2005a), with serious epidemics often localized in tropical areas (WHO, 2005b). With the apparent increase in its emergence, cholera is, at present, one of the most important of the resurgent diseases (Smith et al., 2005; UNEP, 2005). In South America, outbreaks start along the coast (Seas et al., 2000), and epidemiological studies in México have demonstrated that people living in coastal states are at a high risk of contracting the disease, as shown in Figure 4 (Borroto and Martínez-Piedra, 2000).

Genus *Vibrio* includes Gram-negative bacteria indigenous to marine and estuarine waters, and facultative anaerobic, non-spore forming bacilli which are oxidase-positive and halophilic. Furthermore, *V. cholerae* is an autochthonous microbial inhabitant of brackish water, estuarine ecosystems, and coastal areas; the species can remain in a culturable state in the marine environment for years (Colwell, 1996). *V. cholerae* can pose a public health risk when it is ingested via untreated water, contaminated seafood (raw or under-cooked), or exposure of skin wounds to sea water (Whittman and Flick, 1996; Daniels et al., 2002; Potasman et al., 2002; Scott et al., 2002).

The prevalence of pathogenic *Vibrios* appears to be influenced by the physico-chemical features of the environment; studies in coastal and estuarine regions of different parts of the world have highlighted the potential significance of environmental factors to the dynamics of the disease.

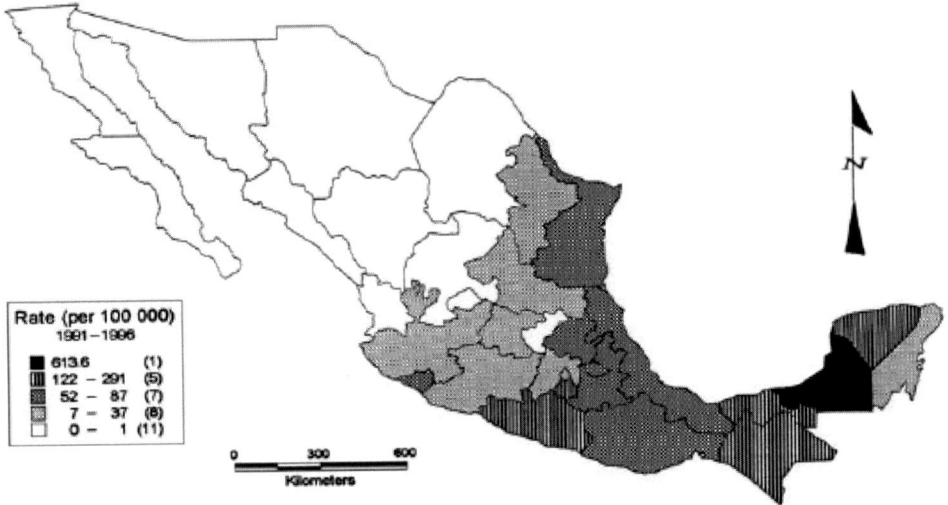

Figure 4. Cholera incidence in States of México (reproduced from Borroto and Martínez-Piedra, 2000).

Previous studies of cholera population dynamics have proposed links between climate, oceanographic environmental conditions such as temperature, phytoplankton productivity, and human cases of cholera morbidity and mortality. It has been reported that climate may affect the dynamics of cholera by shifting pathogen or host reservoir species abundance, population dynamics, and community interactions (Pascual et al., 2002; Koelle et al., 2005a). Nevertheless, the mechanistic basis for climate-cholera connection remains scantily understood. The environmental and climate factors that have been associated with *V. cholerae* are presented in Table 1.

Two of the primary environmental variables influencing this complex ecology of pathogenic *Vibrios* are temperature and salinity which, along with sea surface temperature (SST), is consistent with the role played by sea surface height (Lipp et al., 2002) in combination with elevated pH. The optimum temperature for growth of this organism is 37°C, with possibilities for growth ranging from 16 to 42°C (Colwell, 1996).

Temperature and salinity play roles in the occurrence of *V. cholerae* in the aquatic environment (Singleton et al., 1982a; Singleton et al., 1982b; Barbieri et al., 1999; Jiang, 2001). *Vibrios* are present in the environment even when they cannot be cultured. *V. cholerae* can survive under unfavorable environmental conditions in a dormant state, switching into a viable-but-non-culturable (VBNC) state in response to rapid transitions in environmental conditions such as temperature and osmolarity, and nutrient deprivation. When *V. cholerae* enters the viable-but-non-culturable state, it loses its flagellum and changes to a smaller, spherical form, in a spore-like stage (Xu et al., 1982; McDougald et al., 1998; Huq et al., 2000). This dormant state serves as a survival strategy as cells survive changes in temperature, salinity, or availability of organic matter and remain infectious (Colwell, 1996).

## Table 1. Influence of environment, climate, and weather on cholera and *V. cholerae* dynamics

| FACTOR | CLIMATE AND WEATHER DRIVERS | INFLUENCE(S) |
|---|---|---|
| Temperature | Seasons, interannual variability | Growth of *V. cholerae*, Phytoplankton blooms, infection by temperate phages |
| Salinity | Seasons, monsoons, ENSO, sea level rise | Growth of *V. cholerae*, seroconversion, expression of cholera toxin |
| Sunlight | Seasons, monsoons, interannual variability | |
| pH | Seasons, interannual variability (phytoplankton growth) | Growth of *V. cholerae* |
| $Fe^{3+}$ | Precipitation (runoff), atmospheric deposition (NAO) | Growth of *V. cholerae*, expression of cholera toxin |
| Exogenous products of algal growth | Seasons, monsoons, interannual variability in light, nutrients | Survival of *V. cholerae* |
| Chitin | Seasons, monsoons, zooplankton blooms (following phytoplankton) | Growth of *V. cholerae*, attachment to exoskeletons |

Lipp et al., 2002.

Franco et al. (1997) provided evidence for a positive association between counts of the pathogen *V. cholerae* and river water temperature two months earlier in Perú. The influence of temperature on the timing of the pathogen's appearance was demonstrated with the presence or absence of CT-positive *V. cholerae*. In addition, the number of cholera cases correlated significantly ($r = 0.72$) with CT-positive cholera counts at cleaner sites upriver two to three months earlier, supporting a role of water temperature in its seasonality. Barbieri et al. (1999) observed a positive correlation ($r^2 = 0.559$, $P = 0.038$) between the occurrence of *Vibrio* spp. and temperature in two estuaries along the Italian Adriatic Coast. Pascual et al. (2000) carried out a time-series analysis of an 18-year cholera record from Bangladesh and reported a positive effect of El Niño-Southern Oscillation (ENSO), a major cause of interannual climate variability, on the predictions of cholera incidence in the fall. Their results suggested that higher ambient temperatures would correspond to higher water temperatures in shallow bodies of water, such as ponds and rivers in estuaries, and shallow coastal waters, increasing local temperature and producing the interannual variability of cholera. Lobitz et al. (2000) used satellite data to monitor the timing and spread of cholera. The remote sensing data included sea surface temperature (SST), chlorophyll and sea surface height (SSH). For the time period shown in Figures 5a and 5b, they found a general agreement in the patterns of SST, SSH, and chlorophyll values; moreover, SST showed an annual cycle similar to the cholera case data. The warming of water temperature in ponds and rivers might increase the incidence of cholera through the faster growth rate of the pathogen in aquatic environments.

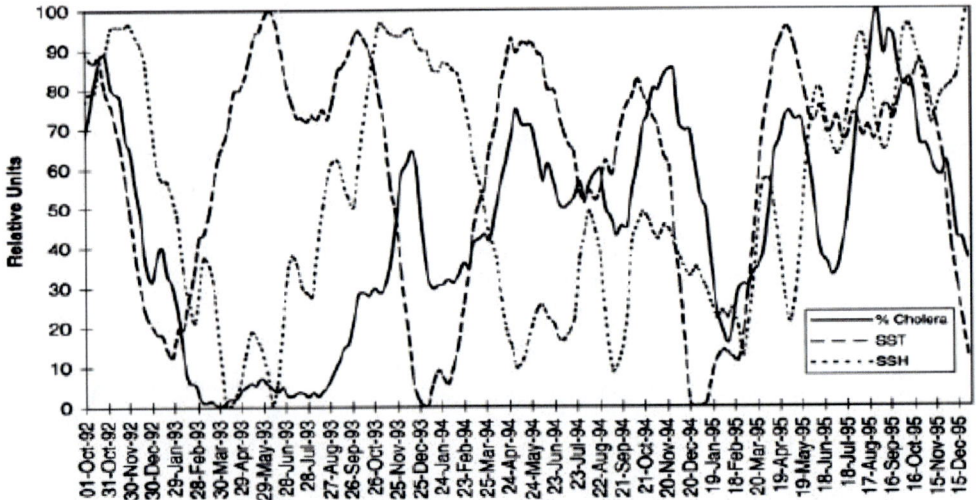

Figure 5a. Cholera cases (solid line), sea-surface temperature SST (dashed), and sea-surface height SSH (dotted), data for September, 1992–1995. In 1994 and 1995, cholera cases followed the SST cycle (see fig. 5b); however, in spring, 1993, SSH was the lowest for this period (reproduced from Lobitz et al., 2000).

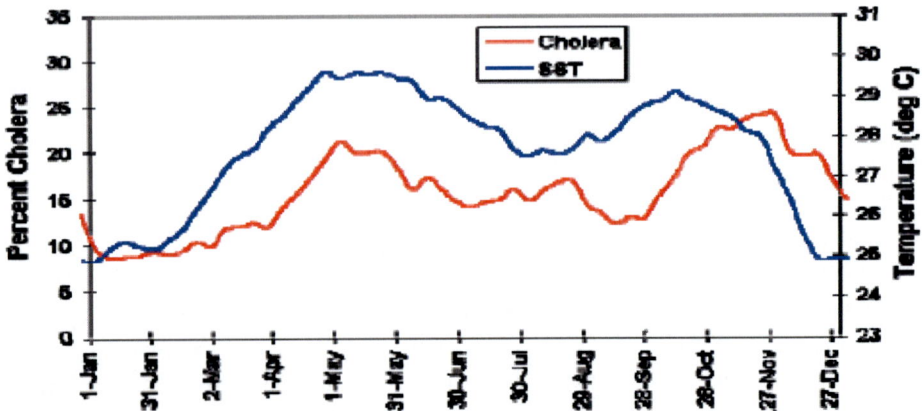

Figure 5b. Remote Sensing of Cholera: 1994 SST and Cholera Plot. Cholera case data in Bangladesh followed sea surface temperatures (SST) in 1994. The SST had an annual cycle similar to the one shown here every year, while the cholera case data did not show a clear pattern. The SST cycle can also be seen in the time series for 1992. Bangladesh has a Northern-hemisphere-summer monsoon, which appears as a decrease in the SST in June through August. Coastal processes are very complex. Without examining the relationship between plankton data and cholera, the link between SST, plankton, and cholera outbreaks cannot be determined (reproduced form Lobitz et al., 2000).

Barnett et al. (2001) reported the warming of the ocean since 1950. The Panel on Climate Change (IPCC) (2001a, 2001b, 2001c) has projected an increase in world average temperature of 1.0°C–3.5°C by the year 2100, and an associated rise in sea level of 50 cm is also expected by the 2080s, with considerable regional variations (Figure 6).

As a result of warming seawater, the world oceans are expanding. Coupled with freshwater input from ice-melt, thermal expansion of the oceans is causing sea levels to rise at $c.$ 2 mm year $^{-1}$. This assessment is derived from projections made by computer-based global climate models that combine, through simultaneous equations within a 3-dimensional global grid, the atmospheric and oceanic processes that occur in response to increased greenhouse gases and the resulting rise in radiative forcing in the lower atmosphere (Haines et al., 2000). Evidence is indicating that these changes in the broad-scale climate system may already be changing the ecology of infectious diseases (Patz et al., 2000; Patz et al., 2005).

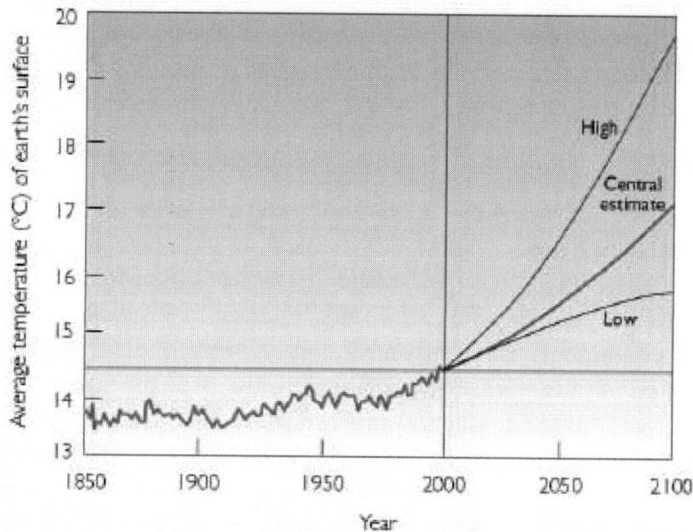

Figure 6. Global temperature record since instrumental recording began in 1860 and projection to 2100, according to the IPCC (2001a, 2001b, 2001c).

Heidelberg et al. (2002) reported an increase of *V. cholerae* in water samples collected from the Choptank River in Chesapeake Bay during summer months, associated with an increase in water temperature ($r^2 = 0.420$-$0.720$) as Louis et al. (2003) reported. According to Rodó et al. (2002), enhanced warming can also affect disease transmission by changing human behavior, with a possible increase in the contact with contaminated water sources under warm conditions immediately before and during the spring, when the first seasonal peak in cholera is typically observed. According to these authors, the effect of more extreme ENSO and climate changing conditions of drought and flood remain to be examined, but are consistent with future global and regional scenarios that also might impact sanitation conditions critical to cholera transmission. A remarkable finding from Huq et al. (2005) study was that water temperature was directly correlated with cases of cholera in Bakerganj, the southernmost site included in the study, having the most direct influence from the Bay of

Bengal, where most of the initial cases of cholera occurred, including the outbreak of newly recognized serotype O139 in 1992.

Another adaptive feature allows *V. cholerae* for prolonged survival when soluble iron is unavailable; *V. cholerae* can produce iron-chelating siderophores to take up insoluble iron from the environment. Survival is thus further improved when insoluble iron ($Fe_2O_3$) is available under alkaline pH environment, as *V. cholerae* also thrives under high-pH conditions (Hayes et al., 2001).

Environmental conditions may also affect expression of virulence genes in *V. cholerae*. Sunlight can induce propagation of the CTX_ phage, and the viability of *V. cholerae* remains stable in full sunlight compared to enteric bacteria such as *E. coli*, which may impart some selective advantage to *vibrios* at tropical latitudes. Furthermore, moderate levels of introduced iron also increase the expression of CT. Therefore, environmental triggers may become epidemiologically important for prevalence of the organism and its virulence (potentially resulting in shorter onset times and lower infectious doses) (Lipp et al., 2002).

Castañeda et al. (2005), reported *V. cholerae* non-O1 occurrence in water samples correlated with water temperature ($r^2=0.468$, $P<0.05$) and *V. cholerae* O1 Inaba occurrence in La Mancha lagoon oysters correlated ($r^2=0.543$, $P<0.05$) with water temperature averaging >30°C. Their results indicated that seasonal salinity variability and warm temperatures, as well as nutrient influx, may influence the occurrence of *V. cholera* non-O1 and O1. The conditions found in the Alvarado (31.12°C, 6.27 psu, pH 8.74) and La Mancha lagoons (31.38°C, 24.18 psu, pH 9.15) during the rainy season in 2002 favored the occurrence of *V. cholera* O1 Inaba enterotoxin positive traced in oysters, indicating that environmental conditions could also affect expression of virulence genes in *V. cholerae*. Mahasneh y Al-Sayed (1997) isolated both *V. cholerae* and *V. parahaemolyticus* in seawater and oysters *(Pinctada radiata)* during the warmer months (June, July, September and October 1992). In oysters, *V. cholerae* was more prominent than *V. parahaemolyticus*. Peak counts in oysters at nearshore and pelagic sites occurred in July and they were five times greater at the pelagic site. In general, most bacteria were isolated from the pelagic waters and oyster samples, and they were highest during the warmer months. Some disappeared totally during February and March. Water temperature was found to have greater influence on distribution of different bacteria in water and oysters, compared with other environmental factors.

Salinities favorable for *V. cholerae* growth are found primarily in inland coastal areas and estuaries, but survival of *V. cholerae* in seawater for more than 50 days has been demonstrated (Munro and Colwell, 1996). Pathogenic *V. cholerae* grows in water with low salinity if the water temperature is relatively high and organic nutrients are present even in small concentrations (less than 1 mg/L), compensating, to a degree, for lack of salt (Singleton et al., 1982a; Singleton et al., 1982b; Colwell, 1996). Although optimal salinity for *V. cholerae* growth is between 5 and 25 psu, *V. cholerae* can tolerate salinities near 45 psu (Singleton et al., 1982). Castañeda et al. (2005) reported that water temperature and salinity were significantly associated ($P<0.05$). *V. cholerae* non-O1 percent isolation in oysters correlated with La Mancha lagoon water temperature and salinity ($r^2=0.603$, $P<0.05$) and Alvarado lagoon water temperature and salinity correlated with *V. cholerae* non-O1 isolation in oyster samples ($r^2=0.576$, $P<0.05$).

Other climatic variables related to water levels such as rainfall have also been invoked to explain cholera patterns since early times. Flooding may result in the contamination of water supplies or the reproduction rate of the pathogen may be influenced by ambient air temperatures (Wilson, 2001). Floods and droughts can affect not only the concentration of the bacterium in the environment, but its survival through the effect of salinity, pH or nutrient concentrations, as well as human exposure to the pathogen, sanitary conditions and susceptibility to disease. Moist and humid conditions have also long been associated with cholera's spatial distribution (Pascual et al., 2002).

Chlorophyll a and/or turbidity have been associated with rainfall events, influence of rainfall and runoff on salinity and nutrients leading to algal blooms, influence of phytoplankton on zooplankton dynamics and succession, influence of temperature and salinity on the growth of *Vibrio* in the copepod, concentration of the bacteria in the copepod, numbers of copepods transmitted upstream and in a glass of water (Epstein, 1993; Colwell and Huq, 2001).

Significant correlation between the cumulative monthly distribution of cholera cases and the monthly distribution of precipitation has been observed in Guam (Borroto and Haddock, 1998). Floods and droughts may affect not only the concentration of the bacterium in the environment, but also its survival, through the effect exerted by these environmental changes on water depth, conductivity, salinity, hours of sunlight, pH, and nutrient concentrations (Islam et al., 1990; Colwell, 1996; Montilla et al., 1996; Bouma and Pascual, 2001; Piarroux and Bompangue, 2006). In addition, water temperature would drive seasonality by direct influence on the abundance and/or toxicity of *V. cholerae* in the environment, or alternatively, through their indirect influence on other aquatic organisms such as zooplankton (Copepods – *Acartia, Cyclops, Diaptomus,* Cladocerans – *Daphnia, Bosmina, Bosminopsis, Ceriodaphnia, Diaphanosoma,* rotifers), phytoplankton (Cyanobacteria – *Anabaena,* Chlorophytes – *Volvox,* desmids, *Rhizoclonium,* Diatoms – *Skeletonema,* Dinoflagellates), macrophytes (Marine taxa – *Ulva, Entermorpha, Ceramium, Polysiphonia,* Freshwater taxa – *Eichhornia* (water hyacinth), *Lemna* (duckweed), Benthos (Prawns – *Penaeus, Metapenaeus, Macrobrachium,* oysters, crabs, chironomid egg masses), and fish (Sea mullet) to which the pathogen is found attached (Islam et al., 1990; Colwell, 1996). Within the marine environment, *V. cholerae* is found attached to surfaces provided by plants, filamentous green algae, copepods (zooplankton), crustaceans, and insects (Huq et al., 1983; Huq et al, 1990). Many pathogenic *Vibrio* spp. are associated with chitinaceous zooplankton and shellfish, and can survive on fish and shellfish as well (Tamplin, 2001). Shellfish feeding on planktonic crustaceans are colonized by *V. cholerae* in natural water systems (Huq et al., 1983; Huq et al., 1984; Islam et al., 1994a; Islam et al., 1994b; Islam et al., 1999). *Vibrio cholerae* survive in association with aquatic vegetation, as well as other zooplankton and crustacean invertebrates in the aquatic environment (Huq et al., 1983; Huq et al., 1984; Huq et al., 1990; Islam et al., 1990; Islam et al., 1994a; Islam et al., 1994b; Montilla et al., 1996; Islam et al., 1999; Jacoby et al., 2000; Bouma and Pascual, 2001; Tamplin 2001; Piarroux and Bompangue, 2006). After concentration by filter-feeding shellfish such as oysters, these bacteria may be present at concentrations that are 100-fold higher than those in the surrounding water (Wright et al., 1996; Morris, 2003).

The algae and zooplankton that feed upon them provide a natural refuge for *Vibrio cholerae*, where, under normal conditions, the bacteria exist in a non-culturable state for long periods. The *Vibrios* appear to enter a non-culturable phase induced primarily by unfavorable conditions such as low temperature (Colwell and Grimes, 2000). An increase in sea surface temperature, along with high nutrient levels (eutrophication) that stimulate algal growth and deplete oxygen, can activate the blooms and in turn, lead to increased numbers of *Vibrios*

(Burke et al., 2001). The culturability of *V. cholerae* remains stable in full sunlight compared to enteric bacteria such as *E. coli*, which may impart some selective advantage to *Vibrios* at tropical latitudes (Mezrioui et al., 1995).

Due to the relationship between zooplankton and phytoplankton populations, dynamics between climate and phytoplankton, although indirect, are important mechanisms that regulate the prevalence of the bacteria in the environment. The specific environmental changes that amplified plankton and associated bacterial proliferation and govern the location and timing of plankton blooms, are nutrient levels, SST and solar radiation (Lipp et al., 2002), pH, currents, winds, and river runoffs (Epstein et al., 1993); wind, tidal forces and other physical processes can affect phytoplankton growth through the transport of deep nutrient-loaded water to the surface; however, heavy precipitation can dilute nutrient levels and decrease phytoplankton production. In addition, sufficient amounts of solar radiation are necessary for photosynthesis among phytoplankton. Favorable combinations of these elements can result in phytoplankton blooms and greater abundance of zooplankton (survival of these small crustaceans depends on the abundance of their food supply, phytoplankton), which ultimately result in greater prevalence of *V. cholerae* in aquatic environments (Colwell and Huq, 2001). Attachment, growth, and multiplication of *V. cholerae* in the aquatic environment, particularly in association with copepods, are influenced by warmer temperatures, elevated pH (8.5), salinity (between 5 and 25 psu, and plankton blooms which are, in turn, controlled by large-scale climate variability (Epstein et al., 1993; Lipp et al., 2002; Louis et al., 2003).

The correlation of cholera outbreaks and the seasonal occurrence of algal blooms has been reported (Epstein, 1993; Islam et al., 1994a; Colwell, 1996); however, there is no direct evidence that such events lead to an enrichment of toxigenic *V. cholerae* strains responsible for cholera epidemics. Because many phyto- and zooplankton organisms may serve as reservoirs for the cholera bacterium (Islam et al., 1994a; Islam et al., 1994b; Colwell, 1996; Islam et al., 1999; Tamplin, 2001), disease outbreaks in humans may be strongly correlated with the population and community dynamics of these hosts (Lobitz et al., 2000; Lipp et al., 2002; Colwell et al., 2003). The appearance of cholera cases in coastal human communities correlated with variations in phyto- and zooplankton abundance in the marine environment induced by variations in rainfall in the Indian Ocean during the monsoon has been reported (Colwell, 1996; Faruque et al., 2005). It has been demonstrated that sea surface temperature in the Bay of Bengal is correlated with algal blooms and outbreaks of cholera in Bangladesh. *V. cholerae* occur in the Gulf of México and along the east coast of North America, where most blooms occur during the summer months and relatively high water temperatures are necessary for algal growth and bloom formation that are directly or indirectly hazardous to human health (Jacoby et al., 2000; Saker and Griffiths, 2001). Climate variability and change may thus influence the introduction of cholera into coastal populations (Haines et al., 2000). According to Huq et al. (2005), environmental sampling and analysis of the environmental and clinical data revealed significant correlations of water temperature, water depth, rainfall, conductivity, and copepod counts with the occurrence of cholera toxin producing bacteria (presumably *V. cholerae*). The lag periods between increases or decreases in units of factors, such as temperature and salinity, and occurrence of cholera correlate with biological parameters, e.g., plankton population blooms. The new information on the ecology of *V. cholerae* is proving useful in developing environmental models for the prediction of cholera epidemics.

Recently, one factor involved in adherence of *V. cholerae* O1 to the chitin of zooplankton - and then in addition acquiring acid tolerance, was identified as the mannose-sensitive hemagglutinin (MSHA) type IV pili. *V. cholerae* O1 El Tor and O139 strains expressed and assembled these pili; nevertheless, O1 classical strains carry the pili (msh) genes but do not assemble functional MSHA pili (Watnick and Kolter, 1999; Chiavelli et al., 2001; Reguera and Kolter, 2005). *V. cholerae* also possess secreted chitinase enzyme(s) which are probably needed to utilize chitin (homopolymer of N-acetylglucosamine) as a carbon source (Connell et al., 1998); the hypothetical chi gene products were identified in the genome sequence (Heidelberg et al., 2000). These findings are suggestive of a close association of *V. cholerae* with chitin structures in the environment (e.g. zooplankton such as copepods) (Reidl and Klose, 2002). *V. cholerae* respond efficiently to different and constantly changing ecosystems due to a variety of transport proteins with broad substrate specificity and catabolic pathways located on its chromosomes. For example, ribose and lactate transport and degradation enzymes are contained on chromosome 2, whereas the trehalose systems reside on chromosome 1; nevertheless, other energy metabolism pathways like the chitin system are split between the chromosomes (Heidelberg et al., 2000). In aquatic environments, chitin represents a source of both carbon and nitrogen. Sequence analysis suggests that *Vibrio cholerae* degrades chitin by a phosphenolpyruvate phosphotransferase system (PTS) for chitobiose transport (Bassler et al., 1991).

Several studies have demonstrated that environmental non-O1 and non-O139 *V. cholerae* strains and *V. cholerae* O1 El Tor and O139 are able to form a three-dimensional biofilm on surfaces which provides a microenvironment (Broza and Halper, 2001; Watnick et al., 2001), increasing bacterial productivity that favors survival and persistence due to increased resistance to various harmful environmental conditions, including chlorination and antibiotics (Davey and O'Toole, 2000; O'Toole et al., 2000). Biofilm formation is likely to be important for the life cycle of *V. cholerae*, facilitating environmental persistence within natural aquatic habitats during interepidemic periods (Reidl and Klose, 2002). For both *V. cholerae* O1 El Tor and O139 strains biofilm formation is dependent on the expression of an exopolysaccharide (*vps*) (Wai et al., 1998; Watnick et al., 2001). Expression of the *vps* (*Vibrio polysaccharide* synthesis) genes, encoded in two gene clusters on the larger chromosome (*vpsA-K* and *vpsL-Q*), is required for synthesis of an exopolysaccharide (EPS) matrix that stabilizes the mature biofilm. Overproduction of EPS results in resistance to osmotic and oxidative stress and bacteriocidal agents (Yildiz and Schoolnick, 1999). Strains containing defects in one of the *vps* genes necessary for exopolysaccharide synthesis fail to make a three-dimensional biofilm, indicating that the exopolysaccharide is used to build a mature biofilm (Watnick et al., 2001). Expression of the *vps* exopolysaccharide causes a rugose colony phenotype, and provides enhanced chlorine and phage resistance (Morris et al., 1996; Nesper et al., 2001). The expression of *vps* inhibited intestinal colonization in infant mouse cholera model, suggesting that the expression of a factor that enhances environmental persistence actually decreases virulence. The lack of flagellar synthesis causes high-level *vps* exopolysaccharide expression (at least in some strains, e.g. O139), indicating that flagellar synthesis is coupled to *vps* expression. This effect is specifically due to the lack of a flagellum, rather than a lack of motility, suggesting that the loss of the flagellum may be a developmental cue during biofilm formation (Watnick et al., 2001).

Kierek and Watnick (2003) identified and characterized a *vps*-dependent *V. cholerae* biofilm and a *vps*-independent *V. cholerae* biofilm. The *vps*-dependent biofilm predominated in growth media containing monosaccharides. They hypothesized that *vps*-dependent biofilm development occurs only in nutrient-rich environments. *vps* transcription is greater in biofilm-associated cells than in planktonic cells, suggesting that *vps* gene transcription is activated by surface association (Haugo and Watnick, 2002). This type of biofilm development may facilitate *V. cholerae*'s colonization of favorable aquatic environments. Conversion of environmental monosaccharides into an exopolysaccharide matrix may even serve as a form of nutrient storage. In contrast, millimolar $Ca^{2+}$ concentrations, which are present in seawater but not in freshwater, are required for *vps*-independent biofilm development. Thus, they hypothesized that *vps*-independent biofilm development probably occurs primarily in marine environments. In contrast to the variability of organic nutrients in aquatic environments, the concentration of $Ca^{2+}$ is uniformly high in the marine environment. Thus, *V. cholerae* surface adhesion in marine environments may be less discriminating. As cholera epidemics have been associated with heavy rainfall and increases in sea surface height (Lobitz et al., 2000; Lipp et al., 2002), both of these conditions are predicted to alter the organic and inorganic compositions of an estuary, which is the primary interface between humans and the marine environment (Reemtsma et al., 1993; Mahadevan and Subramanian, 1999; Padmavathi and Satyanarayana, 1999; Han and Webster, 2002). These changes in the organic and inorganic compositions of the aquatic environment may alter the nature of *V. cholerae*'s association with surfaces and may play some role in the initiation of cholera epidemics.

Evidence indicates that *V. cholerae* toxigenic strains may arise from environmental, non-toxigenic progenitors in coastal areas, as nutrient-rich effluents and warmer sea surface temperatures shift marine ecosystems towards more toxic species (Epstein, 1993; Lobitz et al., 2000; Harvell et al., 2002). Most *V. cholerae* strains, especially those from the environment, lack the genes required to produce CT, but the possibility of genetic exchange in the environment by horizontal transfer of phage transduction with cholera toxin (CT)-encoding phage CTX-Φ mechanism as well as clonal diversity allows the potential emergence of new toxigenic clones (Faruque et al., 1998; Chakraborty et al., 2000). Expression of CT is optimal at salinities between 2 and 2.5 psu. Castañeda et al. (2005) reported that conditions found in Alvarado and La Mancha lagoons (Veracruz, México) might favor *V. cholera* O1 enterotoxin positive occurrence during the rainy season. Evidence for serconversion, in part, may explain this observation. Only serogroup O1 and the newly emerged O139 have been associated with severe disease and cholera pandemics. In contrast, intestinal and/or extraintestinal infections with non-O1 and -O139 serogroups or non-toxigenic O1 strains are rarely found and have been associated with sporadic gastrointestinal diseases and extraintestinal infections (Boyd et al., 2000). The majority of environmental isolates of *V. cholerae* are members of non-O1 and non-O139 serogroups and lack the genes required to produce cholera toxin (CT) (Osorio and Klose, 2000).

The cholera toxin (CT), a protein enterotoxin that elicits profuse diarrhea, causes the signs and symptoms of cholera. CT activates cyclic adenosine monophosphate, leading to increased $Cl^-$ secretion and decreased NaCl-coupled absorption. Glucose, potassium, and bicarbonate absorption, however, remain intact, as does glucose-linked enhancement of sodium and water absorption. For unknown reasons, persons with blood group O are significantly more likely to have severe cholera. Factors that predispose to hypochlorhydria (e.g., malnutrition, gastrectomy, and acid-reducing medications), by decreasing the gastric acid barrier to infection, also increase susceptibility to illness. The atrophic gastritis and hypochlorhydria associated with chronic *Helicobacter pylori* infection has also been associated with an increased risk of severe cholera (Morris, 2003).

## TEMPORAL AND SPATIAL VARIABILITY

The deteriorating conditions of the environment and climate change have led to concerns in the influence of climate on disease dynamics. Seasonality in disease incidence might infer an association with climatic factors. It has been attributed to seasonal changes in pathogen transmission rates, resulting from fluctuations in extrinsic climate factors. Nevertheless, to prove a causal link to climate, non-climatic factors must be considered. Furthermore, in order to assess long-term climate influences on disease trends, data must span numerous seasons and utilize proper statistics to account for seasonal fluctuations (Patz et al., 2003).

Research investigating possible links between temporal and spatial variation of climate and the transmission of infectious diseases can be categorized into one of three conceptual areas (Patz et al., 2003):

1) Evidence for associations between short-term climate variability and infectious disease occurrence in the recent past.
2) Evidence for long-term trends of climate change and infectious disease prevalence.
3) Evidence from climate and infectious disease linkages used to create predictive models for estimating the future burden of infectious disease under projected climatic conditions.

Climate modulates the seasonal and inter-seasonal variability of cholera by affecting both the prevalence and the spatial distribution of the agent in the environment through numerous mechanisms that affect the seasonality of cholera incidence unique for each geographical location. In addition to seasonal variability, endemic areas experience inter-seasonal increases in cholera incidences that have been closely linked to climate variability (Tamerius et al., 2007).

The confirmation that *V. cholerae* occurs in aquatic environments in association with zooplankton and phytoplankton, and the associations found between cholera cases and sea surface temperatures in coastal regions, and sea surface height indicates that cholera dynamics are strongly associated with climate and seasonal variability (Colwell, 1996; Bouma and Pascual, 2001; Colwell and Huq, 2001; Pascual et al., 2002; Koelle et al., 2005a; De Magny et al., 2006). The increase in sea surface heights (SSH) might cause inland incursion of contaminated water containing zooplankton, and human contact with *V. cholerae*. SSH is modulated by seasonal thermal expansion and contraction of sea water, ocean currents, tidal forces, and the piling up of water resulting from persistent winds and other atmospheric phenomenon (Lobitz et al., 2000).

The complex interactions and feedbacks between the dynamics and the ecology of endemic areas result in favorable conditions for the propagation and transmission of the bacteria. The geographical areas once known to have experienced cholera epidemics can be characterized into three levels: (i) Cholera- free communities defined as having no locally acquired infections. (ii) In areas of cholera epidemicity, the disease diminishes after an outbreak. (iii) In regions of cholera endemicity, the disease does not disappear after an epidemic peak and returns in successive waves (Torres-Codeço, 2001). Of particular interest and relevance to identifying environmental or climate factors that may promote epidemics is the understanding of dynamics of the disease in areas of endemicity (Charron et al., 2004).

Where cholera is endemic, cases tend to demonstrate distinct seasonal trends. These patterns are strongly related to the ecology of *V. cholerae* in the environment, where high numbers are observed during times of warm water temperatures and zooplankton blooms (Lobitz et al., 2000; Lipp et al., 2002). Currently, the regions of cholera endemicity, mainly in the tropics and subtropics, include the coasts surrounding the Bay of Bengal, both Bangladesh and the Indian subcontinent, and coastal Latin America. In each of these three geographical regions, patterns of disease frequency follow similar trends and are most likely explained by the same physical or environmental drivers (Lipp et al., 2002). The nature of this phenomenon is not entirely understood but is most likely related to both environmental and socioeconomic factors.

Based on quantitative empirical studies, relationships have been shown to exist between cholera epidemics, land and sea surface temperature anomalies (LSTAs), rainfall, and ENSO, with the latter parameter quantified by the Southern Oscillation Index (SOI), for the Asian and South American subcontinents (Pascual et al., 2000; Pascual et al., 2002; Rodó et al., 2002; Bouma and Pascual, 2001; Checkley et al., 2000; Speelmon et al., 2000; Koelle et al., 2005b). In particular, an increased role of interannual climate variability in cholera interannual dynamics in Bangladesh has been suggested, with strong and consistent signature of ENSO, at least for certain time periods (Rodó et al., 2002). In an epidemiological endemic context, these findings suggest that certain aspects of climate are associated with human cholera incidence in specific areas of the intertropical belt (WHO, 2005a).

The Rodó et al. (2002) study represents the first evidence that warming trends are affecting human disease. Using an extensive cholera database and innovative statistical methods they found quantitative evidence for an increased role of interannual climate variability on the temporal dynamics of cholera. They provided evidence for an increase over time in the frequency and amplitude in the ENSO. According to their findings, they confirmed a robust association of cholera dynamics and ENSO, from the first to the last decades of the 20th century, based on time-series analyses of the relationship between ENSO and cholera prevalence in Bangladesh (formerly Bengal) during two different time periods. Even though debate still remains in reference to the relationship between ENSO intensification and climate change, their observations are consistent with warming models. A strong and consistent signature of ENSO is apparent in the last two decades (1980–2001), while it is weaker and eventually uncorrelated during the first parts of the last century (1893–1920 and 1920–1940, respectively), and the SOI undergoes shifts in its frequency spectrum as well. Their results indicated that a change in remote ENSO modulation alone can only partially serve to substantiate the differences observed in cholera. Regional or basin-wide changes possibly linked to climate changes, mainly warming, seem to facilitate ENSO transmission, as this climate phenomenon accounts for over 70% of disease variance.

Cholera cases can exhibit different seasonal patterns at different locations, including variations in the number of outbreaks and different delays with respect to peaks in rainfall and temperature. These patterns are not well understood because environmental drivers themselves are poorly defined for the seasonal cycle of the disease. The importance of rainfall as a driver of the seasonal cycle of cholera is implied by its waterborne transmission, the dose-dependent nature of infection, and the decline of cases during the rainy season. In endemic regions, cholera should exhibit a negative association with rainfall at zero lag, largely reflecting the dilution effect, and a positive correlation at positive lags reflecting the increase in secondary transmission after the rains. The existence of both positive and negative effects of rainfall in the seasonal cycle is consistent with the previously described associations of this variable with the transmission rate of cholera at longer temporal scales (Pascual et al., 2002). Recent findings suggest that high precipitation and flooding mediate, in part, the effect of El Niño in the approximately 4-year cycle of the disease.

Koelle et al. (2005b) have demonstrated the linkages across time and space scales in the dynamics of the regional system (climate patterns, river basin rainfall variability, river discharge, and flooding), *V. cholerae*, and non-linear human susceptibility levels. According to their study, water temperature in ponds and rivers provides another local mechanism for the remote association of cholera transmission with SST in the Bay of Bengal and the Pacific ENSO, and other factors such as changes in cloud cover, wind stress and evaporation modulate variations in the net heat flux entering the system increasing both the SST in the Bay of Bengal and affecting the surface temperature over land. The resulting warming of water temperature in ponds and rivers might increase the incidence of cholera through the faster growth rate of the pathogen in aquatic environments. Their results showed that a critical interplay of environmental forcing, specifically climate variability, and temporary immunity explained by the interannual disease cycles presented in four-decade cholera time series from Matlab, Bangladesh. By reconstructing the transmission rate, affected by extrinsic forcing over time for the predominant strain (El Tor), a non-linear population model permitted a contributing effect of intrinsic immunity. According to their findings, transmission showed clear interannual variability with a strong correspondence to climate patterns at long periods (over 7 years, for monsoon rains and Brahmaputra river discharge) and at shorter periods (under 7 years, for flood extent in Bangladesh, sea surface temperatures in the Bay of Bengal and the El Niño–Southern Oscillation). According to authors, the complex role of water reflects multiple mechanisms and the spatial and temporal scales involved in the concentration of the pathogen in the environment, and in the human behavior underlying contact with water and the pathogen itself.

De Magny et al. (2006) study was the first to document the existence of an association between climate and cholera outbreaks on the African continent, using the wavelet method to explore periodicity in (i) a long-time monthly cholera incidence in Ghana, West Africa, (ii) proxy environmental variables, and (iii) climatic indices time series, from 1975 to 1995. Cross-analysis was done to explore links between cholera and climate. Results showed strong statistical association from the end of the 1980s, between cholera outbreak resurgences in Ghana and the global climatic index, and two regional climatic parameters, land and/or sea surface temperature anomalies, and rainfall. A periodic common mode of fluctuation of 2–3 years for cholera time series cases and each of the five environmental parameters under scrutiny were observed. Results of coherence revealed the existence of strong associations between, (i) cholera case incidence and rainfall, and (ii) disease incidence and LSTA1 (Land Surface Temperature Anomaly) time series. The influence of warm events may have an impact on the bacterial populations and that of their host reservoirs, providing new favorable environmental conditions such as an increase in temperature in shallow bodies of water, i.e., lagoons, estuaries, and coastal waters.

# CHALLENGES FOR FUTURE RESEARCH

Different scenarios predict health adverse consequences in different regions of the world in the coming decades due to climate change (Hales et al., 2002; Lieshout et al., 2004; Kovats et al., 2005). Associated land use and transformation of resource production (urbanization, agricultural expansion and intensification, and natural habitat alteration) have produced changes in ecological systems, notably in landscapes and, in turn, their natural communities and ultimately in their pathogen, animal host, and human populations. Factors related to public health infrastructure and climate variability, and their interactions with regional environmental change also may contribute significantly to disease emergence (Burke et al, 2001).

In addition to natural climate variability and climate shifts, the climate change contribution may well contribute further to disease emergence (Wilcox and Colwell, 2005). Environmental/ecological changes and economic inequities strongly influence disease patterns. Nevertheless, an ever-increasing role of a warming and unstable climate is driving the global emergence, resurgence and redistribution of infectious diseases (McMichael et al., 2006). Regional environmental change, which is influenced significantly by population growth, resource consumption, and waste generation, plays an important role in the emergence of infectious disease, especially in tropical developing regions. Tropical climate, poor water and food security, low socio-economic status and political instability define the regions that would be most vulnerable to the health effects of climate change (McMichael and Kovats, 2000; Moreno, 2006). Many Latin American countries have these conditions in common. Therefore, there is a need in Latin America to identify areas where populations are vulnerable to the health impacts of climate change, being that populations with the fewest resources would be the most vulnerable to the adverse health effects of climate change. Epidemiological surveillance of areas under risk would allow having better knowledge of how climate change may impact human health. There are areas where diseases are likely to respond to a change in climate and where the population at risk is large with limited capacity to respond to emerging disease threats (Kovats et al., 2001).

Humans have the potential to affect *V. cholerae* abundance and transmission at multiple spatial scales. Human activities at a local scale such as land-use change, pollution (including sewage), aquaculture and fisheries management, have the potential to alter water temperature and nutrient concentrations, which may affect directly conditions for growth, or indirectly, the distribution, abundance, and composition of the plankton. On a broader spatial scale, human-induced climate change could alter *V. cholerae* dynamics by changing the microbe's seasonal regimes or facilitating its spread to new areas (Cottingham et al., 2003).

From an applied perspective, clarifying the mechanisms that link seasonal environmental changes to diseases dynamics will aid in forecasting long-term health risks and in developing strategies for controlling diseases across a range of human and natural systems. This is especially important because longer-term environmental changes caused by climate warming and complex events like ENSO will alter seasonality in ways that influence spread (Harvell et al., 2002; Pascual et al., 2002).

Research on the linkage between climate and infectious diseases must be strengthened in order to examine the consistency of climate/disease relationships in different social contexts and across a variety of temporal and spatial scales. Clarifying the causal pathways linking climate to disease prevalence will require additional knowledge of the ecology of the pathogen and the transmission dynamics of infectious disease. Greater financial support to sponsor research and to strengthen environmental management in developing countries is urgent, in particular to improve water supplies and sanitation, and efficient and integrated global cholera surveillance and reporting systems should be established, for improving risk assessment for potential cholera outbreaks. A clearer understanding of the current role of climate change in disease patterns will enable scientists to improve forecasts of future potential impacts (Kovats et al., 2001). The sooner these consequences are estimated and communicated the better will be the chance of averting future retrograde policy decisions (McMichael, 2001).

## CONCLUSION

*V. cholerae* has long been known as a fecal-oral pathogen; however, the evidence showing that *V. cholerae* is naturally present in warm, brackish environments is overwhelming. Evidence collected over the last 20 years indicates a close association of *V. cholerae* with copepods. The persistence of *V. cholerae* as part of the normal flora in aquatic environments, the lack of an effective vaccine, and increasing antibiotic resistance among strains isolated from cholera patients all suggest that cholera will not be eradicated in the future. As with other tropical diseases, there is growing concern that the combination of climate change, anthropogenic disturbance of local environments, and transport due to travel and trade will expand the range of endemic strains, and consequently create more focal points for cholera outbreaks. The effects of climate variability driven by natural cycles or by anthropogenic activities will expand the range and increase the prevalence of *V. cholerae* and cholera both geographically and temporally, if public health measures are not implemented. As socioeconomic conditions favorable to cholera persist in many countries, revealing the influence of climatic/environmental factors in seasonal patterns is critical to understanding temporal variability of cholera at longer time scales, including sustained monitoring of freshwater and marine ecosystems and trends and interannual variability of climate to improve disease forecasting. An understanding of disease risk related to the environment can also call attention to the need for improving these conditions. A more complete understanding of the ecology of *V. cholerae* is critical to identify and comprehend the seasonality and regional mechanisms as a function of environmental factors for the prediction and management of this disease.

## REFERENCES

Anthes, R. A., Moore, B., Anderson, J. G., Barron, B. E. J., Brown, O. B. Jr., Cutter, S. L., De Fries, R., Gaiel, W. B., Hager, B. H., Hollingsworth, A., Janetos, A. C., Kelly, K. A., Lane, N. F., Lettenmaier, D. P., Traw, B. M., Mika, A. M., Washington, W. M., Wilson, M. L., Zoback, M. L. (2007). Earth Science and Applications from Space: National Imperatives for the Next Decade and Beyond. National Research Council of the National

Academies, Space Studies Board, Division on Engineering and Physical Sciences. Washington, D. C. U.S.A.: The National Academies Press.

Barbieri, E., Falzano, L., Fiorentini, C., Pianitti, A., Baffone, W., Fabbri, A., Matarrese, P., Caisere, A., Katouli, M., Kühn, I., Mölby, F., Bruscolini, F., Donelli, G. (1999). Occurrence, diversity, and pathogenicity of halophilic Vibrio spp. and Non O1 *Vibrio cholerae* from estuarine waters along the Italian Adriatic Coast. *Appl. Environm. Microbiol.*, 65, 2748-2753.

Barnett, T. P., Pierce, D. W. and Schnur, R. (2001). Detection of antropogenic climate change in the world's oceans. *Science*, 292, 270-274.

Bassler, B. L., Yu, C., Lee, Y. C. and Roseman, S. (1991). Chitin utilization by marine bacteria. Degradation and catabolism of chitin oligosaccharides by *Vibrio furnissii. J. Biol. Chem.*, 266, 24276-4286.

Bonilla-Castro, E., Rodríguez, P., Carrasquilla, G. (2000). La Enfermedad de la Pobreza, El Cólera en los Tiempos Modernos. Santafé de Bogotá: Ediciones Uniandes.

Borroto, R.R. and Haddock R.L. (1998). Seasonal pattern of cholera in Guam and survival of *Vibrio cholerae* in aquatic environments. *Journal of Environment, Disease and Health Care Planning*, 3, 1–9.

Borroto, R. J. and Martínez-Piedra, R. (2000). Geographical patterns of cholera in Mexico, 1991-1996. *Int. J. Epidemiol.*, 29, 764-772.

Bouma, M. J., Pascual, M. (2001). Seasonal and interannual cycles of endemic cholera in Bengal 1891–1940 in relation to climate and geography. *Hydrobiologia*, 460, 147–156.

Boyd, E. F., Heilpern, A. J., and Waldor, M. K. (2000). Molecular analyses of a putative CTX_ precursor and evidence for independent acquisition of distinct CTX_s by toxigenic *Vibrio cholerae. J .Bacteriol.*, 182, 5530–5538.

Broza, M. and Halpern, M. (2001). Pathogen reservoirs: Chironomid egg masses and *Vibrio cholerae. Nature*, 412, 40.

Burke, D., Carmichael, A., Focks, D., Gimes, D., Harte, J., Lele, S., Martens, P., Mayer, J., Mearns, L., Pulwarty, R., Leal, L., Ropelewski, Ch., Rose, J., Shope, R., Simpson, J., Wilson, M. (2001). Under the Weather: Climate, Ecosystems, and Infectious Disease. Committee on Climate, Ecosystems, Infectious Diseases, and Human Health, Board on Atmospheric Sciences and Climate, National Research Council, Division on Earth and Life Studies, National Research Council. National Academy of Sciences. Washington, D. C. U.S.A.: The National Academy Press.

Castañeda Chávez, M. R., Pardío, V., Orrantia, E., Lango, F. (2005). Influence of water temperature and salinity on seasonal occurrences of *Vibrio cholerae* and enteric bacteria in oyster-producing areas of Veracruz, México. *Mar. Poll. Bull.*, 50, 1641-1648.

Chakraborty, S., Mukhopadhyay, A. K., Bhadra, R. K., Ghosh, A. N., Mitra, R., Shimada, T., Yamasaki, S., Faruque, S. M., Takeda, Y., Colwell, R. R. and Mair, G. B. (2000). Virulence genes in environmental strains of *Vibrio cholerae. Appl. Environ. Microbiol.*, 66, 4022-4028.

Chan, N. Y., Ebi, K., Smith, F., Wilson, T. F. and Smith, A. E. (1999). An integrated assessment framework for climate change and infectious diseases. *Environmental Health Perspectives*, 107, 329-337.

Charron, D. F., Thomas, M. K., Waltner-Toews, D., Aramini, J. J., Edge, T., Kent, R. A., Maarouf, A. R., Wilson, J. (2004). Vulnerability of waterborne diseases to climate change in Canada: A review. *J .Toxicol. Environm .Health, Part A*, 67, 1667–1677.

Checkley, W., Epstein, L. D., Gilman, R. H., Figueroa, D., Cama, R. I., Patz, J. A. and Black, R. E. (2000). Effects of El Niño and ambient temperature on hospital admissions for diarrhoeal diseases in Peruvian children. *Lancet*, 355, 442–450.

Chiavelli, D. A., Marsh, J. W. and Taylor, R. K. (2001). The mannosesensitive hemagglutinin of *Vibrio cholerae* promotes adherence to zooplankton. *Appl. Environ. Microbiol.*, 67, 3220-3225.

Colwell, R. R. (1996). Global climate and infectious disease: the cholera paradigm. *Science*, 274, 2025-2031.

Colwell, R. R. and Patz, J. A. (1998). Climate, infectious disease and health: an interdisciplinary perspective. Washington, D.C.: American Academy of Microbiology.

Colwell, R. R. and Grimes, D. J. (2000). *Nonculturable Microorganisms in the Environment.* Washington D.C.: ASM Press.

Colwell, R. R., and Huq, A. (2001) Marine ecosystems and cholera. *Hydrobiologia*, 60, 141–145.

Colwell, R. R., Huq, A., Islam, M. S., Aziz, K. M. A., Yunus, M., Khan, N.,H., Mahmud, A., Sack, R. B., Nair, G. B., Chakraborty, J., Sack, D. A., Russek-Cohen, E. (2003). Reduction of cholera in Bangladehi villages by simple filtration. *Proc. Natl. Acad. Sci. USA,* 100, 1051-1055.

Connell, T. D., Metzger, D. J., Lynch, J. and Folster, J. P. (1998). Endochitinase is transported to the extracellular milieu by the eps-encoded general secretory pathway of *Vibrio cholerae. J. Bacteriol.,* 180; 5591-5600.

Cottingham, K. L., Chiavelli, D. A., Taylor, R. K. (2003). Environmental microbe and human pathogen: the ecology and microbiology of Vibrio cholerae. *Frontiers in Ecology and the Environment*, 1, 80-86.

Curriero, F. C., Patz, J. A., Rose, J. B., and Lele, S. (2001). The association between extreme precipitationand waterborne disease outbreaks in the United States, 1948–1994. *Am. J. Public Health,* 91, 1194–1199.

Daniels, N. A., MacKinnon, L., Bishop, R., Altekruse, S., Bay, B., Hammond, R. M., Thompson, S., Wilson, S., Bean, N. H., Griffin, P. M., Slutsker, L. (2002). *Vibrio parahaemolyticus* infections in the United States, 1973-1998. *J. Infect. Dis.*, 181, 1661-1666.

Davey, M. E. and O'Toole, G. A. (2000). Microbial biofilms: from ecology to molecular genetics. *Microbiol. Mol. Biol. Rev.*, 64, 847–67.

De Magny, G. C., Cazelles, B. and Guégan, J-F. (2006). Cholera threat to humans in Ghana is influenced by both global and regional climatic variability. *EcoHealth*, 3, 223-231.

Ebi, K. L. (2007). Healthy people 2100: modeling population health impacts of climate change. *Climatic Change*.

Epstein, P. R. (1993). Algal blooms in the spread and persistence of cholera. *BioSystems*, 31, 209-221.

Epstein, P. R., Ford, T. E., Colwell, R. R. (1993). Health and climate change: Marine ecosystems. *Lancet*, 342, 1216-1219.

Epstein, P. R. (2001). Climate change and emerging infectious diseases. *Microbes and Infection*, 3, 747–754.

Evins, G. M., Cameron, D. N., Wells, J. G., Greene, K. D., Popovic, T., Giono-Cerezo, S., (1995). The emerging diversity of the electrophoretic types of *Vibrio cholerae* in the Western Hemisphere. *J .Infect. Dis.*, 172, 173-179.

Faruque, S. M., Albert, M. J. and Mekalanos, J. J. (1998). Epidemiology, genetics, and ecology of toxigenic *Vibrio cholerae. Microbiol .Mol .Biol. Rev.*, 62, 1301-1314.

Faruque, S. M., Naser, I. B., Islam, M. J., Faruque, A. S., Ghosh, A. N., Nair, G. B., Sack, D. A. and Mekalanos, J. J. (2005). Seasonal epidemics of cholera inversely correlate with the prevalence of environmental cholera phages. *Proc. Natl. Acad. Sci. USA*, 102, 1702–1707.

Franco, A. A., Fix, D., Prada, A., Paredes, E., Palomino, C., Wright, C., Johnson, J. A., McCarter, R., Guerra, H. and Morries, J. E. (1997). Cholera in Lima, Peru correlates with prior isolation of *Vibrio cholerae* from the environment. *Am. J. Epidemiol.*, 146, 1067–1075.

Gleick, P. H. Dirty Water: Estimated deaths from water-related diseases 2000-2020. Pacific Institute for Studies in Development, Environment, and Security [2007 october 7]. Available from: www.pacinst.org

Greene, C. H. and Pershing A. J. (2007). Oceans: Climate drives sea change. *Science* 315, 1084-1085.

Haines, A., McMichael, A. J., Epstein, P. R. (2000). Environment and health: 2. Global climate change and health. *Can. Med. Assoc. J.*, 163, 729-734.

Hales, S., de Wet, N., Maindonald, J., Woodward, A. (2002). Potential effect of population and climate changes on global distribution of dengue fever. *The Lancet*, 360, 830–834.

Han, W. Q. and Webster, P. J. (2002). Forcing mechanisms of sea level interannual variability in the Bay of Bengal. *J. Phys. Oceanogr.*, 32, 216–239.

Harvell, C. D., Mitchell, C. E., Ward, J. R., Altizer, S., Dobson, A. P., Ostfeld, R. S., Samuel, M. D. (2002). Climate warming and disease risk for terrestrial and marine biota. *Science*, 296, 2158-2162.

Haugo, A. J. and Watnick, P. I. (2002). *Vibrio cholerae* CytR is a repressor of biofilm development. *Mol. Microbiol.*, 45, 471–483.

Hayes, M., Bonaventura, J., Mitchell, T. P., Prospero, J. M., Shinn, E. A., Van Dolah, F. and Barber, R. T. (2001). How are climate and marine biological outbreaks functionally linked?. *Hydrobiologia*, 460, 213–220.

Heidelberg, J. F., Elsen, J. A., Nelson, W. C., Clayton, R. J., Gwinn, M. L., Dodson, R. J., Haft, D. H., Hickey, E. K., Peterson, J. D., Umayam, L., Gill, S. R., Nelson, K. E., Read, T. D., Gill, S. R., Nelson, K. E., Read, T. D., Tettlin, H., Richardson, D., Ermolaeva, M. D., Vamathevan, J., Bass, S., Qin, H., Dragoi, I., Sellers, P., Mcdonald, L., Utterback, T., Fleishmann, R. D., Nierman, W. C., White, O., Salzberg, S. L., Smith, H. O., Colwell, R. R., Mekalanos, J. J., Craig, J. and Frase, C. M. (2000). DNA sequence of both chromosomes of the cholera pathogen. *Nature*, 406,477–484.

Heidelberg, J. F., Heidelberg, K. B., Colwell, R. R. (2002). Seasonality of Chesapeake Bay bacterioplankton species. *Appl Environm Microbiol*, 68, 5488-5497.

Huq, A., Small, E. B., West, P. A., Huq, M. I., Rahman, R. and Colwell, R. R. (1983) Ecological relationships between *Vibrio cholerae* and planktonic copepods. *Appl. Environ. Microbiol.*, 45, 275-283.

Huq, A., West, P. A., Small, E. B., Huq, M. I., and Colwell, R. R. (1984) Influence of water temperature, salinity, and pH on survival and growth of toxigenic *Vibrio cholerae* serovar O1 associated with live copepods in laboratory microcosms. *Appl. Environ. Microb.*, 48, 420–24.

Huq, A., Colwell, R. R., Rahmann, R., Ali, A., Chowdhury, M. A. R., Parveen, S., Sack, D. A. and Russek-Chohen, R. (1990). Detection of *Vibrio cholerae* O1 in the aquatic environment by fuorescent-monoclonal antibody and culture methods. *Appl. Environ. Microbiol.*, 56, 2370-2373.

Huq, A. and Colwell, R. R. (1996). Environmental factors associated with emergence of disease with special reference to cholera. *Eastern Medit. Hlth. J.*, 2, 37-45.

Huq, A., Rivera, I. N. G. and Colwell, R. R. (2000). Epidemiological significance of viable but nonculturable microorganisms. In R. R. Colwell and D. J. Grimes (Eds.), *Nonculturable microorganisms in the environment*. Washington, D. C.: American Society of Microbiology Press.

Huq, A., Sack, R. B., Nizam, A., Longini, I. M., Nair, G. B., Ali, A., Morris, J. G. Jr., Khan, M. N., Siddique, A. K., Yunus, M., Albert, M. J., Sack, D. A. and Colwell, R. R. (2005). Critical Factors Influencing the Occurrence of *Vibrio cholerae* in the Environment of Bangladesh. *Appl Environ Microbiol*, 71, 4645-4654.

Hunter, P. R. (2003). Climate change and waterborne and vector-borne disease. *J. Appl. Microbiol.*, 94, 37S–46S.

International Panel on Climate Change (IPCC) (2001a) Climate change 2001: Impacts, Adaptation and Vulnerability. The scientific basis. Contribution of Working Group I to the third assessment report of the Intergovernmental Panel on Climate Change. Canziani O and McCarthy J editors. Cambridge, UK: Cambridge Univ. Press.

International Panel on Climate Change (IPCC) (2001b) Climate change 2001: The scientific basis. In J. T. Houghton, Y. Ding, D. J. Griggs, M. Noguer, P. J. van der Linden, X. Dai, K. Maskell, C. A. Johnson (Eds). *Contribution of Working Group I to the third assessment report of the Intergovernmental Panel on Climate Change*, (pp. 881). Cambridge, UK: Cambridge University Press.

International Panel on Climate Change (IPCC) (2001c) Climate change 2001: Impacts, Adaptation and Vulnerability. In J. J. McCarthy, O. F. Canziani, N. A. Leary, D. J. Dokken, K. S. White (Eds.) *Contribution of Working Group II to the third assessment report of the Intergovernmental Panel on Climate Change.*, (pp. 1032). Cambridge, UK: Cambridge University Press.

Islam, M. S., Drasar, B. S., Bradley, D. J. (1990) Long-term persistence of toxigenic *Vibrio cholerae* O1 in the mucilaginous sheath of a blue-green alga, *Anabaena variabilis*. *J Trop Med Hyg*, 93, 133–139.

Islam, M. S., Drasar, B. S., Sack, R. B. (1994a). The aquatic flora and fauna as reservoirs of Vibrio cholerae: a review. *J. Diarrh. Dis. Res.*, 12, 87–96.

Islam, M. S., Rahim, Z., Alam, M. J., Begum, S., Moniruzzaman, S. M., Umeda, A., Amakao, K., Albert, M. J., Sack, R. B. and Colwell, R. R. (1994b). Detection of non-culturable *Vibrio cholerae* O1 associated with a cyanobacterium from the aquatic environment in Bangladesh. *Trans R. Soc. Trop. Med. Hyg.*, 88, 198-199.

Islam, M. S., Rahim, Z., Alam, M. J., Begum, S., Moniruzzaman, S. M., Umeda, A., Amakao, K., Albert, M. J., Sack, R. B. and Colwell, R. R. (1999). Association of *Vibrio cholerae* O1 with the cyanobacterium, *Anabaena* sp., elucidated by polymerase chain reaction and transmission electron microscopy. *Trans. R. Soc. Trop. Med. Hyg.*, 93, 36-40.

Jacoby, J. M., Collier, D. C., Welch, E. B., Hardy, F. J. and Crayton, M. (2000). Environmental factors associated with a toxic bloom of Microcystis aeruginosa. *Can. J. Fish Aquat. Sci.*, 57, 231–240.

Jiang, S. C. (2001). *Vibrio cholerae* in recreational beach waters and tributaries of Southern California. *Hydrobiologia*, 460, 157-164.

Kierek, K. and Watnick, P. I. (2003). Environmental determinants of *Vibrio cholerae* biofilm development. *App. Environm. Microbiol.*, 69, 5079-5088.

Kiska, D. L. (2000). Global Climate Change: An Infectious Disease Perspective. *Clinical Microbiology* Newsletter, 22, 81-86.

Koelle, K., Pascual, M., Yunus, M. (2005a). Pathogen adaptation to seasonal forcing, climate change. *Proc. Royal Soc. London B*, 272, 971–977.

Koelle, K., Rodo, X., Pascual, M., Yunus, M., Mostafa, G. (2005b). Refractory periods and climate forcing in cholera dynamics. *Nature*, 436, 696–700.

Kovats, R. S., Campbell-Lendrum, D. H., Mc-Michael, A. J., Woodward, A. and Cox, J. S. (2001). Early effects of climate change: do they include changes in vector-borne diseases?. *Philos. Trans. R. Soc. London B*, 356, 1057–1068.

Kovats, S., Campbell-Lendrum, D. H., Matthies, F. (2005). Climate change and human health: estimating avoidable deaths and disease. *Risk Anal.*, 25, 1409–1418.

Levitus, S. (2001). Anthropogenic warming of earth's climate system. Science 292, 267-270.

Lieshout, van M., Kovats, R. S., Livermore, M. T. J., Martens, P. (2004). Climate change and malaria: analysis of the SRES climate and socio-economic scenarios. *Global Environ. Change*, 14, 87–99.

Lipp, E. K., Huq, A., Colwell, R. R. (2002). Effects of global climate on infectious disease: The cholera model. *Clin. Microbiol. Rev.*, 15, 757-770.

Lobitz, B., Beck, L., Huq, A., Wood, B., Fuchs, G., Faruque, A. S. G., and Colwell, R. R. (2000). Climate and infectious disease: use of remote sensing for detection of *Vibrio cholerae* by indirect measurement. *Proc. Natl. Acad. Sci .USA.*, 97, 1438–1443.

Louis, V. R., Cohen, E., Choopun, N., Rivera, I., Gangle, B., Jiang, S., Rubin, A., Patz, J., Huq, A., Colwell, R. R. (2003). Predictability of *Vibrio cholerae* in Chesapeake Bay. *Appl. Environm. Microbiol.*, 69, 2773-2785.

Mahadevan, A. and Subramanian, B. (1999). Seasonal and diurnal variation of hydrobiological characters of coastal water of Chennai (Madras), Bay of Bengal. *Indian J. Mar. Sci.*, 28, 429–433.

Mahasneh, A. M. and Al-Sayed, H. A. (1997). Seasonal incidence of some heterotrophic aerobic bacteria in Bahrain pelagic and nearshore waters and oysters. *Intern. J. Environ. Studies,* 51, 301-12.

McDougald, D., Rice, S. A., Weichart, D., and Kjelleberg, S. (1998). Nonculturability: adaptation or debilitation? *FEMS Microbiol. Ecol.*, 25, 1–9.

McMichael, A. J. and Kovats, R. S. (2000). Climate change and climate variability: adaptations to reduce adverse health impacts. *Environ. Monitor Assess*, 61, 49-64.

McMichael, A. J. (2001). Global environmental change as ''Risk Factor'': can epidemiology cope?. *Am. J. Public Health*, 91, 1172–1174.

McMichael, A. (2004). Environmental and social influences on emerging infectious diseases: past, present and future. *Phil. Trans. R. Soc. Lond. B,* 359, 1049–1058.

McMichael, A., Woodruff, R. E., Hales, S. (2006). Climate change and human health: present and future risks. *Lancet*, 367, 859-869.

Mezrioui, N., Oufdou, K. and Baleux, B. (1995). Dynamics of non-O1 *Vibrio cholerae* and fecal coliforms in experimental stabilization ponds in the arid region of Marrakesh,

Morocco, and the effect of sunlight on their experimental survival. *Canadian J. Microbiol.*, 41, 489-498.

Montilla, R., Chowdhury, M. A., Huq, A., Xu, B., Colwell, R. R. (1996). Serogroup conversion of *Vibrio cholerae* non-O1 to *Vibrio cholerae* O1: effect of growth state of cells, temperature, and salinity. *Can. J. Microbiol.*, 42, 87–93.

Moren-Abat, M., Quevauviller, P., Feyen, L., Heiskanen, A. S., Noges, P., Solheim, A. L. and Lipiatou, E. (2007). *Climate Change Impacts on the Water Cycle, Resources and Quality.* European Commission.

Moreno, A. R. (2006). Climate change and human health in Latin America: drivers, effects, and policies. *Reg. Environ. Change.*, 6, 157–164.

Morris, J. G., Sztein, M. B., Rice, E. W., Nataro, J. P., Losonsky, G. A., Panigrahi, P., Tacket, C. O. and Johnson, J. A. (1996). *Vibrio cholerae* O1 can assume a chlorine-resistant rugose survival form that is virulent for humans. *J. Infect. Dis.*, 174, 1364-1368.

Morris, J. G. Jr. (2003). Cholera and other types of vibriosis: A story of human pandemics and oysters on the half shell. *Clin. Infect. Dis.*, 37, 272-280.

Munro, P. M. and Colwell, R. R. (1996). Fate of Vibrio cholerae O1 in seawater microcosms. *Water Res*, 1, 47-50.

Nesper, J., Lauriano, C. M., Klose, K. E., Kapfhammer, D., Kraib, A. and Reidl, J. A. (2001). Characterization of *Vibrio cholerae* O1 El Tor *galU* and *galE* Mutants: Influence on Lipopolysaccharide Structure, Colonization, and Biofilm Formation. *Infection and Immunity*, 69, 435-445.

Osorio, C. R. and Klose, K. E. (2000). A region of the transmembrane regulatory protein toxR that tethers the transcriptional activation domain to the cytoplasmic membrane displays wide divergence among *Vibrio* species. *J. Bacteriol.*, 182, 526-528.

O'Toole, G., Kaplan, H. B., and Kolter, R. (2000). Biofilm formation as microbial development. *Annu. Rev. Microbiol.*, 54, 49–79.

Padmavathi, D. and Satyanarayana, D. (1999). Distribution of nutrients and major elements in riverine, estuarine and adjoining coastal waters of Godavari, Bay of Bengal. *Indian J. Mar. Sci.*, 28, 345–354.

Patz, J. A., McGeehin, M. A., Bernard, S. M., Ebi, K. L., Epstein, P. R., Grambsch, A., Gubler, D. J., Reiter, P., Romieu, I., Rose, J. B. Samet, J. M. and Trtanj, J. (2000). The potential health impacts of climate variability and change for the United States: Executive summary of the report of the health sector of the U.S. national assessment. *Env. Hlth. Perspec.*, 108, 367-376.

Patz, J. A., Githeko, A. K., McCarty, J. P., Hussein, S., Confalonieri, U., de Wet, N. (2003). Climate change and infectious diseases. In A.J. McMichael, D.H. Campbell-Lendrum, C.F. Corvalán, K.L. Ebi, A.K. Githeko, J.D. Scheraga, A. Woodward (Eds.) *Climate Change and Human Health. Risk and Response.* World Health Organization: Geneva.

Patz, J. A., Campbell-Lendrum, D., Holloway, T., Foley, J. A. (2005). Impact of regional climate change on human health. *Nature*, 7066, 310–317.

Patz, J. A., Olson, S. H. and Gray, A. (2006). Climate change, oceans, and human health. *Oceanography*, 19, 52-59.

Pascual, M., Rodo, X., Ellner, S. P., Colwell, R. R., and Bruma, M. J. (2000). Cholera dynamics and El Niño-southern oscillation. *Science*, 289, 1766-1769.

Pascual, M., Bouma, M. J., Dobson, A. P. (2002). Cholera and climate: revisiting the quantitative evidence. *Microbes and Infection*, 4, 237-245.

Piarroux, R., and Bompangue, D. (2006). Needs for an integrative approach of epidemics: the example of cholera. In M. Tibayrenc (Ed.), *Encyclopedia of Infectious Diseases: Modern Methodologies*. Chichester, UK: John Wiley and Sons Ltd.

Pidwirny, M. Causes of climate change. (27/03/2007). Available from:    http://www. eoearth.org/article/Causes_of_climate_change.

Potasman I, Paz A, Odeh M 2002. Infectious otbreaks associated with bivalve shellfish consumption: a worldwide perspective. *Clin. Infect. Dis*. 35, 921-928.

Prüss, A. and Havelaar, A. (2001). The global burden of disease study and applications in water, sanitation and hygiene. In L. Fewtrell and J. Bartram (Eds.), *Water Quality: Guidelines, Standards and Health*. (pp. 43–59). London: IWA Publishing.

Reemtsma, T., Ittekkot, V., Bartsch, M., and Nair, R. R. (1993). River inputs and organic-matter fluxes in the northern Bay of Bengal. *Chem. Geol.*, 103, 55–71.

Rees, R. (1996). Under the weather: climate and disease, 1700-1900. *History Today*, 46, 35-42.

Reidl, J. and Klose, K. E. (2002). *Vibrio cholerae* and cholera: out of the water and into the host. *FEMS Microbiol. Rev.*, 26, 125–39.

Reguera, G. and Kolter R. (2005). Virulence and the environment: a novel role for *Vibrio cholerae* toxin-coregulated pili in biofilm formation on chitin. *J. Bacteriol.*, 187, 3551-3555.

Rodó, X., Pascual, M., Fuchs, G., Faruque, A. S. G. (2002). ENSO and cholera: A nonstationary link related to climate change. *Proc. Natl. Acad. Sci. USA*, 99, 12901-12906.

Rose, J. B., Epstein, P. R., Lipp, E. K., Sherman, B. H., Bernard, S. M., and Patz, J. A. (2001). Climate variability and change in the United States: potential impacts on water- and foodborne diseases caused by microbiological agents. *Environ. Helth Perspect*, 109, 211-222.

Rivera Tapia, J. A. (2003). Environment and Health. *Anales Médicos Hospital ABC*, 48, 223-227.

Saker, M. L. and Griffiths, D. J. (2001). Occurrence of blooms of the cyanobacterium *Cylindrospermopsis raciborskii* (Woloszynska) Seenayya and *Subba Raju* in a north Queensland domestic water supply. *Marine and Freshwater Research*. 52, 905–915.

Scott, G. I., Fulton, M. H., Wirth, E. F., Chandler, G. T., Key, P. B., Daugomah, J. W., Bearden, D., Chung, K. W., Strozier, E. D., DeLorenzo, M., Siversten, S., Dias, A., Sanders, M., Macauley, J. M., Goodman, L. R., LaCroix, M. W., Thayer, G. W., Kucklick, J. (2002). Toxicological studies in tropical ecosystems: an ecotoxicological risk assessment of pesticide runoff in south Florida estuarine ecosystems. *J. Agric. Food Chem.*, 50, 4400-4408.

Seas, C., Miranda, J., Gil, A. I., Leon-Baru, R., Patz, J. A., Huq, A., Colwell, R. R. and Sack, R. B. (2000). New insights on the emergence of cholera in Latin America during 1991. The Peruvian experience. *Am. J. Trop. Med. Hyg.*, 62, 513-517.

Speelmon, E. C., Checkley, W., Gilman, R. H., Patz, J., Calderon, M., Manga, S. (2000). Cholera incidence and El Niño related higher ambient temperature. *JAMA*, 283, 3072–3073.

Simanjuntak, C. H., Larasati, W., Arjoso, S., Putri, M., Lesmana, M., Oyofo, B. A., Sukri, N., Nurdin, D., Kusumaningrum, R. P., Punjabi, N., Subekti, D., Djelantik, S., Lubis, A., Siregar, H., Mas'Ud, B., Abdi, M., Sumardiati, A., Wibisana, S., Wanto, H., Wan, B. S.,

Santoso, W., Putra, E., Sarumpaet, S., Ma'Ani, H., Lebron, C., Soeparmanto, S. A., Campbell, J. R. and Corwin, A. L. (2001). Cholera in Indonesia in 1993–1999. *Am. J. Trop. Med. Hyg.*, 65, 788–797.

Smith, K. F., Dobson, A. P., McKenzie, F. E., Real, L. A., Smith, D. L., Wilson, M. L. (2005). Ecological theory to enhance infectious disease control and public health policy. *Frontiers in Ecology and the Environment*, 3, 29–37.

Singleton, F. L., Attwell, R., Jangi, S., Colwell, R. R. (1982a). Effects of temperature and salinity on *Vibrio cholerae* growth. *App. Environm. Microbiol.*, 44, 1047-1058.

Singleton, F. L., Attwell, R., Jangi, S., Colwell, R. R. (1982b). Influence of salinity and organic nutrient concentration on survival and growth of *Vibrio cholerae* in aquatic microcosms. *Appl. Environm. Microbiol.*, 43, 1080-1085.

Tamerius, J. D., Wise, E. K., Uejio, C. K., McCoy, A. L. and Comrie, A. C. (2007). Climate and human health: synthesizing environmental complexity and uncertainty. Stoch Environ Res Risk Assess. On line first.

Tamplin, M. L. (2001). Coastal Vibrios: identifying relationships between environmental condition and human disease. *Human and Ecological Risk Assessment*, 7, 1437 – 1445.

Torres-Codeço, C. (2001). Endemic and epidemic dynamics of cholera: the role of the aquatic reservoir. *BMC Infectious Diseases*, 1, 1-14.

UNEP. (2005). Emerging challenges—new findings. In *Geo Year Book 2004–2005: an Overview of Our Changing Environment.* (pp. 71–79), New York: United Nations Publications.

Villalpando, S., Eusebio, M. G., Aviles, D. (2000). Detection of *Vibrio cholerae* O1 in oysters by the visual colorimetric immunoassay and the cultura technique. *Rev Latinoamer Microbiol*, 42, 63-68.

Wai, S. N., Mizunoe, Y., Takade, A., Kawabata, S-I. and Yoshida, S-I. (1998). *Vibrio cholerae* O1 strain TSI-4 produces the exopolysaccharide materials that determine colony morphology, stress resistance, and biofilm formation. *Appl. Environ. Microbiol.*, 64, 3648-3655.

Walther, G. R., Post, E., Convey, P., Menzel, A., Parmesank, C., Beebee, T. J. C., Fromentin, J-M., Guldberg, O. H., and Bairlein, F. (2002). Ecological responses to recent climate change. *Nature*, 416, 389-395.

Watnick, P. I. and Kolter, R. (1999). Steps in the development of a *Vibrio cholerae* El Tor biofilm. *Mol. Microbiol.*, 34, 586-595.

Watnick, P. I., Lauriano, C. M., Klose, K. E., Croal, L. and Kolter, R. (2001). The absence of a flagellum leads to altered colony morphology, biofilm development and virulence in Vibrio cholerae 0139. *Molec. Microbiol.*, 39, 223-235.

Whittman RJ, Flick GJ (1996). Microbial contamination of shellfish prevalence risk to human health and control strategies. *Annual Rev. Public Hlth.*, 16, 123-140.

Wilcox, B. A. and Colwell, R. R. (2005). Emerging and reemerging infectious diseases: Biocomplexity as an interdisciplinary paradigm. *EcoHealth*, 2, 244–257.

Wilson, M.L. (2001). Ecology and infectious disease. In J. L. Aron and J.A. Patz (Eds.) *Ecosystem change and public health:a global perspective* (pp. 283–324), Baltimore, USA: John Hopkins University Press.

Wright, A. C., Hill, R. T., Johnson, J. A., Roghman, M. C., Colwell, R. R., Morris, J. G. Jr. (1996). Distribution of *Vibrio vulnificus* in the Chesapeake Bay. *Appl. Environ. Microbiol.*, 162, 17–24.

World Health Organization. (2002).The World Health Report 2002. Geneva: WHO.

World Health Organization. (2005a). Available: http://www.who.int.

World Health Organization. (2005b). Cholera, 2004. *Weekly Epidemiological Record,* 80,261–268.

WHO Weekly epidemiological record No. 31, 81, 297–308 (2007 octubre 5). Available from: http://www.who.int/wer .

Xu, H. S., Roberts, N. C., Singleton, F. L., Attwell, R. W., Grimes, D. J., and Colwell, R. R. (1982). Survival and viability of nonculturable *Escherichia coli* and *Vibrio cholerae* in the estuarine and marine environment. *Microb. Ecol.*, 8, 313–323.

Yildiz, F. H. and Schoolnik, G. K. (1999). *Vibrio cholerae* O1 El Tor: identification of a gene cluster required for the rugose colony type, exopolysaccharide production, chlorine resistance, and biofilm formation. *Proc. Natl. Acad. Sci. USA*, 96, 4028–4033.

In: Aquatic Ecosystem Research Trends
Editor: George H. Nairne

*Chapter 11*

# AQUATIC POLLUTANT ASSESSMENT
# ACROSS MULTIPLE SCALES

## *Clint D. McCullough*[*]

Centre for Ecosystem Management and Centre of Excellence for Sustainable Mine Lakes,
Edith Cowan University, 100 Joondalup Drive, Perth, WA 6027, Australia

## ABSTRACT

Aquatic pollutant testing using biological assays is useful for ranking the toxicity of different chemicals and other stressors, for determining acceptable concentrations in receiving systems and for elucidating cause and effect relationships in the environment. This 'ecotoxicological' testing approach supplants previous approaches that indirectly estimated toxicity using chemical and physical surrogate measurements alone.

Nevertheless, many published aquatic pollution studies are restricted to examining the effects of a single toxicant on only a single species. Moreover, laboratory-based ecotoxicity tests often intrinsically suffer from a number of limitations due to their small-scale. For example, a major criticism of single-species bioassays is their failure to integrate and link toxicants (and other associated abiotic components) with higher scales of biological and ecological complexity (predation, competition, etc.). Many researchers have suggested that single-species toxicant testing has become so widely entrenched that it has hindered the development and greater use of testing at more ecologically-relevant scales. An improvement to single-species laboratory tests are microcosm and mesocosm studies using more complex and relevant measures to aquatic biotic communities.

Nevertheless, mesocosms still do not entirely simulate the ecosystem they come from, rather they mirror its general properties. As a result, there is increasing interest in correlating pollution measures from field surveys with measures of aquatic biotic community structure to determine a toxicant's scale of effect. However, field assessments, although extremely useful in determining site-specific impacts, may be limited by lack of experimental controls, too few or poorly-positioned regional reference sites and by confounding effects from impacts unrelated to the disturbance of concern.

A hegemony on the evolution of ecotoxicological science and practice is that the primary application of ecotoxicological data is regulatory. As ecological systems do not

---

[*] *Clint D. McCullough: e-mail: c.mccullough@ecu.edu.au*

have a single characteristic of scale, "validation" of single-species toxicity assessments by higher ecological level assessments remains the highest standard for aquatic pollution studies. As such, multi-scale assessments at all of these scales are now being recognised as providing the highest reliability for environmental protection of lake ecosystems.

# INTRODUCTION

Toxicity assessment of chemicals and other stressors is useful for ranking their toxicity to aquatic biota, for determining their acceptable concentrations in receiving systems and for elucidating cause and effect relationships in the environment [34]. A fundamental premise of toxicity assessment (toxicology) is that any chemical occurring at a high enough concentration may cause toxicity. Indeed, this original analysis can be traced back to Paracelsus, often recognised as the father of Toxicology [49].

"Poison is in everything, and no thing is without poison. The dose makes it either a poison or a remedy."

Paracelsus (1492–1541)

Assessing toxicity to the environment is often formalised through practice of Ecological Toxicology or 'Ecotoxicology'. Ecotoxicology is the science of investigating the effects of toxicants on, and their relationship with, the ecology of a receiving ecosystem. The ecosystem in question may be either the atmosphere, the soil or, as will be the focus of this discussion, an aquatic system such as a lake. Aquatic ecotoxicology, although a relatively new science, compares well with the advancement of other branches of ecotoxicology such as those of air and soil pollutant impacts [27]. As such, ecotoxicology provides an objective and mensurative basis for making decisions about the likely impact of a chemical or physical change on the ecosystem of a given receiving environment [34]. The essence of ecotoxicology is that risk evaluation and regulation of toxic discharges is incomplete unless biological organisms are also used as indicators of the presence of toxic effects [124]. Biological assessment of water quality, and ecotoxicological testing as a subset of this, arose from the recognition that measurement of physico-chemical variables alone does not allow for an assessment of the endpoint for most waterbody monitoring, that is, the suitability of such waters for sustaining biological communities [32; 111].

Biological methods of assessing pollution in aquatic environments have the capacity to integrate effects through continuous exposure and because they measure directly the level of change at which a particular stressor becomes toxic. This approach supplants earlier efforts at indirectly estimating toxicity, using chemical and physical surrogate measurements alone [5; 76; 77]. Comprehensive and effective assessment and management of water quality relies on integrating biological approaches with the more traditional chemical and physical-based approaches, where chemical data provide explanatory variables for trends observed for biota ("cause for consequence") [33].

# SINGLE-SPECIES LABORATORY TESTING

Ecotoxicology is most typically used to determine at what concentration a detrimental effect will become apparent for a given toxicant. In this regard many ecotoxicological tests occur as single-species laboratory bioassays in the screening of complex effluents to waterways, or in cases where there is little knowledge of the toxicity and mechanism of toxicity of a discharged substance [21; 34]. Therefore, ecotoxicological testing can also identify the absence of toxic effects with regards to the environmental safety of a discharge [124]. By definition there is no disturbance (pollution) to a biological community without an observable change in the structure/function or other definable characteristics of that community [100]. The use of toxicity testing involving biological organisms (bioassays) is therefore of prime importance in anticipating the likelihood of environmental impacts [114].

Nonetheless, many studies are restricted to examining the effects of a single toxicant on only a single species [114]. Acute (short-term) studies have also dominated the ecotoxicological literature, although there is an increasing focus on chronic (long-term) and sub-lethal (e.g., population growth rate) assays [88; 4; 13]. Data from single-species tests are typically derived using established standard protocols, where tests are maintained in a relatively simple form, and at small scale under highly controlled ambient conditions. Such procedures enable tests to be more easily reproduced within and amongst different laboratories, and in a given laboratory at different times [25]. These features, together with use of standard test species (e.g., rainbow trout, *Oncorhynchus mykiss*) also enable comparisons of quantitative results from different studies and toxicants. Repeatability (precision) for chronic ecotoxicity tests may be quite high [3], and mode of toxicity information can typically be readily inferred from one toxicant to another.

Despite these advantages, laboratory-based ecotoxicity tests suffer from a number of limitations. These may include inappropriate test species selection, inherent confounding of toxicity responses in the experimental design, inappropriate test species selection, and a failure to accommodate all modes of toxicant exposure to an organism. As a result, there has been some published criticism that responses in laboratory-scale studies are not representative of responses in natural ecosystems [98; 83; 18; 22].

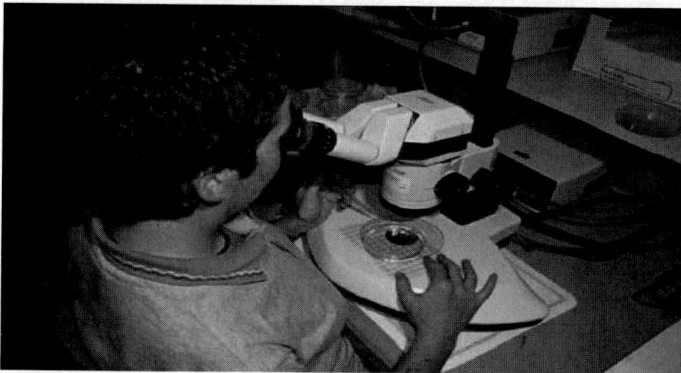

Figure 1. Counting *Hydra* abundances in a single-species, laboratory-based bioassay testing a mine water discharge. (Photo: Clint McCullough).

It is often not recognised by researchers that different species may respond differently in their tolerance to different pollutants [23]. However, species chosen for testing are generally those that are most easily maintained in captivity [83; 109]. This explicit selection represents a major bias toward the relatively few robust taxa occurring in an ecosystem that are amenable to laboratory testing. More often than not, the few taxa that are amenable to laboratory testing do not meet other important criteria relating their ecological significance to key ecosystem processes or to their toxicant sensitivity being representative of other receiving system taxa [6]. Laboratory-based, single-species tests have been found to best predict the field effects of strong toxicants upon *common* species [10; 81]. However, this is likely to be because taxa selected for single-species ecotoxicological testing are typically common species [18]. By definition, these species are unlikely to be of as much conservation significance as rare species which may already be constrained in abundance and distribution by narrower environmental tolerances. Yet these latter species may be the least protected by single-species testing [28; 29; 89]. Similarly, although laboratory bioassay tests are used to predict ecological effects on natural populations of species, laboratory test data are often based upon captive stocks; the gene pool of which may differ from wild stocks, yielding erroneous toxicology conclusions [113; 92].

The extent to which differing races/locales of the same species may differ in their response to toxicants is an important consideration for ecotoxicological science. This is especially so when environmental conditions from where test populations are sourced may be appreciably different [128]. Thus, a relative ranking of toxicant toxicity risk will depend very much upon which strain(s) are tested [58]. Testing for comparability of response of laboratory-maintained stock and wild stock is therefore also a good way of achieving high laboratory quality control [58; 122] as concerns have been voiced for asexually reproducing laboratory populations of test species in general [58]. It is possible that a clone with a slight selective advantage under specific laboratory conditions may successfully exclude other genotypes, resulting in less variability of test results within labs because of reduced genetic diversity but increased variability of test results between labs culturing populations of separate genotypes [113].

Experimental design limitations may also lead to confounding of toxicity responses. For example, test organism deaths during the period of culture and/or maintenance may remove weaker test individuals, thereby potentially underestimating toxicity for that population when tested [17; 23]. Alternatively, stress associated with the culture and testing procedures may reduce genetic variability of a test species population and thereby render individuals more susceptible to the stress of the test toxicant. This increased susceptibility may in turn lead to overestimation of toxicity compared with the unstressed populations in the actual toxicant receiving populations [58]. Repercussions expected from this altered gene pool following transfer from a natural population include predictions of increased magnitude of the effect concentration of a toxicant and at the same time reduced variation of their response [7]. Measures of toxicity obtained from a subsample of a natural population with reduced variability are therefore likely underestimate the variability of the response of the more genetically diverse natural populations, although by how much is difficult to quantify [87; 97] (Figure 2).

However, if a population receiving elevated toxicants is more resistant to the given toxicant than natural, reference populations, then this may be taken as direct evidence that the

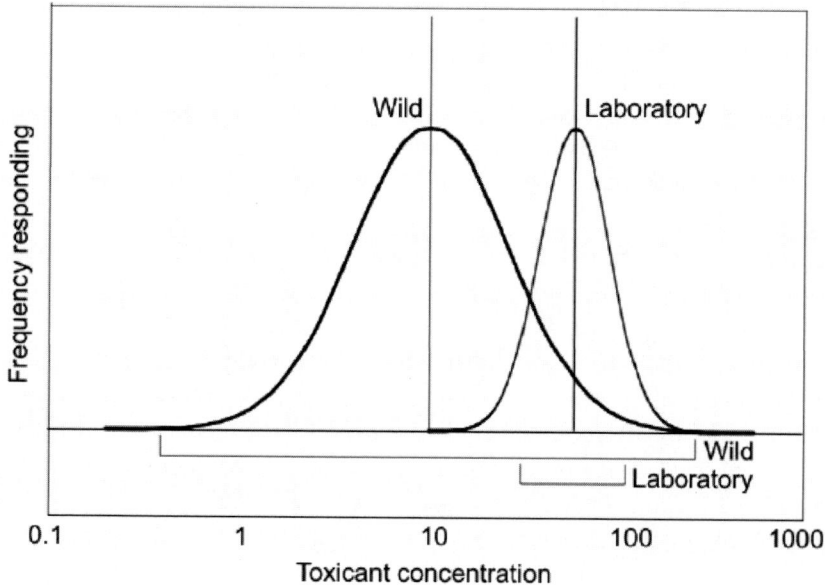

Figure 2. Expected response to a captive population of a test species relative to its original wild population (after Baird [7]).

toxicant concentration has been sufficient to elicit biological community changes [92]. Furthermore, this effect on the population of one species in the receiving system strongly implies that other species are likely to be affected by the toxicant in similar fashion [87]. The presence of tolerant forms may therefore be used as an indicator that a community is being subjected to an environmental stress [61; 24]. Furthermore, the ecological consequence of favouring obligatory pollution-tolerant genotypes may result in species being rendered less fit in other areas which are necessary for their success [70; 122]. Because of this physiological "cost", resistant strains have typically been found only in areas of sufficient toxicant concentration. The level of resistance to a toxicant has also been found to be directly related to the concentration of the toxicant to which the population is exposed [58].

Short-term, single-species laboratory tests for pollution may also fail to address issues of community complexity, as they have no way of incorporating factors such as the role of species interactions including predation and competition, or other population processes such as immigration and emigration [109]. Consequently, when there is differing sensitivity of test species to different toxicants, reliance on too few test species without regard to the test species' relevance to higher scales of ecological complexity may lead to a lack of agreement with field responses [93].

The extent to which one species can act as a proxy and provide predictive information on the effect of a toxicant to the receiving ecosystem is clearly a critical issue [88]. However, the ability of single-species assays to accurately predict the impact on an ecosystem, even when the species chosen for assay are indigenous to the receiving waters, has frequently been called into question both empirically [21; 99; 50] and philosophically [94]. Nonetheless, practicality and convenience remain the dominant influences on the choice of ecotoxicological testing. This is why single-species tests still predominate and are likely to do so for some time [42; 43].

Single-chemical laboratory toxicity tests also continue to play a role in pollution assessment for derivation of water quality standards or guidelines. In these assessments, the direct toxicity assessment (DTA) approach, using chronic (or sub-chronic) tests on appropriate species and using local dilution waters, is now regarded as the preferred, more realistic methodology for predicting effects on ecosystems [4]. DTA has been shown to result in stronger agreement between field validations and laboratory studies (see review by deVlaming and Norberg-King [50]).

There have, however, been few field validations of the conclusions of laboratory-based, single-species bioassays [44]. Nevertheless, reviews of studies comparing predictions of laboratory-based, single-species bioassays to field studies indicate good agreement when toxicity is high, such that only a single study of either laboratory or field approach may be required to characterise the pollutant's risk [10; 54; 51]. Nonetheless, there has also been criticism that responses of laboratory studies are not representative of responses in natural ecosystems where local differences in species composition and community complexity are integrated [18; 22]. Thus, studies which have compared predicted "ecosystem" responses (generally using community characteristics as a surrogate for ecosystem responses) by way of single-species bioassays have reached mixed conclusions.

In summary, the major criticism of single-species bioassays is their failure to integrate and link toxicants (and other associated abiotic components) with higher scales of biological and ecological complexity (predation, competition, etc.) [122; 125]. Nevertheless, the direct cause and effect nature and repeatability of a single-species test is also very important in maintaining the credibility of conclusions with scientists, regulators and lay-persons. Indeed, across various test types, measurement repeatability (precision) for chronic bioassays may be quite high [3]. Furthermore, more sensitive responses to stress have been predicted with smaller organisms that have shorter generation times [109]. Consequently, although there is more of a trend toward multi-species and community scale experimentation, single-species tests, often on non-indigenous species, remain the most frequent ecotoxicological approach [20; 88; 32; 34].

## MICROCOSMS/MESOCOSMS

Compared with laboratory approaches, proponents of community-scale testing have argued that single-species toxicant testing has become so widely entrenched that it has hindered the development and greater use of community-scale testing [83]. Thus, despite continuing development of ecologically-relevant, population-based test endpoints in the laboratory, such as reproduction, growing concerns have been for a *complete* risk analysis of a pollutant requiring studies at the scales of populations, communities and finally ecosystems themselves [12].

Shortcomings of single-species studies have therefore been increasingly addressed through use of various multi-species tests, which may range in scale from laboratory studies using mixed-flask cultures, to field studies using algae and macroinvertebrates on artificial substrates, microcosm and mesocosm communities, and model streams and lake enclosures . Such mesocosm methods may be employed to complement existing single-species data as part of a multi-scale risk assessment.

There has been a growing interest amongst the ecotoxicological research community in the use of these larger scale (and consequently, higher scale of biological organisation) studies for pollution assessments. A compromise between the two extremes of spatial and organismal scales of assessment that both single-species tests and field studies represent, is intermediate scale microcosm and mesocosm systems. The concept of mesocosms was an extension of early experiments at the scale of farm ponds [78; 79]. The attractiveness of experimental mesocosms is that they retain a strong element of environmental realism and applicability, whilst permitting laboratory-like replication and manipulations to be performed. The greater realism of mesocosm environments is achieved through their capacity to include higher scale components of ecosystem responses such as micro/macroinvertebrate communities as well as basic ecosystem interactions.

A useful definition of the term "mesocosm" is provided by Odum [98], describing it as a middle-sized experimental environment falling between a laboratory-based (microcosms) and a full field-scale (macrocosm) multi-species study. Suitable enclosures representative of receiving water communities of micro- and macroinvertebrates have been variously described as artificial "ecosystems" and "multispecies toxicity tests". Mesocosm experiments may use artificial enclosures to achieve independent units, or they may make use of naturally isolated water bodies such as pools or backwaters. The most frequently cited advantages of these studies over single-species bioassays, are improved environmental realism and greater predictive abilities [88; 126; 31].

Figure 3. Large mesocosms may be labour-intensive and expensive to establish. (Photo: Clint McCullough).

Pollutant exposure durations in mesocosm pollutant exposure experiments are also frequently greater than those in single-species laboratory tests [50; 126]. The predominant reason for greater exposure times in mesocosm experiments is a function of their more expensive setup costs and cheaper ongoing costs than for the smaller-scale but more intensive single-species laboratory tests [98].

However, disadvantages to using mesocosms remain. Mesocosms typically exclude predators which may structure some communities e.g., zooplankton, macroinvertebrates [110]. For example, fish are generally excluded from the enclosures in natural water bodies and populations of test organisms are assessed by conventional field sampling techniques [114]. Replication is also often poor while variability amongst the communities of replicate mesocosms is often very high [85; 110].

It has also long been proposed by some that there is a direct link between ecosystem complexity and its stability [41; 55; 119]. This proposal suggests that, for an ecosystem already near a threshold level of tolerance to a stressor, the loss of even a small portion its species diversity may reduce both the resistance and resilience of the ecosystem to further perturbation [48]. Therefore, the loss of any taxa from complex ecosystems has been interpreted as having an effect of reducing the stability of the system to perturbations and stressors [119]. Consequently, deletion of a single species from a natural community may cause a simplification of the biological system and that this may then lead to species losses, possibly including losses of significant species such as keystone, flagship or economically important taxa [45; 101; 46]. Suitability of environmental conditions for the continued existence of a species can, therefore, only strictly be evaluated by assessing an impact at a community scale. Testing of more than one community type within mesocosms is also therefore recommended.

Ecological communities of organisms that are potentially useful for assessing water quality in artificial (experimental) and natural settings include the following.

Although single-species toxicity testing datasets used to derive guideline values are typically under-represented in aquatic insects [66], macroinvertebrates are a popular biological group chosen to assess effects of aquatic pollution at the community scale. Internationally, analysis of benthic macroinvertebrate communities has been the foremost tool for biological assessment of aquatic ecosystems due to the availability of good taxonomy and extensive literature of pollutant effects [111; 4]. Macroinvertebrates are also often the most speciose community in aquatic ecosystems [65]. Their typical characteristics of relatively long life cycles, ecological relevance, low motility and comparatively simple identification are good qualities for assessing the impacts of pollutants [105; 108; 106].

Periphyton is often the greatest source of primary production for communities of both littoral and pelagic habitats [65] and may represent the major source of energy to secondary and higher trophic scales [86]. Along with macroinvertebrates, periphyton have historically received the most attention in community-scale assessments of environmental quality [42; 43]. Periphytic diatoms (family Bacillariophyceae) have been especially suitable for the biomonitoring of aquatic ecosystems [52; 104]. Diatoms occur ubiquitously in high numbers and diversity, are generally relatively sensitive to changes in water chemistry, are easily collected, analysed and preserved and can be readily identified to species scale [104]. Good preservation in sediments also enhances the usefulness of these algae in both temporal and historical studies such as restoration and palaeolimnology [96; 38]. The distributions and associated water chemistry of diatom taxa are cosmopolitan and well documented and there is typically good information available on their environmental requirements. Furthermore, diatoms often have narrow ranges of tolerance to pH, nutrients and salinity, which have often been widely studied and defined [52; 104]. Nevertheless, as for microinvertebrates, the application of diatoms may be limited because of their small size and requirements for preparation prior to sorting and expert identification [74; 104; 75].

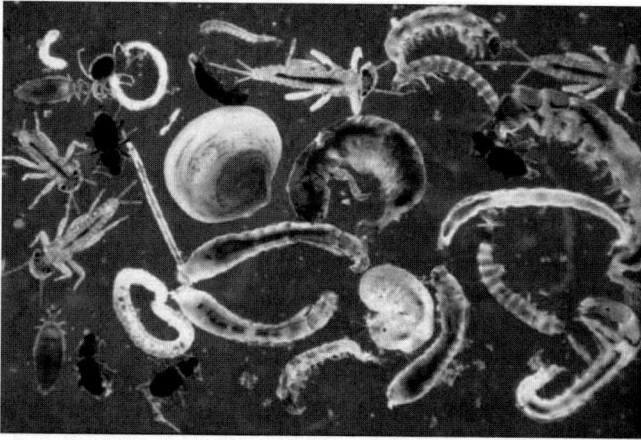

Figure 4. Macroinvertebrates are a diverse and easily identified biotic community of lake littoral margins (Photo: Chris Humphrey).

Sampling of diatom communities on natural rather than artificial substrates has been recommended by some authors, although many have realised the limited statistical power that the intrinsic variability of these methods allows for in experimental studies. However, artificial substrates are generally unrepresentative analogues for naturally occurring diatom communities. This may occur through a "founder effect" with differing inter-replicate seeding potential for the algae themselves, or through differences at higher trophic scales such as zooplankton grazers or their fish predators [2].

Figure 5. Artificial substrates, such as this glass slide-based 'periphytometer' allow easy quantification of periphyton abundances. (Photo: Clint McCullough).

Another group of biota suitable for water quality assessment is the aquatic microinvertebrates. Conventionally, microinvertebrates are invertebrates less than 250 μm in body length that share many of the desired characteristics of macroinvertebrates, but with shorter life cycles and thus faster community responsiveness to environmental change [123]. Commonly encountered microinvertebrates are zooplankters, either littoral or pelagic. Microinvertebrates are important components of aquatic ecosystems, grazing on detritus, bacteria and phytoplankton and often forming an important link between lower organisational levels of energy (primary producers) and those of higher trophic scales such as the numerous fish species [57]. A significant disadvantage of this group for water quality assessment is their smaller size which may make their enumeration and taxonomic identification difficult and consequently limit their use to more specialised applications.

As primary producers, unicellular algae phytoplankton are also often used as indicators of water quality because of their high sensitivity to environmental change and short generation time. Indeed, van Dam *et al.* [121] considered phytoplankton studies as potentially the most promising indicators of shallow lake and wetland degradation. Consequently, phytoplankton are often employed in single-species pollutant assay tests. Phytoplankton are also useful indicators of high nutrient conditions due to their ability to reproduce rapidly under ideal conditions. They provide fundamental information on an important trophic scale and act as an interface between the water chemistry and a significant component of the aquatic food web. In primarily autotrophic-based communities, phytoplankton are of great importance to ecosystem functioning [116].

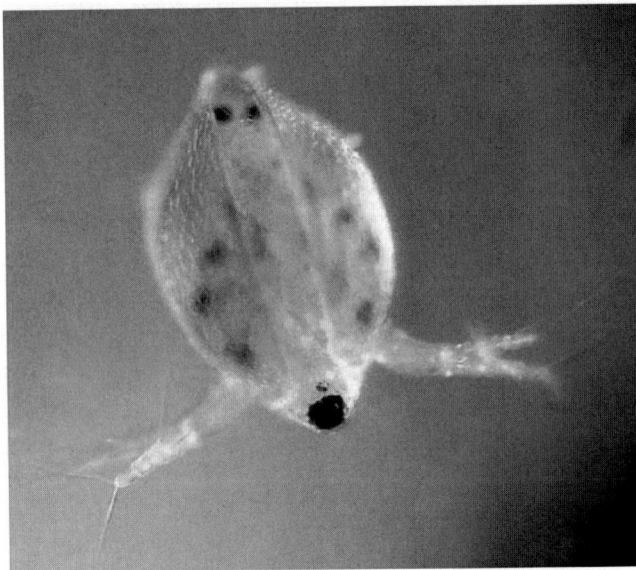

Figure 6. Microinvertebrates, such as this cladoceran, are small and may be hard to identify but they form a crucial trophic linkage in most lake food webs. (Photo: Clint McCullough).

Water body chlorophyll *a*, *b* and *c* concentrations can be used as proxies for phytoplankton biomass and general composition. The concentrations of these different photosynthetic pigments from various receiving conditions of mine waters may, therefore, be expected to act as proxies for validating the (single-species) laboratory *Chlorella* sp. bioassay results in an ecosystem-scale field setting [72].

However, higher-level endpoint criteria for manipulative ecotoxicity experiments are still unclear [26]. Moreover, there is a general consensus amongst researchers that the results of community and ecosystem studies are often complex, highly variable, and therefore, difficult to interpret [42; 68]. Natural spatial and temporal variability of communities may render detection of effects of stressors difficult at community scale in all but extreme cases [97]. For example, there may be difficulty in ascribing the change seen in a community to the toxicant in question when different concentrations and different community endpoints are simultaneously assessed [42]. Thus, historically, one of the most challenging tasks for ecologists is determining whether or not a stressor is detrimentally affecting the biological communities of a receiving aquatic ecosystem [51]. Mesocosms do not entirely simulate the ecosystem they come from, rather they mirror only the general properties that characterise that system [120].

The responses of biological communities to a toxicant often lends a different, more informative insight into the expected effects of perturbation at the scale of an ecosystem than single-species data. However, the scale and multivariate nature of their data also requires a different statistical approach. Many ecological applications of multivariate techniques are readily extended to ecotoxicological field studies. As a consequence, caveats and considerations for ecological study design, such as optimum scale of taxonomic resolution, type of community summary and replication, must also be provided for during the design phase of experiments involving communities. Low amount of replication is often associated with both mesocosm and field studies due to the relatively high intrinsic costs of construction and establishment [90]. Low replication and high intra-treatment variability frequently encountered in mesocosm studies may lead to low reliability of community data caused by artificial enclosures diverging for reasons unrelated to that of the dosed toxicant, such as confounding by unconstrained variables and founder effects [60]. Hence, replication and associated variability must be fundamental in consideration of study design [90]. Nevertheless, the more environmentally realistic scale of data gives field community studies more power to predict expected ecological effects than single-species studies [4]. Consequently, mesocosm studies increasingly contribute to aquatic pollutant studies by a variety of approaches which are particularly relevant for the management of large and unique ecological systems [112].

# FIELD STUDIES

As discussed, estimates of toxicant risk to ecosystems are generally extrapolated from single-species bioassays without substantial validation of the accuracy of the specific response of the toxicant in the field [67]. This lack of field validation represents a limitation of both the real and perceived value of these toxicological data. Field validation of single-species studies of aquatic pollution has, therefore, been found to be a useful approach in assessing the ultimate effects of stressors and also in determining the confidence in the prediction of simpler ecotoxicological models [51; 122]. Since pollutant studies are typically aimed at determining effects of a stressor on higher scales of organisation, it is logical that analysis and interpretation should be performed at the same scale [90].

Manipulative field exposure experiments are typically of similar cost to mesocosm tests with high initial but lower ongoing costs. The reason for greater exposure times in field studies is by reason of their observational nature, where existing exposures are assessed retrospectively. These greater exposure durations of the large-scale experiments would be expected to increase the sensitivity of these studies. Further differences in sensitivity between single-species toxicity testing and the community assessments of the mesocosm and field studies may also be expected to be due to the absence or, at best, greatly reduced contribution of the single-species bioassay test species themselves to the mesocosm communities (i.e., possible lack of relevance of a single-species test species).

Observational field exposure tests, by their nature, typically involve testing over long durations regardless of their initial setup expenses. However, the exposure of entire ecosystems in field studies means that these studies are frequently reactive rather than proactive. Similarly to mesocosm studies, the reliability of laboratory studies to predict effects in the field is also often inconsistent. Some comparisons have found underestimations of effects whilst others have found overestimations [37]. To this end, it highly desirable to validate single-species laboratory tests by studying effects in a functioning community [114]. While laboratory bioassays may be useful as initial screening tools, more comprehensive studies must form the basis of ecosystem management [109]. Consequently, it may not be sufficient to simply detect a change in a sensitive or "early detection" indicator, because such a change cannot easily be linked to prediction of a change at the population, community or ecosystem scale in the field. Instead, responses must be sought in the field from suitable surrogates for these higher scales of organisation and complexity [70] using, for example:

species richness, community composition or structure [9],

patterns of abundance and distribution of species of high conservation value or ecological significance [4],

physical, chemical or biological processes e.g., production:respiration ratios, primary production, energy flow pathways [14; 47; 15; 11; 84].

Many studies have used results from field surveys showing correlations of pollutants with measures of biotic community structure to determine a toxicant's scale of effect [63; 1; 115]. Nevertheless, one of the great difficulties in evaluating ecosystem-scale risks is replicating treatments adequately enough to account for high inter-treatment variability. Every water body serving as a replicate is also different in many more ways than just the treatment in question e.g., physical morphometrics, micro-climatic patterns, water retention time etc. [109].

Such field assessments, although extremely useful in determining site-specific impacts, are still frequently limited through a lack of proper experimental control including too few or poorly-positioned reference sites and confounding effects from impacts unrelated to the disturbance of concern [66]. Consequently, the scale of biological organisation producing the most ecosystem-relevant dataset (e.g., biological communities) conversely also provides the least reliable database for regulators [83; 91]. Nonetheless, although there is a need to trade-off between test simplicity, costs and environmental relevance; relevance may still warrant greatest consideration in many pollutant studies and broader risk assessments [39]. Furthermore, in addition to having more ecological relevance, evidence also suggests that pollutant studies that address biotic effects at higher-scales of ecological organisation are at least as powerful in detecting the biological effects of pollutants as are single-species approaches [19; 90]. This is especially so when multivariate analyses are applied to these community data [39].

## MULTIPLE SCALES OF EVIDENCE STUDIES

As discussed, the failure of smaller-scale, single-species bioassay designs to address all the complexities of actual receiving ecosystems has long been recognised [20; 99; 50]. Large-scale system behaviour cannot generally be predicted from individual sub-units [83] and, indeed, in many situations it may be that laboratory-based ecotoxicological testing measures the *wrong* variables *more precisely* than less repeatable and replicable (but more environmentally realistic) field-based experiments [26]. For example, although population abundance and growth may be measured, this measurement is made in the absence of many factors that limit abundance such as direct predation or competition and consequent resource limitation. Ecological field studies are sometimes initiated when laboratory tests have indicated the existence of a potential risk [79]. An integrated assessment approach has been especially recommended for situations where the effect of the toxicant is subtle [43]. The integration of different scales of organisation in this manner may provide complementary information and, ultimately, a better understanding of both the scales at which a stressor is likely to affect a community and in what manner this stress will reveal itself [43].

Ecotoxicological assessment of aquatic pollution is already, and needs to be more widely recognised, as a multi-disciplinary subject [8]. As such, there is always scope for criticisms of particular individual discipline's methods; all have different advantages and disadvantages, and consequently there is a need for multiple approaches to an ecosystem-scale toxicity assessment [80]. Such a more holistic ecotoxicological approach would ideally employ laboratory and field-based approaches as complementary methods. What is important is that their limitations and context are realised and accounted for. Ideally, ecotoxicological assessments should include endpoints from many different scales ranging from cellular and physiological processes at the individual level through to ecosystem changes such as functional feeding group relationships and food web structure [12] (Figure 7).

Although numerous endeavours have been made, there still appears to be no parsimonious way of marrying data from very different scales of study together into a single holistic trigger value derivation, without significant loss of information specific to each scale. An example of such an holistic assessment method, the "weight-of-evidence" approach (also know as "multiple lines of evidence" or a "meta-analysis" approach) [40; 53] seeks concordance between controlled experimental findings and actual field results [117; 71]. The weight-of-evidence approach has been recommended where different types of site-specific data provide partial information on different aspects of a stressor's action at those sites [95]. The weight-of-evidence approach is achieved at its simplest through such testing at a variety of ecological scales [83]. The essence of this multi-scalar process is that, where necessary, toxicant information from water physico-chemistry, single-species tests, multi-species tests, and ecosystem processes (e.g., changes in trophic relationships as indicated by functional feeding groups) are all considered in the ecotoxicological assessment [83]. As such, a multi-scalar type weight-of-evidence approach has proven to be a reliable complex risk assessment strategy [127].

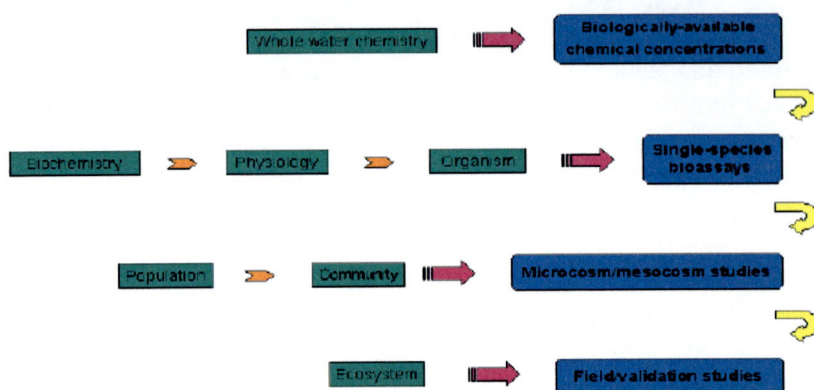

Figure 7. Increasing environmental relevance at increasing scale of complexity of pollution assessment typically undertaken and their relationship to each other in order.

## CONCLUSION

As a result of the continuing evolution of "ecotoxicology", the emphasis on extrapolation of laboratory results to field expectations is considered an increasingly important requirement [35; 69]. A large part of the change in emphasis is, although toxicological data derived from laboratory bioassays may provide repeatable results, they may still not provide an accurate assessment of actual receiving system ecotoxicity.

Inaccurate pollutant assessments may either fail to sufficiently protect receiving ecosystems or, conversely, may contribute to overly conservative guidelines which unnecessarily restrict industry more than is required for environmental protection [35; 30]. Consequently, a hegemony on the evolution of ecotoxicological science and practice is that the primary application of pollutant toxicity data is regulatory [20]. Scientific results from the various forms of ecotoxicological testing available are frequently held in higher esteem by regulatory bodies when they are more precise and repeatable. This is the case even if these data may not be relevant or sufficiently comprehensive for the receiving system of concern.

Given that the two activities of science and policy are distinct [20], some authors suggest that such pollutant toxicology data should not provide guidelines alone. Instead ecotoxicological data are best incorporated into a risk assessment framework which accounts for the confidence in quality and relevance of the data to indicate whether environmental harm will occur, and if so, the acceptability of this harm to society [64]. Formal risk assessments provide an ideal framework for integrating ecotoxicological and other data [16] as economic, political and sociological considerations are also important in ecological risk-management [68], but are often not considered in risk-assessment [107]. As a result, single-species toxicity tests have been, are, and most likely also will remain, the mainstay of toxicological assessment [4]. Many of the criticisms of the validity of single-species, laboratory-test data to pollutants in a receiving environment therefore relate more to the *regulatory* application of the science within its limitations, than to the way the science itself is executed [20; 62].

Risk predictions of environmental impacts arising from pollutants in aquatic environments are likely to remain an inexact science for some time yet. A broader criticism of single-species testing is therefore that it is relatively rare for a single line of evidence to reach a definitive risk conclusion [56]. Indeed, some authors consider that single-species studies alone cannot be used to assess environmental toxicant risk e.g., Joern and Hoagland [73]. While extrapolation of results from small-scale studies to larger scales may be sufficient for generic or screening assessment, small-scale study data on their own are likely to be unreliable for water bodies with site-specific and high-reliability requirements for environmental protection [36; 59]. Furthermore, there is a danger that these smaller-scale studies can easily become the over-interpreted object of study, rather than the toxicological issue they are designed to address [30; 36]. Studies at smaller, proxy scales to that of the receiving system in question often encourage a "Type III error" of the original definition *sensu* Kimball [82], where an irrelevant question is answered. In this type of error, the questions "is there an impact on the proxy?" and "what mode of action on the proxy does the toxicant take?", etc. become the sole questions of analysis, with little regards to the context of scale or management application which initiated them. Given that ecological systems do not have a single characteristic of scale, the context and relevance of toxicant guidelines are now considered more reliably derived from studies at more appropriate, higher scales [30].

Nevertheless, although differences between assessment scales are expected, the alternative approach of employing more relevant field-scale data is still rarely used to derive water quality guidelines for aquatic pollutants [102]. Some authors have even questioned whether existing field study design can give managers enough information about the risks of potential pollutants on which to base management decisions [103]. Nevertheless, on the rare occasion that they have been completed, studies that have combined data from laboratory tests with data from field observations have often been found to provide the greatest information about a toxicant and its likelihood of ecosystem-scale effects [118]. This greater understanding of a toxicant's risk increases both the scientific validity of the water quality criteria derivation process, and also improves the confidence for managers and stakeholders. Consequently, "validation" (*sensu* Cairns [19]) of single-species toxicity assessments of this sort (or from tests of similar lower levels of complexity) by high level assessments, remains the highest standard for ecotoxicological trigger value derivation.

# REFERENCES

[1]     Agard, JBR; J Gobin and R Warwick (1993). Analysis of marine macrobenthic community structure in relation to pollution, natural oil seepage and seasonal disturbance in a tropical environment (Trinidad, West Indies). *Marine Ecology Progress Series*. 92: 233-243.

[2]     Aloi, JE (1990). A critical review of recent freshwater periphyton field methods. Canadian *Journal of Fisheries and Aquatic Sciences*. 47: 656-670.

[3]     Anderson, SL (1991). Letter to the Editor: Precision of short-term chronic toxicity tests in the real world. *Environmental Toxicology and Chemistry*. 10: 143-145.

[4]     ANZECC/ARMCANZ (2000). Australian and New Zealand guidelines for fresh and marine water quality. National Water Quality Management Strategy Paper No 4.

Canberra, Australian and New Zealand Environment and Conservation Council and Agriculture and Resource Management Council of Australia and New Zealand.

[5]   Auer, CM; JV Nabholz and KP Baetcke (1990). Mode of action and the assessment of chemical hazards in the presence of limited data: use of structure activity relationships (SARS) under TSCA Section 5. *Environmental Health Perspectives.* 87: 183-197.

[6]   Bacher, GJ; JC Chapman and RP Lim (1992). The impact of agriculture on inland temperate waters: field validation. *Australian Biologist.* 5: 196-202.

[7]   Baird, DJ (1992). Predicting population response to pollutants; in praise of clones. A comment on Forbe and Depledge. *Functional Ecology.* 6: 616-617.

[8]   Bartell, SM (1997). Charlatan or Sage-A dichotomy of views on ecological risk assessment. *Environmental Management.* 21: 822-824.

[9]   Baskin, Y (1994). Ecosystem function of biodiversity. *Bioscience.* 44: 657-660.

[10]  Birge, WJ; JA Black; TM Short and AG Westerman (1989). A comparative ecological and toxicological investigation of a secondary wastewater treatment plant effluent and its receiving stream. *Environmental Toxicology and Chemistry.* 8: 437-450.

[11]  Boulton, AJ (2003). Parallels and contrasts in the effects of drought on stream macroinvertebrate assemblages. *Freshwater Biology.* 48: 1173-1185.

[12]  Bradbury, SP (1995). Ecological risk assessment for chemical stressors: challenges in predictive ecotoxicological research. *Australasian Journal of Ecotoxicology.* 1: 3-9.

[13]  Bunce, NJ and RBJ Remillard (2003). Haber's Rule: the search for quantitative relationships in toxicology. *Human and Ecological Risk Assessment.* 9: 1547-1559.

[14]  Bunn, SE (1995). Biological monitoring of water quality in Australia: workshop summary and future directions. *Australian Journal of Ecology.* 20: 220-227.

[15]  Bunn, SE and PM Davies (2000). Biological processes in running waters and their implications for the assessment of ecological integrity. *Hydrobiologia.* 422: 61-70.

[16]  Burgman, MA (2005). Risks and decisions for conservation and environmental management. Cambridge, UK, Cambridge University Press.

[17]  Cairns, J, Jr. (1983). Are single species toxicity tests alone adequate for estimating environmental hazard? *Hydrobiologia.* 100: 47-57.

[18]  Cairns, J, Jr. (1986). The myth of the most sensitive species. *Bioscience.* 36: 670-672.

[19]  Cairns, J, Jr. (1986). What is meant by validation of predictions based on laboratory toxicity tests? *Hydrobiologia.* 137: 271-278.

[20]  Cairns, J, Jr. (1988). Should regulatory criteria and standards be based on multispecies evidence? *Environmental Professional.* 10: 157-165.

[21]  Cairns, J, Jr. (1995). Future trends in ecotoxicology. Ecological toxicity testing: Scale, complexity and relevance. J. Cairns Jr and B. R. Niederlehner. Boca Raton, Florida, Lewis Publishing: 217-222.

[22]  Cairns, J, Jr. and BR Niederlehner (1987). Problems associated with selecting the most sensitive species for toxicity testing. *Hydrobiologia.* 153: 87-94.

[23]  Cairns, J, Jr. and JR Pratt (1986). On the relation between structural and functional analyses of ecosystems. *Environmental Toxicology and Chemistry.* 5: 785-786.

[24]  Calow, P (1992). Can ecosystems be healthy? Critical consideration of concepts. *Journal of Aquatic Ecosystem Health.* 1: 1-5.

[25]  Calow, P (1992). The three Rs of ecotoxicology. *Functional Ecology.* 6: 617-619.

[26]  Calow, P (1994). Ecotoxicology; what are we trying to protect? *Environmental Toxicology and Chemistry.* 13: 1549.

[27]  Calow, P (1995). Risk assessment: principles and practice in Europe. *Australasian Journal of Ecotoxicology.* 1: 11-13.

[28]  Cao, Y and DD Williams (1999). Rare species are important in bioassessment (reply to the comment by Marchant). *Limnology and Oceanography.* 44: 1841–1842.

[29]  Cao, Y; DD Williams and NE Williams (1999). How important are rare species in aquatic community ecology and bioassessment? *Limnology and Oceanography.* 43: 1403–1409.

[30]  Carpenter, SR (1996). Microcosm experiments have limited relevance for community and ecosystem ecology. *Ecology.* 77: 677-680.

[31]  Caux, P-Y and RA Kent (2001). Exploring future directions in environmental quality guideline development in Canada. *Australasian Journal of Ecotoxicology.* 7: 13-30.

[32]  Chapman, JC (1995). The role of ecotoxicity testing in assessing water quality. Australian *Journal of Ecology.* 20: 20–27.

[33]  Chapman, PM (1990). The Sediment Quality Triad approach to determining pollution-induced degradation. *Science of the Total Environment.* 98/98: 815-825.

[34]  Chapman, PM (1995). Bioassay testing for Australia as part of water quality assessment programmes. *Australian Journal of Ecology.* 20: 7-19.

[35]  Chapman, PM (1995). Ecotoxicology and pollution - key issues. *Marine Pollution Bulletin.* 31: 167-177.

[36]  Chapman, PM (2002). Integrating toxicology and ecology: putting the "eco" back into ecotoxicology. *Marine Pollution Bulletin.* 44: 7-15.

[37]  Chapman, PM; A Fairbrother and D Brown (1998). A critical evaluation of safety (uncertainty) factors for ecological risk assessment. *Environmental Toxicology and Chemistry.* 99–108.

[38]  Clark, JS; T Hussey and PD Royall (1996). Presettlement analogs for Quarternary fire regimes in eastern North America. *Journal of Paleolimnology.* 16: 76-96.

[39]  Clarke, KR (1999). Non-metric multivariate analysis of changes in community-level ecotoxicology. *Environmental Toxicology and Chemistry.* 18: 118-127.

[40]  Clarke, KR and RM Warwick (2001). Change in marine communities: an approach to statistical analysis and interpretation. Plymouth, Plymouth Marine Laboratory.

[41]  Clements, FE (1916). Plant succession: an analysis of the development of vegetation. Washington D.C., Carnegie Institution of Washington: 512.

[42]  Clements, WH (1994). Editorial: Assessing contaminant effects at higher levels of biological organisation. *Environmental Toxicology and Chemistry.* 13: 357-359.

[43]  Clements, WH and PM Kiffney (1994). Integrated laboratory and field approach for assessing impacts of heavy metals at the Arkansas River, Colorado. *Environmental Toxicology and Chemistry.* 13: 397-404.

[44]  Connell, D; P Lam; B Richardson and R Wu (1999). Introduction to Ecotoxicology. Oxford, United Kingdom, Blackwell Science.

[45]  Connell, JH (1978). Diversity in tropical rainforests and coral reefs. Science 199: 1302-1310.

[46]  Davic, RD (2003). Linking keystone species and functional groups: a new operational definition of the keystone species concept. *Conservation Ecology.* 7: r11.

[47]  Davies, PM (1997). Assessment of river health by the measurement of community metabolism. Canberra.

[48]    de March, BGE (1988). Acute toxicity of binary mixtures of five cations ($Cu^{2+}$, $Cd^{2+}$, $Zn^{2+}$, $Mg^{2+}$, and $K^+$) to the freshwater amphipod *Gammarus lacustris* (Sars): alternative descriptive models. *Canadian Journal of Fisheries and Aquatic Sciences.* 45: 625-633.

[49]    Deichmann, WB; D Henschler; B Holstedt and G Keil (1986). What is there that is not a poison? *Archives of Toxicology.* 58: 207-213.

[50]    deVlaming, V and T Norberg-King (1997). A review of single species toxicity tests: are the tests reliable predictors of aquatic ecosystem community responses?, U.S. Environmental Protection Agency Office of Research and Development.

[51]    Dickson, KL; WT Waller; JH Kennedy and LP Ammann (1992). Assessing the relationship between ambient toxicity and instream biological response. *Environmental Toxicology and Chemistry.* 11: 1307-1322.

[52]    Dixit, SS; JP Smol and JC Kingston (1992). Diatoms: powerful indicators of environmental change. *Environmental Science and Technology.* 26: 23-33.

[53]    Downes, BJ; LA Barmuta; PG Fairweather; DP Faith; MJ Keough; PS Lake; BD Mapstone and GP Quinn (2002). Monitoring ecological impacts: concepts and practice in flowing waters. Cambridge, Cambridge University Press.

[54]    Eagleson, KW; DL Lenat; LW Ausley and FB Winborne (1990). Comparison of measured instream biological responses with response predicted using the *Ceriodaphnia dubia* chronic response toxicity test. *Environmental Toxicology and Chemistry.* 9: 1019-1028.

[55]    Elton, CS (1958). The ecology of invasions by plants and animals. London, UK, Methuen.

[56]    Fairbrother, A (2003). Lines of evidence in wildlife risk assessments. *Human and Ecological Risk Assessment.* 9: 1475-1491.

[57]    Fernando, CH (1994). Zooplankton, fish and fisheries in tropical waters. *Hydrobiologia.* 272: 105-123.

[58]    Forbes, VE and MH Depledge (1992). Predicting population response to pollutants: the significance of sex. *Functional Ecology.* 6: 376-381.

[59]    Freckleton, RP (2004). The problems of prediction and scale in applied ecology: the example of fire as a management tool. *Journal of Applied Ecology.* 41: 599-603.

[60]    Futuyma, DJ (1998). Evolutionary Biology. Sunderland, USA, Sinauer Associates, Inc.

[61]    Grant, BF (1976). Endrin toxicity and distribution in freshwater: a review. *Bulletin of Environmental Contamination and Toxicology.* 15: 283-290.

[62]    Gray, JS (1994). Science and the environment. *Marine Pollution Bulletin.* 28: 270-271.

[63]    Gray, JS; KR Clarke; R Warwick and G Hobbs (1990). Detection of initial effects of pollution on marine benthos: an example from the Ekofisk and Eldfisk oilfields, North Sea. *Marine Ecology Progress Series.* 66: 285-299.

[64]    Hart, BT; MA Burgman; M Grace; C Pollino; C Thomas and JA Webb (2006). Risk-based approaches to managing contaminants in catchments. *Human and Ecological Risk Assessment.* 12: 66-73.

[65]    Havens, KE; LA Bull; GL Warren; TL Crisman; EJ Phlips and JP Smith (1996). Food web structure in a subtropical lake ecosystem. *Oikos.* 75: 20-32.

[66] Hickey, CW and E Pyle (2001). Derivation of water quality guideline values heavy metals using a risk-based methodology: a site specific approach for New Zealand. *Australasian Journal of Ecotoxicology.* 7: 137-156.

[67] Holdway, DA (1997). Truth and validation in ecological risk assessment. *Environmental Management.* 21: 816-819.

[68] Horwitz, P and P Nichols (2002). Located toxicology: the need for alternative methodologies to address toxicological significance. *Australasian Journal of Ecotoxicology.* 8: 45-50.

[69] Hose, GC and PJ Van den Brink (2004). Confirming the species-sensitivity distribution concept for Endosulfan using laboratory, mesocosm and field data. *Archives of Environmental Contamination and Toxicology.* 47: 511-520.

[70] Humphrey, CL; DP Faith and PL Dostine (1995). Baseline requirements for assessment of mining impact using biological monitoring. *Australian Journal of Ecology.* 20: 150–166.

[71] Humphrey, CL; L Thurtell; RWJ Pidgeon; RA van Dam and CM Finlayson (1999). A model for assessing the health of Kakadu's streams. *Australian Biologist.* 12: 33-42.

[72] Jeffrey, SW; SW Wright and M Zapata (1999). Recent advances in HPLC pigment analysis of phytoplankton. *Marine and Freshwater Research.* 50: 879-896.

[73] Joern, A and KD Hoagland (1996). In defense of whole-community bioassays for risk assessment. *Environmental Toxicology and Chemistry.* 15: 407-409.

[74] John, J (1993). The use of diatoms in monitoring the development of created wetlands at a sand-mining site in Western Australia. *Hydrobiologia.* 269/270: 427-436.

[75] John, J (2000). A guide to diatoms as indicators of urban stream health. Perth, Curtin University of Technology.

[76] Karr, JR and EW Chu (1997). Biological monitoring: essential foundation for ecological risk assessment. *Human and Ecological Risk Assessment.* 3: 993-1004.

[77] Karr, JR and EW Chu (1999). Restoring life in running waters: better biological monitoring. Washington, D.C., Island Press.

[78] Kedwards, TJ; SJ Maund and PF Chapman (1999). Community level analysis of ecotoxicological field studies: I. Biological monitoring. *Environmental Toxicology and Chemistry.* 18: 149-157.

[79] Kedwards, TJ; SJ Maund and PF Chapman (1999). Community level analysis of ecotoxicological field studies: II. Replicated-design studies. *Environmental Toxicology and Chemistry.* 18: 158-171.

[80] Kefford, BJ; PJ Papas; D Crowther and D Nugegoda (2002). Are salts toxicants. *Australasian Journal of Ecotoxicology.* 8: 63-68.

[81] Kefford, BJ; PJ Papas; L Metzeling and D Nugegoda (2004). Do laboratory salinity tolerances of freshwater animals correspond with their field salinity? *Environmental Pollution.* 129: 355-362.

[82] Kimball, AW (1957). Errors of the third kind in statistical consulting. *Journal of the American Statistician Association.* 52: 133-142.

[83] Kimball, KD and SA Levin (1985). Limitations of laboratory bioassays: the need for ecosystem-level testing. *Bioscience.* 35: 165-171.

[84] Kremen, C (2005). Managing ecosystem services: what do we need to know about their ecology? *Ecology Letters.* 8: 468-479.

[85]   Lawrence, JR and MJ Hendry (1995). Mesocosms for subsurface research. *Water Quality Research Journal of Canada*. 30: 493-512.

[86]   Lewis, W, M. Jr.; SK Hamilton; MA Rodríguez; J Saunders, F. III and MA Lasi (2001). Foodweb analysis of the Orinoco floodplain based on production estimates and stable isotope data. *Journal of the North American Benthological Society*. 20: 241-254.

[87]   Luoma, SN (1977). Detection of trace contaminant effects in aquatic ecosystems. *Journal of Fisheries Research Board Canada*. 34: 436-439.

[88]   Maltby, L and P Calow (1989). The application of bioassays in the resolution of environmental problems: past, present and future. *Hydrobiologia*. 188/189: 65-66.

[89]   Marchant, R (1999). How important are rare species in aquatic community ecology and bioassessment: A comment on the conclusions of Cao et al. Limnology, *Oceanography*. 44: 1840–1841.

[90]   Maund, SJ; PF Chapman; TJ Kedwards; L Tattersfield; P Matthiessen; R Warwick and E Smith (1999). Editorial: Application of multivariate statistics to ecotoxicological field studies. *Environmental Toxicology and Chemistry*. 18: 111-112.

[91]   McArdle, BH; KJ Gaston and JH Lawton (1990). Variation in the size of animal populations: patterns, problems and artefacts. *Journal of Animal Ecology*. 59: 439-454.

[92]   McCullough, CD; AC Hogan; CL Humphrey; RA van Dam and MM Douglas (in press). Failure of *Hydra* populations to develop tolerance, indicates absence of toxicity from a whole-effluent. Proceedings of the 30th Congress of the International Association of Theoretical and Applied Limnology Montreal, Canada.

[93]   McPherson, CA and PM Chapman (2000). Copper effects on potential sediment test organisms: the importance of appropriate sensitivity. *Marine Pollution Bulletin*. 40: 656-665.

[94]   Mentis, M (1988). Hypothetico-deductive and inductive approaches in ecology. *Functional Ecology*. 2: 5-14.

[95]   Menzie, C; MH Henning; J Cura; K Finkelstein; J Gentile; J Maughan; D Mitchell; S Petron; B Potocki; S Svirsky and PA Tyler (1996). Special report of the Massachusetts Weight-of-Evidence Workgroup: a weight-of-evidence approach to evaluating ecological risks. *Human and Ecological Risk Assessment*. 2: 277-304.

[96]   Millspaugh, SH and C Whitlock (1995). A 750 year fire history based upon lake sediment records in central Yellowstone National Park. *Holocene*. 5: 283-292.

[97]   Millward, RN and A Grant (1995). Assessing the impact of copper on nematode communities from a chronically metal-enriched estuary using pollution-induced community tolerance. *Marine Pollution Bulletin*. 30: 701-706.

[98]   Odum, EP (1984). The mesocosm. *Bioscience*. 35: 419-422.

[99]   Parkhurst, BR (1995). Are single species toxicity test results valid indicators of effects to aquatic communities. Ecotoxicological testing: scale, complexity and relevance. J. Cairns Jr and B. R. Niederlehner. Boca Raton, Florida, Lewis Publishing: 105-121.

[100]  Parliamentary Commissioner for the Environment. (2006). "Glossary." Retrieved July, 2006, from http://www.pce.govt.nz/reports/pce_reports_glossary.shtml#p.

[101]  Pimm, SL (1979). Complexity and stability: another look at MacArthur's original hypothesis. Oikos 33: 351-357.

[102]   Pollino, C and BT Hart (2005). Bayesian approaches can help make better sense of ecotoxicological information in risk assessments. *Australasian Journal of Ecotoxicology.* 11: 57-58.

[103]   Power, M and LS McCarty (1997). Fallacies in ecological risk assessment practices. *Environmental Science and Technology.* 31: 370-375.

[104]   Reid, MA; JC Tibby; Penny and PA Gell (1995). The use of diatoms to assess past and present water quality. *Australian Journal of Ecology.* 20: 57–64.

[105]   Resh, VH and JK Jackson (1993). Rapid assessment approaches to biomonitoring using benthic macroinvertebrate. Freshwater biomonitoring and benthic macroinvertebrates. D. M. Rosenberg and V. H. Resh. New York, Capman and Hall: 195-233.

[106]   Resh, VH; RH Norris and MT Barbour (1995). Design and implementation of rapid assessment approaches for water resource monitoring using benthic macroinvertebrates. *Australian Journal of Ecology.* 20: 108-121.

[107]   Roelofs, W; WAJ Huijbregts; T Jager and AMJ Ragas (2003). Prediction of ecological no-effect concentrations for initial risk assessment: combining substance-specific data and database information. *Environmental Toxicology and Chemistry.* 22: 1387-1393.

[108]   Rosenberg, DM and VH Resh, Eds. (1993). Freshwater biomonitoring and benthic macroinvertebrates. New York, Chapman and Hall.

[109]   Schindler, DW (1987). Detecting ecosystem responses to anthropogenic stress. Canadian *Journal of Fisheries and Aquatic Sciences.* 44: 6–25.

[110]   Schmidt, K; M Koski; J Engström-öst and A Atkinson (2002). Development of Baltic Sea zooplankton in the presence of a toxic cyanobacterium: a mesocosm approach. *Journal of Plankton Research.* 24: 979-992.

[111]   Schofield, NJ and PE Davies (1996). Measuring the health of our rivers. Water May-June 1996: 39-43.

[112]   Schrader-Frechette, KS and ED McCoy (1993). Method in ecology: strategies for conservation. London, UK, Cambridge University Press.

[113]   Snyder, TP; KM Switzer and RE Keen (1991). Allozymic variability in toxicity-testing strains of *Ceriodaphnia dubia* and in natural populations of *Ceriodaphnia.* *Environmental Toxicology and Chemistry.* 10: 1045-1049.

[114]   Sprague, JB (1990). Aquatic toxicology. Methods for fish biology. C. B. Schreck and P. B. Moyle. Bethesda, USA, American Fisheries Society: 491-528.

[115]   Stark, JS (1998). Heavy metal pollution and macrobenthic assemblages in soft sediments in two Sydney estuaries, Australia. *Marine and Freshwater Research.* 49: 533-540.

[116]   Stauber, JL (1995). Toxicity testing using marine and freshwater unicellular algae. *Australasian Journal of Ecotoxicology.* 1: 15-24.

[117]   Suter II, GW (1996). Abuse of hypothesis testing statistics in ecological risk assessment. *Human and Ecological Risk Assessment.* 2: 331-347.

[118]   Thompson, SA and GG Thompson (2004). Adequacy of rehabilitation monitoring practices in the Western Australian mining industry. *Ecological Management and Restoration.* 5: 30-31.

[119]   Tilman, D (1999). The ecological consequences of changes in biodiversity: a search for general principles. *Ecology.* 18: 1455-1474.

[120]   Tsirtsis, G and M Karydis (1997). Aquatic microcosms: a methodological approach for the quantification of eutrophication processes. *Environmental Monitoring and Assessment.* 48: 193-215.

[121]   van Dam, RA; C Camilleri and CM Finlayson (1998). The potential of rapid assessment techniques as early warning indicators of wetland degradation: a review. *Environmental Toxicology and Water Quality.* 13: 297-312.

[122]   van Dam, RA and JC Chapman (2001). Direct toxicity assessment (DTA) for water quality monitoring guidelines in Australia and New Zealand. *Australasian Journal of Ecotoxicology.* 7: 175-198.

[123]   Van den Brink, PJ; J Hattink; F Bransen; E Van Donk and TCM Brock (2000). Impact of the fungicide carendazim in freshwater microcosms. II. Zooplankton, primary producers and final conclusions. *Aquatic Toxicology.* 48: 251-264.

[124]   Wall, TM and RW Hanmer (1987). Biological testing to control toxic water pollutants. *Journal of the Water Pollution Control Federation.* 59: 7-12.

[125]   Ward, JV and K Tockner (2001). Biodiversity: towards a unifying theme for river ecology. *Freshwater Biology.* 46: 807-819.

[126]   Warne, MS (1998). Critical review of methods to derive water quality guidelines for toxicants and a proposal for a new framework. Canberra, Supervising Scientist.

[127]   Wickwire, WT and C Menzie, A. (2003). New approaches in ecological risk assessment: expanding scales, increasing realism, and enhancing causal analysis. *Human and Ecological Risk Assessment.* 9: 1411-1414.

[128]   Wu, R (1996). Editorial: Ecotoxicology: problems and challenges in Australasia. Australasian *Journal of Ecotoxicology.* 2: 1.

[129]   Reviewed by Associate Professor Mark Lund, School of Natural Sciences, Edith Cowan University, Australia.

In: Aquatic Ecosystem Research Trends                    ISBN 978-1-60692-772-4
Editor: George H. Nairne                          © 2009 Nova Science Publishers, Inc.

*Chapter 12*

# PECULIARITIES OF RADIONUCLIDE DISTRIBUTION IN THE MAIN COMPONENTS OF AQUATIC ECOSYSTEMS WITHIN THE CHERNOBYL ACCIDENT EXCLUSION ZONE

*D. I. Gudkov[1]\*, S. I. Kireev[2], M. I. Kuzmenko[1], A. B. Nazarov[2], A. E. Kaglyan[1], L. N. Zub[3] and V. G. Klenus[1]*

[1] Institute of Hydrobiology, National Academy of Sciences of Ukraine, Kiev
[2] State Specialised Scientific and Production Enterprise
"Chernobyl Radioecological Centre",
Ukraine Ministry for Emergency Situations, Chernobyl
[3] I.I. Shmal'gauzen Institute of Zoology,
National Academy of Sciences of Ukraine, Kiev

## ABSTRACT

The investigation results for dynamics of $^{90}$Sr, $^{137}$Cs, $^{238}$Pu, $^{239+240}$Pu and $^{241}$Am content and distribution in components of the aquatic ecosystem located in the Chernobyl accident exclusion zone. The main data massif was obtained in the period of 1992 – 2004. Specific radionuclide activity data are present for bottom sediments, water, seston, macrozoobenthos (including bivalve molluscs), gastropods, higher aquatic plants, and fishes. The species specificity of radionuclide concentration by hydrobionts is studied, and the role of various groups of aquatic organisms in distribution of radioactive substances by the main components of the lake biocoenoses is estimated.

**Keywords:** $^{90}$Sr, $^{137}$Cs, $^{238}$Pu, $^{239+240}$Pu, $^{241}$Am, the Chernobyl accident exclusion zone, freshwater ecosystems, bottom sediments, seston, macrozoobenthos, gastropods, higher aquatic plants, fishes

---

\* Main addressee: 12 Geroyev Stalingrada Avenue, 04210 Kiev, Ukraine; Institute of Hydrobiology, National Academy of Sciences of Ukraine, Department of Freshwater Radioecology; Tel: (044) 419-8437; Fax: (044) 418-2232; E-mail: digudkov@svitonline.com

Radioactive substances emitted to the atmosphere during the Chernobyl accident have hit the areas of watersheds and precipitated directly on the surface of water objects, in which radioactive fission fragments and neutron activation products were partly dissolved, where chemically interacted with natural water and mineral organic components in it. In accordance with geochemical subordination in the sequence of elementary landscapes connected by migration flows of the substances, aquatic landscapes represent peculiar "receivers" of the majority of chemical compounds, including radionuclides, delivered by the overland discharge to the hydrological network and secluded reservoirs, located in internal-drainage lowlands. The latter reservoirs retain radioactive substances for long, where they distribute and migrate by the components of aquatic ecosystems.

Discussed in the present review are features of concentration and distribution dynamics for $^{90}$Sr, $^{137}$Cs, $^{238}$Pu, $^{239+240}$Pu and $^{241}$Am in aquatic ecosystems of the Chernobyl accident exclusion zone possessing different hydrological conditions and levels of radionuclide contamination. The species specificity of radionuclide concentrating in hydrobionts and the values of various groups of aquatic organisms at radioactive substance distribution in hydrobiocoenosis are analyzed.

## MATERIALS AND METHODS

The investigations were performed in the period of 1989 – 2004 in Azbuchin Lake, Yanovsky (Pripyatsky) crawl, Chernobyl NPP cooling pond, Pripyat' River left-bank flood plain lakes – Glubokoe and Dalekoe-1, and rivers Uzh (Cherevach village) and Pripyat' (Chernobyl town) (Figure 1). Water samples (20 liters each) were taken from the surface layers of water reservoirs. Seston (small plankton-forming organisms and inorganic and organic particles suspended in water) were filtered by a "blue belt" filter. Bottom sediments were sampled in 1995 – 1999, in winter, from ice with 20 – 70 m step using sampling unit PD-100. Sampling points were adjusted and fixed with the help of surveying tools.

Radionuclide concentration in tissues of biological objects was determined and analyzed for 28 species of higher aquatic plants, 6 species of mollusks and 18 species of fishes. Floristic and geobotanic descriptions, as well as sampling of macrophytes, mollusks and macrozoobentos for the purpose of estimating biomass and determining radionuclide content were performed using methods, described in refs. [1 – 5]. The results were measured in Bq/kg at natural humidity.

The radionuclide distribution by components of lake ecosystems was estimated by measuring specific activity of radionuclides in water, seston and hydrobionts, as well as using quantitative data on biomass obtained in summer periods (late July – August) of 1998 – 1999. Currently shown parameters of radioactive substance concentration in hydrobionts (concentration factor – CF) were calculated from specific radionuclide activity in tissues related to their average specific activity in habitat waters within one year for aquatic animal organisms and within the vegetation period for higher aquatic plants. The mistake of estimated radionuclide concentration in the ecosystem components fell within 15 – 20%.

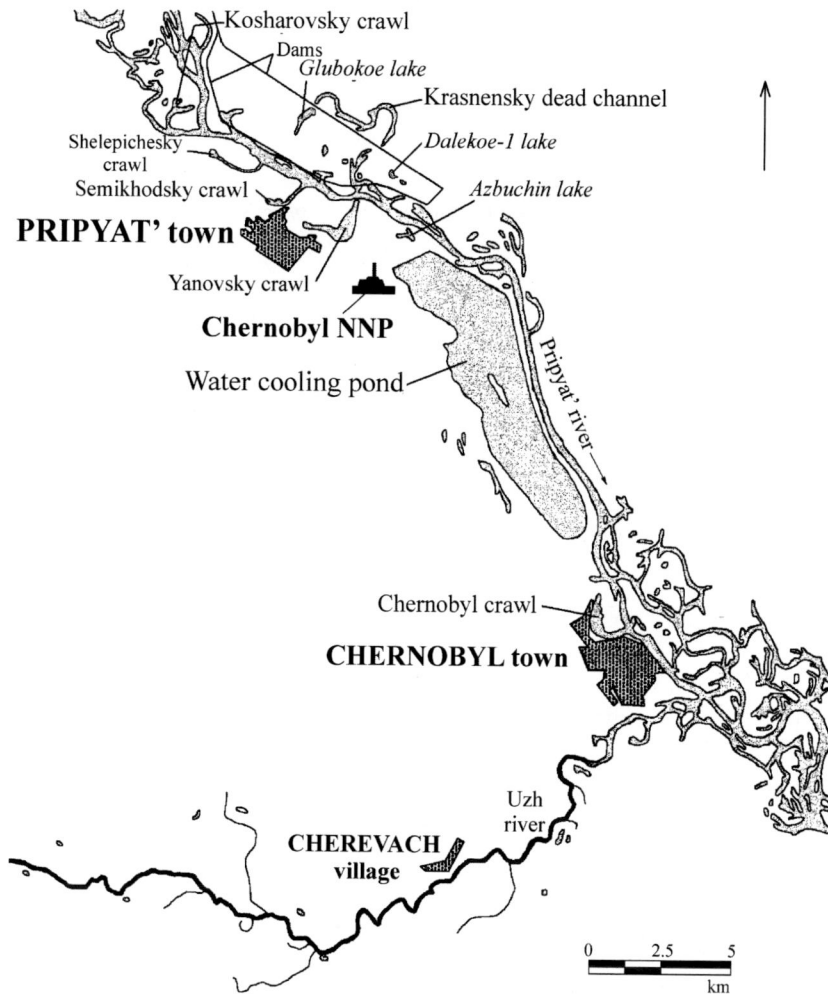

Figure 1. The map of main water bodies in the Chernobyl NPP exclusion zone

## RESULTS AND DISCUSSION

The modern level and composition of radionuclide contamination of aquatic ecosystems in the Chernobyl accident exclusion zone is primarily provided by the intensity of radioactive substance delivery to a watershed area and hydrodynamic processes proceeding in water reservoirs. Hence, transformation of radionuclides in soils of the watershed areas and bottom sediments of water reservoirs, and their migration with water flows are of special importance.

The lowest specific radioactive activity is observed in river ecosystems, where bottom sediments were subject to natural decontamination (specifically, in high waters and spring floods) and, therefore, during years passed since the accident, these factors became insignificant as the secondary source of channel flow contamination. At present, most radionuclides are delivered to rivers from overflooding and inflow from higher contaminated water objects. At the same time, closed water reservoirs and, especially, lakes located in the nearest exclusion zone, possess considerably higher radiation levels, provided by the water exchange limitation and comparatively high concentration of radionuclides, accumulated in bottom sediments. Hence, the concentration of radionuclides, determined for water of the majority of closed basins, is defined by intensity of mobile radionuclide form exchange between bottom sediments and water masses and, as well, external washout from the bailing area. Therefore, flooded landscapes of Pripyat' River become of the greatest significance. Impacted by the Chernobyl accident, these areas were seriously contaminated with radionuclides. Currently, they are the largest source of radionuclides, delivered with the overland flow to river systems of the exclusion zone, then connected to the Dnepr basin.

## Water

Maximal values of total β-activity of water in the mouth of Pripyat' river were registered during initial two weeks after the Chernobyl accident, reaching 10 kBq/l [6]. In this period, the main contribution was made by $^{131}$I, which gave up to 70 – 90% of specific activity. By middle of May, 1986, after the end of aerosol emission, total radioactivity of water in Pripyat' river reduced to several thousand and then in June to several hundred Bq/l. In June, the contribution of $^{131}$I into total radioactivity of water was below 30%. Since the end of October 1986 and till the beginning of 1987 total radioactivity of water in Pripyat' river infrequently exceeded 40 Bq/l. Hence, the main contribution into radionuclide contamination of aquatic objects was made by $^{90}$Sr and $^{137}$Cs radionuclides. Since 1988, the fraction of $^{90}$Sr in total radioactivity of river water was significantly higher at the background of reduction of $^{137}$Cs specific activity. At present, the ratio of these radionuclides in water bodies located in the Chernobyl NPP exclusion zone is significantly different. For Uzh and Pripyat' rivers the average $^{90}$Sr specific activity is 2.5 – 4-fold higher than that of $^{137}$Cs, whereas in closed lakes this parameter may reach the level of 20-fold higher specific activity. Among water bodies located in the Chernobyl NPP exclusion zone, the only exclusion is presented by the cooling pond of the Chernobyl NPP, in which $^{137}$Cs concentration in water exceeds that of $^{90}$Sr. As suggested, this is related to predominant delivery of high $^{137}$Cs concentrations with contaminated reactor waters to the cooling pond during the accident period via discharge channel, but not atmospheric precipitations [7].

For all water objects, studied since the early 1990, located in the Chernobyl accident exclusion zone, the general tendency in variations of radionuclide concentration in water of these objects is a decrease of $^{90}$Sr and $^{137}$Cs specific activity, which dynamics is firstly related to intensity of water exchange processes, with the only exclusion for closed water bodies of the left-bank flood plain of Pripyat' river, located at the territory of dammed area of Krasnenskaya flood plain, where recently, at the background of $^{137}$Cs specific activity stabilization, a tendency to $^{90}$Sr concentration increase was observed.

At present, the mean annual concentration in waters of test site water bodies is recorded in the wide ranges: from 0.11 Bq/l in Uzh river to 135 Bq/l or more in Glubokoe lake for $^{90}$Sr and from 0.05 Bq/l in Pripyat' river to 8 Bq/l in Azbuchin lake for $^{137}$Cs (Table 1). For the period of 1998 – 2003, the specific activity of transuranic elements, observed in water of some water objects is shown in Table 2.

**Table 1. Dynamics of the mean annual concentration of radionuclides in water of some tested water bodies in the Chernobyl NPP exclusion zone, Bq/l**

| Year | Cooling pond | | Azbuchin lake | | Glubokoe lake | | Dalekoe-1 lake | | Pripyat' river (Chernobyl town) | | Uzh river (Cherevach vil.) | |
|---|---|---|---|---|---|---|---|---|---|---|---|---|
| | $^{90}Sr$ | $^{137}Cs$ | $^{90}Sr$ | $^{137}Cs$ | $^{90}Sr$ | $^{137}Cs$ | $^{90}Sr$ | $^{137}Cs$ | $^{90}Sr$ | $^{137}Cs$ | $^{90}Sr$ | $^{137}Cs$ |
| 1992 | 4.8 | 5.2 | –* | – | 333 | 16 | – | – | 0.44 | 0.21 | 0.35 | 0.22 |
| 1993 | 4.4 | 6.3 | – | – | 281 | 18 | – | – | 0.85 | 0.21 | 0.81 | 0.20 |
| 1994 | 3.4 | 5.6 | – | – | 215 | 15 | – | – | 0.93 | 0.20 | 0.56 | 0.17 |
| 1995 | 3.2 | 4.8 | 229 | 22 | 159 | 11 | – | – | 0.33 | 0.11 | 0.31 | 0.13 |
| 1996 | 2.7 | 4.1 | 85 | 15 | 137 | 14 | – | – | 0.34 | 0.13 | 0.34 | 0.13 |
| 1997 | 2.2 | 2.8 | 85 | 13 | 100 | 13 | 45 | 4.5 | 0.25 | 0.16 | 0.26 | 0.08 |
| 1998 | 1.8 | 3.1 | 120 | 17 | 120 | 14 | 50 | 3.4 | 0.30 | 0.14 | 0.32 | 0.14 |
| 1999 | 1.9 | 3.1 | 190 | 23 | 120 | 14 | 45 | 2.8 | 0.50 | 0.15 | 0.25 | 0.10 |
| 2000 | 1.7 | 2.7 | 133 | 13 | 103 | 8 | 48 | 1.7 | 0.22 | 0.11 | 0.16 | 0.10 |
| 2001 | 1.5 | 2.1 | 110 | 10 | 79 | 7 | 35 | 2.6 | 0.23 | 0.12 | 0.18 | 0.09 |
| 2002 | 1.4 | 2.5 | 52 | 6 | 74 | 8 | 29 | 2.0 | 0.17 | 0.08 | 0.08 | 0.07 |
| 2003 | 1.7 | 2.4 | 49 | 8 | 102 | 7 | 40 | 2.3 | 0.15 | 0.05 | 0.11 | 0.06 |
| 2004 | 1.6 | 2.1 | 56 | 7 | 135 | 7 | 55 | 2.2 | 0.18 | 0.05 | 0.17 | 0.06 |

* – no measurements were performed

**Table 2. The average specific activity of transuranic elements in water of some water bodies in the exclusion zone in the period of 1998 – 2003, Bq/l**

| Water object | $^{238}Pu$ | $^{239+240}Pu$ | $^{241}Am$ |
|---|---|---|---|
| Glubokoe lake | 0.0050 | 0.0107 | 0.0061 |
| Dalekoe-1 lake | 0.0025 | 0.0068 | 0.0042 |
| Chernobyl NPP cooling pond | 0.00025 | 0.00071 | 0.00027 |
| Pripyat' river (Chernobyl) | 0.00017 | 0.00022 | –* |

* – no measurements were performed

Enhanced $^{90}Sr$ migration activity in the soils of contaminated areas comparing with that for $^{137}Cs$ provides for higher $^{90}Sr$ delivery to the surface discharge to river networks and subsequent withdrawal outside the exclusion zone. Hence, firstly, $^{90}Sr$ discharge is defined by the annual discharge of river networks (Figure 2). In 2003, over 70% of $^{90}Sr$ discharged by Pripyat' river, was delivered to it by the surface runoff from the areas of the Chernobyl accident exclusion zone. So far as concerns $^{137}Cs$, in recent years, its specific activity in Pripyat' river upstream the exclusion zone equaled that in the aligning of Chernobyl town. This testifies about that the most of $^{137}Cs$ is in low-soluble form. It is strongly fixed by soils that does not provide for its intensive discharge from drainage basin surfaces and occurrence in water flows. Totally, in the period of 1986 – 2004 Pripyat' river has discharged doses equal 158 TBq for $^{90}Sr$ and 126 TBq for $^{137}Cs$.

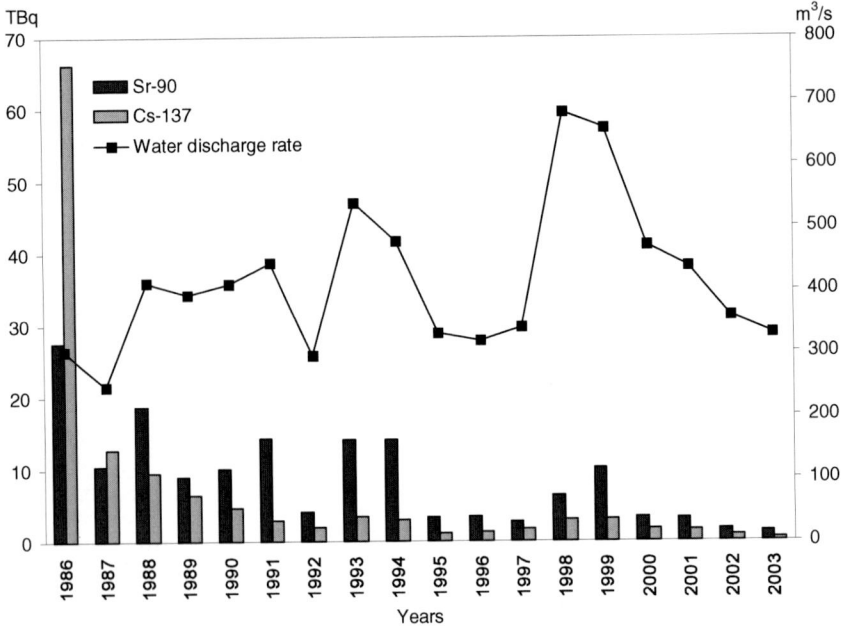

Figure 2. Radionuclides $^{90}$Sr and $^{137}$Cs withdrawal with Pripyat' river waters in through Chernobyl town section

## Bottom Sediments

The main water objects of the exclusion zone, in bottom sediments of which a noticeable quantity of radioactive substances is deposited, are closed or low flowage water bodies – lakes, cutoff meanders, crawls, abandoned irrigation channels, etc. Many crawls and cutoff meanders display so high contamination level of bottom sediments that to prevent radionuclide drainage during flood runoffs and spring tides were staked off from the truck of Pripyat' river by dead filling dams. Determinations of bottom sediment contamination level in dead water bodies, performed in the period of 1995 – 1999, allowed for specifying morphometry of the objects under study, their flat water square and water and water mass volume, as well as calculating radionuclide contamination density in bottom sediments (Table 3).

Among studied water objects, maximal contamination densities in bottom sediments were detected in Yanovsky (Pripyatsky) crawl, 2.5 km distant from the epicenter and on the formation way of the Northwest radioactive trail. In the current water body, localities with radionuclide concentration above 36 TBq/km$^2$ for $^{90}$Sr and $^{137}$Cs and above 1 TBq/km$^2$ for $^{238+239+240}$Pu were detected. Within the zone of silt deposits, several areas (covering about 13% of the crawl area) are detected, where total density of radionuclide contamination equals 41 – 70 TBq/km$^2$, and total reserves of radionuclide activity are about 46%. The contamination densities for $^{90}$Sr and $^{137}$Cs exceeding 4 TBq/km$^2$ are usually confined to depths of 6 – 10 m, whereas concentrations above 37 TBq/km$^2$ – beneath 10 m. There is an area in the western region of the crawl, where anomalously high concentration of radionuclides was detected according to data, obtained in 1995: 307, 252, and 53 TBq/km$^2$ for $^{90}$Sr, $^{137}$Cs and $^{238+239+240}$Pu, respectively, which are approximately 20 times higher than the average for the whole crawl. Total reserve of radionuclides deposited in the bottom sediments of Yanovsky crawl equaled about 13.5 TBq for $^{90}$Sr, 12.1 TBq for $^{137}$Cs and 0.3 TBq for $^{238+239+240}$Pu, respectively.

During the accident, the cooling pond of Chernobyl NPP was also subject to intensive radionuclide contamination. Here, at depths beneath 11 m (up to 35% of the bottom square) reaches 100 cm at $^{137}$Cs contamination density within the range of 19 – 133 TBq/km$^2$. At the depths of 3 – 11 m, the bottom is composed of primary soils, covered by silt 1 – 6 cm thick, where $^{137}$Cs contamination density varies from 2 to 6 TBq/km$^2$ [8].

**Table 3. Morphometric characteristics and concentration densities of radionuclides in bottom sediments of some water bodies in the exclusion zone**

| Water body characteristics | | Yanovsky crawl | Semikhodsky crawl | Shepelichesky crawl | Kosharovsky crawl | Azbuchin lake | Chernobyl crawl | Dalekoe-1 lake | Glubokoe lake |
|---|---|---|---|---|---|---|---|---|---|
| Distance from Chernobyl NPP* | azimuth, ° | 330 | 320 | 320 | 310 | 30 | 130 | 5 | 340 |
| | km | 2.5 | 4.5 | 7.0 | 9.5 | 2.0 | 13.0 | 4.5 | 6.5 |
| Volume, million m$^3$ | | 3.70 | 0.84 | 0.61 | 0.28 | 0.78 | 1.20 | 0.02 | 0.46 |
| Water plane, km$^2$ | | 0.84 | 0.41 | 0.23 | 0.16 | 0.27 | 0.61 | 0.01 | 0.17 |
| Maximal depth, m | | 13.4 | 5.0 | 6.2 | 3.9 | 5.6 | 10.0 | 6.6 | 7.1 |
| Average concentration densities of radionuclides, $\cdot 10^{10}$ Bq/km$^2$ | | | | | | | | | |
| $^{137}$Cs | | 1,498.5 | 310.8 | 85.1 | 62.9 | 1,147.0 | 38.5 | 339.6 | 558.7 |
| $^{90}$Sr | | 1,628.0 | 225.7 | 70.3 | 28.9 | 666.0 | 14.1 | 301.8 | 259.0 |
| $^{238+239+240}$Pu | | 37.0 | 7.8 | 1.7 | 0.9 | 24.4 | 0.4 | 7.8 | 7.4 |
| $^{241}$Am | | 31.1 | 7.0 | 1.4 | 7.4 | 21.8 | 0.4 | 7.7 | 6.3 |

* – to the point closest to Chernobyl NPP

According to assessments made in 2001 [7], bottom sediments of the cooling pond contains up to 160 TBq of $^{137}$Cs, 24 TBq $^{90}$Sr and 0.5 TBq $^{239+240}$Pu, respectively. About 30 – 50% of total radionuclides are accumulated in bottom sediments, precipitated at depths above 7 m.

At present, the radionuclide concentrations in bottom sediments of trunks in Uzh and Pripyat' rivers are just insignificantly higher than the pre-accident level. Much higher specific concentration of radionuclides is still observed in open crawls, cutoff meanders and other low flowage areas of rivers.

## Higher Hydrophytes

Higher hydrophytes possess high production potential and an ability to actively absorb radioactive substances. Occupying littoral and partly sublittoral zones in the majority of water bodies and, hence, being one of predominant components of biocoenoses by biomass, hydrophyte coenoses may be of importance for self purification of aquatic ecosystems. The content of radionuclides in higher hydrophytes from water bodies of the Chernobyl NPP exclusion zone is mostly determined by the radiation contamination level of water objects and neighboring lands, and the features of hydrochemical regime of water bodies affecting the forms of radionuclide presence in said water bodies and their biological accessibility.

Radionuclide $^{90}$Sr and $^{137}$Cs accumulation by macrophytes dwelling in the exclusion zone possess signified species specificity. Species with relatively high $^{137}$Cs concentration include aerial aquatic plants of Cyperaceae family, common reed (*Phragmites australis* (Cav.) Trin ex. Steud.), reed sweet-grass (*Glyceria maxima* (Hartm.) Holmb.), lesser reedmace (*Typha angustifolia* L.), and inherent aquatic plants – spiked milfoil (*Myriophyllum spicatum* L.) and water-soldier (*Stratiotes aloides* L.). The following species with floating leaves accumulate $^{137}$Cs in low concentration: first of all, they are representatives of Nymphaeaceae family, which are white water lily (*Nymphaea candida* L.) and yellow water-lily (*Nuphar luteum* (L.) Smith), as well as frogbit (*Hydrocharis morsus-ranae* L.) [9]. Among species with comparatively high $^{90}$Sr concentration, Potamogetonaceae family representatives are of special importance. It is obvious that this is related to Potamogetonaceae ability to intensively absorb calcium on their surface during photosynthesis processes, and this calcium is not washed off at standard sampling procedure [10]. Hence, calcium carbonate separated from may contain 7 – 20-fold high concentration of radioactive strontium than directly in tissues. The data on $^{137}$Cs and $^{90}$Sr concentration in macrophytes growing in water bodies of Chernobyl NPP exclusion zone are shown in Table 4. The majority of higher aquatic plant species were characterized by increasing specific activities of $^{137}$Cs and $^{90}$Sr at the vegetation peak (the end of July – August) compared with spring and autumn periods.

In higher aquatic plants of lakes, located in the left flood plain of Pripyat' river, concentrations of $^{238+239+240}$Pu and $^{241}$Am were registered in the range of 0.4 – 65.6 (11.2) Bq/kg with *CF* within 23.6 – 4,175.2 and the range of 0.3 – 44.8 (11.0) Bq/kg with *CF* within 82.5 – 7,458.3, respectively. The highest *CF* was observed for lesser reedmace (*Typha angustifolia* L.). For this species, the values of this index were 5 – 7 times higher than average *CF* for other species of plants under study that allows for considering lesser reedmace as a specific accumulator of transuranic elements under conditions of water bodies, located in Chernobyl NNP exclusion zone.

Within the period of investigation since 1989 till 2004, dynamics of the main radionuclide concentrations in tissues of the higher aquatic plants from river ecosystems of the exclusion zone indicated a decrease of specific activity of $^{90}$Sr and $^{137}$Cs (Figure 3a, b). So far as concerns closed of low flowage water bodies, the most representative retrievals obtained in the period of 1993 – 2004 have shown that since the late 1990ies the higher aquatic plants related to different ecological groups are indicating a frank tendency to $^{90}$Sr content increase in tissues (Figure 3c–f). Primarily, this tendency was observed for test materials, sampled in the period of 1993 – 1998 in Glubokoe lake [11], and then was confirmed by tests performed in 1998 – 2004 for Dalekoe-1 lake. So far as concerns $^{137}$Cs, its specific activity in higher aquatic plants of the lakes under study either decreases or remains practically constant. Hence, in the middle 1990ies specific activity of $^{137}$Cs in tissues of higher aquatic plants from Krasnenskaya flood plain was much higher than specific activity of $^{90}$Sr in them, whereas in the late 1990ies these values became comparable, and at present specific activity of $^{90}$Sr is much higher than that of $^{137}$Cs, the concentration of which either decreasing or remaining practically constant.

In the cooling pond of Chernobyl NPP, due to more intensive water exchange and, at the same time, high contamination density of bottom sediments with $^{137}$Cs, which defines the ratio between $^{90}Sr$ and $^{137}Cs$ contents in water, anomalous for water bodies of the exclusion zone, dynamics of $^{90}Sr$ specific activity in tissues of macrophytes is not so clear, and $^{137}Cs$ specific activity varies within a broad range showing no clear linear dependence (Figure 3e).

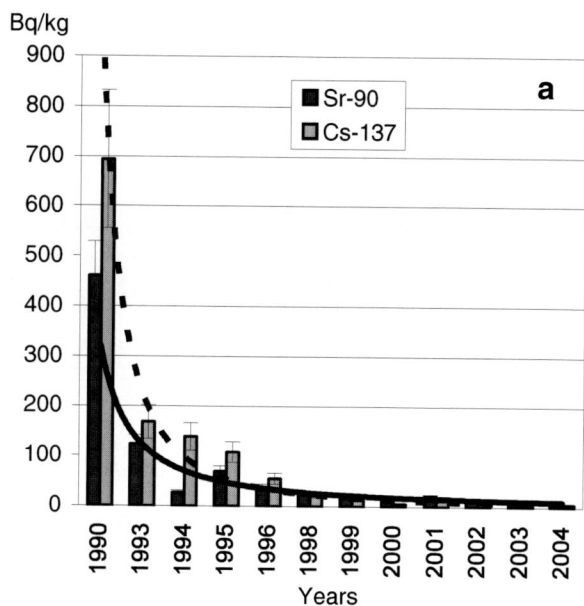

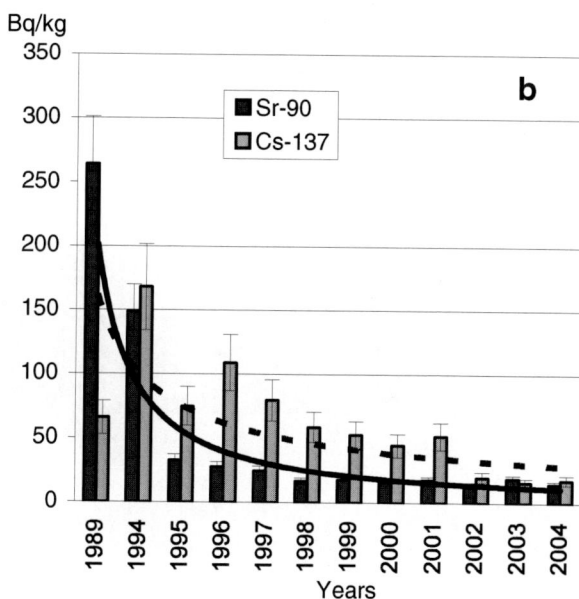

Figure 3(a, b)

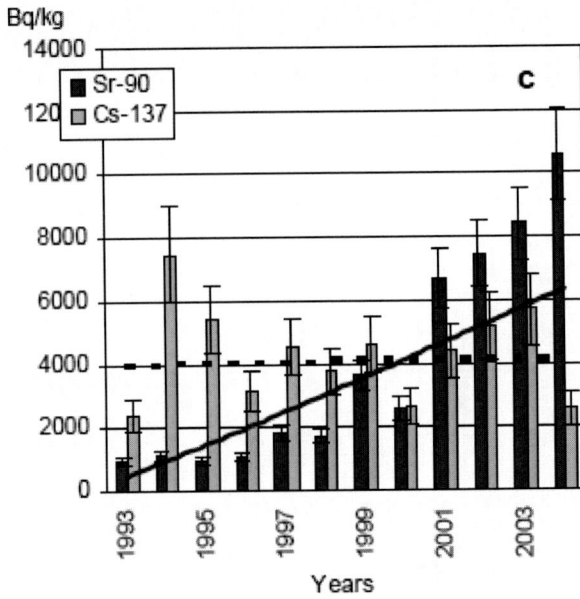

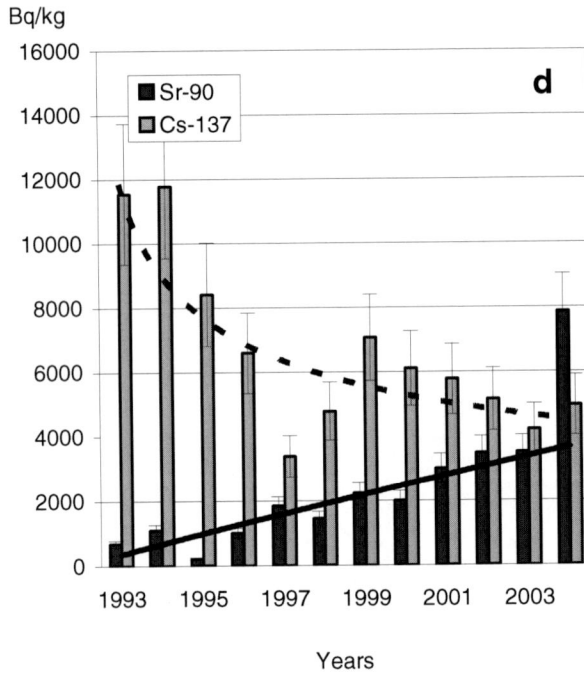

Figure 3(c, d)

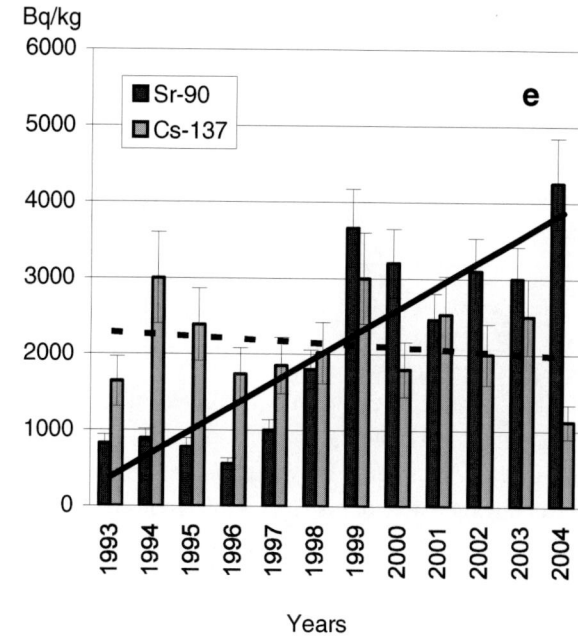

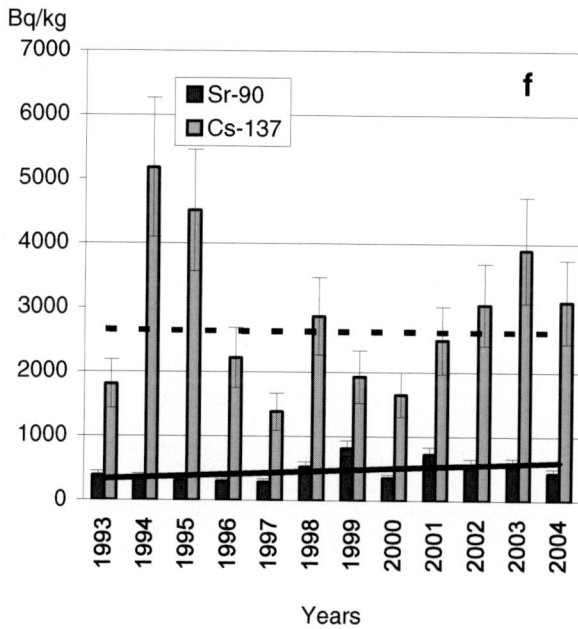

Figure 3. Dynamics of radionuclide content in higher aquatic plants of Chernobyl NPP exclusion zone: a – lesser reedmace (Pripyat' river); b – fennel pondweed (Pripyat' river); c – water-soldier (Glubokoe lake); d – yellow water lily (Glubokoe lake); e – yellow water lily (Dalekoe-1 lake); f – common reed (Chernobyl NPP cooling pond). Continuous line denotes [90]Sr specific activity trend; dashed line denotes [137]Cs specific activity trend

**Table 4. Specific activity of radionuclides in higher aquatic plants from water bodies in the exclusion zone of Chernobyl NPP in the period of 1997 – 2001, Bq/kg**

| Water object | $^{90}$Sr | | | $^{137}$Cs | | |
|---|---|---|---|---|---|---|
| | max | min | average | max | min | average |
| Glubokoe lake | 14,060 | 67 | 2,210 | 36,500 | 1,220 | 8,730 |
| Dalekoe-1 lake | 5,100 | 200 | 1,810 | 19,470 | 1,170 | 5,170 |
| Azbuchin lake | 24,210 | 730 | 4,900 | 23,860 | 220 | 2,030 |
| Cooling pond | 1,600 | 36 | 218 | 3,130 | 310 | 1,380 |
| Yanovsky crawl | 2,200 | 110 | 779 | 1,702 | 38 | 814 |
| Uzh river (Cherevach vil.) | 234 | 3 | 10 | 185 | 4 | 42 |
| Pripyat' river (Chernobyl) | 357 | 5 | 15 | 164 | 5 | 30 |

It is suggested that $^{90}$Sr specific activity in tissues of macrophytes from Krasnenskaya flood plain increases due to dynamics of radionuclide transformation in soils of water catchment areas and bottom sediments of water bodies. Since after the Chernobyl accident Krasnenskaya flood plain appeared one of the most radionuclide-contaminated areas of the exclusion zone, in 1991 – 1995 a complex of flood control dams were constructed here, which changed the hydrological regime of flood plain flows during floods and preventing radioactive substance washing off the soils of contaminated areas (refer to Figure 1). In its turn, this became the reason for intensification of waterlogging and swamping of dam-locked areas. As a result, at the background of general tendencies of $^{90}$Sr mobile form increase in the soils of water catchment areas and bottom sediments of the exclusion zone water bodies; in swamped soils of Krasnenskaya flood plain fulvic and humic acid concentration increases that decreases pH value in water and intensifies a denuding of water-soluble forms of $^{90}$Sr forming soluble complexes with fulvic acids. Hence, an increase of concentrations of mobile forms of the radionuclide and their inclusion to biotic turnover of aquatic ecosystems are observed. This also confirms the increase of $^{90}$Sr specific activity, observed in recent years in Krasnenskaya flood plain lakes at the background of stabilization of this index for $^{137}$Cs.

The tendency of the root contamination of plant tissues by $^{90}$Sr was also observed for terrestrial plants in the exclusion zone [12 – 14]. Currently, some authors [15] suggest that $^{90}$Sr mobility in soils of the exclusion zone is maximal, and this will last during the nearest decade. Thereafter, the rate of radionuclide decomposition will exceed the rate of its mobilization.

## Mollusks

Freshwater mollusks are frequently considered as the species-indicators of radionuclide contamination of water objects. These invertebrates accumulate practically all radionuclide detectable in water, and owing to high biomass mollusks play very important role in bioaccumulation and radionuclide redistribution processes in freshwater ecosystems. The data on $^{137}$Cs and $^{90}$Sr concentration in freshwater mollusks from water bodies of Chernobyl NPP exclusion zone are shown in Table 5.

**Table 5. Specific activity of radionuclides in freshwater mollusks from the exclusion zone of Chernobyl NPP in 1998 – 2003, Bq/kg**

| Water object | $^{90}$Sr | | | $^{137}$Cs | | |
|---|---|---|---|---|---|---|
| | max | min | average | max | min | average |
| Glubokoe lake | 170,300 | 39,770 | 61,830 | 27,150 | 3,160 | 9,070 |
| Dalekoe-1 lake | 62,830 | 17,120 | 32,310 | 2,410 | 847 | 1,523 |
| Azbuchin lake | 87,670 | 34,010 | 51,120 | 4,750 | 2,704 | 3,620 |
| Cooling pond | 2,500 | 1,600 | 2,133 | 2,900 | 830 | 1,450 |
| Yanovsky crawl | 13,350 | 7,720 | 10,350 | 590 | 430 | 510 |
| Uzh river (Cherevach vil.) | 216 | 98 | 177 | 30 | 14 | 25 |
| Pripyat' river (Chernobyl) | 101 | 77 | 86 | 61 | 31 | 35 |

Maximal *CF* values for both $^{90}$Sr and $^{137}$Cs characterize bivalve mollusks – zebra mussel (*Dreissensia polymorpha* Pall.) and *Unio* sp., which are the most active filtration organisms. Maximal *CF* values of $^{90}$Sr exceeding 1,100 were observed for zebra mussel and of $^{137}$Cs for *Unio* sp. tissues – about 500. Much lower *CF* of radionuclides were determined for gastropods: *Lymnaea stagnalis* L., *Stagnacolia palustris* L., ram's-horn snail (*Planorbarius corneus* L.), and viviparous pond-snail (*Viviparus vivparus* L.). Minimal *CF* of $^{90}$Sr and $^{137}$Cs were determined for *Lymnaea stagnalis*: 440 and 137, respectively [16].

The average specific activities of transuranic $^{238}$Pu and $^{239+240}$Pu elements in tissues of mollusks inhabiting in Glubokoe and Dalekoe-1 lakes were the following: minima were observed for *Lymnaea stagnalis* – 0.1 and 0.2 Bq/kg in Dalekoe-1 lake and 2.7 and 6.4 Bq/kg in Glubokoe lake, respectively. Higher concentration was determined for *Lymnaea stagnalis* from Glubokoe lake: 14.0 and 36.0 Bq/kg, respectively. Among gastropods of Krasnenskaya flood plain lakes, maximal activities were detected for ram's-horn snail: 0.92 and 2.1 Bq/kg in Dalekoe-1 lake and 24.7 and 53.0 Bq/kg in Glubokoe lake, respectively. Zebra mussel from cooling pond of Chernobyl NPP contained $^{238}$Pu and $^{239+240}$Pu in amounts of 2.8 and 6.2 Bq/kg, respectively.

**Table 6. Specific activity of radionuclides in fish from the exclusion zone of Chernobyl NPP for the period of 1997 – 2004, Bq/kg**

| Water object | $^{90}$Sr | | | $^{137}$Cs | | |
|---|---|---|---|---|---|---|
| | max | min | average | max | min | average |
| Glubokoe lake | 6,800 | 660 | 3,011 | 14,000 | 1,200 | 8,250 |
| Dalekoe-1 lake | 13,060 | 410 | 3,129 | 27,020 | 16,110 | 17,500 |
| Cooling pond | 1,520 | 18 | 246 | 13,260 | 1,100 | 3,950 |
| Uzh river | 24 | 1 | 8 | 290 | 3 | 63 |
| Pripyat' river | 87 | 1 | 18 | 800 | 6 | 113 |

The specific activity of $^{241}$Am in *Lymnaea stagnalis* tissues was minimal, determined within the range of 4.6 – 30.0 (15.2) Bq/kg in Dalekoe-1 lake and within the range of 6.3 – 51.0 (26.5) Bq/kg in Glubokoe lake, respectively. For *Lymnaea stagnalis* from Glubokoe lake concentrations about 75.0 Bq/kg were detected. Maximal values were determined for ram's-horn snail: 18.0 – 29.0 (23.5) Bq/kg in Dalekoe-1 lake and 80.0 – 310.0 (170.0) Bq/kg in Glubokoe lake, respectively. The concentration of $^{241}$Am in zebra mussel from the cooling pond of Chernobyl NPP equaled 7.9 Bq/kg.

## Fishes

For radioecological investigations of aquatic ecosystems fishes occupying the upper levels of freshwater biocoenosis trophic pyramid and representing one of man's feeding objects are of certain interest. The concentrations of $^{90}$Sr and $^{137}$Cs radionuclides in fishes of various water bodies in the exclusion zone are shown in Table 6.

The concentrations of transuranic $^{238}$Pu, $^{239+240}$Pu and $^{241}$Am elements were measured in fishes from Glubokoe and Dalekoe-1 lakes. The specific activity of $^{238}$Pu in fish tissues equaled 0.4 – 0.5 (0.4) Bq/kg with $CF$ equal 72 – 98 (83), $^{239+240}$Pu – 0.7 – 0.9 (0.8) Bq/kg with $CF$ equal 68 – 87 (75), and $^{241}$Am – 2.2 – 10.0 (6.2) Bq/kg with $CF$ equal 367 – 1,667 (1.028).

In all cases, the content of $^{90}$Sr and $^{137}$Cs radionuclides in fishes from the lakes of Pripyat' river left-bank flood plain far exceeded maximum permissible concentrations (MPC) pursuant to the fish industry standards of Ukraine [17]: on average, 90-fold (at MPC equal 35 Bq/kg) by $^{90}$Sr and 120-fold (at MPC equal 150 Bq/kg) by $^{137}$Cs, respectively. Maximal recorded values exceeded MPC by 370 and 180 times, respectively. The content of $^{90}$Sr in fished inhabiting in the cooling pond also exceeded MPC, by 7 time, on average, maximally reaching 40 times. In all tests, $^{137}$Cs concentration also exceeded MPC by 30 times, on average, and by 90 times as a maximum.

Despite the fact that the average concentrations of radionuclides in fishes of Pripyat' river did not exceed MPC, the cases of $^{137}$Cs concentration elevated above MPC during the period of tests were observed for 20% of all caught species. Maximal $^{137}$Cs concentrations were 5-fold above MPC. In Uzh river $^{137}$Cs concentrations exceeding MPC were detected in singular cases.

**Table 7. Total concentration and ratio of radionuclides in the main components of the ecosystem of Dalekoe-1 lake**

| Object | $^{90}$Sr | | $^{137}$Cs | | $^{238+239+240}$Pu, $^{241}$Am | |
|---|---|---|---|---|---|---|
| | MBq | % | MBq | % | MBq | % |
| Bottom sediments | 37,000 | 95.35 | 51,800 | 99.11 | 1,100 | 99.90 |
| Water | 1,650 | 4.25 | 236 | 0.45 | 0.27 | 0.03 |
| Seston | 58 | 0.15 | 155 | 0.30 | –* | – |
| Biota | 96 | 0.25 | 73 | 0.14 | 0.81 | 0.07 |

* – no measurements were performed

Since the requires selection of even-aged individuals was not obtained, reliable differences in concentrations of radionuclides within groups of predatory and nonpredatory fishes can be hardly determined and compared with ecological specialization of one species or another, because even nonpredatory species, characterized by similar trophic orientation, possess $CF$ for $^{90}$Sr and $^{137}$Cs varying within broad ranges. Nevertheless, the averaged data indicate elevated $^{137}$Cs concentration for fishes of high trophic levels, whereas MPC exceeding by $^{137}$Cs concentrations in the exclusion zone were mostly observed for predatory fish species.

**Table 8. Total concentration and ratio of radionuclides in the main components of the ecosystem of Glubokoe lake**

| Object | $^{90}Sr$ | | $^{137}Cs$ | | $^{238+239+240}Pu, ^{241}Am$ | |
|---|---|---|---|---|---|---|
| | MBq | % | MBq | % | MBq | % |
| Bottom sediments | 444,000 | 89.02 | 962,000 | 98.64 | 25,900 | 99.80 |
| Water | 50,900 | 10.21 | 6,200 | 0.64 | 10 | 0.04 |
| Seston | 800 | 0.16 | 2,471 | 0.25 | –* | – |
| Biota | 3,035 | 0.61 | 4,598 | 0.47 | 42 | 0.16 |

\* – no measurements were performed

## Radionuclide Distribution in Basic Components of Lake Ecosystems

Distribution and further migration of radionuclides in the components of aquatic ecosystems is defined by biogeochemical repeating pattern of substance motion in the nature. Hence, besides predominant sedimentation and sorption processes of radionuclide depositing to bottom sediments and precipitation in suspensions, their accumulation in living organisms and subsequent invoking in the biotic turnover is also urgent.

The studies of $^{90}Sr$, $^{137}Cs$, $^{238}Pu$, $^{239+240}Pu$ and $^{241}Am$ distributions in basic biotic and abiotic components of Dalekoe-1 and Glubokoe lakes ecosystems, located at dam-fenced site of Krasnenskaya flood plain, indicated depositing of the main quantity of radionuclides to bottom sediments. Hence, concentrations of radionuclides equaled as follows: by $^{90}Sr$ – 89 and 95% for Glubokoe and Dalekoe-1 lakes, respectively; by $^{137}Cs$ – 99%; by transuranic $^{238+239+240}Pu$ and $^{241}Am$ elements – almost 100% of total content in ecosystems (Tables 7 and 8). The difference in $^{90}Sr$ contribution to bottom sediments of the lakes and contents of other radionuclides is stipulated by elevated migration activity of $^{90}Sr$ compared with $^{137}Cs$ and transuranic elements. This is the determining reason for higher concentration of dissolved $^{90}Sr$ in waters of both lakes (10.2 and 4.3%) compared with $^{137}Cs$ (0.6 and 0.5%) and transuranic elements (0.04 and 0.03%) and, vice versa, lower $^{137}Cs$ concentration in seston (0.16 and 0.15%) compared with $^{137}Cs$ (0. 25 and 0.30%), respectively.

The content of transuranic elements in biotic components of ecosystems was minimal 0.07 and 0.16%, respectively; for $^{137}Cs$ it equals 0.14 and 0.47%, and for $^{90}Sr$ – 0.25 and 0.61%, respectively. Maximal values were determined for Glubokoe lake, where dissolved $^{90}Sr$ concentration in water was also higher. As suggested, such differences in $^{90}Sr$ distribution in ecosystems of lakes is, firstly, related to high biomass of the higher aquatic vegetation, intensively overgrowing Glubokoe lake (covering about a half of water plane). It is known that as growths of macrophytes increase and become denser, the oxygen regime becomes worse, significant quantities of organic substances and biogenic elements are accumulate, and pH of the water medium decreases. As pH decreases, radionuclides are desorbed more actively and transit into the dissolved state, $^{90}Sr$, first of all [13, 14]. This is also confirmed by lower average pH in Glubokoe lake (8.2) compared with Dalekoe-1 lake (8.8) and the ratio between specific activity of $^{90}Sr$ and $^{137}Cs$ in water equal, on average, 12.5 in Glubokoe lake and 7.3 in Dalekoe-1 lake, respectively. Contamination densities of bottom sediments by different radionuclides are also typical: the average specific activities of transuranic elements in both lakes were practically equal, whereas the ratio between average concentrations of $^{137}Cs$ and $^{90}Sr$ in Dalekoe-1 and Glubokoe lakes equaled 1.3 and 2.2, respectively. Therefore, affecting hydrochemical regime of water bodies, the intensity of closed lake overgrowth by associations of higher aquatic plants may also change the type of distribution of radionuclides in the components of ecosystems. It is also obvious that a part of radionuclides transit to water from biomass of dead plants and return to the biotic turnover of the water body.

**Table 9. Total concentration of radionuclides in basic biotic components of Dalekoe-1 lake, kBq**

| Hydrobionts | $^{90}$Sr | $^{137}$Cs | $^{238+239+240}$Pu | $^{241}$Am |
|---|---|---|---|---|
| Macrozoobenthos: | | | | |
| Oligochaetes | 48 | 2,128 | –* | – |
| Midges | 3 | 1,637 | – | – |
| Bivalve mollusks | 85,160 | 1,238 | 284 | 320 |
| Total | 85,211 | 5,003 | 284 | 320 |
| Gastropods: | | | | |
| *Lymnaea stagnalis* | 860 | 12 | 0.04 | 0.5 |
| ram's-horn snail | 476 | 8 | 0.03 | 0.2 |
| Total | 1,336 | 20 | 0.07 | 0.7 |
| Macrophytes: | | | | |
| yellow water-lily | 800 | 3,600 | 1 | 2 |
| hornwort | 3,400 | 12,900 | 38 | 24 |
| common reed and sedge | 4,100 | 41,900 | 79 | 58 |
| Total | 8,300 | 58,400 | 118 | 84 |
| Fishes | 1,407 | 5,573 | 0.7 | 2 |
| TOTAL | 96,254 | 68,996 | 403 | 407 |

* – no measurements were performed

Total concentrations of radionuclides in the main biotic components of the lakes under study are shown in Tables 9 and 10. In zoobenthos associations of Dalekoe-1 lake the overwhelming majority of $^{90}$Sr (99%) is concentrated in bivalve mollusks. Total $^{90}$Sr concentration in these mollusks equals about 89% of its content in the lake biota. About 8.5% is accumulated in higher hydrophytes, 1.5% – in fishes, about 1% – in gastropods, and less than 1% – in representatives of "soft" zoobenthos (oligochaetes and midges). Similar situation is observed for distribution of transuranic elements. However, in this case, quantities accumulated in bivalve mollusks are decreased (down to 70 – 80%), as well as these accumulated in gastropods and fishes (to fractions of percent), whereas the value of macrophytes increases to 21 – 29% (Figure 4).

In the benthos of bottom invertebrates the part of $^{137}$Cs accumulated in bivalve mollusks is 25% of its total amount or lower. This relates to high *CF* values for this radionuclide in "soft" zoobenthos representatives, in which, despite lower biomass compared with bivalve mollusks (10-fold lower), $^{137}$Cs content gives about 65% of its total quantity. About 85% of $^{137}$Cs present in the lake biota is accumulated in high aquatic plants, 7 and 8% – in zoobenthos and fish, respectively, and less than 0.1% – in gastropods.

**Table 10. Total concentration of radionuclides in basic biotic components of Glubokoe lake, MBq**

| Hydrobionts | $^{90}$Sr | $^{137}$Cs | $^{238+239+240}$Pu | $^{241}$Am |
|---|---|---|---|---|
| Gastropods: | | | | |
| *Lymnaea stagnalis* | 155 | 12 | 0.04 | 0.29 |
| *Stagnicolia polustris* | 25 | 3 | 0.04 | 0.07 |
| ram's-horn snail | 115 | 11 | 0.17 | 0.37 |
| Total | 295 | 26 | 0.25 | 0.73 |
| Macrophytes: | | | | |
| yellow water lily and white water | 85 | 178 | 0.1 | 0.2 |
| lily | 656 | 1,473 | 2.5 | 2.8 |
| water-soldier | 108 | 477 | 0.3 | 0.5 |
| hornwort | 454 | 840 | 12.1 | 8.2 |
| lesser reedmace | 26 | 28 | 0.1 | 0.1 |
| broad leaved pondweed | | | | |
| joint associations of common reed, | 260 | 1,457 | 3.4 | 2.8 |
| sedge, reedmace and bur-reed | 1,589 | 4,453 | 18.5 | 14.6 |
| Total | | | | |
| Macrozoobenthos | 1,126 | 80 | 3 | 5 |
| Fishes | 25 | 37 | 0.01 | 0.03 |
| TOTAL | 3,035 | 4,596 | 22 | 20 |

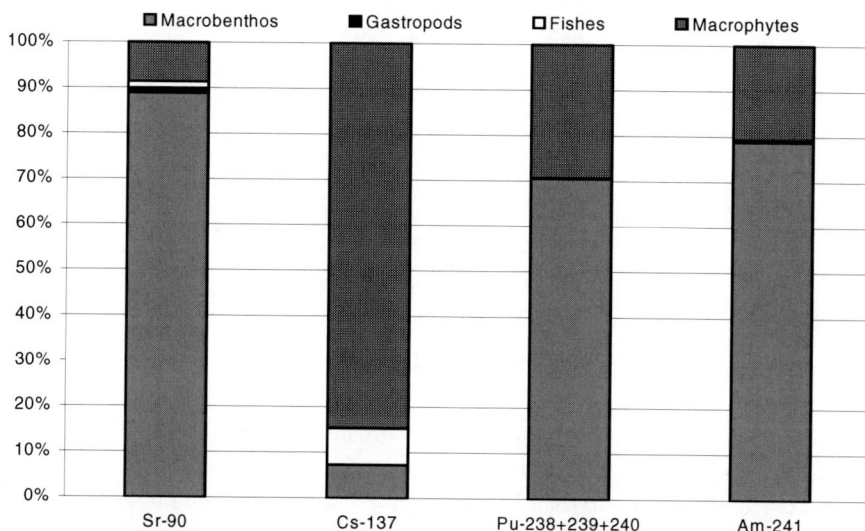

Figure 4. Distribution of radionuclides by main groups of hydrobionts in the biotic component of Dalekoe-1 lake ecosystem

In Dalekoe-1 lake biota higher aquatic plants occupying less than 5% of the water plane dominate exclusively in $^{137}$Cs distribution, whereas in Glubokoe lake comparative biomass of macrophytes is so high that they accumulate about 90% of all radionuclides present in the biotic component of the lake (Figure 5). The exception is $^{90}$Sr, for which the contribution of aquatic plants is slightly more than 50%, and the rest of it is mostly concentrated in gastropod and bivalve mollusk valves.

The main factors determining the contribution of various ecological groups of hydrobionts to total content of radionuclides in the biotic component of lake ecosystems are *CF* and aquatic organism biomass. Moreover, *CF* values may be so significant that sometimes the biomass factor is displayed in the only cases, when the difference in this parameter between different groups of hydrobionts reaches two orders of magnitude.

The basic data used for estimation of radionuclides distribution in the main components of lake ecosystems were obtained in 1998, when specific activity of $^{90}$Sr in higher aquatic plant tissues did not significantly increase. In this relation, it can be suggested that, currently, $^{90}$Sr reserves in the vegetation of tested lakes are increasing. This tendency will be preserved with intensifying swamping processes on the territories under study and corresponding overgrowth of water bodies with macrophytes.

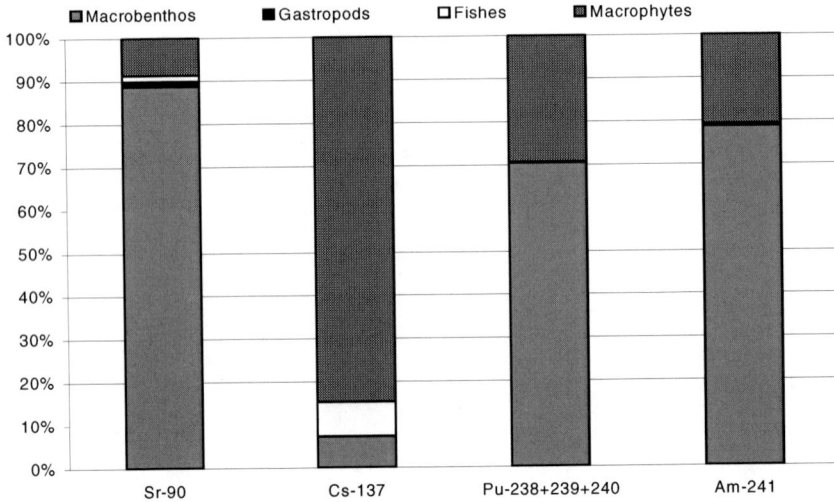

Figure 5. Distribution of radionuclides by main groups of hydrobionts in the biotic component of Glubokoe lake ecosystem

# CONCLUSION

Despite the years passed since the Chernobyl accident, the exclusion zone area is still the open source of radionuclide contamination with a complex distribution structure in different landscapes and characterized by dynamic character of the state of radioactive substance forms, which affect their migration and redistribution by components of ecosystems. The basic questions of radioactive safety of the exclusion zone concern radionuclides wash-off with surface drainage water to river network, their export outside the exclusion zone and affection of the water quality in Dnepr River.

Self purification of closed water bodies in the Chernobyl NPP exclusion zone is extremely slow process. Therefore, ecosystems of the majority of lakes, dead channels and crawls possess high level of radionuclide contamination of all the components. The main quantity of radionuclides in aquatic landscapes of the exclusion zone are deposited to bottom sediment of closed water bodies, hence, the distribution of radioactive substances in biotic and abiotic components of hydrobiocoenoses is defined by biogeochemical regularities and transformations of radioactive substances in bottom sediments of water bodies and in soils of adjacent territories.

The content of radionuclides in tissues of higher aquatic plants from the exclusion zone is characterized by high species specificity. Maximal $^{137}$Cs concentrations were observed for aerial-aquatic species and minimal – for Nymphaeaceae family. Maximal $^{90}$Sr concentrations were determined for Potamogetonaceae family and minimal – for Cyperaceae and Nymphaeaceae families. Maximal *CF* values of transuranic elements were observed for lesser reedmace (*Typha angustifolia* L.). On average, the values of this parameter are 6-fold higher than *CF* for the rest studied species of plants. This allows for considering lesser reedmace as a hydrophyte specifically accumulating transuranic elements in conditions of water bodies in the Chernobyl NPP exclusion zone.

The species specificity of $^{90}$Sr and transuranic elements concentrating by freshwater mollusks is, firstly, determined by specific weight of the valve, whereas for $^{137}$Cs – by the features of functional ecology and the type of feeding. Among hydrobionts, mollusks dominate in $^{90}$Sr biosedimentation processes.

The average values of $^{90}$Sr and $^{137}$Cs concentration in fishes from rivers of the exclusion zone does not exceed MPC pursuant to the fish product standards of Ukraine. However, in the period of 1997 – 2004, $^{137}$Cs concentration exceeded MPC in 20% of fish species caught in Pripyat' river, whereas for Uzh river these were singular cases. For $^{90}$Sr, MPC exceeding was not observed at all. All fishes sampled in closed or low flowage water bodies showed high concentration of radionuclides. The maximal values were observed in Krasnenskaya flood plain lakes, where average specific activities of radionuclides exceeded MPC by two orders of magnitude. Maximal specific activities of radionuclides in all water bodies of the exclusion zone are mostly observed for predatory species.

The construction of a complex of flood retarding dams and degradation of existing melioration systems at the site of the left-bank flood plain of Pripyat' river implied a change of hydrological regime and the character of water object overgrowth. The absence of flowage in water bodies, stagnation effects during spring flooding and seasonal runoffs intensified waterlogging and swamping of dam-fenced territories. As a result, at the background of general tendencies of $^{90}$Sr mobile forms increase in the soils of water catchment areas and bottom sediments of the exclusion zone water bodies located at dam-fenced sites, as well as increasing intensity of this radionuclide concentrating by higher aquatic plants and, obviously, other autotrophic organisms. For some species of macrophytes, $^{90}$Sr concentration compared with the early 1990ies has increased by more than an order of magnitude and exceeds the specific activity of $^{137}$Cs. Obviously the specific activity of $^{90}$Sr will also increase at higher trophic levels, however, at present, such dynamics was not reliably detected. It is suggested that for radioecological monitoring of aquatic ecosystems in the exclusion zone, higher aquatic plants possessing high *CF* values of radionuclides are the most sensitive test-objects for recording increasing specific activity of mobile forms of radionuclides in the water of test site water bodies.

The results of tests performed confirm the tendency to further deterioration of the radiation situation in aquatic ecosystems of the exclusion zone. Swamping of contaminated territories leads to acceleration of $^{90}$Sr deposited form mobilization processes and their migration and redistribution in closed aquatic landscapes. Hence, an original "depot" of mobile forms of radioactive substances is formed, which in high-flood periods may become a source of increasing $^{90}$Sr drainage to Pripyat' river and then outside the exclusion zone. In this connection, the necessity to implement hydraulic engineering procedures preventing underflooding of territories with high densities of radionuclide contamination; optimization and enhancement of radioecological monitoring system, and further development of investigations of radionuclide behavior in aquatic ecosystems of the exclusion zone – the important components in the complex of measures related to forecasting and minimization of the Chernobyl disaster consequences.

## REFERENCES

[1]    Aleksandrova V.D., *Flora Classification*, Leningrad, Nauka, 1969, 275 p. (Rus)

[2]    Katanskaya V.M., *Higher Aquatic Vegetation of Continental Water Bodies of the USSR*, Leningrad, Nauka, 1981, 1185 p. (Rus)

[3]    Kuz'menko M.I., Romanenko V.D., Derevets V.V. *et al.*, *Radionuclides in Aquatic Ecosystems of Ukraine. The Influence of Radionuclide Contamination on Hydrobionts of the Exclusion Zone*, Kiev: Chernoibylinterinform, 2001, 318 p. (Ukr)

[4]    Kulikov N.V. and Molchanova I.V., *Continental Radioecology: Soil and Freshwater Ecosystems*, Moscow, Nauka, 1975, 185 p. (Rus)

[5]    Romanenko V.D., Kuz'menko M.I., Evtushenko N.Yu. *et al.*, *Radioactive and Chemical Contamination of Dnepr River and Its Water Reservoirs After the Accident at the Chernobyl NPP*, Kiev, Naukova Dumka, 1992, 194 p. (Rus)

[6]    Derevets V.V., Ivanov Yu.P., Kazakov S.V. *et al.*, *Bulletin of Ecological State of the Exclusion Zone and the Zone of Compulsory (Mandatory) Evacuation*, 1999, No. 13, pp. 9 – 19. (Ukr)

[7]    Kashparov V.A., Khomutin Yu.V., Glukhovsky A.S. *et al.*, *Bulletin of Ecological State of the Exclusion Zone and the Zone of Compulsory (Mandatory) Evacuation*, 2003, No. 1 (21), pp. 67 – 74. (Ukr)

[8]    Kanivets V.V. and Voitsekhovich O.V., *Thes. Rep. Intern. Conf. "Radioactivity at Nuclear Explosions and Accidents"*, 2000, Moscow – Saint-Petersburg, Hydrometeoizdat, 2000, p. 128. (Rus)

[9]    Gudkov D.I., Zub L.N., Savitsky A.L. *et al.*, *Hydrobiol. J.*, 2002, vol. 38 (5), pp. 116 – 132.

[10]   Lainerte M.P. and Seisuma Z.K., *Higher Aquatic and Littoral Aquatic Plants, Thes. Rep. 1st All-Union Conference*, 7 – 9 September, 1977, Borok, pp. 117 – 119. (Rus)

[11]   Kaglyan A.E., *The Features of $^{90}$Sr and $^{137}$Cs Accumulation by Hydrobionts and Abiotic Components of Water Bodies in the Exclusion Zone of Chernobyl NPP*, Candidate Dissertation Thesis on Biology, Kiev, 2003, 24 p. (Ukr)

[12]   Kashparov V.A., *Bulletin of Ecological State of the Exclusion Zone and the Zone of Compulsory (Mandatory) Evacuation*, 1998, No. 12, pp. 67 – 74. (Ukr)

[13]   Kashparov V.A., *Formation and Dynamics of Environmental Radioactive Contamination During the Accident at the Chernobyl NPP and After the Accident*, Ed. V.G. Barjakhtar, Kiev, Naukova Dumka, 2001, pp. 11 – 46. (Ukr)

[14]   Ivanov Yu.A., 'Dynamics of radionuclide redistribution in soils and vegetation', In Book: *Chernobyl – the Exclusion Zone*, Ed. V.G. Barjakhtar, Kiev, Naukova Dumka, 2001, pp. 47 – 76. (Ukr)

[15] Sobotovich E.V., Bondarenko G.N., Kononenko L.V. *et al.*, *Geochemistry of Production Induced Radionuclides*, Kiev, Naukova Dumka, 2002, 332 p. (Ukr)

[16] Gudkov D.I., Derevets V.V., Kuz'menko M.I., and Nazarov A.B., *Radiats. Biol. Radioekol.*, 2001, vol. 41(3), pp. 326 – 330. (Rus)

[17] *Maximum Permissible Concentrations of $^{137}$Cs and $^{90}$Sr Radionuclides in Foods and Drinking Water (MPC-97)*, Kiev, Ministry of Health Protection of Ukraine; Committee on Questions of Hygienic Regulation; NKRZU, 1997, 38 p. (Ukr)

# INDEX

## C

## D

## E

## F

**G**

## H

# I

## N

## O

161, 162, 186, 208, 220, 232, 252, 258, 277, 337, 370, 383, 385, 386, 387, 391, 392, 393, 394, 398, 400, 401, 402, 405
seed, 38, 63
seeding, 371
seedlings, 211
seeds, 35, 213
selecting, 250, 378
selectivity, 123, 304, 305, 323
selenium, 254
self, 102, 393, 405
senescence, 12
sensing, 341, 358
sensitivity, 2, 20, 96, 123, 226, 229, 230, 259, 336, 366, 367, 372, 374, 381, 382
separation, 155
sequencing, 322
series, ix, x, 124, 144, 189, 195, 197, 341, 343, 350, 351
serine, 291, 297
serum, 148, 154, 235
serum albumin, 235
services, iv, 381
sesquiterpenoid, xi, 279
severity, xi, 329, 330
sewage, x, xii, 7, 15, 16, 109, 125, 128, 154, 155, 156, 157, 158, 160, 161, 163, 166, 219, 231, 232, 243, 255, 261, 266, 329, 331, 336, 337, 338, 339, 352
sex, vii, viii, 105, 106, 107, 109, 110, 122, 123, 124, 125, 126, 128, 129, 130, 134, 136, 139, 141, 144, 146, 149, 150, 151, 152, 154, 157, 166, 225, 226, 227, 232, 233, 237, 238, 242, 243, 249, 251, 380
sex hormones, vii, viii, 105, 106, 107, 109, 110, 122, 123, 124, 125, 126, 128, 129, 130, 134, 136, 141, 144, 149, 150, 151, 152, 157, 166
sex ratio, viii, 105, 106
sex steroid, 154
shade, 43
shape, 12, 106, 207, 242
shares, 317
shellfish, 345, 360, 362
shipping, x, 265, 266
shock, 223, 257
shoot, 11, 12, 14, 191
shores, 28
short-term, 5, 68, 256, 349, 365, 367, 377
sigmoid, 55, 56, 82, 83, 86
sign, 172
signaling, 221, 312, 314, 315, 316, 318, 320
signaling pathway, 314, 320
signalling, 284, 286, 291, 297, 315
signals, 106, 226

significance level, 236
signs, 229, 348
silica, 108, 115, 116, 117, 118, 119, 121, 127
silver, 20, 222
similarity, 64, 65, 200, 210, 212, 213, 268, 272, 273, 274
simulation, viii, 49, 63, 64, 66, 94, 100
simulations, 60, 61, 62, 69, 79, 94, 95
singular, 400, 405
sister chromatid exchange, 224
sites, viii, xi, xiii, 3, 5, 6, 7, 15, 28, 30, 31, 32, 34, 36, 37, 39, 44, 105, 126, 127, 129, 133, 135, 139, 140, 198, 210, 220, 226, 254, 256, 265, 268, 269, 278, 282, 284, 293, 341, 344, 363, 374, 375, 405
skeleton, 110, 142
skimming, 147
skin, 325, 339
sludge, 128, 141, 155, 157, 158, 159, 160, 164, 170, 232, 261
slums, 334
smelters, 18
social context, 353
social factors, 332
social influence, 359
social influences, 359
society, 376
socioeconomic, 331, 350, 353
socioeconomic conditions, 331, 353
sociological, 376
sodium, 137, 196, 349
software, 236, 268
soil, 5, 15, 21, 28, 127, 128, 130, 131, 132, 133, 134, 141, 142, 143, 144, 145, 146, 147, 150, 151, 156, 160, 161, 162, 163, 164, 186, 208, 209, 257, 364
soil particles, 130
soils, 126, 127, 130, 131, 134, 139, 141, 142, 144, 145, 146, 147, 150, 156, 159, 160, 161, 162, 163, 165, 260, 387, 390, 392, 398, 405, 407
solar, 149, 165, 317, 337, 346
solid phase, 123, 130, 156
solid tumors, 293
solid waste, 125
solubility, 108, 110, 131, 134, 135, 136
solvent, 131, 325
somatic cell, 109
sorption, ix, 106, 107, 108, 124, 128, 130, 131, 132, 133, 134, 135, 136, 138, 139, 140, 141, 150, 151, 152, 160, 161, 162, 163, 401
sorption experiments, 132
sorption isotherms, 140
sorption process, 133, 401
sorting, 371
South Africa, 289, 318